Advances in Cardiac Signal Processing

Rajendra Acharya U, Jasjit S. Suri, Jos A.E. Spaan and S.M. Krishnan

Advances in Cardiac Signal Processing

With 268 Figures, 13 in color and 59 Tables

Springer

Dr. Rajendra Acharya U
Visiting Faculty
Dept. of ECE
Ngee Ann Polytechnic
Singapore 599489

E-mail: aru@np.edu.sg

Prof. Jos A.E. Spaan
Editor Chief MBEC
Dept. of Medical Physics
Academic Medical Center
University of Amsterdam
P.O. Box 22700
1100 DE Amsterdam
The Netherlands

E-mail: j.a.spaan@amc.uva.nl

Prof. Jasjit Suri
Idaho State Biomedical Research Institute
ID, USA and Biomedical Technologies, CO, USA

E-mail: jsuri@comcast.net

Mr. Shankar M. Krishnan
8 Whittier Place
BOSTON MA 02114
USA

E-mail: smkrishnan@gmail.com

Library of Congress Control Number: 2006931013

ISBN-10 3-540-36674-1 Springer Berlin Heidelberg New York
ISBN-13 978-3-540-36674-4 Springer Berlin Heidelberg New York

Springer is a part of Springer Science+Business Media.

springer.com

Typesetting by authors and SPi using a Springer LaTeX macro package
Cover design: deblik, Berlin

Printed on acid-free paper SPIN 11731283 62/3100/SPi 5 4 3 2 1 0

Preface

Various disciplines have been benefited by the advent of high-performance computing in achieving practical solutions to their problems and the area of health care is no exception to this. Signal processing and data mining tools have been developed to enhance the computational capabilities so as to help clinicians in diagnosis and treatment.

The electrocardiogram (ECG) is a representative signal containing information about the condition of the heart. The shape and size of the P-QRS-T wave and the time intervals between various peaks contains useful information about the nature of disease afflicting the heart. However, the human observer cannot directly monitor these subtle details. Besides, since biosignals are highly subjective, the symptoms may appear at random in the timescale. The presence of cardiac abnormalities are generally reflected in the shape of ECG waveform and heart rate. However, by the very nature of biosignals, this reflection would be random in the timescale. That is, the diseases may not show up all the time, but would manifest at certain irregular (random) intervals during the day. Therefore the study of ECG pattern and heart rate variability has to be carried out over extended periods of time (i.e., for 24 hours). Naturally the volume of the data to be handled is enormous and its study is tedious and time consuming. As a consequence, the possibility of the analyst missing (or misreading) vital information is high. Hence, the electrocardiogram and heart rate variability signal parameters, extracted and analyzed using computers, are highly useful in diagnostics. This book deals with the acquisition, extraction of the various morphological features, classification and analysis of the cardiac signals. This book comprises of twenty chapters. It deals with the acquisition, extraction of the various features, classification, cardiac function during different phases, arterial pressures and analysis of fluid perfusion through soft tissues of left ventricular myocardium heart muscle.

The Chapter 1 of the book explains the recording of ECG and electrophysiologic principles underlying cardiac electrical activity. The terms used in clinical electrocardiography are defined, and the depolarization sequence of the heart is reviewed. Different modes of lead placement and the various arrhythmia are discussed in depth.

Generally the various characteristics features of ECG are extracted and used for decision making purposes. This makes the decision making and diagnosis process simpler and faster. Hence appropriate feature description and extraction becomes the most important component in cardiac health diagnostics. In Chapter 2, various techniques for feature extraction are described.

In Chapter 3, a detailed discussion is done on the prediction of heart rate signals using Auto Regressive (Burg's method), and Recursive Neural Network (Elman's method). The performance of the prediction methods is evaluated using the Normalized Root Mean Square Error.

Visualization of ECG data is an important part of the display in life threatening state. The symptoms of disease may occur at random in the timescale, and the physician has to view enormously large data to diagnose the affliction. The tedium of reading large data can be considerably reduced by using the computer for a hierarchical visualization technique in Chapter 4.

Heart Rate Variability (HRV) is a reliable reflection of the many physiological factors modulating the normal rhythm of the heart. It is a powerful means of observing the interplay between the sympathetic and parasympathetic nervous systems. Chapter 5 discusses the various applications of heart rate variability and different linear, non-linear techniques used for the analysis of the HRV.

The fusion of ECG, blood pressure, blood oxygen saturation and respiratory data for achieving improved clinical diagnosis of patients in cardiac care units. Computer based analysis and display, of the heterogeneous signals for the detection of life threatening states is demonstrated using fuzzy logic based data fusion. A new parameter called deterioration index is proposed to evaluate the severity of the disease and results are tabulated for various cases in Chapter 6.

In Chapter 7, a new method is developed using artificial neural network (ANN) techniques for classification of the states of patients in intensive care unit (ICU) from the electrocardiogram (ECG) obtained from the patients. The states are classified into normal, abnormal and life threatening classes using different neural network techniques.

In Chapter 8, the AR modeling technique is proposed to classify six types of cardiac arrhythmias. Quadratic discriminant function (QDF) based classification algorithm was performed in various stages. The AR modeling technique can be utilized for parameter extraction and are subsequently used for classification and diagnosis in telemedicine system.

Chapter 9, presents a comparative approach for classification of eight cardiac diseases using classifiers as, neural network, fuzzy and neuro-fuzzy. The features are extracted from the raw heart rate signals using the non-linear signal processing techniques and fed to the classifiers for classification.

Digital watermarking technique is adapted in Chapter 10, for interleaving patient information and cardiac signals with medical images, to reduce storage and transmission overheads. The text data is encrypted and the graphical signals like heart rate signals are compressed and subsequently interleaved with the error correcting codes to increase the reliability during transmission and storage of the images.

In Chapter 11, we developed a mathematical and clinical evaluation for a quantitative understanding of the cardiac function and dysfunction during diastolic-filling and ejection phases. The computation of clinical-diagnostic measures of (i) left ventricular (LV) volume-dependent passive elastance and active elastance, (ii) LV maximal change rate of normalized wall stress with respect to intracavitary pressure are discussed.

Chapter 12 discusses the determination of arterial pressure pulse wave propagation velocity and arterial properties. It represents the analytical basis for determining arterial elasticity parameters from pulse wave velocity (PWV). It is shown that the PWV can be used as an index of arterial distensibility.

Morphological filtering algorithm using modified morphological operators is proposed for baseline correction and noise suppression in ECG signals. The performance of the proposed algorithm was evaluated by using simulated signals and clinically acquired ECG data from a standard set in Chapter 13.

ECG, blood pressure and respiratory signals can provide important information on the pathophysiology of the cardiovascular regulatory mechanisms. Spectral and cross-spectral analysis of these signals gives quantitative information which can be of potential interest in clinical studies. This discussed in Chapter 14 using a methodology based upon multivariate autoregressive identification and parametric power spectral density estimation.

Considering the heart as a nonlinear complex system and processing various cardiovascular signals like ECG, HRV and ABP seems to provide very useful information for detection of abnormalities in the condition of the heart than is possible by conventional means. In Chapter 15, a new multidimensional phase space analysis of these cardiovascular signals using weighted spatial filling index has been proposed for detecting cardiac dysfunction.

Chapter 16, deals with the linear, non-linear and wavelet analysis of eight kinds of cardiac abnormalities. It also, presents the ranges of linear and nonlinear parameters calculated for them with a confidence level of more than 90%. A unique visual pattern of scalogram and phase space plot of heart rate signal, which may provide considerable insight into the nature and pattern of the disease are discussed.

Myocardial perfusion is the flow or forced passage of blood through the coronary arteries and to the LV myocardium. A review of the biomechanical models focused on the analysis of fluid perfusion through soft tissues of left ventricular myocardium heart muscle in particular and its fundamental constitutive relations was included in Chapter 17.

Conventional Fourier Transform techniques are not suitable for analysis of nonstationary signals. Wavelet analysis, on the other hand, provides a better insight into both the timing and intensity of transient events. Chapter 18 deals with the theory of the wavelet transforms and few applications related to the cardiac signals are dealt exhaustively in this.

Several phenomenon show the 1/f fluctuation in nature which is of interest. Recently electrical appliances have a control system included 1/f fluctuation. Heart rate in daily life also show the 1/f fluctuation. After suffering several diseases, the patients still show the 1/f fluctuation. In Chapter 19, the 1/f fluctuation in brain-death patients is discussed with results.

Stress measurement by heart rate variability is popular not only in daily life but in hospital ward. In ICU, the spaghetti syndrome gives rise to stress in patients and easily estimated by heart rate monitor. After stroke, several people suffer the aphasia. In Chapter 20, the stress during speech therapy using the method of heart rate variability is investigated.

It is our humble hope that this book will assist those who seek to enrich their lives and those of others with the wonderful powers of cardiac signal processing. Electrical, Computer and Biomedical Engineering are great fields, contributing immensely to the service of humanity.

Rajendra Acharya U
Jasjit S. Suri
Jos A E Spaan
Shankar M Krishnan

Cardiac Signal Processing

Rajendra Acharya U
Jasjit S. Suri
Shankar M. Krishnan
Jos A. E. Spaan
Editors

Rajendra Acharya U, PhD is a visiting faculty in Ngee Ann Polytechnic, Singapore. He received his Doctorate National Institute of Technology Karnataka, Surathkal, India in 2001. He served as Assistant Professor, in Bio-Medical Engineering department in Manipal Institute of Technology, Manipal, India till 2001. His current interests are Visualization, Bio-signal Processing and Image Processing.

Jasjit S. Suri, PhD is an innovator, scientist, and an internationally known world leader and has spent over 20 years in the field of biomedical engineering/ sciences and its management. He received his Doctorate from University of Washington, Seattle and Management Sciences from Weatherhead, Case Western Reserve University, Cleveland. Dr. Suri was crowned with President's

Gold medal in 1980 and the Fellow of American Institute of Medical and Biological Engineering for his outstanding contributions.

Jos A.E. Spaan is Full Professor of Medical Physics at the faculty of Medicine, Academic Medical Center and the Faculty of Science, University of Amsterdam. He is Editor in Chief of Medical & Biological Engineering & Computing, the official journal of the International Federation of medical and Biological Engineering, the IFMBE.

Shankar M. Krishnan received the Ph.D. degree in Electrical Engineering from the University of Rhode Island and is conducting research at the Rhode Island Hospital. He is presently a consultant and advisor to various international projects in the biomedical and clinical engineering field.

Contents

List of Contributors

Johnny Chee
Biomedical Engineering Centre
School of Engineering, Ngee Ann Polytechnic, Singapore

Seow Swee Chong
Cardiologist, National University Hospital, The Heart Institute
Singapore

Kannathal N, Rajendra Acharya U, Lim Choo Min
Department of Electronics and Computer Engineering, Ngee Ann Polytechnic, Singapore

Jayachandran E S, Paul Joseph K
Department of Electrical Engineering, National Institute of Technology
Calicut, India

Jasjit S. Suri
Idaho's Biomedical Research Institute, ID, USA and Biomedical Technologies, Inc., CO, USA

and

Research Professor
Biomedical Research
Institute, ISU, USA

P Subbanna Bhat
Department of Electronics & Communication Engineering, National Institute of Technology Karnataka
Surathkal, India

Niranjan U C, Sathish Kumar
Department of Biomedical Engineering, Manipal Institute of Technology
Manipal, India

E. Y. K. Ng
School of Mechanical and Aerospace Engineering, Nanyang Technological University, Singapore

Jos A E Spaan
Department of Medical Physics
Academic Medical Center, University of Amsterdam, Netherlands

Liang Zhong, Eddie Y.K Ng
School of Mechanical and Aerospace Engineering, Nanyang Technological University, Singapore

Dhanjoo N Ghista
School of Chemical & Biomedical Engineering, Nanyang Technological University, Singapore

Ru San Tan, Soo Teik Lim, Terrance S. J. Chua
Department of Cardiology, National Heart Center, Singapore

S. M. Krishnan, Kwoh Chee Keong, Sun Yan, Chan Kap Luk, Narayanan Srinivasan
Biomedical Engineering Research Centre, Research TechnoPlaza, 50 Nanyang Drive, XFrontiers Block Nanyang Technological University Singapore, 6th Storey, Singapore 637553

D Narayana Dutt
Indian Institute of Science Bangalore, India

Y W Chan, V. Anantharaman
Singapore General Hospital Singapore

Dingfei Ge
School of Information and Electronics Engineering, Zhe Jinag University of Science and Technology, Hangzhou 310012, China

Narayanan Srinivasan
Indian Institute of Science Bangalore, India

Nakajima K, Sasaki K
Division of Bio-Information Engineering, Faculty of Engineering University of Toyama, Japan

Tamura T
Department of Medical System Engineering, Faculty of Engineering Chiba University, Japan

T Maekawa
Division of Stress and Bio-response Medicine, School of Medicine, Yamaguchi University; Advanced Medical Emergency and Critical Care Center, Yamaguchi University Hospital, Japan

Toshiyo Tamura, Masaki Sekine
Department of Medical System engineering, School of Engineering Chiba University

Ayako Maeda, Yuji Higashi, Toshiro Fujimoto
Rehabilitation Centre, Fujimoto-Hayasuzu Hospital, Japan

R.C. Jegathese
College of Engineering, Nanyang Technological University, 50 Nanyang Avenue Singapore 639798

1

The Electrocardiogram

Johnny Chee and Swee-Chong Seow

The electrocardiogram (ECG) is a graphical recording of the electrical signals generated by the heart. The signals are generated when cardiac muscles depolarise in response to electrical impulses generated by pacemaker cells. Upon depolarisation, the muscles contract and pump blood throughout the body. The ECG reveals many things about the heart, including its rhythm, whether its electrical conduction paths are intact, whether certain chambers are enlarged, and even the approximate ischemic location in the event of a heart attack (myocardial infarction).

A typical ECG recording from a normal person (Fig. 1.1) and human heart (Fig. 1.2) are shown below.

The ECG is described by waves, segments and intervals:

- Waves are labelled using the letters P, QRS, T and U. The typical normal ECG may not show a U wave.
- Segments are time durations between waves, e.g. P-R segment is the duration between the P and R waves (or P and Q waves, when Q wave is present).
- Intervals are time durations that include waves and segments, e.g. P-R interval is made up of the P-wave and the P-R segment.

In the absence of an S-wave, the junction where the R-wave joins the S-T segment is described as the J-point (Fig. 1.3).

The significance of the waves (Fig. 1.4) may be broadly described as follows:

- P-wave corresponds to the depolarisation of the atrial myocardium (muscles of upper chambers of the heart), and indicates the start of atrial contraction that pumps blood to the ventricles.
- The Q, R, and S waves are usually treated as a single composite wave known as the QRS-complex. The QRS-complex reflects the depolarisation of ventricular myocardium, and indicates the start of ventricular contraction that pumps blood to the lungs and the rest of the body.

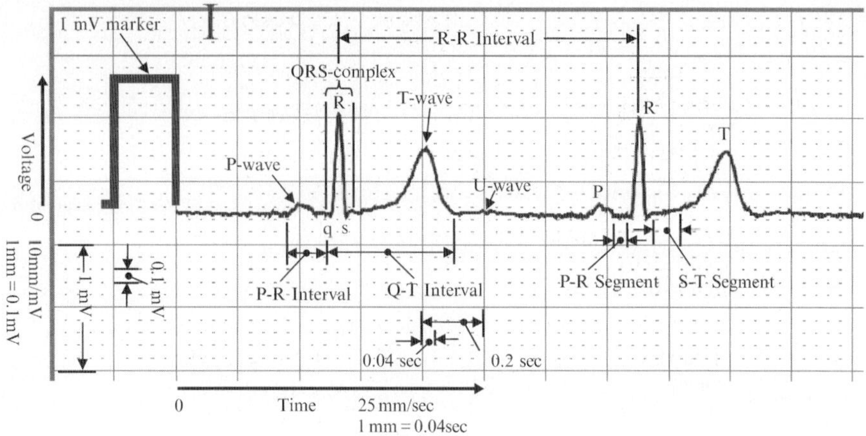

Fig. 1.1. Typical ECG of a normal person

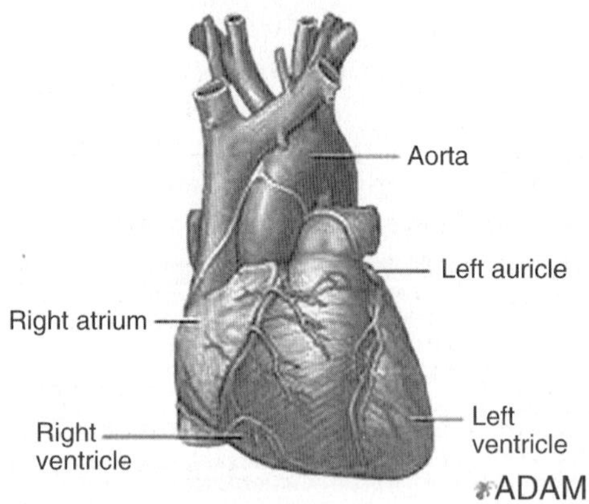

Fig. 1.2. Human heart

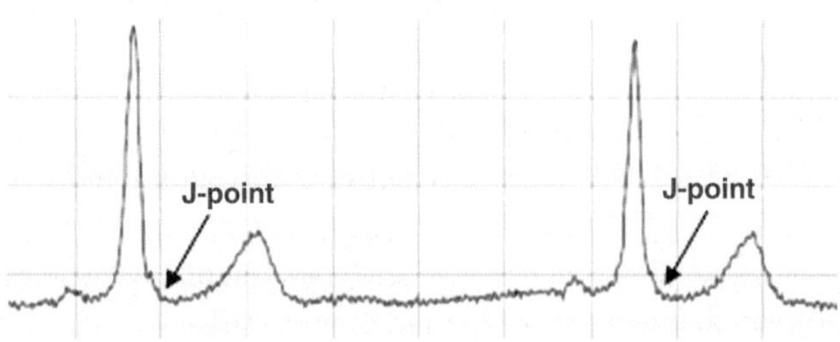

Fig. 1.3. In the ECG of some normal persons, the S-wave is not seen. Instead a J-point appears in its place and is defined as the intersection of the S-T segment with the ending of the QRS-complex

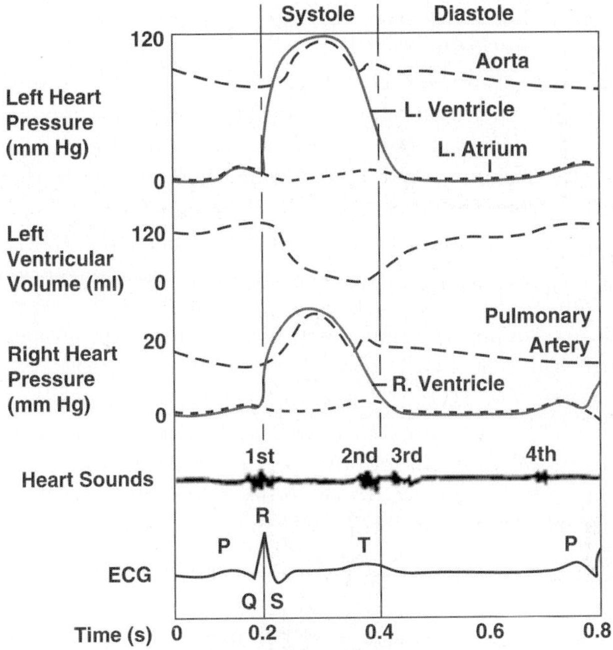

Fig. 1.4. When the ventricles depolarise, the ventricular muscles contract and rapidly builds up systolic pressure. When left ventricular pressure exceeds that of the aorta, the aortic valve opens and blood ejects out to circulate in the body

- The T-wave corresponds to the repolarisation of the ventricular myocardium, which is a necessary recovery process for the myocardium to depolarise and contract again. The end of the T-wave coincides with the end of ventricular contraction. Atrial depolarisation (T_a-wave) is usually not visible as it normally coincides with the QRS-complex (and is buried in the larger waveform).
- The origin of the U-wave is uncertain, and is thought to represent repolarisation of endocardial structures or late depolarisation of the ventricular myocardium. U waves may be seen in a normal ECG, but are <10% of the height of the QRS complex. They become prominent under abnormal conditions such as electrolyte imbalance and drug toxicity.

More detailed discussion of the waves, segments and intervals in relation to arrhythmias follows in Sect. 1.3 of this chapter, after a brief study of the heart's anatomy and its electrical conduction system.

1.1 Anatomy of the Heart

The heart is an efficient muscular organ that pumps blood throughout the whole body. Blood brings needed nutrients and oxygen to tissue, and carries away metabolic waste and carbon dioxide for excretion through the kidneys and the lungs, respectively.

The heart is made up of 4 chambers (Fig. 1.5). The two upper chambers are called the left and right atria or auricles, while the lower two chambers are called the left and right ventricles. The atria are attached to the ventricles by fibrous, non-conductive tissue that keeps the ventricles electrically isolated from the atria [1–4]. A thin membranous wall called the interatrial septum separates the left atrial chamber from the right while a thicker muscular wall called the interventricular septum separates the left ventricular chamber from the right.

The right atrium and the right ventricle together form a pump to the circulate blood to the lungs. Oxygen-poor blood is received through large veins called the superior and inferior vena cava and flows into the right atrium. The right atrium contracts and forces blood into the right ventricle, stretching the ventricle and maximising its pumping (contraction) efficiency. The right ventricle then pumps the blood to the lungs where the blood is oxygenated. Similarly, the left atrium and the left ventricle together form a pump to

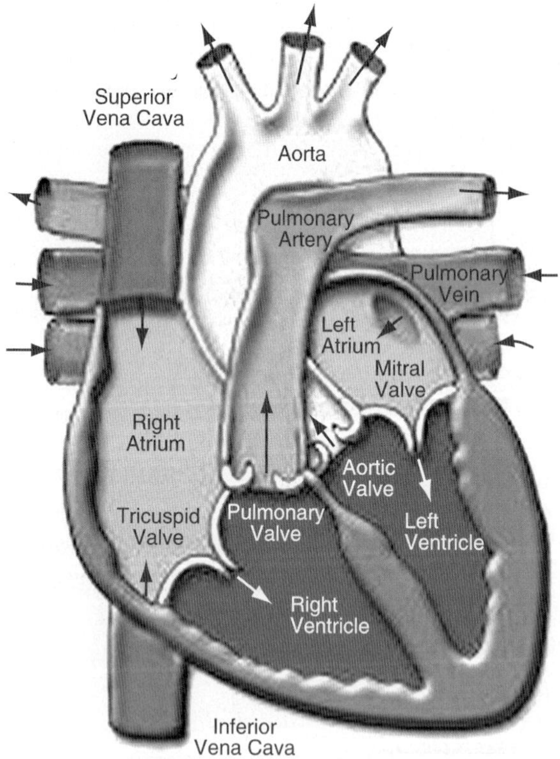

Fig. 1.5. A diagrammatic structure of the human heart

circulate oxygen-enriched blood received from the lungs (via the pulmonary veins) to the rest of the body.

Pumping is performed with synchronised motion. The right and left atria contract in unison to force-fill the ventricles, and then the right and left ventricles contract in unison to forcefully pump blood to the lungs and other parts of the body, respectively. The time duration during which the ventricles contract is known as "systole" while the time duration during which the ventricles relax to receive blood is called "diastole".

The left ventricle typically has a muscular wall about three times as thick as that of the right ventricle (Fig. 1.6) because of the heavier workload to circulate blood to the rest of the body, as compared to that required to circulate blood to the lungs.

The muscle wall of the heart is made up of three layers. The inner layer, called the endocardium, lines the chambers of the heart. The centre layer is the myocardium, which forms the bulk and provides the contractile force for pumping. This layer of myocardium is further divided into the subendocardial area which is the inner half of the myocardium, and the subepicardial area, the outer half. The outermost layer of the heart wall overlying the myocardium is called the epicardium. The entire heart is encased in a thin membrane called the serous pericardium, which is made up of 2 layers viz the visceral (inner) and parietal (outer) pericardium. Pericardial fluid between these 2 layers minimises friction against heart movements as the heart beats.

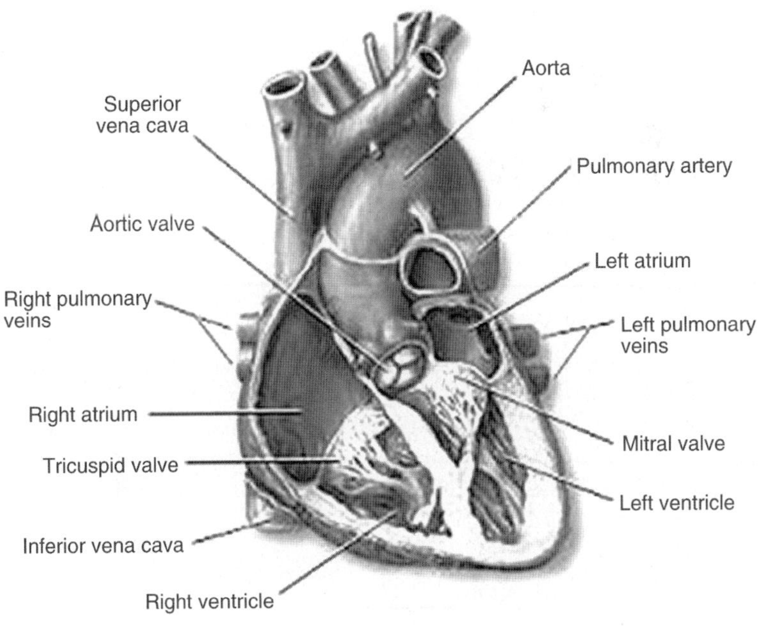

Fig. 1.6. Cutaway section of the human heart showing the myocardium

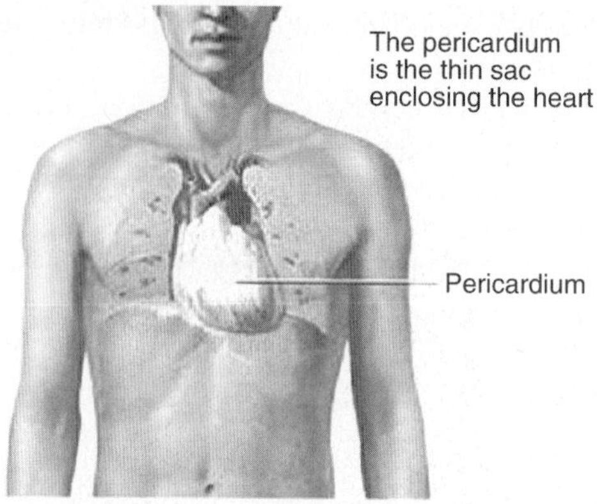

Fig. 1.7. The pericardium anchors the heart within the body

External to the serous pericardium is an outer tough, fibrous sac called the fibrous pericardium which holds the heart in place within the body (Fig. 1.7). Ligaments attach the fibrous pericardium inferiorly to the centre of the diaphragm, anteriorly to the sternum, and posteriorly to the oesophagus, trachea, and main bronchi.

Valves regulate the flow of blood by ensuring that blood flows only in the desired direction, and prevent backflow. If the valves are incompetent (i.e. "leaky") or stenotic (i.e. narrowed), the heart has to work harder. This additional workload can, over time, cause the heart to compensate by developing a thicker myocardium (hypertrophy); and in severe cases eventually lead to heart failure.

1.2 Electrical Conduction System

Pumping is only efficient when the heart contracts in a coordinated manner. Blood must first fill the atria, and then be pumped into the ventricles before being forcefully ejected. This coordination is achieved by an elaborate electrical conduction system that controls the precise timing for depolarising the substantial mass of electrically excitable myocardium. This delicate control starts with an intrinsic self-excitable cardiac pacemaker which sets the rate at which the heart beats. The pacemaker spontaneously generates regular electrical impulses, which then spread through the conduction system of the heart and initiate contraction of the myocardium. This pacemaker is called the Sino-atrial (S-A) node.

1.2.1 The Sino-Atrial (S-A) Node

The S-A node lies in the upper wall of the right atrium, near the entrance of the superior vena cava (Fig. 1.8). It is normally the initial source of electrical

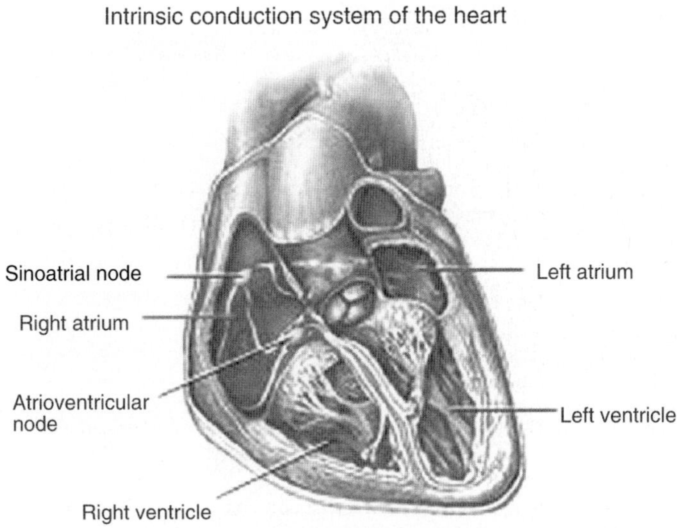

Fig. 1.8. The intrinsic conduction system

excitation. The S-A node is a network of pacemaker cells – excitable tissue that exhibits automaticity. Automaticity is a property of the cell to periodically generate an electrical impulse even without the presence of an external stimulus. Automaticity of the S-A node may be modulated by the balance of sympathetic and parasympathetic inputs, or by drugs. Sympathetic stimulation (from nerves connected to the brain) speeds up the rate of impulse generation while parasympathetic stimulation slows the rate. This variability in rate allows the heart to respond to demands for higher or lower cardiac output (i.e. faster or slower heart rate).

The electrical impulse from the S-A node spreads through the myocardium of the right atrium, stimulating its contraction (Fig. 1.9). At the same time, the interatrial conduction tract (Bachman's Bundle) between the S-A node and the left atrium carries the impulse quickly to the left atrium, spreading it through the left atrial myocardium so that the contraction of the left atrium occurs almost simultaneously with that of the right. In addition, three internodal conduction tracts carry the impulse from the S-A node to Atrioventricular (A-V) node, the gateway to the ventricular conduction system.

1.2.2 Depolarisation

Propagation of an electrical impulse through excitable tissue is achieved through a process called depolarisation. At rest, a potential difference (voltage) exists across the cell membrane, between the fluid external to the cell and its internal cellular fluid as the cell membrane separates the ions of both fluids. When a trigger occurs (e.g. an electrical impulse exceeding a threshold

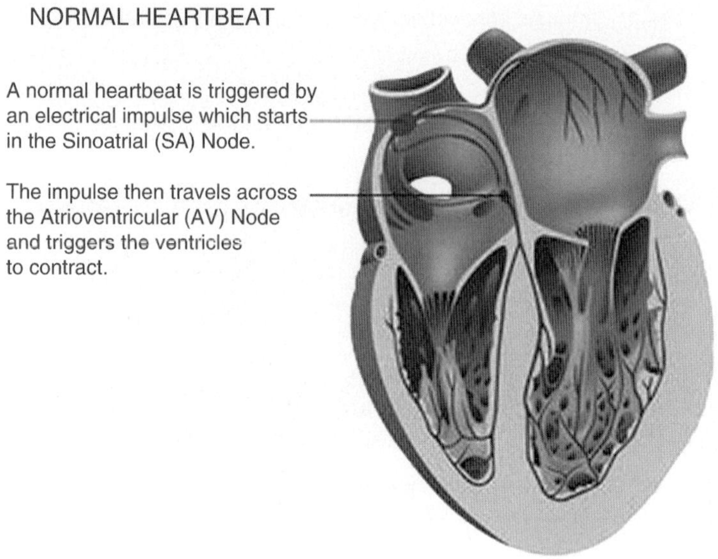

Fig. 1.9. Sino-atrial node in the right atrium connects electrically via conduction tracts to the atrioventricular node

voltage) the cell membrane suddenly becomes permeable. The ions that were previously held at bay across the membrane cross the membrane generating an ionic current flow.

The depolarisation of a myocardial cell will lead to a physical contraction within milliseconds. The contraction would then last for tens of milliseconds. At the end of depolarisation, the cell membrane once again becomes impermeable. Repolarisation begins as the ion channels within the cell membrane pump out unwanted ions and brings the ionic balance of the cell back to its resting (equilibrium) state.

Myocardial cells that make up the heart muscles are interconnected to adjacent myocardial cells through specialised cellular membranes known as intercalated disks (Fig. 1.10). These disks contain areas of low electrical resistance called gap junctions, which permit rapid conduction of electrical impulses from one cell to another [5].

With its large muscle mass, depolarisation of the heart muscles collectively generates a strong ionic current. This current flows through the resistive body tissue generating a voltage drop. The magnitude of the voltage drop is sufficiently large to be detected by electrodes attached to the skin (typically 3 mV-peak at the chest). ECGs are thus recordings of voltage drops across the skin caused by ionic current flow generated from myocardial depolarisations. Although nerve depolarisations also generate ionic currents, the magnitudes of such currents are too small to be detected by skin electrodes.

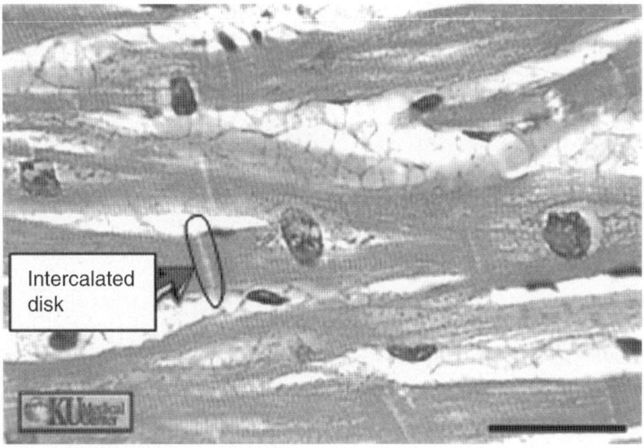

Fig. 1.10. Myocardial cells are interconnected to one another electrically and physically by the intercalated disks

Atrial depolarisation results in the spreading of the electrical impulse through the atrial myocardium and appears as the P-wave. This wave is normally less than 120 ms wide and corresponds with the start of atrial muscular contraction. The P-R interval, which is measured from the onset of the P-wave to the onset of the QRS-complex, is normally within 120–200 ms. (Note that if the Q-wave is present, the P-R interval should terminate on the onset of Q-wave although it would still be labelled as P-R interval.) Atrial contraction typically lasts longer than the P-R interval.

Similarly, ventricular depolarisation results in the spreading of the electrical impulse throughout the ventricular myocardium. Depolarisation is triggered when the pacemaker impulse from the S-A node comes through the atrioventricular node [1–4] and spreads through the ventricular conduction system to the myocardium.

1.2.3 The Atrioventricular (A-V) Node

The A-V node lies partly in the right side of the interatrial septum and partly in the interventricular septum. Since the ventricles are separated from the atria by a fibrous layer of non-conducting tissue, no electrical impulse from the S-A node will reach the ventricles except through the A-V node (Fig. 1.10). This allows the A-V node to control the impulse received from the S-A node. Each impulse, received through the atrial conduction tracts, is propagated at a slow rate ($\sim$0.05 m/s) through the A-V node, generating a time delay. This time delay is to allow blood from the atria to fully fill the ventricles before the latter contract.

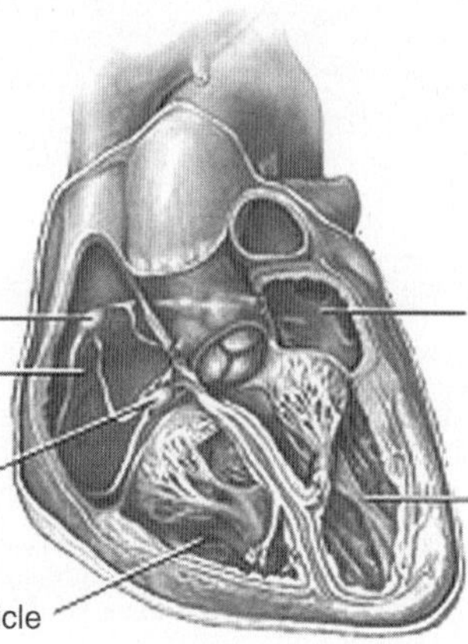

Fig. 1.11. The ventricular conduction system

The conduction system of the ventricles is more elaborate than that of the atria. The A-V node connects to the Bundle of His which in turn connects to two bundle branches – the Right Bundle Branch and the Left Bundle Branch (Fig. 1.11). The right bundle branch travels down the right side of the interventricular septum to subdivide again and again before connecting to the Purkinje Network of the right ventricle. The Purkinje network of fine conducting fibres is embedded in subendocardium of the ventricles. In like manner, the left bundle branch divides into the left posterior fascicle and left anterior fascicle which travel down the left side of the interventricular septum to further subdivide repeatedly before connecting with the Purkinje Network of the left ventricle. In addition, the left anterior fascicle also branches to form septal fibres which innervate the interventricular septum.

Thus from the A-V node, the impulse propagates through the bundle of His, spreads through the bundle branches and sub-branches until it reaches the Purkinje network. The spread of impulses to depolarise the ventricular myocardium is nearly simultaneous despite the large mass of myocardium because of the high speed of propagation through the conduction fibres.

1.2.4 Automaticity

Cells that form different parts of the conduction system exhibit different rates of automaticity. The S-A node has the fastest inherent rate of $60 \sim 100$ pulses per minute. It becomes the dominant pacemaker in the heart, setting the rhythm at which the heart contracts by overriding (resetting) potential pacemakers running at slower rates. The A-V node and the Bundle of His are the next fastest with an inherent rate of $40 \sim 60$ pulses per minute. Finally the Bundle Branches and Purkinje Network exhibit the slowest rates, at $30 \sim 40$ pulses per minute.

Propagation speeds of electrical impulses along the conduction paths are:

- A-V node: ~0.05 m/s
- Bundle of His and Bundle Branches: ~2 m/s
- Purkinje Fibres: ~4 m/s
- Myocardium: ~0.5 m/s

When the S-A node fails to fire or the impulse from the S-A node fails to reach the A-V node, the A-V node takes over as the dominant pacemaker since it has the next fastest rate of automaticity. The heart rate will then be slower at 40 ~ 60 pulses per minute. If there is a conduction block below the level of the A-V node, then the bundle branches, the Purkinje network or the ventricular muscles will take over as the pacemaker for ventricular contraction. The ventricular contraction rate will then be even slower at 30 ~ 40 pulses per minute.

A conduction disorder is often observable in an ECG. For instance, a block in the interatrial tract will lead to normal depolarisation of the right atria but a delayed depolarisation of the left atria, resulting in a broadened, notched P-wave. This occurs because the impulse can only propagate slowly by way of myocardial depolarisation to reach the left atrium rather than through the fast conduction pathway.

1.2.5 Accessory Pathways

Normally, the only electrical connection between the atria and the ventricles is the A-V node. But in some hearts, abnormal conduction (accessory) pathways may be present. These are electrically conducting fibres that short circuit the normal route that a depolarising impulse would be forced to take. The shorter route leads to an early start in the depolarisation of the ventricles causing the QRS complex of the ECG to be abnormal in morphology (shape).

Commonly found accessory pathways are:

(1) Accessory A-V pathways (Bundles of Kent) crossing the insulating fibres between the atria and the ventricles.
(2) Atrio-His Fibres (James Fibres) – between the atria and the A-V node.
(3) Mahaim Fibres, comprising:
   (a) Atriofascicular fibres – between the right atrium and the right bundle branch.
   (b) Nodoventricular fibres – between the A-V node and the right ventricle.
   (c) Fasciculoventricular fibres – between a fascicle of the bundle branch and the ventricle.

1.2.6 Excitable Tissue and Generation of Ionic Currents

Muscle is a form of contractile tissue. When triggered by an electrical impulse (or other chemical, mechanical stimulus, etc.) it depolarises and contracts. Nerves also depolarise when triggered, but they do not contract.

When an excitable tissue is at rest, its cell membrane is impermeable to ions. Ions in the interstitial fluid external to the cell thus remain separated from those within the cell. If we take zero millivolts to be the potential voltage of the interstitial fluid, then the potential of the intracellular fluid will be found to be −90 mV (Fig. 1.12). When a stimulus such as an ionic current raises the cell internal potential beyond its threshold voltage of −70 mV, the cell depolarises. The cell membrane suddenly becomes permeable and an exchange of ions takes place across the membrane (phase 0). The potential within the cell rises rapidly and momentarily reaches to +20 mV at which point the membrane becomes impermeable to ions again. Repolarisation starts to take place as charge pumps within the cell force out unwanted ions, restoring the ion concentration of the cell back to resting equilibrium (phases 1–3). The action potential eventually drops to its resting potential of −90 mV (phase 4).

The cardiac pacemaker cell has a differently shaped action potential from the rest of the myocardial cells. Its resting potential does not remain constant but increases steadily with time (phase 4 of Fig. 1.13). The gradual increase in action potential is due to the slow "leak" of ions across the cell membrane at rest, leading to a gradual diminution of the voltage across the cell membrane. Once the cell potential crosses the threshold voltage, the cell depolarises. If the slope of the resting potential is increased, the automaticity of the pacemaker cell is increased (i.e. faster rate of impulse generation).

Both the S-A node and A-V nodes are innervated by efferents from the brain and the nervous system. Parasympathetic nerves from the

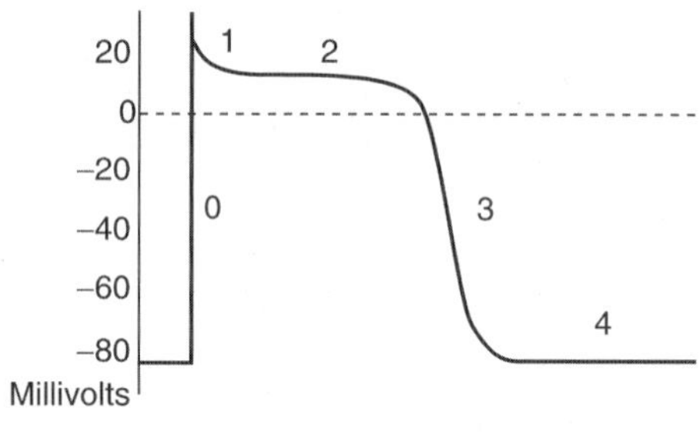

Fig. 1.12. Action potential of a myocardial cell

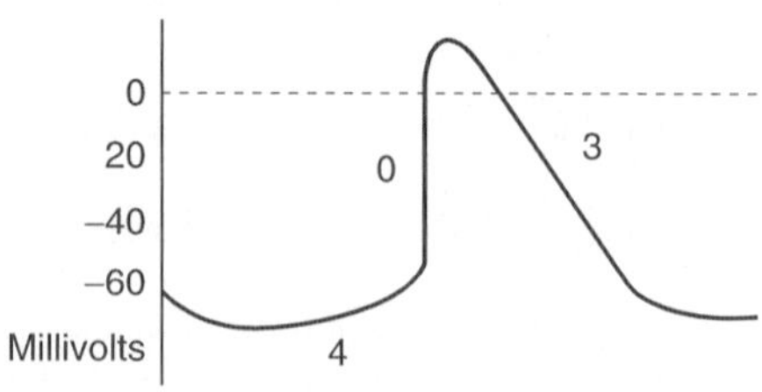

Fig. 1.13. Action potential of a pacemaker cell

cardio-inhibitor centre of the brain slow the intrinsic rate of the S-A node and delay the conduction at the A-V node; while sympathetic nerves from the cardio-accelerator centre of the brain accelerate the S-A nodal rate and augment conduction at the A-V node.

1.2.7 ECG Signal Detection

Surface (skin) electrodes are used to detect the depolarisation of excitable myocardium (Fig. 1.14). When depolarisation propagates towards the positive electrode of the amplifier, the voltage detected is seen as positive and is represented by a upward deflection in the ECG. For example, if the net electrical depolarisation is directed towards lead I, this will be represented graphically by an upward deflection on the ECG tracing for lead I.

Atrial depolarisation is seen as a positive (upright) P-wave in lead I because the impulse propagates from the right atrium towards the left (i.e. towards lead I). Repolarisation (T_a-wave) is seen as a negative (inverted) P-wave in the same lead. However for most ECGs, atrial repolarisation is masked by the QRS-complex and the T_a-wave is not discernable.

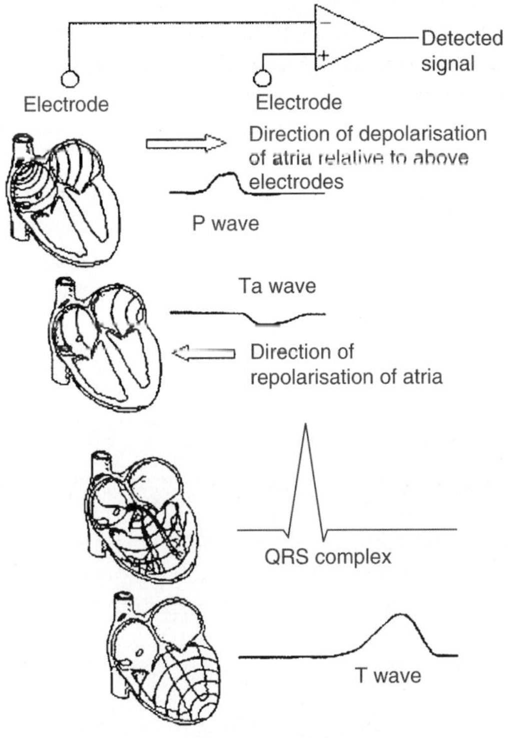

Fig. 1.14. Depolarisation of atria and ventricles generate different components of ECG

The depolarisation of ventricular myocardium starts with the interventricular septum from left to right, generating the Q-wave of the ECG in the left-sided leads. This is followed by the near-simultaneous depolarisation of the right and left ventricular muscle walls. Depolarisation starts from the subendocardial myocardium and propagates to the subepicardial myocardium. Although the right and left ventricular walls are depolarising in opposite directions, the net (sum) direction along the horizontal axis is to the left of the heart, because of the much thicker left ventricular wall generates a larger electrical potential [5].

Repolarisation of the ventricles starts in the opposite direction – from the subepicardial myocardium and proceeds to the subendocardial myocardium resulting in an upright T-wave. (If repolarisation were to take place in the same sequence as depolarisation, viz from endocardium to epicardium, the T wave would be inverted.)

1.2.8 ECG Lead Placements

The graphical recording of electrical heart signals or the electrocardiogram (ECG) is obtained by using electrodes attached to the skin across different areas of the heart. The placement of the electrodes determines the directional viewpoint of the heart. Each viewpoint is called a "lead". The standard ECG as recorded by clinicians is the 12-lead ECG, which uses 10 electrodes. A single-lead ECG recorder would typically have three electrodes: the positive electrode, the negative electrode and an indifferent electrode (ground or "right-leg" drive electrode).

Electrodes detect ionic current flow within the body by detecting the potential difference between them as current flows through resistive tissue. An electrode is labelled as positive when the ECG recording shows a positive signal (upward deflection) corresponding to depolarisation propagating towards it. Conversely, if the direction of depolarisation is propagating away from it, a negative signal would result and will be represented on the ECG as a downward deflection.

In nearly all ECG instruments, one electrode is always an indifferent electrode. The indifferent electrode is the "ground" reference of the instrument. While an indifferent electrode could be connected to the instrument ground, most often it would be connected to the output of an amplifier that generates a "right-leg drive" signal. This amplifier inverts and amplifies the common-mode signal (50 or 60 Hz power-line interference) detected at the input electrodes. By feeding back this signal into the body, the common-mode signal at the input is reduced. This connection scheme is commonly referred to as right-leg drive as the electrode is usually attached to the right leg or an electrically equivalent location.

Surface electrodes are attached to the skin of the patient and are labelled according to their location on the patient's anatomy. For example an electrode placed on his left arm will be called the left-arm electrode (LA), while those on the chest are called chest (ventral) electrodes V1, V2...V6.

LA = Left Arm Electrode
RA = Right Arm Electrode
LL = Left Leg Electrode
RL = Right Leg Electrode (indifferent electrode)
Vx = Chest Electrodes: V1 to V6

1.2.9 The Limb Leads (Bipolar) – Leads I, II, III

The limb leads are called as such because the electrodes are attached to the limbs (Fig. 1.15). Three views are immediately obtained (Table 1.1):

Leads I, II and III are commonly referred to bipolar leads as they use only two electrodes to derive a view. One electrode acts as the positive electrode while the other as the negative electrode (hence bipolar).

Note that the RL electrode is never used for obtaining the ECG. It is a ground reference electrode and is there to help the ECG instrument reduce common mode interference.

Electrically, the placement of a limb electrode at any position along the arm is the same. The minor difference is the extra impedance of tissue resistance if the electrode is placed further from the heart. Thus, an electrode can readily be attached either to the wrist or to the chest near that same arm.

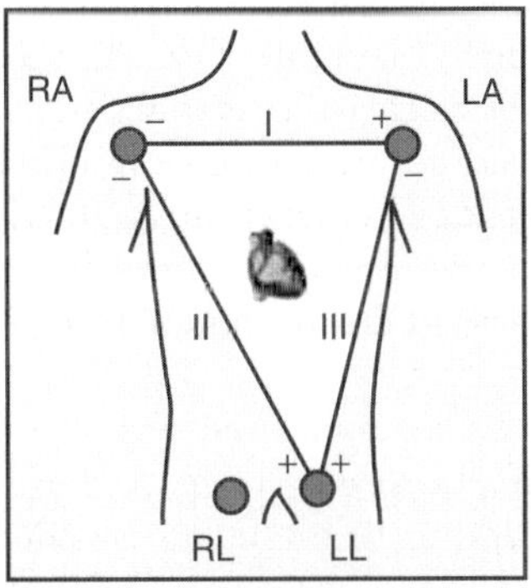

Fig. 1.15. Electrode placements for the limb leads

Table 1.1. Bipolar lead combination

Lead	Electrode+ (real)	Electrode− (real)	Signal combination	Medical angle	Mathematical angle
I	LA	RA	LA–RA	0°	0°
II	LL	RA	LL–RA	+60°	−60°
III	LL	LA	LL–LA	+120°	−120°

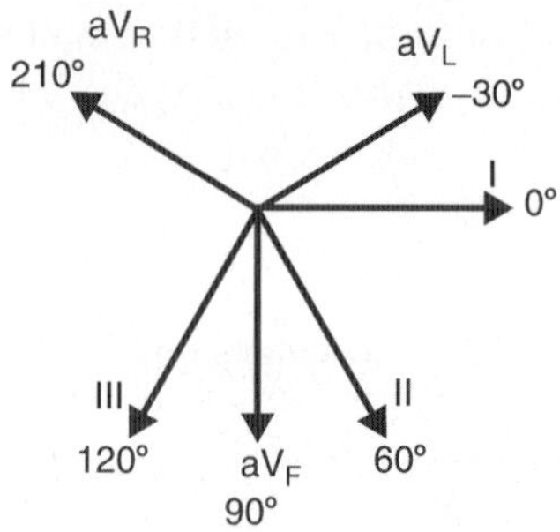

Fig. 1.16. Different directional views of the heart along the frontal plane are obtained from the limb leads and augmented limb leads

Table 1.2. Augmented limb lead combination

Lead	Electrode+ (real)	Electrode− (virtual)	Signal combination	Medical angle	Mathematical angle
aVL	LA	RA, LL	LA–1/2(RA + LL)	−30°	+30°
aVF	LL	RA, LA	LL–1/2(RA + LA)	+90°	−90°
aVR	RA	LA, LL	RA–1/2(LA + LL)	−150°	+150°

1.2.10 The Augmented Limb Leads (Unipolar) – Leads aVL, aVR, aVL

The signals from the limb electrodes can be combined to give further views called the augmented leads (Fig. 1.16). One of the limb electrodes serves as the positive electrode. The negative electrode is virtual, being the average of the signals from the remaining two limb electrodes. In contrast to Leads I, II and III, the augmented leads are known as unipolar leads (Table 1.2).

In total there are six views obtained from the limb leads (the hexaxial system) and they all view the heart signals from different angles along the frontal (anterior) plane.

1.2.11 The Precordial Leads (Unipolar) – Leads V1, V2, V3, V4, V5, V6

There are six chest electrodes V1, V2...V6 (Fig. 1.17) giving rise to six views of the heart signals across the front (ventral aspect) of the chest (Fig. 1.18). The views fall along the transverse (cross-sectional, i.e. looking into the chest) plane. The positive electrode is the chest electrode. The negative electrode is a virtual electrode commonly called the Wilson Central Terminal (WCT). This virtual electrode is realised by electrically averaging the signals from the three electrodes LA, RA and LL. The WCT is thus the electrical centre of the heart. These six leads are known as precordial leads and are unipolar leads (Table 1.3).

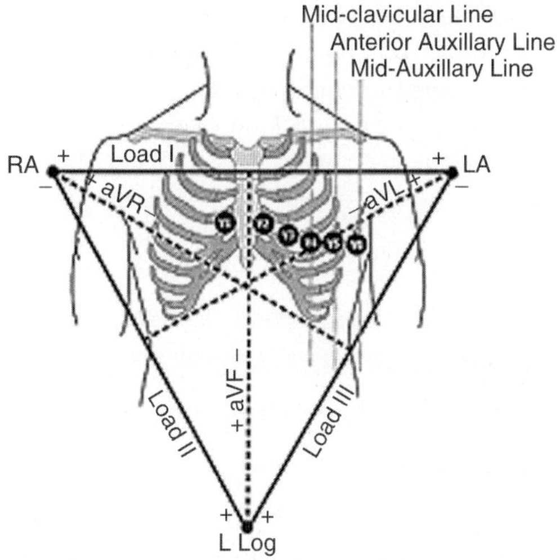

Fig. 1.17. Precordial chest electrodes are normally placed on the left side of the chest (V1...V6) allowing

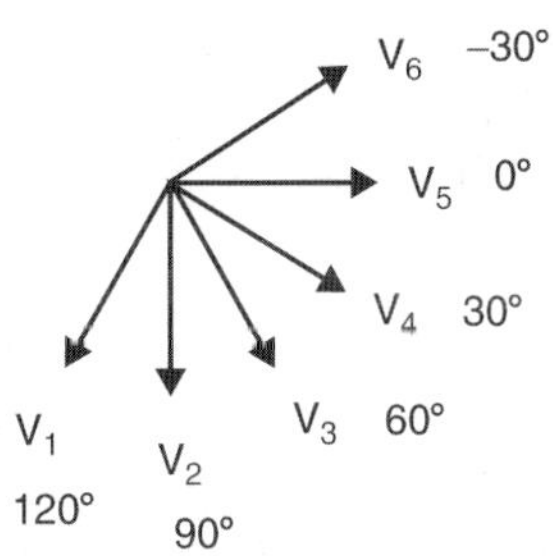

Fig. 1.18. Precordial leads are views that are in the transverse plane (i.e. cuts across the chest in the horizontal plane)

Table 1.3. Precordial limb lead combination

View	Electrode+ (real)	Electrode− (virtual)	Signal combination	Approx medical angle	Mathematical angle
V1	V1	LA, RA, LL	V1–1/3(LA + RA + LL)	+120°	−120°
V2	V2	LA, RA, LL	V2–1/3(LA + RA + LL)	+90°	−90°
V3	V3	LA, RA, LL	V3–1/3(LA + RA + LL)	+60°	−60°
V4	V4	LA, RA, LL	V4–1/3(LA + RA + LL)	+30°	−30°
V5	V5	LA, RA, LL	V5–1/3(LA + RA + LL)	+0°	+0°
V6	V6	LA, RA, LL	V6–1/3(LA + RA + LL)	−30°	+30°

All the six unipolar chest leads view the heart signals from different angles along the transverse plane. Together with the limb leads, a total of 12 views are used for diagnosis – resulting in the standard 12-lead ECG. These leads are designed to primarily look at the left side of the heart. If investigations are needed to look at the right side of the heart, then the chest leads may be mirrored and applied in a similar manner on the right side of the chest (instead of the left). The views are then designated as $V_{1R}, V_{2R}, \ldots, V_{6R}$.

In addition, there are other leads which can be used for the monitoring the patient's ECG. These are:

- Modified Chest Lead 1 (MCL1) – gives a view similar to V1 but uses bipolar leads.
- Modified Chest Lead 6 (MCL6) – gives a view similar to V6 but uses bipolar leads.

For these leads, the positive electrode is placed at the normal positions for V1 or V6, while the negative electrode is placed at a location that approximates the electrical centre of the heart. This position is approximated to be near the mid-clavicular line of the patient's chest, just below the clavicle.

1.3 Arrhythmias

The manner in which the heart contracts over time determines the rhythm of the heart. Normal sinus rhythm (NSR) is the normal rhythm of the heart when there is no disease or disorder affecting it. NSR is characterised by a heart rate of 60 to 100 beats per minute. The regularity of the R-R interval varies slightly with the breathing cycle, typically shortening slightly during inspiration (Fig. 1.19). The source of the rhythm is the Sino-atrial node, which is the normal pacemaker of the heart. Hence another characteristic feature of NSR is a normal P-wave followed by a normal QRS-complex [1–4].

When the heart rate increases beyond 100 beats per minute, the rhythm is known as sinus tachycardia (Fig. 1.20). This is not an arrhythmia but a normal response of the heart to higher demand for blood circulation.

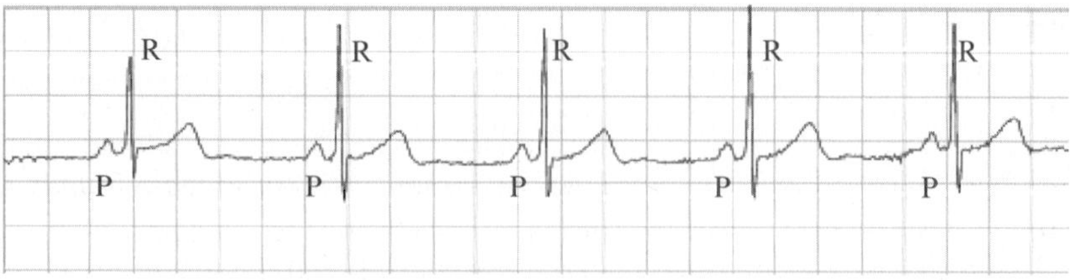

Fig. 1.19. The normal sinus rhythm has regular, repeating waveforms of P, QRS, T waves and stable time segments between the waves

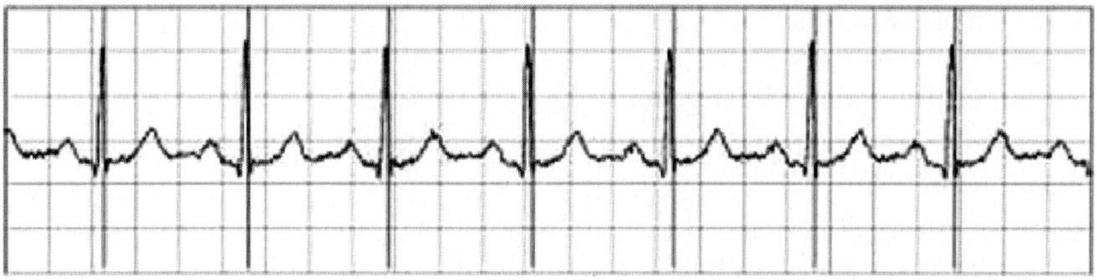

Fig. 1.20. In sinus tachycardia, the heart beats quickly, resulting in very short R-R intervals

Rhythms that deviate from NSR are called arrhythmias (or dysrhythmias) since they are abnormal and dysfunctional. Arrhythmias can be life-threatening. If the heart rate is too slow as in bradycardia, perfusion may be insufficient and this can adversely affect vital organs. Similarly, if the heart rate is too fast, the ventricles are not completely filled before contraction and pumping efficiency drops, adversely affecting perfusion. Arrhythmias may be more readily understood by categorising them in the following manner:

1. Sinus Node Arrhythmias – pacemaker in the Sino-atria node.
2. Atrial Arrhythmias – pacemaker in the atria.
3. Junctional Arrhythmias – pacemaker in A-V junction.
4. Ventricular Arrhythmias – pacemaker in the bundle branches, Purkinje network, or ventricular myocardium.
5. Atrioventricular Blocks – impulse blockage in the A-V junction.
6. Bundle Branch and Fascicular Blocks – impulse blockage in the bundle branches and sub-branches (fascicles).

1.3.1 Sinus Node Arrhythmias

The arrhythmia arises from the S-A node for this group of disorders. Since the electrical impulse is generated from the normal pacemaker, the consistent characteristic feature of these arrhythmias is that P-wave morphology (wave shape) of the ECG is normal.

Sinus Arrhythmia

This is not a disorder or a true arrhythmia, but a normal, physiologic variation in the sinus rate with the phases of respiration. The slowest instantaneous heart beat may be less than 60 beats per minute, while the highest may exceed 100 beats per minute. It is caused by variation in the vagal tone during the breathing cycle. The tone decreases during inspiration causing the heart rate to increase, but increases during expiration causing the heart rate to decrease. Sinus arrhythmia has no clinical significance.

Sinus Bradycardia

In sinus bradycardia, the rhythm originates from the S-A node but at a rate of less than 60 beats per minute (Fig. 1.21). The ECG appears normal except for the slow heart rate. Mild sinus bradycardia (50–59 beats per minute) is usually asymptomatic, while marked sinus bradycardia (30–45 beats per minute) may lead to hypotension and result in insufficient perfusion of the brain and other vital organs. Treatment is indicated if the bradycardia is symptomatic (e.g. giddiness, fainting, shortness of breath or chest pain).

Sinus Arrest

In sinus arrest, the S-A node intermittently fails to fire. There is no P-wave and therefore no accompanying QRS-complex and no T-wave (Fig. 1.22). Bradycardia (slow average heart rate) may result if the occurrence of sinus arrests is frequent. Sinus arrest results from a marked depression of the automaticity of the S-A node. Since the automaticity of the S-A node is abnormal, the longest P-P interval (i.e. the "pause" on the ECG) will not be a multiple of the shortest P-P interval; unlike Sino-atrial exit block.

Sino-Atrial Exit Block

Sino-atrial exit block is similar to sinus arrest, except that the S-A node retains its automaticity but the generated electrical impulse is unable to exit from the S-A node and to propagate. This is caused by an intermittent conduction block in the tissue surrounding the S-A node. Occasionally, conduction manages to occur through the perinodal tissue. Since the automaticity of the S-A node is normal, the longest P-P interval (or the "pause" on the ECG) is a multiple of the shortest P-P interval (the underlying rhythm).

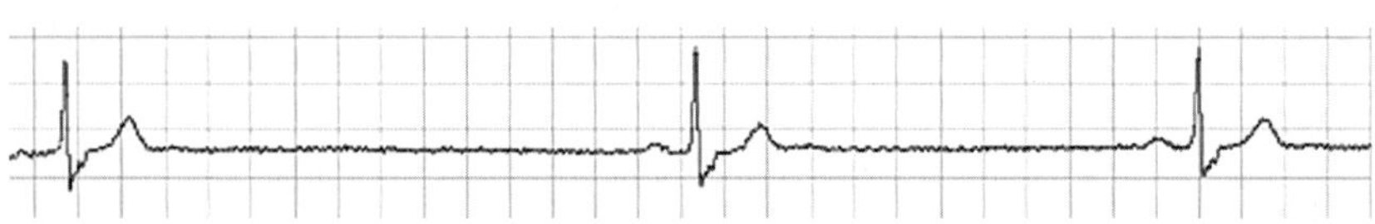

Fig. 1.21. In bradycardia, the heart beats at a slow rate, resulting in long R-R intervals

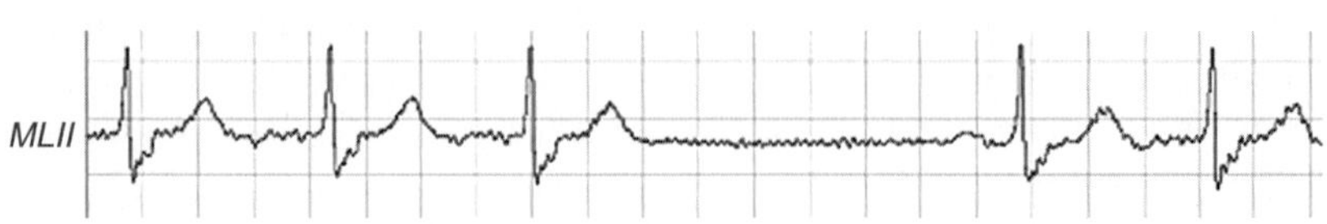

Fig. 1.22. Sinus arrest is characterised by a missing beat (P-QRS-T) that occurs repeatedly

1.3.2 Atrial Arrhythmias

Atrial arrhythmias result from electrical impulses that originate outside the S-A node but within the atria. Since the origin is not from the S-A node, the P-wave inscribed is different in morphology from the sinus P-wave. The ensuing QRS-T complex however appears as normal since the ventricles receive their impulses through the A-V node and in the usual manner.

Wandering Atrial Pacemaker (WAP)

In this condition, instead of the S-A node being the dominant pacemaker, other parts of the atria fire at a rate faster than the S-A node and usurp control of the heart rate from the S-A node. A few ectopic foci take turns doing so, depending on which fires at a faster rate. As the pacemaker site changes, the P′-wave size and shape also vary. The P-R interval varies depending on how close the ectopic focus is from the A-V node; shortening as the pacemaker site gets closer to the A-V node, and lengthening for foci farther from the A-V node (Fig. 1.23). Heart rate is usually around 60 to 100 beats per minute, becoming slower as the pacemaker site shifts away from the S-A node. WAP is often caused by the inhibitory vagal (parasympathetic) effect of respiration on the S-A node and the A-V junction. Usually WAP is not clinically significant and appears in the very young, the elderly and in athletes.

Premature Atrial Contractions (PAC)

Premature atrial contraction results in an earlier than expected occurrence of a (non-sinus) P′-wave followed by a QRS-complex and a T-wave (Fig. 1.24).

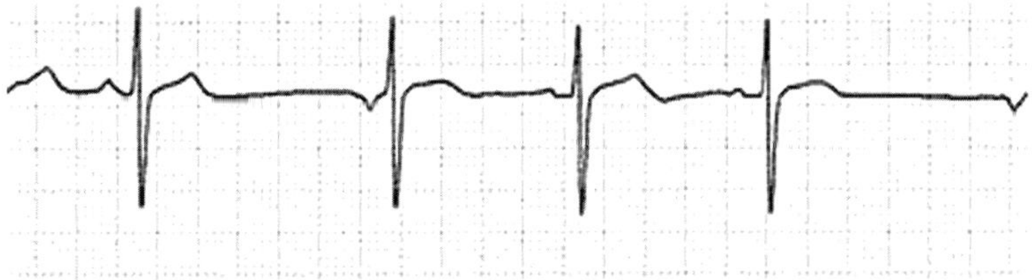

Fig. 1.23. The rhythm in wandering atrial pacemakers

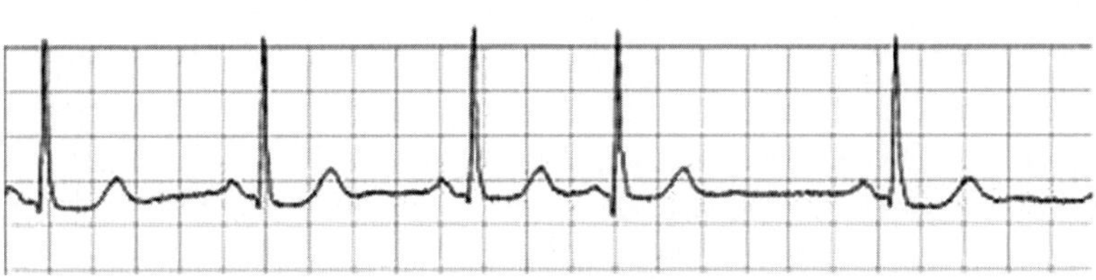

Fig. 1.24. A premature atrial contraction is an atrial contraction that appears early in time with an abnormal P′-wave morphology (shape)

This happens because of an ectopic pacemaker firing before the S-A node does. The ectopic pacemaker may reside in any part of the atria outside the S-A node. Multiple ectopic pacemakers may be involved. PACs may occur as a couplet whereby another PAC consecutively follows the first PAC. Or it may occur as atrial bigeminy, whereby a PAC occurs alternately for every normal atrial contraction. In atrial trigeminy, a PAC occurs for every two normal atrial contractions. The ensuing QRS-complex remains normal since the A-V conduction path is intact. If a PAC occurs very prematurely before the bundle branches have fully repolarised, it may result in a broad QRS due to bundle branch block. When three or more consecutive PACs occur, the rhythm is considered to be atrial tachycardia.

Atrial Tachycardia (Ectopic and Multifocal)

In atrial tachycardia the heart rate is fast and ranges from 160 to 240 beats per minute. When a single ectopic pacemaker is involved, the rhythm is called ectopic atrial tachycardia (Fig. 1.25). When three or more ectopic pacemakers are involved, the rhythm is called multifocal atrial tachycardia. The P′-waves from different ectopic pacemakers differ in size and shape since the location of pacemakers affects the direction of depolarisation of the atria (Fig. 1.26).

The AV conduction ratio is commonly 1:1 when atrial tachycardia is less than 200 beats per minute and 2:1 (every two P′-waves for one QRS-complex) above that. Although the ventricular rate may fall within the normal range because of A-V block, cardiac output is reduced as the atria are not completely filled during diastole. Frequently atrial tachycardia is accompanied by feelings of palpitations, nervousness, or anxiety.

Atrial Flutter

In atrial flutter, the atrial rate is very fast, ranging from 240 to 360 per minute. The P′-waves occur regularly and so quickly that they take on a characteristic saw-tooth waveform known as flutter (F) waves (Fig. 1.27). Atrial flutter usually occurs as a result of a rapid re-entry circuit in the atria. Usually the ventricular rate is much slower, with an AV conduction ratio of 2:1 (2 F-waves

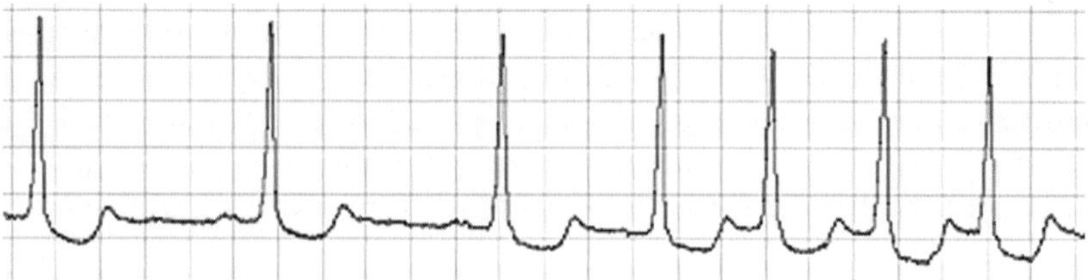

Fig. 1.25. Ectopic atrial tachycardia occurs when more than 3 premature atrial contractions (PACs) occur consecutively

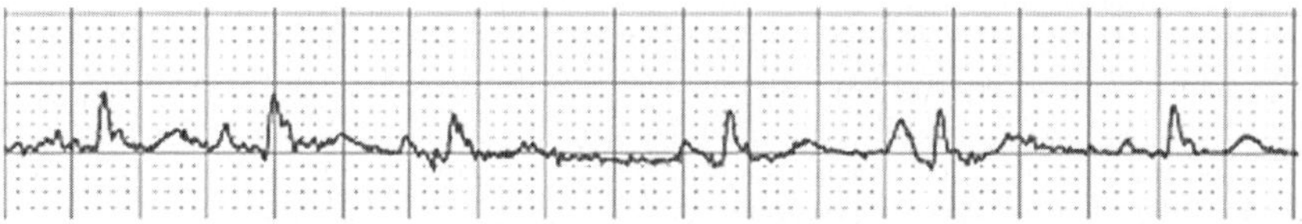

Fig. 1.26. When multiple ectopic pacemakers are involved, the rhythm is called multifocal atrial tachycardia. The distinguishing feature is that there are different P′-wave morphologies

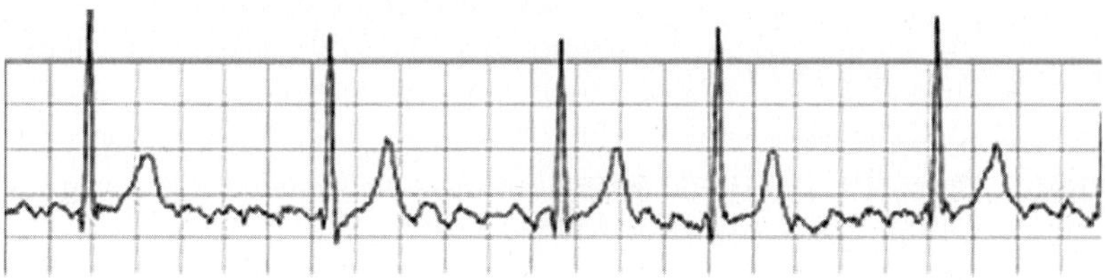

Fig. 1.27. 4:1 Atrial Flutter. Atrial rate is very fast and has a consistent repetitive waveform with a saw-tooth like appearance

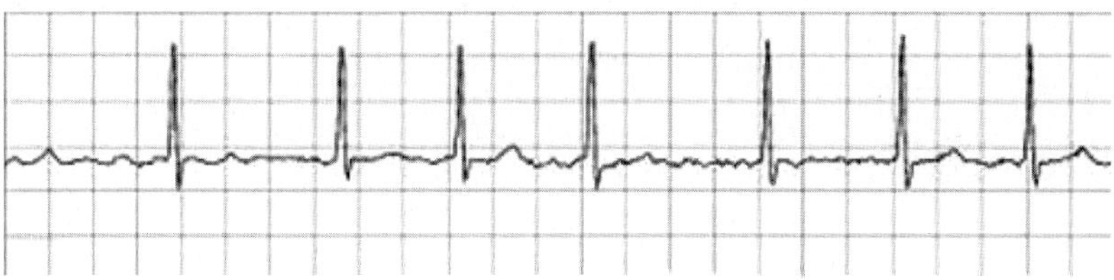

Fig. 1.28. When atrial fibrillation occurs, the QRS appears erratically. There is no distinct P′-wave

to 1 QRS-complex), 3:1 or 4:1. Cardiac output may drop by as much as 25% because of incomplete filling of the ventricles before contraction. Ventricular rate is correspondingly regular in rhythm [1–4].

Atrial Fibrillation

In atrial fibrillation, the atrial rate exceeds 350 per minute. This arrhythmia occurs because of uncoordinated activation and contraction of different parts of the atria (Fig. 1.28). Multiple sites of re-entry in the atria fire rapidly in a haphazard fashion resulting in chaotic atrial contraction. The rapid atria rate and uncoordinated contraction leads to ineffective pumping of blood into the ventricles. Cardiac output falls by as much as 25%. Ventricular contraction occurs irregularly and is commonly around 160 to 180 beats per minute when untreated (fast atrial fibrillation), and about 60 to 70 beats per minute when treated (slow atrial fibrillation). If the atrial fibrillation waves (f-waves) are small (less than 0.1 mV) they are called fine fibrillatory waves, and if larger they are called coarse fibrillatory waves. Atrial fibrillation may be intermittent, occurring in paroxysms (short bursts) or chronic (persistent).

1.3.3 Junctional Arrhythmias

In junctional arrhythmias, the origin of the impulse is within the A-V junction, comprising the A-V node and the Bundle of His. The ensuing QRS-complex and T-wave appear normal since this is the normal pathway that triggers the depolarisation of the ventricles. However, retrograde conduction to the atria may occur and its depolarisation results in P′-wave with abnormal morphology and timing. The polarity of the P′-wave would be opposite to that of the normal sinus P-wave since depolarisation is propagated in the opposite direction – from the A-V node towards the atria ("retrograde"). The wave also appears after the onset of the QRS-complex.

Premature Junctional Contractions (PJC)

A premature junctional contraction is a ventricular contraction initiated by an ectopic pacemaker in the atrioventricular (A-V) node. The atria may be depolarised if retrograde conduction to the atria occurs. The resulting P′-waves may precede, be buried in the QRS complex or follows the QRS complex and differ in size, shape and direction (Fig. 1.29). In addition, if the ectopic pacemaker discharges too soon, the bundle branches may not have fully repolarised to conduct the impulse normally. Should this occur, the electrical impulse may only be conducted in one bundle branch, usually the left one, producing a wide bizarre-looking QRS-complex resembling a right bundle branch block. The S-A node is frequently not depolarised by the PJCs (i.e. not "reset" by the PJCs). Consequently the P-waves that follow after PJCs appear normal and with the expected timing as if the PJC had not occurred.

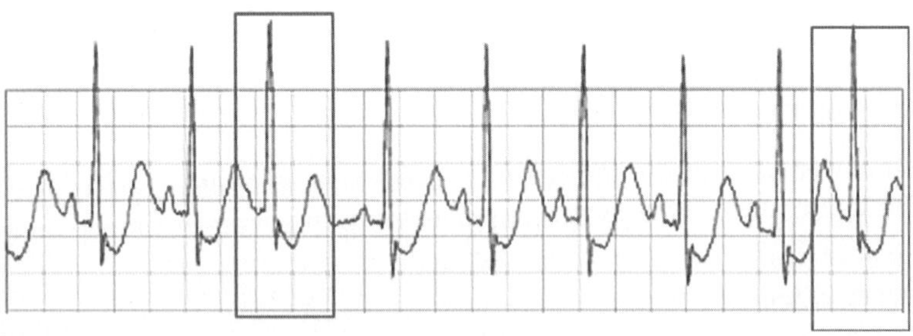

Fig. 1.29. In premature junctional escape contraction, a normal-looking QRS-complex prematurely appears, but without a preceding P-wave. The accompanying T-wave is normal

Junctional Escape Rhythm

If the impulse in the S-A node is blocked, or is not generated, or has a rate that is too slow, a junctional escape rhythm may occur. In this rhythm, a pacemaker site in the A-V junction takes over the role of the S-A node pacemaker. Since the rhythm is an escape from the absence of depolarisation impulses, the heart rate is slow and varies from 40 to 60 beats per minute [1–4]. If the atrial depolarisation occurs through retrograde conduction, the P′-waves will be abnormal in appearance (shape, size and direction) and timing.

Junctional Escape Rhythm, Non-Paroxysmal Junctional Tachycardia (Accelerated Junctional Rhythm, Junctional Tachycardia)

In non-paroxysmal (sustained) junctional tachycardia, an ectopic pacemaker in the A-V junction takes over the role of the S-A node and generates a regular rhythm of 60 to 150 beats per minute. If the rate is between 60 to 100 beats per minute, it is commonly called Accelerated Junction Rhythm. If greater than 100 beats per minute, it is called Junctional Tachycardia (Fig. 1.30).

Paroxysmal Supraventricular Tachycardia (PSVT)

In paroxysmal supraventricular tachycardia, the heart rate ranges from 160 to 240 beats per minute (Fig. 1.31). PSVT may occur as a result of a re-entry circuit in the A-V junction (Atrioventricular nodal re-entry tachycardia or AVNRT). It may also occur as a result of a re-entry circuit involving an accessory pathway between the atria and ventricles (Atrioventricular re-entry

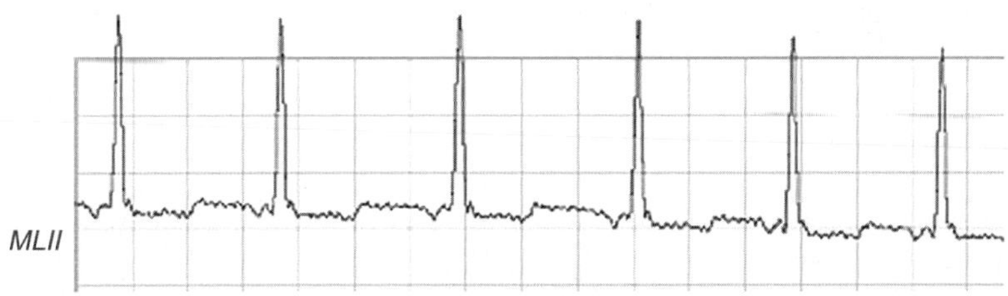

Fig. 1.30. Non-paroxysmal junctional tachycardia with retrograde conduction results in inverted P′-waves. The atrial and ventricular rate falls in the normal range

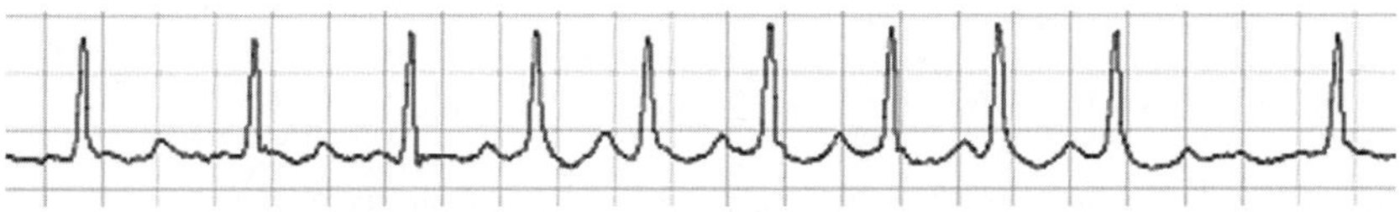

Fig. 1.31. Paroxysmal supraventricular tachycardia results in very fast heart rate ranging from 160 to 240 beats per minute for a short period of time

tachycardia or AVRT). The onset and termination of PSVT is abrupt, and may occur in repeated episodes (paroxysms) that last for seconds, hours or days. In AVNRT, P′-waves are usually buried in the QRS complex and hence not visible; whereas for AVRT the P′ waves may be visible.

1.3.4 Ventricular Arrhythmias

In ventricular arrhythmias, the impulses originate from the ventricles and spread outwards to the rest of the heart. These arrhythmias may occur as a result of enhanced automaticity of the ventricular myocardium or because of a re-entry circuit. While conduction of impulses down the normal conduction pathway to the ventricles from the atria results in a narrow, normal QRS-complex (since all the ventricular myocardium depolarise about the same time), for ventricular arrhythmias, the QRS-complex is wide and bizarre in shape because the impulse is not propagated through the normal pathway but through non-specialised myocardium that conducts more slowly (hence a wide QRS-complex) and in a different direction (hence a bizarre looking QRS-complex).

Premature Ventricular Contractions (PVC)

A premature ventricular contraction is an extra (abnormal) ventricular contraction originating from the ventricles. The site of the pacemaker may be in the bundle branches, Purkinje network, or ventricular myocardium (Fig. 1.32). More than one pacemaker may be involved, each generating its own bizarre-shaped QRS-complex.

PVCs usually do not depolarise the atria or the S-A node and hence the P-waves maintain their underlying rhythm and occur at the expected time (i.e. the "pause" is compensatory). PVCs are not preceded by (ectopic) P-waves and may occur anywhere in the heart beat cycle. If a PVC appears on the ending part (downslope) of the T-wave (near completion of repolarisation of different conduction pathways), a re-entry mechanism may occur resulting in ventricular tachycardia or ventricular fibrillation, which can be lethal.

PVCs are described as isolated if they occur singly, and as couplets if two consecutive PVCs occur (Fig. 1.33). If three or more consecutive PVCs

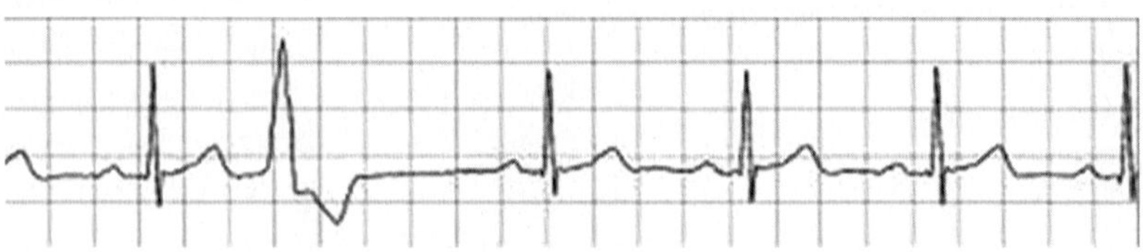

Fig. 1.32. An isolated premature ventricular contraction occurs without a preceding P-wave. The QRS morphology is bizarre in shape since depolarisation of the ventricles do not follow the normal sequence

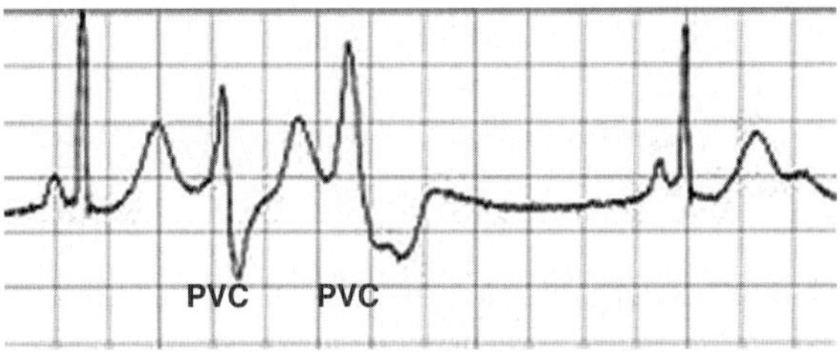

Fig. 1.33. In multifocal paired PVCs, the bizarre-shaped QRS complexes appear in pairs but each with a different morphology

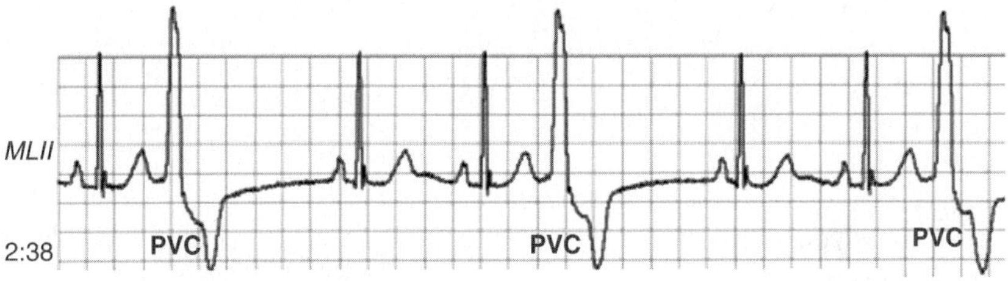

Fig. 1.34. In trigeminy unifocal PVCs, one out of every three beats has a bizarre-shaped QRS-complex which does not vary in morphology

occur at a rate of more than 100/min, the rhythm is known as ventricular tachycardia. If PVCs alternate with normal beats, the rhythm is ventricular bigeminy; if with every two normal beats, ventricular trigeminy (Fig. 1.34); and if with every three normal beats, ventricular quadrigeminy.

Ventricular Tachycardia (VT)

In ventricular tachycardia, the heart rate is 110 to 250 beats per minute. The QRS complex is abnormally wide (usually > 0.14 s), bizarre in shape, and of a different direction from the normal QRS complex. If the ventricular tachycardia (3 or more PVCs) lasts less than 30 seconds, it is known as paroxysmal VT or non-sustained VT, and if more than 30 seconds, sustained VT. Depending on the pacemakers involved, VT can take several forms:

- Monomorphic VT: A single ectopic pacemaker drives the VT and consequently the QRS-complexes look alike (Fig. 1.35).
- Bidirectional VT: Two ectopic pacemakers alternate with each other in depolarising the ventricles resulting in two distinct forms of QRS-complexes, alternating with each other in occurrence.
- Polymorphic VT: Multiple ectopic pacemakers are involved resulting in more than two forms of QRS-complexes (Fig. 1.36).
- Torsade de pointes: This is a special form of polymorphic VT where the QRS complexes gradually change back and forth from one shape, size and direction to another over a series of beats (Fig. 1.37).

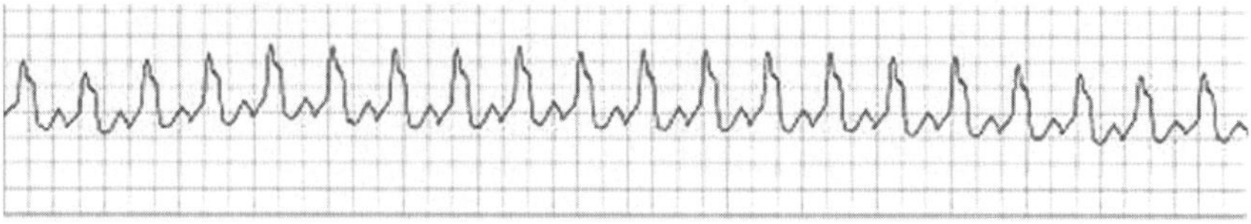

Fig. 1.35. Monomorphic ventricular tachycardia beats at a very fast rate and with a consistent bizarre-shaped QRS morphology

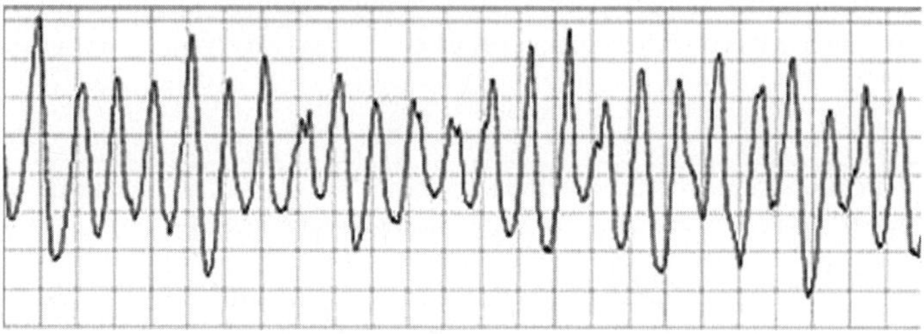

Fig. 1.36. Polymorphic ventricular tachycardia not only exhibits a very rapid ventricular beat but also a variety of bizarre-shaped QRS-complexes

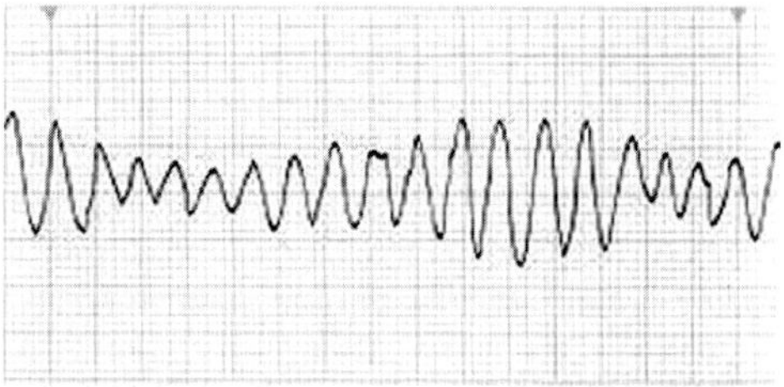

torsade de pointes

Fig. 1.37. Torsade de pointes is a special form of polymorphic VT. It exhibits a gradual change in morphology of the QRS′-complex from one form to another and back again

- Ventricular Flutter: An ectopic pacemaker rapidly drives the depolarisation of the ventricles such that the QRS-complexes appear saw-tooth-like in appearance (Fig. 1.38).

VT is considered life-threatening as the rapid rate may prevent effective ventricular filling and result in a drop in cardiac output. It can also degenerate into ventricular fibrillation, which is lethal.

Ventricular Fibrillation

Ventricular fibrillation occurs when numerous ectopic pacemakers in the ventricles cause different parts of the myocardium to contract at different times

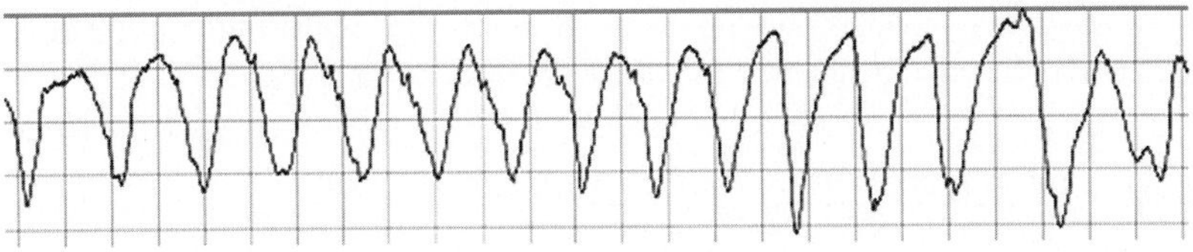

Fig. 1.38. Ventricular flutter exhibits a very rapid ventricular rate with a saw-tooth like ECG waveform

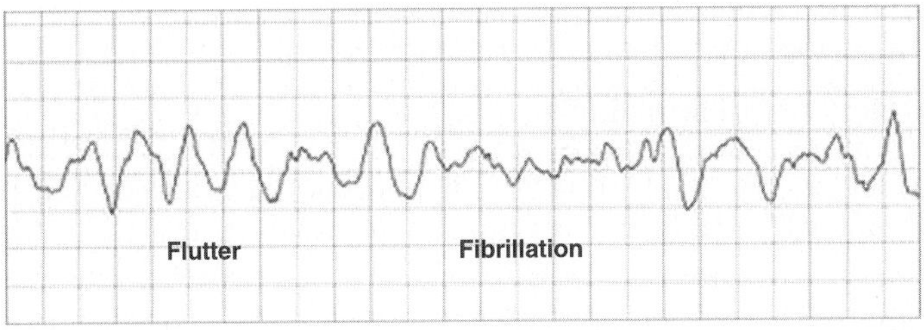

Fig. 1.39. Coarse ventricular fibrillation mixed with ventricular flutter. Fibrillation carries very irregular periodicity

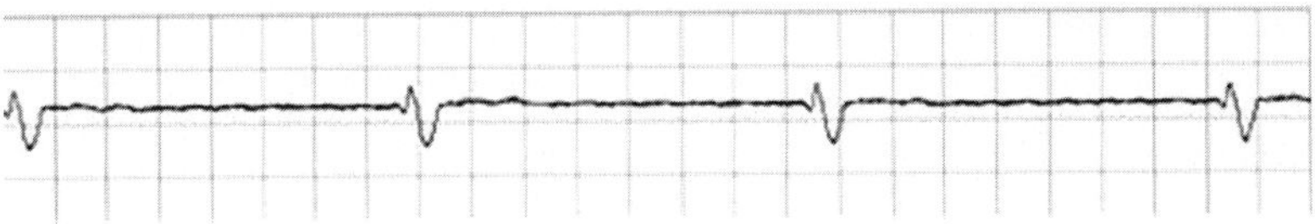

Fig. 1.40. Ventricular escape rhythm has very long R-R intervals

in a non-synchronised fashion (Fig. 1.39). Ventricular contraction is uncoordinated, resulting in insignificant or no cardiac output. Perfusion to vital organs is compromised and death ensues in minutes. Defibrillation (DC shock) is used to abort ventricular fibrillation.

Ventricular Escape Rhythm (Idioventricular Rhythm)

In ventricular escape rhythm, an ectopic pacemaker in the bundle branches, Purkinje network or ventricles dominates at a rate of less than 40 beats per minute (Fig. 1.40). Cardiac output is low as a result of the slow heart rate, giving rise to hypotension and decreased perfusion of the brain and other vital organs. This may result in syncope (fainting), shock and congestive heart failure.

Accelerated Idioventricular Rhythm

In accelerated Idioventricular rhythm, the ectopic ventricular pacemaker fires at an accelerated rate that lies between 40 and 100 beats per minute.

The rhythm may also be a known as Accelerated Ventricular Rhythm, Idioventricular Tachycardia, or Slow Idioventricular Tachycardia.

Ventricular Asystole (Cardiac Standstill)

In ventricular asystole, the ventricles fail to contract. The pacemaker sites in the ventricles fail to generate a life-sustaining escape rhythm. If P-waves exist, they are not conducted through to the ventricles. There is no heart beat, no rhythm and no cardiac output. Death quickly follows.

1.3.5 Atrioventricular Blocks

Atrioventricular blocks disrupt the normal propagation of the electrical impulse along the conduction pathways to the ventricles. The block may delay or completely prevent propagation of the impulse to the rest of the conduction system, and may do so intermittently.

A first-degree AV block is said to occur when all the P-waves are conducted to the ventricles, but the PR-interval is prolonged.

Second-degree AV blocks are said to occur when some of the P-waves fail to conduct to the ventricles. They are subdivided into:

1. Wenkebach (or Mobitz Type 1)
2. Mobitz Type II
3. 2:1 AV block
4. High grade AV block (>3 : 1)

A third degree or complete heart block is said to occur when all the P-waves fail to conduct down to the ventricles. Pacemakers are used to treat this condition and bring the heart rate back to normal.

First-Degree AV Block

In first-degree atrioventricular block, the conduction of electrical impulses is delayed longer than normal at the level of the A-V node. P-R intervals are longer than 0.2 second but constant in duration (Fig. 1.41). Except for a lengthened P-R interval, the ECG appears normal since depolarisation of the atria and ventricles follow the normal route. This is usually asymptomatic, but may deteriorate to a higher order AV block.

Second-Degree, Type I AV Block (Wenckebach)

In second-degree type I AV block (Wenckebach), the conduction delay in the AV-node increases progressively over a few beats until the conduction is completely blocked. Consequently the P-R interval correspondingly lengthens until the P-wave has no accompanying QRS-complex (Fig. 1.42). Then the cycle repeats.

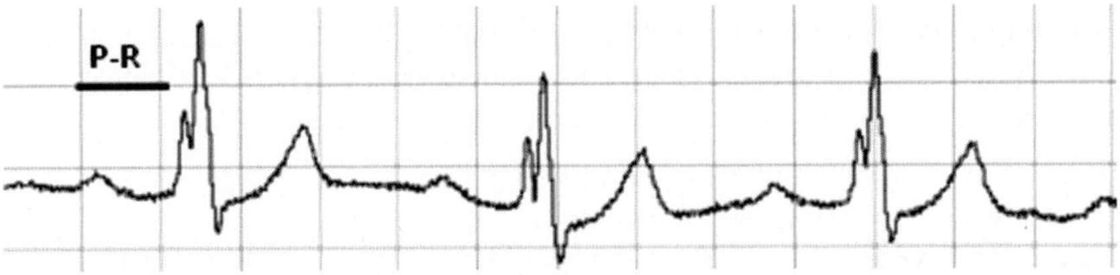

Fig. 1.41. First-degree AV blocked is characterised by a lengthened P-R interval. In addition the QRS-complex of this particular ECG (above) also exhibits a left bundle branch block

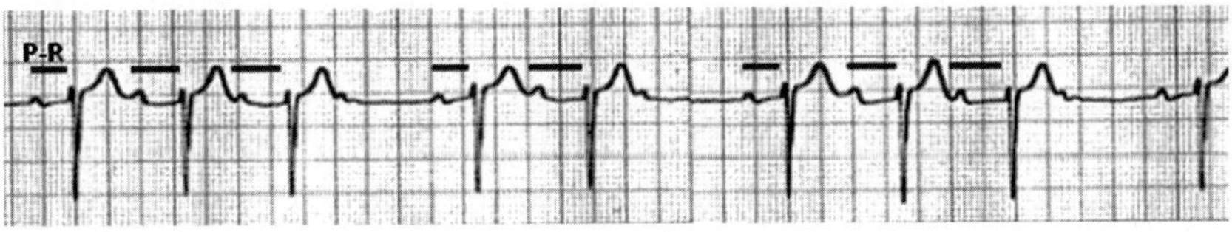

Fig. 1.42. Second-degree Wenckebach AV block is characterised by a cyclical progressively lengthening of the P-R interval until a QRS complex fails to appear

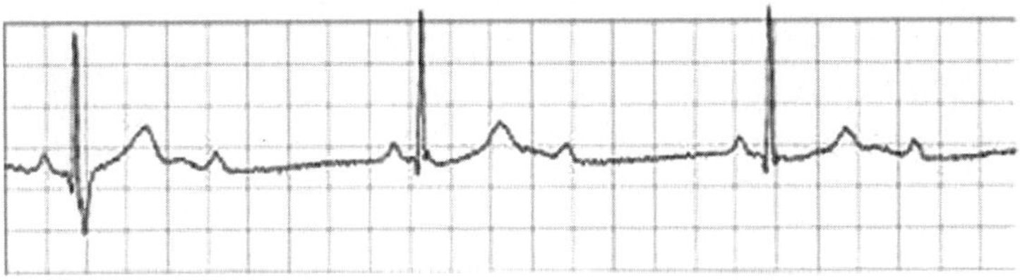

Fig. 1.43. 2:1 second-degree Mobitz AV block. The QRS complexes appear once for every two P-waves

Second-Degree, Type II AV Block

In second-degree, type II AV block (also known as Mobitz Type II AV block), conduction intermittently fails to go down the bundle of His and bundle branches. There is usually a pre-existing complete block in one bundle branch, such that when an intermittent block occurs in the remaining bundle branch, conduction down to the ventricles fails. As a result of the pre-existing bundle branch block, the QRS-complex is commonly wide. Unlike Wenckebach AV block, the PR interval in Mobitz Type II AV block is constant (Fig. 1.43). Commonly the conduction ratio is 4:3 or 3:2 P-waves to QRS-complexes. This type of AV block is dangerous as complete heart block can occur unpredictably, and is an indication for implantation of an artificial pacemaker [1–4].

Second-Degree 2:1 and Advanced AV Block

This AV block is caused by defective conduction of electrical impulses through the AV node, or the bundle branches or both. Conduction ratios of P-waves to QRS-complex range from 2:1, 3:1, or greater. QRS-complex may appear abnormal because of a bundle branch block (Fig. 1.44). Conduction ratios greater than 2:1 are considered as a High-grade AV Block.

Third-Degree AV Block (Complete AV Block)

In third-degree AV block, all of the electrical impulses from the atria are not conducted through the AV node. The atria and the ventricles beat independently, each at their own rate. The pacemaker for the ventricles is in the ventricles itself below the AV block and may reside in the AV junction, bundle branches, Purkinje network or the ventricular myocardium. If the escape pacemaker is in the AV junction the heart rate will be around 40 to 60 beats per minute, but if located lower in the conduction path such as Purkinje networks, or in the ventricular myocardium, it will be slower at 30 to 40 beats per minute (Fig. 1.45).

Pacemaker Rhythm (Implant)

Hearts implanted with an artificial pacemaker will generally beat around 60 to 70 beats per minute depending on the setting of the artificial pacemaker. The pacing electrode is commonly attached to the apex of the right ventricular cavity (ventricular pacemaker) or the right atrium (atrial pacemaker), or both (dual chamber pacemaker). The artificial pacemaker produces a narrow, often biphasic spike. A pacemaker lead positioned in an atrium produces a

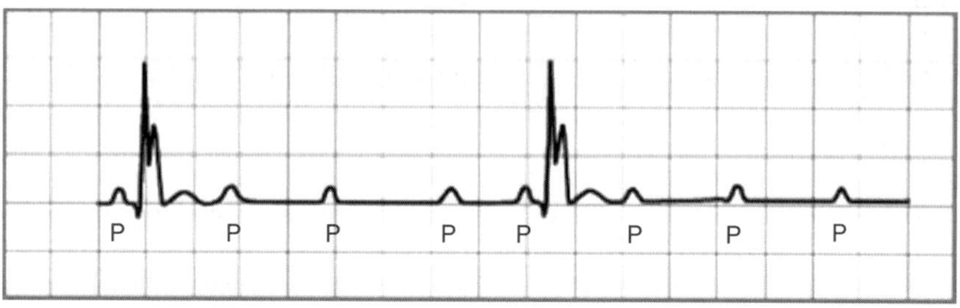

Fig. 1.44. 3:2 AV block with bundle branch block

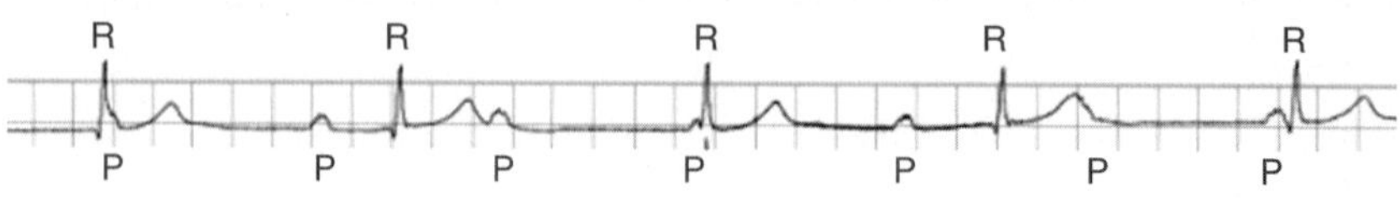

Fig. 1.45. In third-degree AV block, the rhythm of the P-waves is completely dissociated from the rhythm of the QRS-complexes. Each beat at their own rate. QRS morphology may appear normal if the site of the ventricular pacemaker is at the AV junction

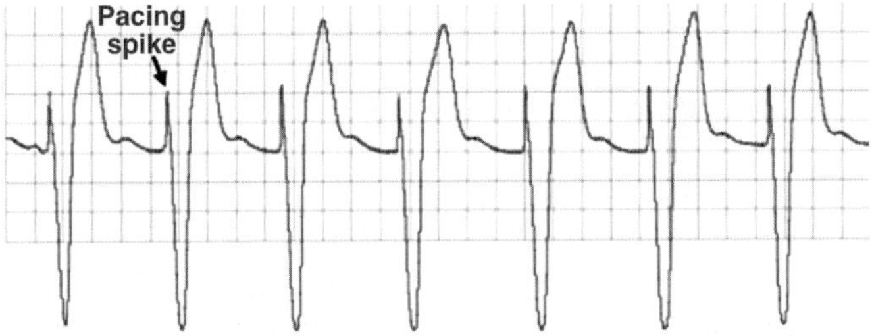

Fig. 1.46. A ventricular pacemaker generates a normal heart rate but with a bizarre-shaped QRS-complex. The pacemaker spike precedes each QRS-complex

pacemaker spike followed by a P′-wave. A pacemaker lead positioned in a ventricle produces a pacemaker spike followed by a wide, bizarre QRS-complex (Fig. 1.46).

1.3.6 Bundle Branch and Fascicular Blocks

The electrical conduction system for the ventricles starts with the A-V node, connects to a short stem called the Bundle of His, and then divides into the right and left bundle branches. The right bundle travels down the right ventricle along the interventricular septum, then subdivides into smaller and smaller branches until they join the Purkinje network of fine conducting fibres that are embedded in the endocardium. The left bundle branch divides into the Left Posterior Fascicle and the Left Anterior Fascicle which similarly travel down the left ventricle along the septum. Both fascicles subdivide repeatedly before connecting to the Purkinje network that innervates, respectively, to the posterior walls of the left ventricle, and to the anterior and lateral walls. Septal fibres also branch out from the main left bundle branch to innervate the septal myocardium.

A block in the conduction of the impulse from the AV-node may affect the whole conduction system, one of the bundle branches, or only one fascicle (sub-branch) of the left bundle branch. When an area of myocardium does not receive a depolarising impulse, its depolarisation is delayed. It has to wait for the depolarisation of neighbouring myocardial cells to trigger its depolarization. Consequently the depolarisation process of the entire ventricles takes longer, and the QRS-complex appears wider in duration, and abnormal in shape.

Right Bundle Branch Block (RBBB)

When the right bundle branch is blocked, the electrical impulse from the AV node is not able propagate to the Purkinje network to depolarise the right ventricular myocardium. Instead, the impulse propagates in a convoluted

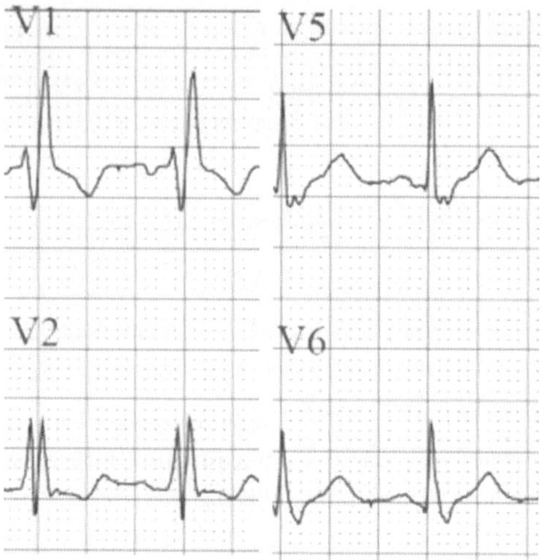

Fig. 1.47. Right bundle branch block. The convoluted path of depolarisation caused by a block in the conduction path leads to a change in QRS morphology

manner through the left ventricular myocardium to reach the right ventricular myocardium. Since propagation through the myocardium is much slower than through the specialised conducting tissue, the QRS-complex becomes widened. The morphology of the QRS-complex also changes and takes on a bizarre appearance because of the different direction of depolarisation (Fig. 1.47).

The conduction block may be complete or incomplete. A complete block will widen the QRS complex to greater than 0.12 seconds, while an incomplete block to 0.10 to 0.11 seconds. ECG waveforms in the facing leads show waveform changes most prominently [1–3].

Left Bundle Branch Block (LBBB)

Similar to right bundle branch block, a block in the left bundle branch will prevent the electrical impulses from the A-V node from depolarising the left ventricular myocardium in the normal way. For LBBB, the right ventricle is depolarised first, generating an electrical wavefront that eventually spreads to the left ventricular myocardium causing the myocardium to depolarise.

A complete conduction block will widen the QRS complex to 0.12 second or more, while an incomplete block will widen it to 0.10 to 0.11 second. ECG waveforms in the facing leads show waveform changes most prominently (Fig. 1.48).

Left Anterior Fascicular Block (Left Anterior Hemiblock)

A conduction block in the left anterior fascicle affects the normal depolarisation of the anterior and lateral walls of the left ventricle (Fig. 1.49). The QRS

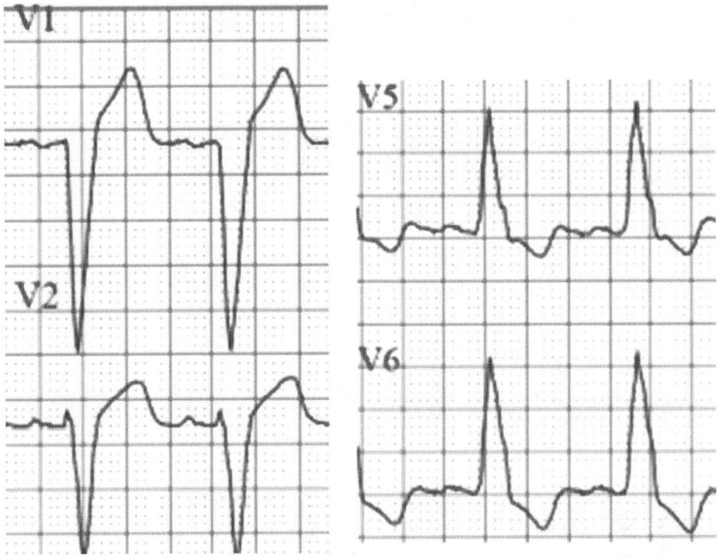

Fig. 1.48. Complete Left bundle branch block (LBBB)

axis is typically deviated to the left (i.e. left axis deviation of −30° to −90°) because the impulse now has to travel from the left posterior fascicle leftwards and anteriorly towards the left anterior fascicle, consequently shifting the summed electrical vector of the depolarisation of the heart. The duration of the QRS-complex is normal, within 0.10 second.

Left Posterior Fascicular Block (Left Posterior Hemiblock)

A conduction block in the left posterior fascicle prevents normal depolarisation of the interventricular septum and posterior wall of the left ventricle (Fig. 1.50). QRS duration remains normal within 0.10 second, while QRS axis is around +110° to +180°).

1.4 Miscellaneous Electrocardiogram Changes

Other than conduction blocks, the morphology of ECG may be affected by other conditions. These include enlargement of the myocardium, inflammation of the pericardium (pericarditis), electrolyte imbalance, pulmonary disease, drug effects, hypothermia and accessory conduction pathways.

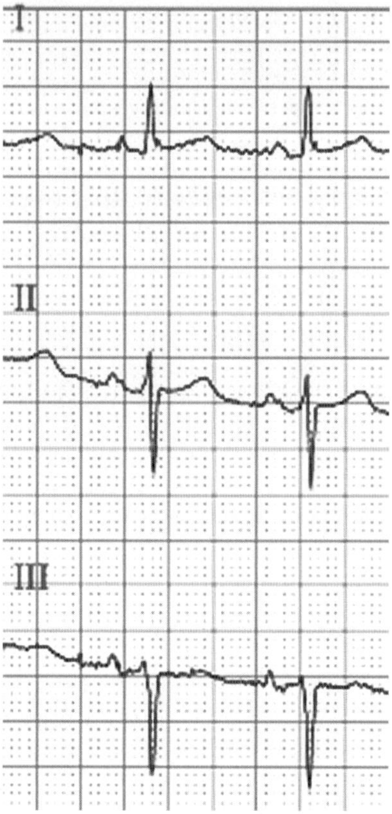

Fig. 1.49. Left anterior fascicular block

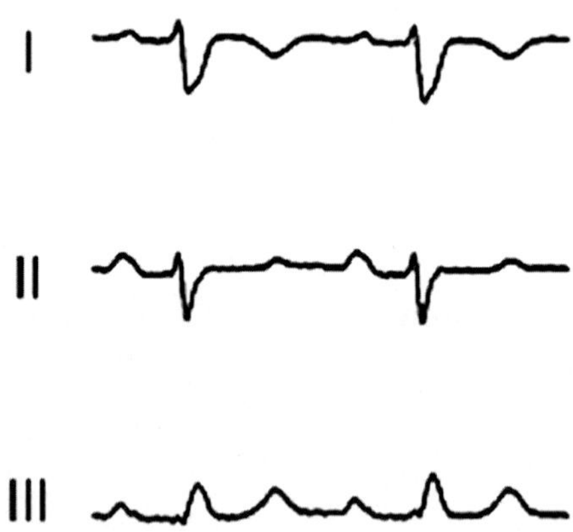

Fig. 1.50. Left posterior fascicular block

1.4.1 Enlargement of the Myocardium

Right and Left Atrial Enlargement

Heart diseases often force the atria and ventricles to become enlarged as they accommodate to higher pressures or volumes demanded of them. Enlargement may involve dilatation (distension of the heart chamber) or hypertrophy (increased thickness of the myocardium in response to chronic increased workload) (Fig. 1.51).

Right Ventricular Hypertrophy

With thicker myocardium, the right ventricle depolarisation changes the QRS axis to deviate away from the normal to +90° or more. T-wave inversion is often present. Ventricular action time (VAT) is prolonged beyond 0.035 second. R-waves are abnormally tall in lead V1, which lies over the right ventricle. Usually the right atrium is correspondingly enlarged, giving rise to enlarged P-waves (Fig. 1.52).

Left Ventricular Hypertrophy (LVH)

An increased thickness of the left ventricle usually does not deviate the QRS axis significantly since the left ventricle normally accounts for the majority of the electrical vector in the normal heart. But in some cases, the deviation to

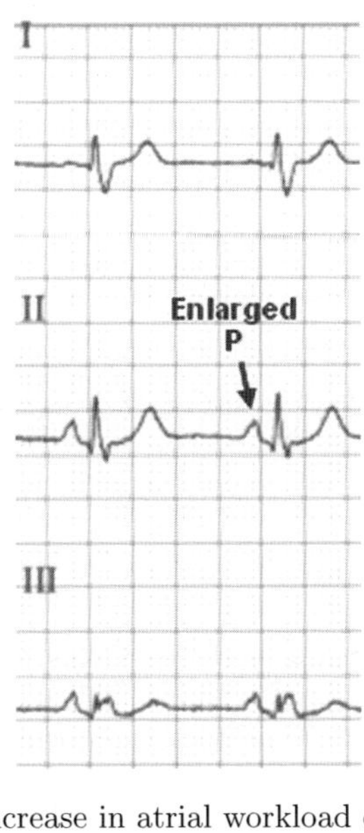

Fig. 1.51. Prolonged increase in atrial workload causes atrial hypertrophy

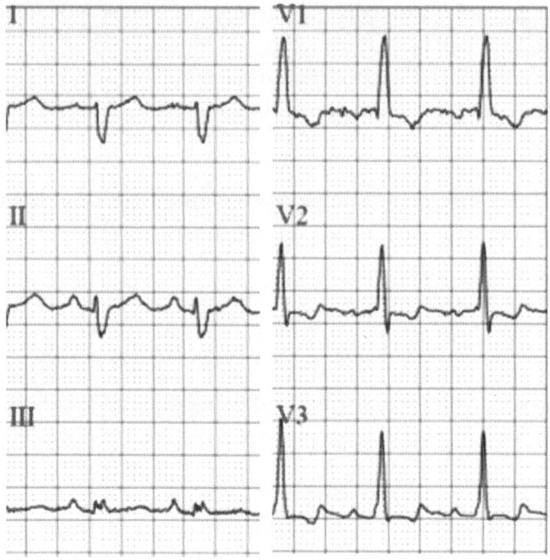

Fig. 1.52. Right ventricular hypertrophy shows ST-segment shift and T-wave inversion in V1-V3, consistent with right ventricular overload

the left may be up to $-30°$. LVH is manifested on ECG as an increase in the height of the R-wave in the left-sided leads (V5-6, I, aVL). The T-waves may be inverted in these leads (Fig. 1.53).

1.4.2 Pericarditis

In pericarditis, the pericardium is inflamed with a variable amount of serous, fibrous, purulent or haemorrhagic exudates within the pericardial sac. S-T segment is elevated and concave (similar to acute myocardial infarction), but unlike infarction, the S-T elevation would appear in most ECG leads (instead of just the facing leads) since pericarditis usually affects the entire myocardial surface of the heart (Fig. 1.54(a)).

If accompanied by large pericardial effusion, the surface ECG voltages may be small due to the additional intervening layer of fluid between the heart and the skin (Fig. 1.54(b)).

1.4.3 Electrolyte Imbalance

Electrolyte imbalances affect the depolarisation and repolarisation of the cardiac cells by their effect on the flow of ions across the cell membrane. These result in changes in the morphology and width of the ECG complexes.

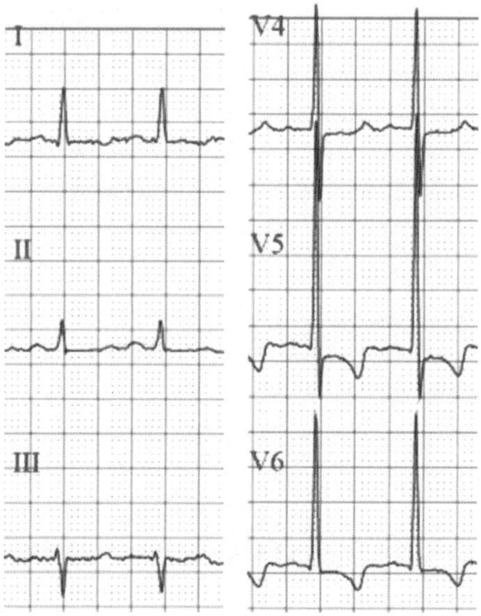

Fig. 1.53. Left ventricular hypertrophy shows a tall R-wave with "strain" pattern of inverted T-wave in leads V5 and V6

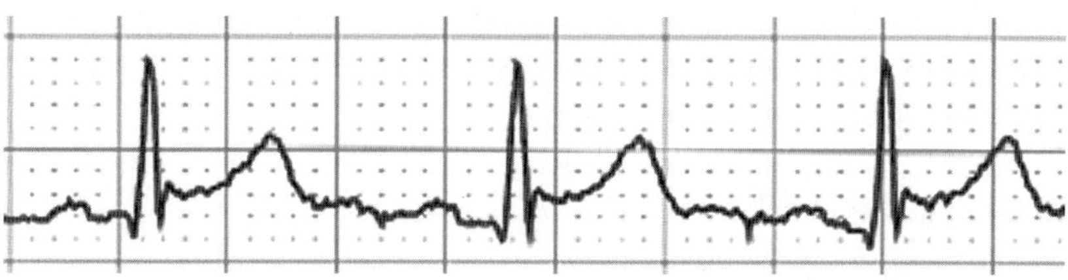

Fig. 1.54(a). In pericarditis, the QRS amplitude drops and the ST-segment becomes elevated and concave

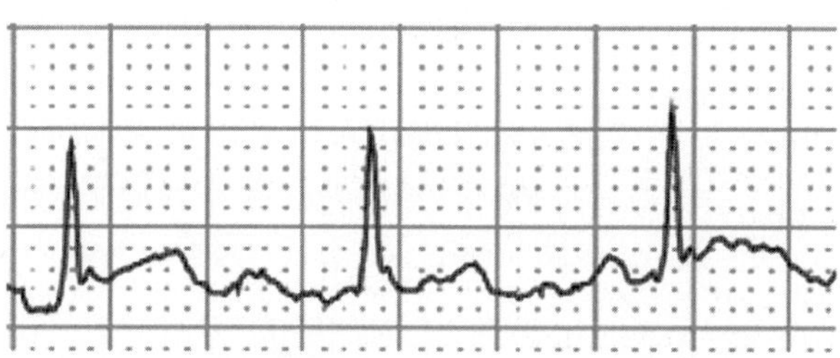

Fig. 1.54(b). When accompanied by large pericardial effusion, the amplitude of the ECG drops substantially

Hyperkalemia

In hyperkalemia there is an excess of serum potassium above the normal levels of 3.5 to 5.0 milliequivalents per litre (mEq/l). The most common cause is kidney failure and certain diuretics. The P-wave begins to flatten and disappears at higher levels. P-R interval may be prolonged beyond 0.2 seconds [1–3]. The QRS-complex widens and starts to merge with the T-wave (Fig. 1.55(a)). The T-waves become narrow, tall and peaked eventually exceeding the R-wave (Fig. 1.55(b)).

Hypokalemia

In hypokalemia, there is a deficiency of serum potassium below the normal levels of 3.5 to 5.0 mEq/l. The most common cause is vomiting, gastric suction and excessive use of diuretics. The P-waves become tall and symmetrically peaked beyond 0.25 mV, appearing similar to P-pulmonale. The QRS-complex begins to widen, and ST-segments become depressed by 0.1 mV or more. U-waves begin to increase in size and may reach the size of normal

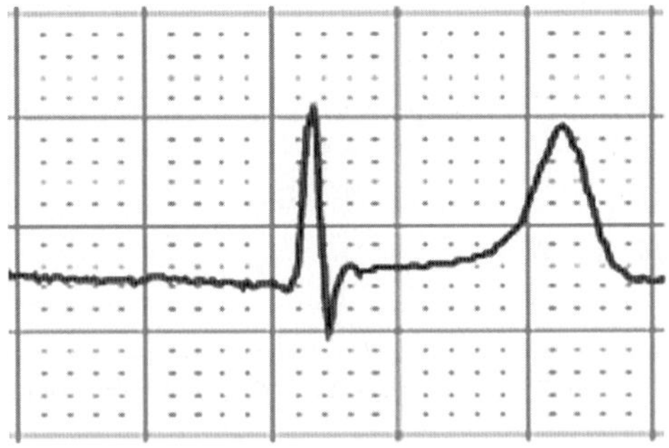

Fig. 1.55(a). Hyperkalemia's strongly prominent feature is a tall T-wave (left) as a result of mild excess of serum potassium. The QRS-complex also widens and begins merging with the T-wave

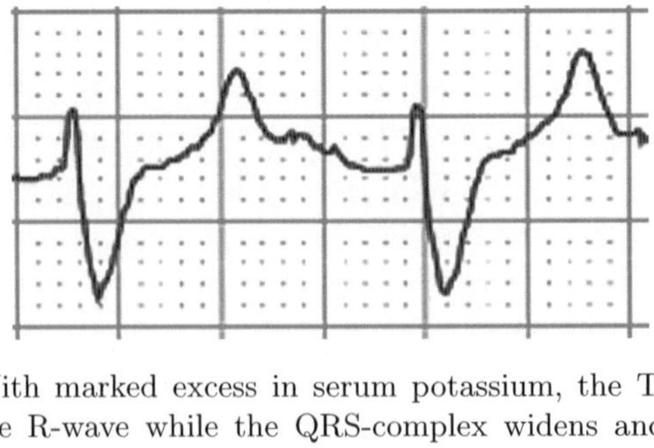

Fig. 1.55(b). With marked excess in serum potassium, the T-wave increases in height beyond the R-wave while the QRS-complex widens and merges with the T-wave

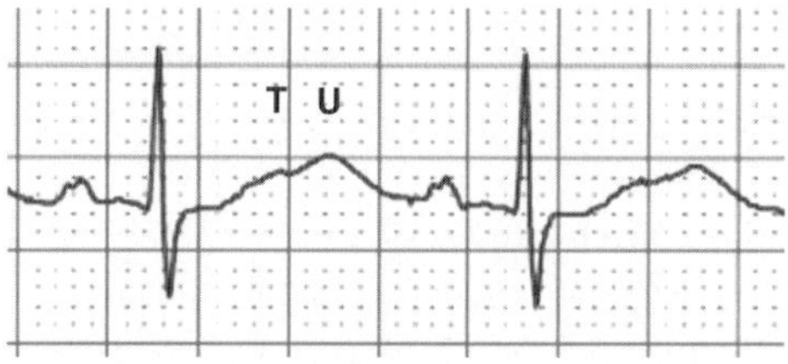

Fig. 1.56. In hypokalemia, the T-wave flattens and U-wave increases in size as the deficiency of serum potassium gets worse

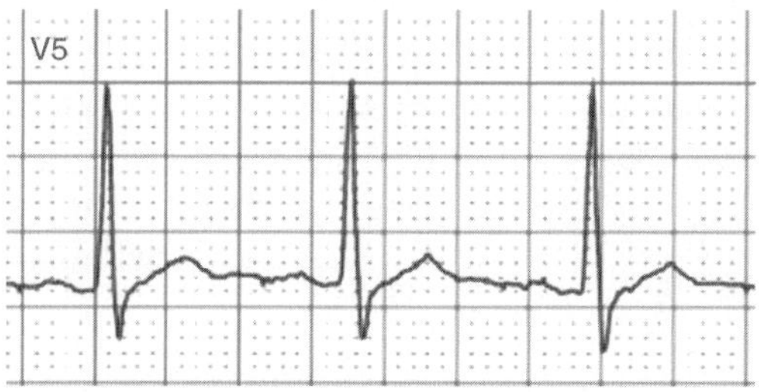

Fig. 1.57. In hypercalcemia, the QT-interval shortens

T-waves. The T-waves begin to flatten and may invert or merge with U-waves (Fig. 1.56).

Hypercalcemia

In hypercalcemia, there is an excess of serum calcium above the normal levels of 2.1 to 2.6 mEq/l. The QT-intervals are shorter than normal (Fig. 1.57).

Hypocalcemia

In hypocalcemia, there is a shortage of serum calcium below the normal levels of 2.1 to 2.6 mEq/l. The ST-segments are prolonged and the QT-intervals are lengthened beyond normal (Fig. 1.58).

1.4.4 Drug Effects

Digitalis

Digitalis administered within the therapeutic range may produce characteristic changes in the ECG. PR-intervals may be prolonged beyond 0.2 second and ST-segments depressed by 0.1 mV or more with a characteristic scooped-out effect (Fig. 1.59). T-waves may be flattened, inverted or biphasic. QT-intervals are shorter than normal for a given heart rate.

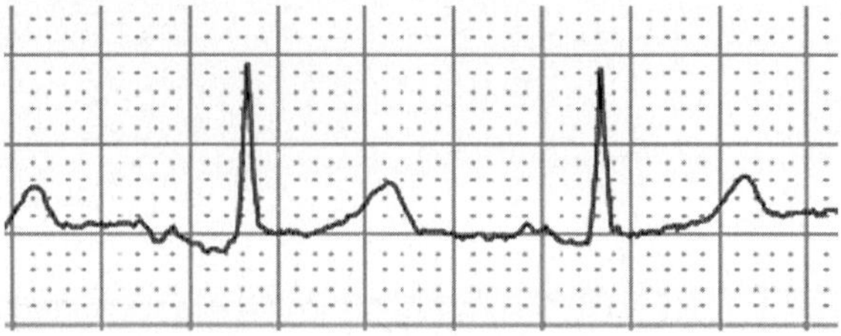

Fig. 1.58. In hypocalcemia, the QT-interval lengthens beyond normal

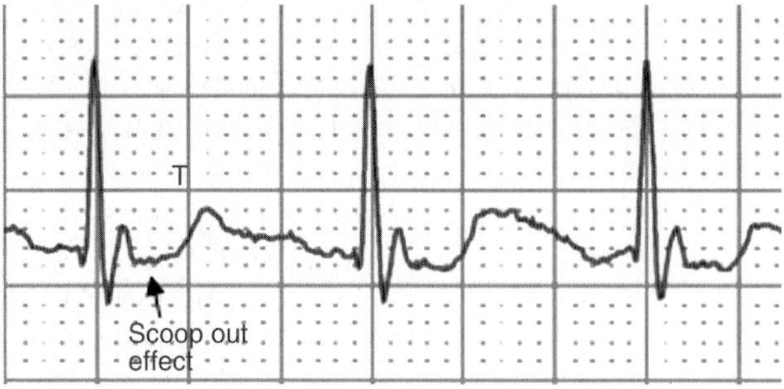

Fig. 1.59. Digitalis produces a shortening of the QT-interval with a scooped-out ST-segment

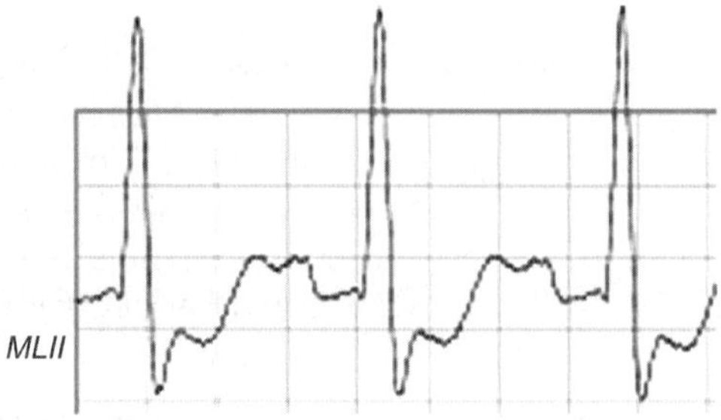

Fig. 1.60. Therapeutic dosages of procainamide results in ECG waveform changes including a widening of the QRS-complex, decrease in T-wave amplitude, depression of the ST-segment and prolonging of the PR-interval

Procainamide

Procainamide administered within the therapeutic range produces characteristic changes in ECG. QRS duration is widened beyond 0.12 second. The R-waves may decrease in amplitude. T-waves may also decrease in amplitude. Occasionally T-waves may widen and have a notch because of the appearance of U-waves. PR-intervals may be prolonged. ST-segments may be depressed by 0.1 mV or more. QT-intervals may occasionally be prolonged beyond normal for a given heart rate (Fig. 1.60).

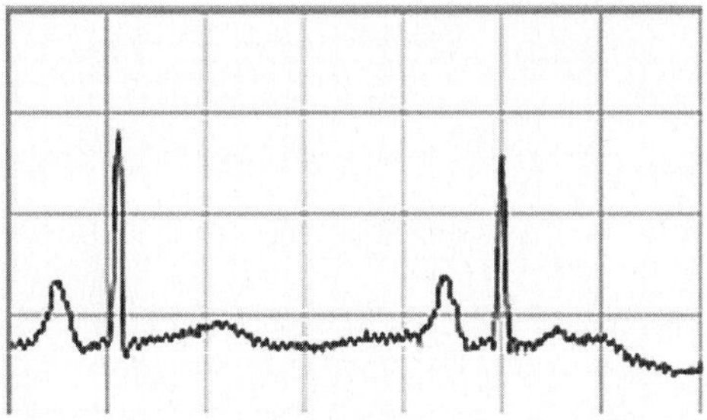

Fig. 1.61. Therapeutic dosages of quinidine results in ECG waveform changes (similar to those of procainamide) including a widening of the QRS-complex, decrease in T-wave amplitude, depression of the ST-segment and prolonging of the PR-interval

Quinidinc

Quinidine administered within the therapeutic range produces characteristic changes in ECG. P-waves may be wide, often notched. QRS duration may widen beyond 0.12 second. T-waves may decrease in amplitude, widened and notched by the appearance of U-waves. Or the T-waves may be inverted. PR-intervals may be prolonged. ST-segments may be depressed by more than 0.1 mV. QT-intervals may be prolonged beyond normal for a given heart rate (Fig. 1.61).

1.4.5 Pulmonary Disease

Chronic Obstructive Pulmonary Disease (COPD)

This chronic disease of the lungs encompasses chronic bronchitis and emphysema, which cause the lungs to be over-distended and enlarged. The primary cause of COPD is prolonged heavy cigarette smoking. The diseased lungs force the right ventricle and right atrium to work harder in pumping blood to them. As a result, there is often right atrial and right ventricular enlargement. R-waves may be low in voltage because the enlarged lungs increase the distance of the chest wall from the heart. QRS-axis may be greater than +90° because of right ventricular hypertrophy as well as rotation of the heart (Fig. 1.62).

Acute Pulmonary Embolism

Pulmonary embolism occurs when a blood clot (thromboembolus) lodges in a pulmonary artery and occludes blood flow to a segment of the lungs. This obstruction to forward flow places a strain on the right side of the heart. Hence the ECG changes mimic that of right ventricular and right atrial enlargement. Tall, peaked P-waves may be seen. An S-wave in lead I, a Q-wave in lead III,

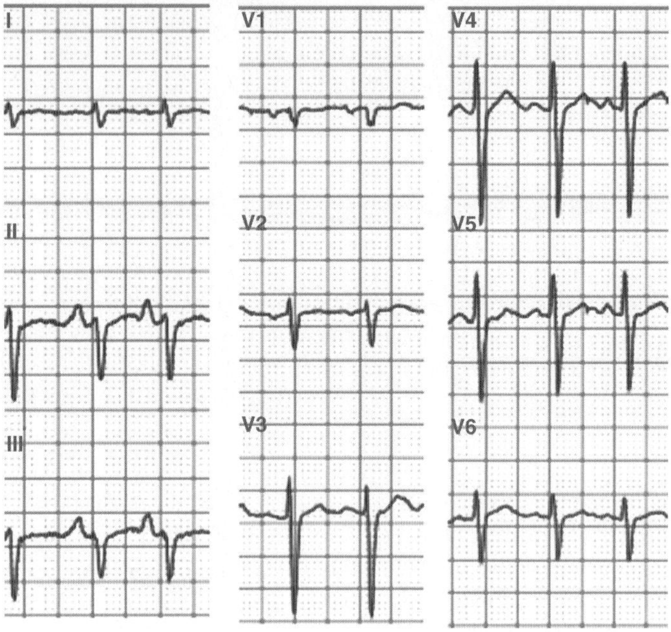

Fig. 1.62. COPD causes right atrial enlargement (tall P-wave) and low-amplitude QRS-complex from increased thorax and lung volume as a result of emphysema (over-inflation of alveoli)

and an inverted T-wave in lead III may occur acutely (Fig. 1.63). The QRS axis is greater than +90°. The right ventricle may show a "strain" pattern (inverted T-waves in leads V1-V3) as it works harder to pump blood to the rest of the lungs.

Chronic Cor Pulmonale

Chronic cor pulmonale is the enlargement of the right ventricle commonly accompanied by right heart failure. It is usually the end stage of prolonged pulmonary hypertension that occurs with many lung diseases such as COPD. P-waves are tall and peaked indicating right atrial enlargement. QRS-complexes are tall in lead V1 and V2 because of right ventricular hypertrophy (Fig. 1.64). QRS axis is greater than +90°. The right ventricle may show a "strain" pattern (inverted T-waves in leads V1–V3).

1.4.6 Early Repolarisation

This is a normal ECG variant (pattern). In early repolarisation, the ST-segment is elevated from 0.1 to 0.3 mV above the baseline. The ST segment is concaved upwards and tends to be most prominent in the mid-precordial

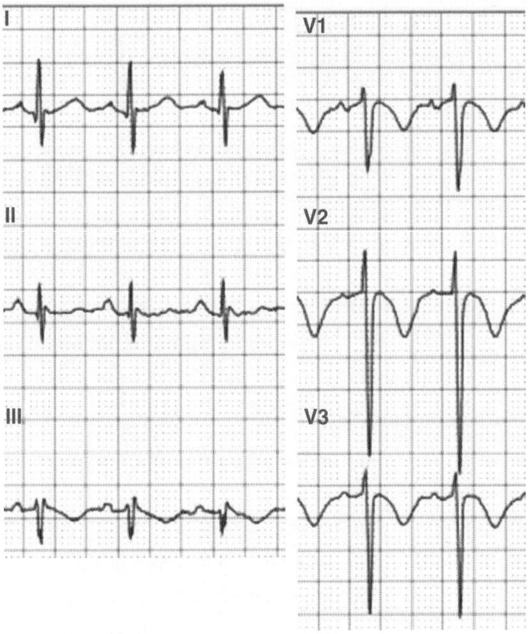

Fig. 1.63. Acute pulmonary embolism causes right atrial enlargement (tall P-waves) characteristics

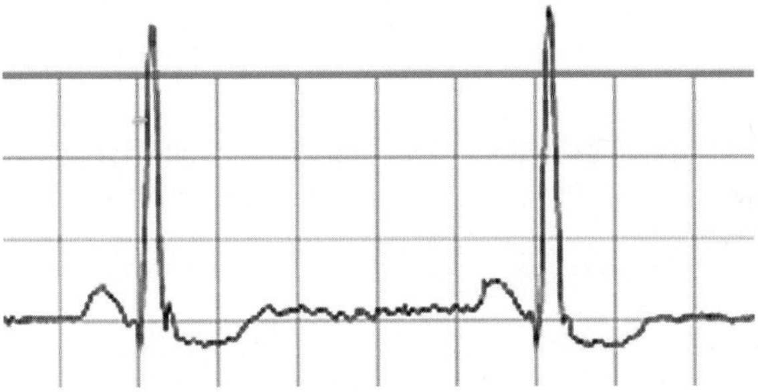

Fig. 1.64. Chronic cor pulmonale exhibits characteristics of right atrial enlargement (tall P-waves) and right ventricular hypertrophy (QRS-axis oriented to the right)

leads (V3–V4). Often there may be a notch at the J point. Importantly, there are no reciprocal depressions or elevations in opposite leads (Fig. 1.65). In addition, the ST-segment elevations persists and do not return to baseline over time.

1.4.7 Hypothermia

When core temperature of the body drops below $35°C (< 95°F)$, a distinctive narrow, positive wave referred to as the "J-wave" or Osborne-wave appears

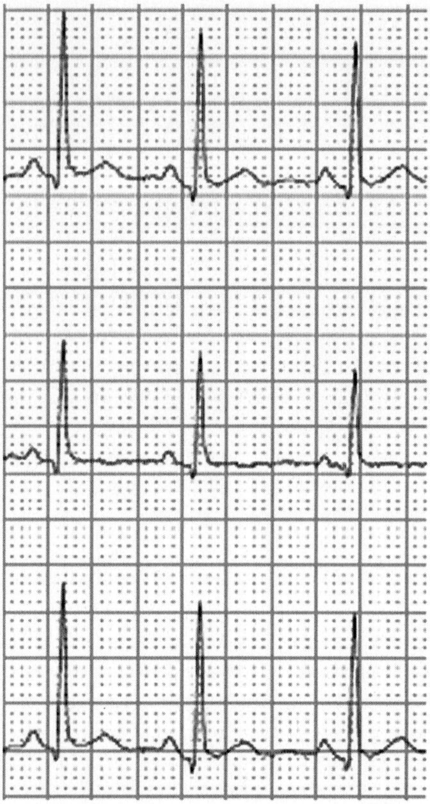

Fig. 1.65. Early repolarisation is indicated by ST-segment elevation or depression without reciprocal depressions or elevations in opposite leads

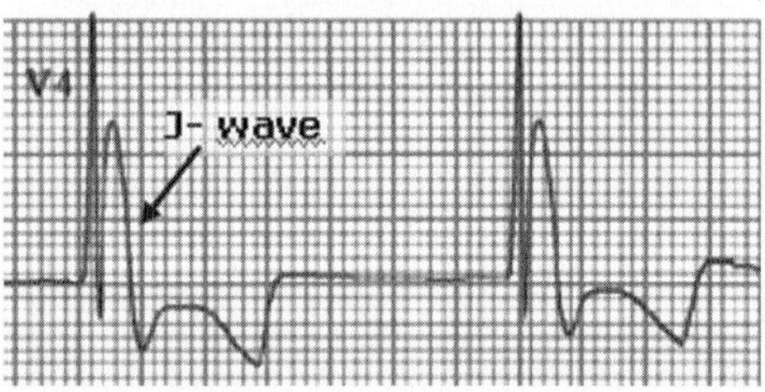

Fig. 1.66. Hypothermia generates an unusual wave called the J-wave or the Osborne wave which appears added onto the QRS-complex just before the ST-segment

(Fig. 1.66). The PR-interval may occasionally be longer than 0.2 seconds. The QRS duration may widen beyond 0.12 seconds. The QT interval may occasionally be prolonged.

1.4.8 Preexcitation Syndromes

When accessory conduction pathways from the atria to the ventricles exist, the electrical impulse propagating down the accessory pathway will excite the ventricles ahead of time, before the impulse propagating through the A-V node (normal pathway) arrives. This results in a shorter PR-interval and a "slur" (delta wave) at the onset of the QRS-complex. The QRS-complex that is seen is really the result of a fusion of the depolarisations triggered by the impulse coming down the accessory pathway with that triggered by the impulse going down the normal route via the AV-node [1–4]. Thus the QRS-complex is wider than normal.

Accessory pathways are abnormal strands of myocardial fibres that conduct the electrical impulses from the atria to the ventricles or to the AV-junction, or from the AV-junction to the ventricles. These pathways not only conduct the impulses forward (anterograde) but backward (retrograde) as well.

Ventricular Preexcitation

In Ventricular Preexcitation (Accessory AV pathways) PR-interval is usually shortened to between 0.09 and 0.12 second. QRS duration is longer than 0.10 second and abnormal in shape (with a delta wave preceding the QRS-complex). This distortion occurs because of the early excitation occurring before the main excitation (Fig. 1.67).

Atrio-His Preexcitation

In Atrio-His preexcitation (Atrio-His fibres), the PR-interval is usually shortened to less than 0.12 seconds. The QRS-complex remains normal.

Nodoventricular/Fasciculoventricular Preexcitation

In Nodoventricular/Fasciculoventricular preexcitation (Nodoventricular/Fasci culoventricular fibres), the PR-interval is usually normal (>0.12 seconds) but the QRS duration is widened with a delta wave preceding the QRS complex.

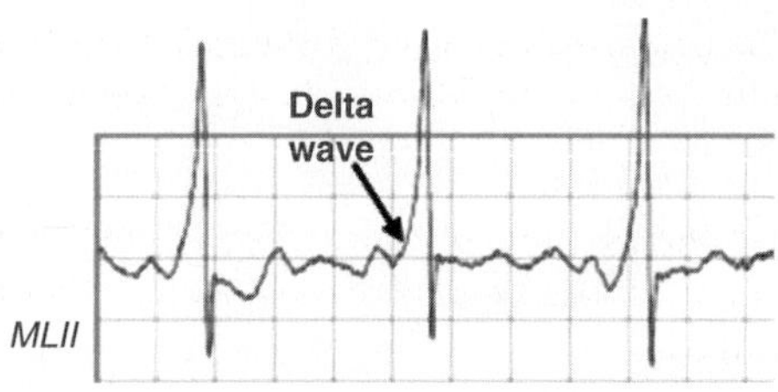

Fig. 1.67. Preexcitation syndromes occur when an accessory pathway starts ventricular depolarisation ahead of time (seen as a delta wave)

1.5 Myocardial Ischemia, Injury and Infarction

Myocardial ischemia, injury and infarction (tissue death) occur when blood circulation to a segment of the myocardium is decreased or disrupted. This may be caused by an occlusion of the local coronary artery as a result of a blood clot or by a spasm of that local artery. The lack of oxygen (anoxia) causes a delay in the depolarisation and repolarisation of the cells leading to a change in ECG morphology. Mild or moderate anoxia (myocardial ischemia) can be tolerated for a short time and upon return to adequate perfusion, the cells usually recover to a normal or near normal condition [1–4].

1.5.1 Zones of Ischemia, Injury and Infarction

Blood perfusion may be reduced by a stenotic artery, an arterial spasm or by a thrombotic (blood clot) occlusion. If perfusion is severely reduced, acute myocardial infarction (tissue death) occurs at the central zone served by the artery. Myocardial injury occurs in the outlying zone bordering the infarcted area, and myocardial ischemia occurs in a further outlying zone after that (Fig. 1.68).

The size of the ischemic zone of the heart depends on which coronary artery is blocked or stenotic (narrowed) and whether it is proximal (near the beginning of the vessel) or distal (distant to the beginning). In general, a proximal occlusion of a coronary blood vessel will affect a larger zone, while a distal occlusion will affect a smaller zone. If the occlusion occurs at a small distal blood vessel, the myocardium may remain adequately perfused by neighbouring blood vessels (called collateral vessels) and no significant ischemia occurs.

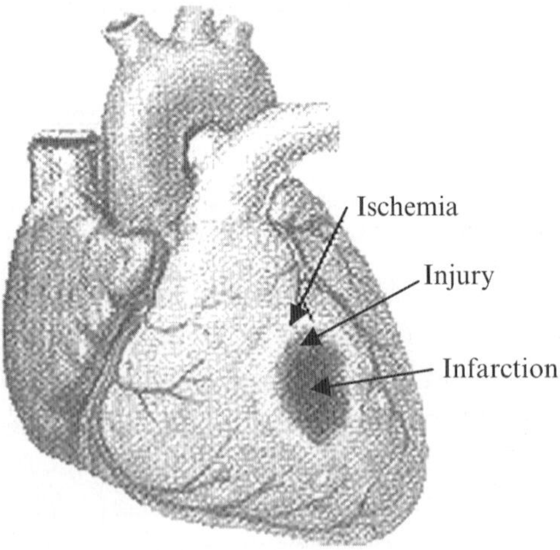

Fig. 1.68. An occlusion of the coronary blood vessel may cause a region of the heart to suffer ischemia, injury or even infarction (necrosis)

1.5.2 Myocardial Injury

If myocardial ischemia is severe or prolonged, the myocardial cells sustain injury and stop functioning normally. The cells are unable to contract or conduct electrical impulses. However the damage to the cell is still reversible and the injured cells may return to normal or near-normal after the return of adequate blood flow and reoxygenation.

1.5.3 Acute Myocardial Infarction (AMI)

If severe myocardial ischemia continues, the cells die (necrosis) and acute myocardial infarction occurs. Over time, the dead tissue will be replaced by scar tissue, which has no contractile ability. Myocardial infarction may be non-transmural or transmural. In a non transmural infarction, the zone of infarction involves a partial thickness of the ventricular wall, usually the subendocardial area of the myocardium. In transmural infarction, the zone of infarction is more, extending outward from the subendocardial area to the entire thickness of the ventricular wall.

1.5.4 ECG Changes

Changes in ECG appear as morphological changes in T-wave, ST-segment and (eventually Q-wave) as the different regions of the heart goes through varying degrees of ischemia, injury and infarction.

Abnormal T-waves appear within seconds of an acute myocardial transmural infarction for leads that face the ischemic zone of the heart. The T-waves become abnormally tall and peaked. The QT-intervals are usually prolonged. The ST-segment becomes highly elevated because of injury current generated by the leakage of ions across the cell membranes. Over a period of days to weeks, as healing (or the opposite – necrosis) takes place, the region of injury reduces in size. The ST-segment elevation correspondingly returns to baseline while the T-wave becomes inverted from the altered repolarisation pattern (Fig. 1.69).

ST-segment elevation indicates severe and extensive myocardial ischemia and injury. The segment is usually considered elevated if the segment is 0.1 mV or greater above baseline, as measured from 40 ms after the J-point of the ECG. While leads facing the zone of ischemia record an elevation, opposite (reciprocal) leads record a depression [1–4]. ST-segment elevation is often accompanied by increased amplitude in the QRS-complex during the early stages of AMI.

ST-segment depression in the facing leads indicates the occurrence of subendocardial ischemia and injury, and may appear downsloping, horizontal or upsloping. Usually no Q-wave results from this less severe form of AMI (Fig. 1.70).

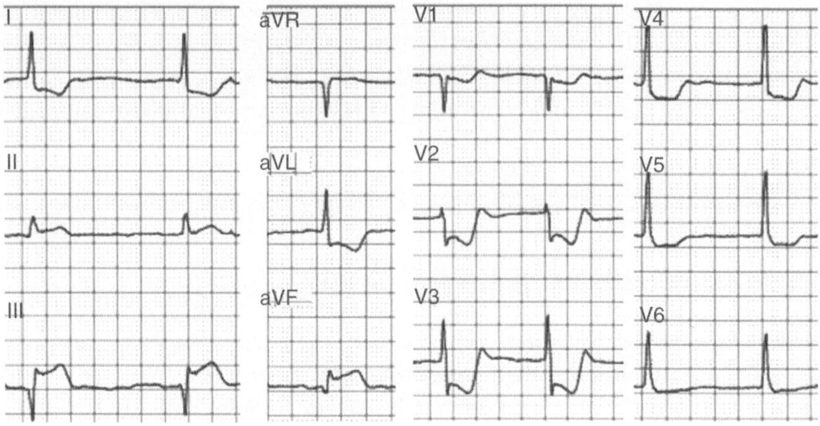

Fig. 1.69. In acute myocardial infarction, the ECG in the facing lead shows abnormally tall T-waves with elevated ST-segment while those in the reciprocal leads show inverted T-waves with depressed ST-segment

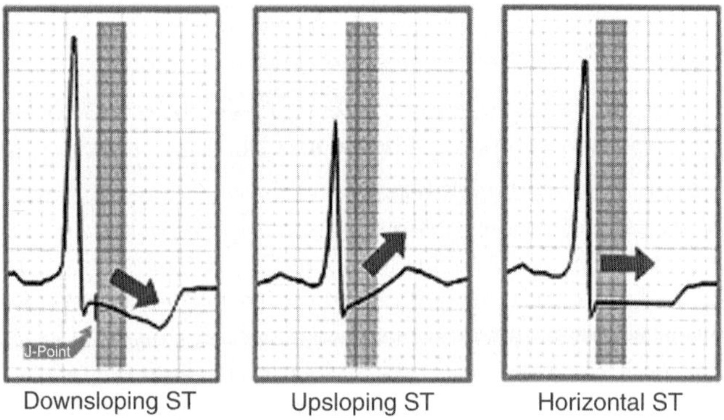

Fig. 1.70. ST-segment depressions may be downsloping, horizontal or upsloping

1.5.5 Evolution of the Deep Q-Wave

As the infarcted ventricular wall is gradually replaced by fibrous tissue, an electrical "window" forms since fibrous tissue does not generate any ionic current. A lead facing this window will see the electrical depolarisation of the far (opposite) non-infarcted ventricular wall. (This is analogous to a window in the near wall of the heart allowing the far wall of the heart to be seen.) Since the direction of depolarisation seen will be moving away from the lead, a negative deflection is observed by that lead. This is observed as a deep Q-wave in the ECG (for that lead). The T-wave that was inverted during early infarction remains inverted (Fig. 1.71).

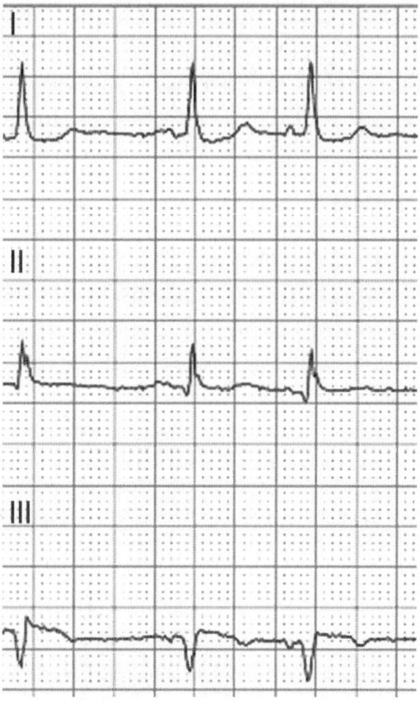

Fig. 1.71. A deep Q-wave appears, accompanied by an inverted T-wave as the heart recovers from transmural myocardial infarction

1.5.6 Silent Ischemia

Myocardial ischemia may be "silent" (or asymptomatic). The patient experiences no chest pain. The ST-segment is depressed during exertion and accompanied by T-wave inversion. The ECG returns to normal at rest. It is normally caused by atheromatous plague forcing significant stenosis (narrowing) of a coronary artery. The stenosis results in ischemia when myocardial demand is increased during exertion, but remains adequate while at rest.

1.5.7 Stable Angina

Stable angina exhibits ST-segment depression during exertion. The T-wave may be inverted. The patient often experiences chest pain. The ECG returns to normal at rest but reappears with the same changes when the patient exerts again. It is normally caused by a stable atheromatous plague causing significant stenosis (narrowing) of a coronary artery.

1.5.8 Unstable Angina

The ST-segment is depressed during exertion. The T-wave may be inverted. The patient often experiences chest pain lasting 10 to 20 minutes or more.

It is normally caused by a rupture of the plaque followed by the formation of a thrombus that further occludes the already stenotic artery. Erosion of the existing plague and partial spontaneous thrombolysis (dissolution of the blood clot) may occur and reduce the severity of the occlusion.

Dilated Cardiomyopathy

Various conditions can damage the myocardium, resulting in impaired contractile function and eventually an enlarged, dilated, poorly contracting heart (dilated cardiomyopathy). The commonest cause is atherosclerotic coronary artery disease. Multiple infarctions result in myocardial cell death and replacement of myocytes by non-functioning scar tissue. Dilated cardiomyopathy due to coronary artery disease is sometimes termed "ischaemic cardiomyopathy". Other causes of dilated cardiomyopathy include viral myocarditis, toxicity due to drugs such as doxorubicin/daunorubicin, peripartum cardiomyopathy, and hereditary cardiomyopathy. These patients often have atrial fibrillation (due to elevated left atrial pressures) and rapid ventricular rates as the heart tries to compensate for the poor contractile function (Cardiac output = Heart rate × Stroke volume). Figure 1.72 shows the ECG tracing of a patient with dilated cardiomyopathy in atrial fibrillation.

Sick Sinus Syndrome (SSS) (Brady-Tachy) Type

In this condition, the Sino-atrial (SA) node, which normally regulates the heart rate, is "sick". It fires at a rate that is slower than normal or fails to fire at all; resulting in bradycardia and/or sinus pauses. Other pacemakers in the heart (eg the AV node) may take over as the dominant rhythm. SSS is seen mainly in the elderly, and as such, is often accompanied by atrial fibrillation or atrial flutter (which are more common in old age). With atrial fibrillation/flutter, the ventricular rate is often rapid (tachycardia). Yet, when the atrial fibrillation/flutter terminates spontaneously, the SA node fails to

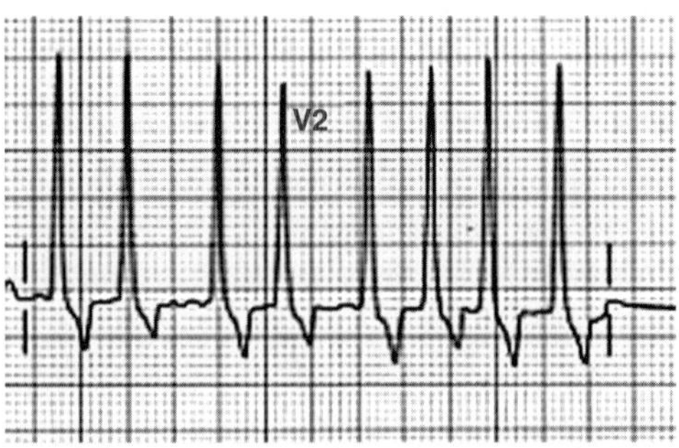

Fig. 1.72. In dilated cardiomyopathy, the R-waves are tall and ventricular rate is fast to compensate for poor contractile function

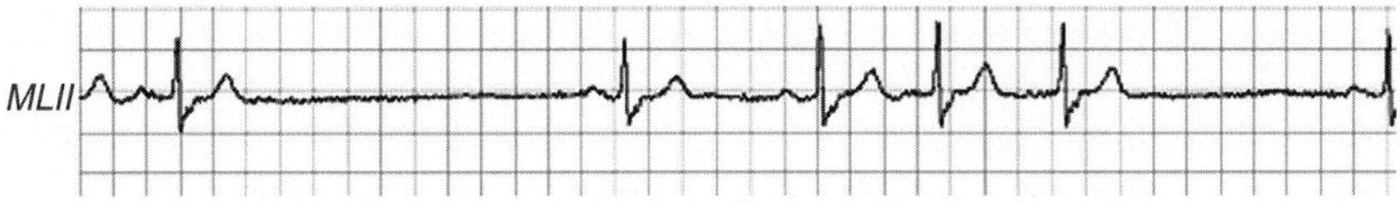

Fig. 1.73. In sick sinus syndrome (brady-tachy), the contractions of heart oscillates between fast & slow rates

fire, resulting in asystole or bradycardia. This leads to cycles of tachycardia followed by bradycardia and hence the term "tachy-brady syndrome". An example of this is shown in Fig. 1.73.

Acknowledgements

Most figures on ECG waveforms are reproduced with permission from PhysioBank (`http://www.physionet.org/physiobank/`), Prof. Alan Lindsay (`http://library.med.utah.edu/kw/ecg/`), ECG Wave-Maven (`http://ecg.bidmc.harvard.edu/maven`)

References

1. Jenkins, R.D., and Gerred, S.J. (1997): 'ECGs by Example', *Churchill Livingstone.*
2. Golschlager, N., Golman, M.J. (1989): 'Principles of Clinical Electrocardiography', Thirteenth Edition, *Appleton and Lange.*
3. Robert, J.H. (2001): 'Basic Dysrhythmias – Interpretation and Management', Third Edition, *Mosby, Inc.*
4. Webster, J.G. (1992): 'Medical Instrumentation – Principles and Applications', Third Edition, *John Wiley & Sons.*
5. Malmivuo, J., and Plonsey, R. (1995): 'Bioelectromagnetism: Principles and Applications of Bioelectric and Biomagnetic Fields', *Oxford University Press.*

2

Analysis of Electrocardiograms

N. Kannathal, U. Rajendra Acharya, Paul Joseph, Lim Choo Min
and Jasjit S. Suri

The electrocardiogram (ECG) representing the electrical activity of the heart is the key biosignal for aiding the clinical staff in disease diagnosis. ECG has been always chosen as the basic signal for diagnosing the cardiac abnormalities and detecting the patients' states. Generally the various characteristic features of ECG are extracted and used for decision making purposes. This makes the decision making and diagnosis process simpler and faster. Hence appropriate feature description and extraction becomes the most important component in cardiac health diagnostics. In this chapter various techniques for feature extraction are described.

2.1 Steps in ECG Analysis

The major steps in the analysis of the ECG signals are:

Noise elimination from ECG using noise filtering techniques
Cardiac cycle detection by detecting QRS complex
Detection of significant characteristic points in ECG signal
Formulation of characteristic feature set

Noise filter removes and reduces the noise components from various sources in the ECG signal.

Cardiac cycle detection involves detecting the QRS complex peak corresponding to each beat. QRS Complex detection is implemented using Tompkins QRS complex detection algorithm.

ECG characteristic points detection involves determining of significant points on the ECG for feature extraction. It includes the detection of QRS complex onset and offset, ST segment detection and T peak detection.

Feature set formulation includes formulation and selection of characteristic features such that they significantly relate to the abnormalities. Additional features are extracted by performing complexity analysis on the signal.

2.2 Preprocessing of ECG

The ECG consists of three basic waves, P, QRS and T. These waves correspond to the far field induced by specific electrical phenomena on the cardiac surface, namely the atrial depolarization (P wave), the ventricular depolarization (QRS complex), and the ventricular repolarization (T wave). The ECG does not look the same in all the leads of the standard 12 lead system used in clinical practice. The polarity and the shape of the ECG constituent waves are different depending on the lead that is used. A sample ECG wave measured at lead II is shown in Fig. 2.1.

In a normal cardiac cycle, the P wave occurs first, followed by the QRS complex and the T wave. The sections of the ECG between the waves and complexes are called segments. The ECG is characterized by three segments namely the PR segment, the ST segment and the TP segment. The characteristic time periods in the ECG wave are the PR interval, the RT interval, and the R-R interval.

Usually ECG signals are contaminated by various kinds of noise. Various noise contaminating the ECG are described in the following section.

Power Line Interference

Power line interference consists of 60/50 Hz pickup and harmonics that can be modeled as sinusoids and combination of sinusoids. According to Friesen *et al* [1], the frequency content of this kind of noise is 60/50 Hz with harmonics and the amplitude is 50% of peak-to-peak ECG amplitude.

Let us consider the presence of a periodic artifact with the fundamental frequency of 60 Hz and odd harmonics at 180, 300, and 420 Hz. Let the sampling frequency (fs) be 1000 Hz, and assume the absence of any aliasing error. Zeros are then desired at 60, 180, 300, and 420 Hz, which translate to $\pm 21.6°$, $\pm 64.8°$,

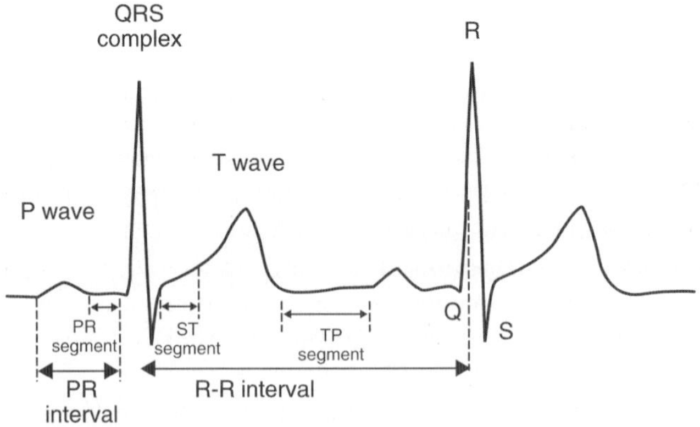

Fig. 2.1. The ECG signal

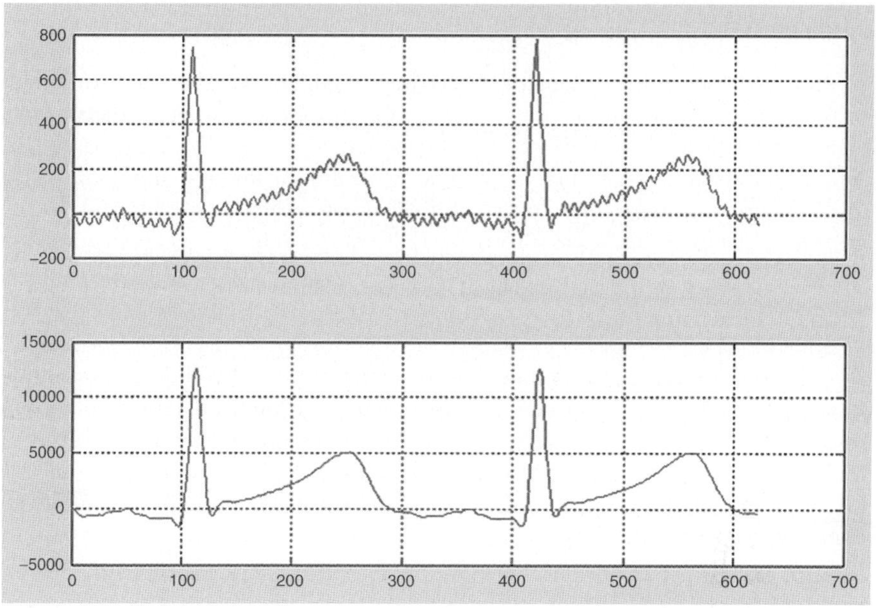

Fig. 2.2. Result of power line interference. (**a**) Original signal with power line interference. (**b**) Output of power line interference filter

$\pm 151.2°$, with 360° corresponding to 1000 Hz. The coordinates of the zeros are $0.9297 \pm j0.3681$, $0.4257 \pm j0.9048$, $-0.3090 \pm j0.9510$, and $-0.8763 + j0.4817$. The transfer function of the filter is $H(z) = G(1 - 1.8595\,z^{-1} + z^{-2})(1 - 0.8515\,z^{-1} + z^{-2}) \times (1 + 0.6180\,z^{-1} + z^{-2})(1 + 1.7526\,z^{-1} + z^{-2})$, where G is the desired gain or scaling factor. With G computed so as to set the gain at DC to be unity, the filter transfer function becomes $H(z) = 0.6310 - 0.2149\,z^{-1} + 0.1512\,z^{-2} - 0.1288\,z^{-3} + 0.1227\,z^{-4} - 0.1288\,z^{-5} + 0.1512\,z^{-6} - 0.2149\,z^{-7} + 0.6310\,z^{-8}$. Figure 2.2 shows the performance of the power line elimination.

Electrode Contact Noise

Electrode contact noise is transient interference caused by loss of contact between the electrode and the skin, which can be permanent or intermittent. The switching action can result in large artifacts since the ECG signal is usually capacitively coupled to the system. This type of noise can be modeled as a randomly occurring rapid baseline transition that decays exponentially to the base line and has a superimposed 60 Hz component. According to Friesen *et al* [1], the duration of the noise signal is 1 sec and the amplitude is the maximum-recorded output with the frequency of 60 Hz.

Motion Artifact

Motion artifacts are transient base line changes in the electrode skin impedance with electrode motion. The shape of the base line disturbance caused by the motion artifacts can be assumed to be a biphasic signal resembling one cycle of a sine wave. The peak amplitude and duration of the artifacts are variables. The duration of this kind of noise signal is 100–500 ms with amplitude of 500% peak-to-peak ECG amplitude.

Muscle Contraction

Muscle contraction causes generation of artifactual millivolt level potentials. It can be assumed to be transient burst of zero mean band limited Gaussian noise. The variance of the distribution may be estimated from the variation and duration of the bursts. Standard deviation of this kind of noise is 10% of peak-to-peak ECG amplitude with duration of 50 ms and the frequency content being dc to 10 kHz.

Base Line Wander

The baseline wander of the ECG signals causes problems in the detection of peaks. For example, due to the wander, the T peak could be higher than R peak, and it is detected as an R peak instead. Low frequency wander of the ECG signal can be caused by respiration or patient movement. The drift of the baseline with respiration can be represented as a sinusoidal component and the frequency of respiration added to the ECG signal. The variation could be reproduced by amplitude modulation of the ECG by the sinusoidal component that is added to the base line. The amplitude variation is 15% of peak-to-peak ECG amplitude and the base line variation is 15% of ECG amplitude at 0.15 to 0.3 Hz.

These noise should be removed from ECG before extracting the characteristic features. Noise removal is accomplished by passing the cardiovascular signals through filter whose cutoff frequency is a function of the noise frequency.

To solve baseline wander, median filtering can be used. The steps involved in the implementation of baseline wander are shown in Fig. 2.3. First, the first 200 ms of samples were extracted and sorted out in ascending order, then its median was calculated. Then for every 200 ms of samples till the end of the ECG signal, the same procedure was carried out. Now, these samples are fed as input to the 600 ms window median filtering. Later, the median value is evaluated for every 600 ms of samples. Then these median values were subtracted from the original waveform to remove the baseline wander of the ECG signal. Figure 2.4 shows the result of the baseline wander filter.

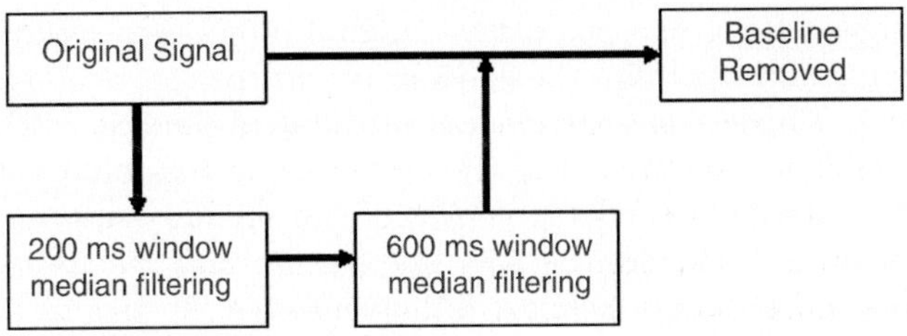

Fig. 2.3. Algorithm to remove the baseline wander

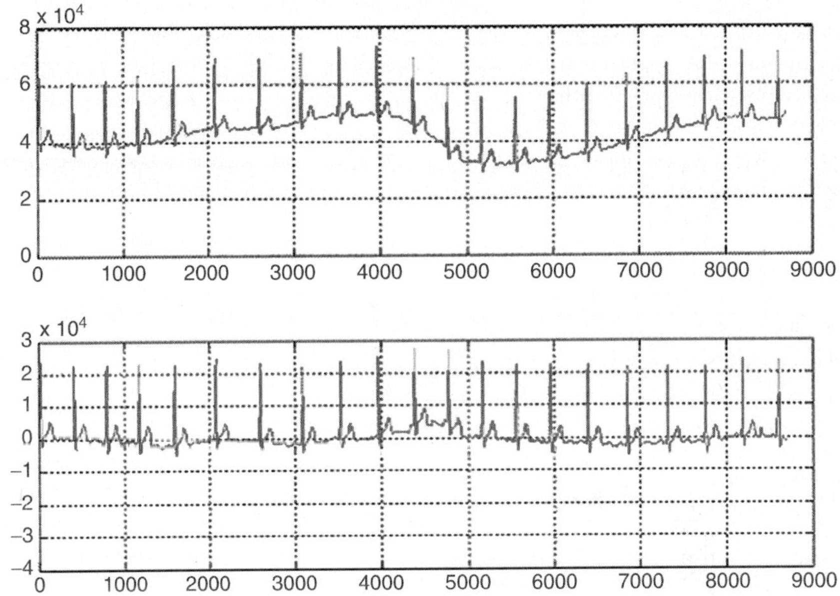

Fig. 2.4. Results of baseline wander algorithm. (**a**) Original signal. (**b**) Output of baseline wander filter

2.2.1 Noise Filtering Technique

The first step in noise filtering is determining the frequency corresponding to the significant characteristics of the ECG signal and the noises. From the Fourier transform of human ECG signal it has been found that P and T wave frequency generally lie between 0.5 and 10 Hz and QRS complex frequency ranges between 4 and 20 Hz [2]. The P or T wave sometimes coincides with the baseline noise having a low frequency range of 0–0.8 Hz. Hence it is very essential to eliminate the baseline noise from ECG signals.

In order to attenuate noise, the signal is passed through a band pass filter composed of cascaded high-pass and low-pass integer filters. The band pass filter is designed from a special class of digital filters that require only integer coefficients. This permits the software to do the signal processing using only integer arithmetic and thereby permitting real-time processing. Since it was not possible to directly design the desired band pass filter with this special approach, the design actually consist of cascaded low-pass and high-pass filter sections. For removing the base line wander and the AC power noise, the algorithm of Alste and Schilder is implemented [3]. In this algorithm, a non recursive finite impulse response (NRFIR) filter has been used with the reduced number of taps. The basic principle behind this filtering process is that the designed frequency response has been defined with small stop band notches at 0 Hz to remove base line wander, as well as 50 Hz and its higher harmonics to remove power line interference. In this algorithm, in order to reduce the long computational time caused by the large number of multiplication involved in the filtering in the time domain the property of the discrete Fourier transform and symmetrical nature of the impulse response has been used. This is achieved by the property that the phase is the linear function of frequency that corresponds with an exact delay time. The convolution sum for the NRFIR is

$$\begin{aligned} y(n\mathrm{T}) = &\sum_{i=1}^{\frac{(\mathrm{M}-3)}{2}} (h(i.k\mathrm{T}) \times ((n-ki)\mathrm{T}) + x((n-k(\mathrm{M}-1)+ki)\mathrm{T})) \\ &+h\left(\frac{(\mathrm{M}-1)}{2}k\mathrm{T}\right) \times \left(n-k\frac{(\mathrm{M}-1)}{2}\mathrm{T}\right) \end{aligned} \tag{2.1}$$

where

T = sampling interval of both the input and output signal
$k = 5$
$k\mathrm{T}$ = time interval between successive impulse response coefficients
$x(n\mathrm{T})$ = input signal
$y(n\mathrm{T})$ = output signal
$h(ik\mathrm{T})$ = filter impulse response coefficients
M = number of filter coefficients

This filter removes the baseline wander and AC power frequency noise. But some high frequency noises like muscle noise or EMG are still present in the ECG signal. These high frequency noises are eliminated by using another low pass FIR filter with a cutoff frequency higher than the frequency of the QRS complex.

2.3 QRS Complex Detection

All the required features from ECG are extracted from the filtered ECG signal. The basic and essential component for feature extraction is the detection of the QRS complex i.e. locating the R point for each beat of the signal. Once R point is determined, all other characteristic points on the wave are determined with reference to the R point. Thus an accurate detection of the QRS complex of the ECG is an important task in ECG analysis.

2.3.1 QRS Detection Algorithm

The algorithm used to detect QRS complexes is an adaptation of the commonly used real-time QRS detection algorithm developed by Pan *et al* [4] and further described by Hamilton *et al* [5]. It recognizes QRS complexes based on analysis of the slope, amplitude and width.

Figure 2.5 shows the various processes involved in the analysis of the ECG signal. In order to isolate the portion of the wave where QRS energy is predominant, the signal is passed through a band pass filter composed of cascaded high-pass and low-pass integer filters. Then the signal is subjected to differentiation, squaring, time averaging and finally peak is detected by applying threshold.

The band pass filter is designed from a special class of digital filters that require only integer coefficients. Since it was not possible to directly design the desired band pass filter with this special approach, the design actually consist of cascaded low-pass and high-pass filter sections.

The next processing step is differentiation, a standard technique for finding the high slopes that normally distinguish the QRS complexes from other ECG waves. To this point in the algorithm, all the processes are accomplished by linear digital filters.

The differentiated waveform is subjected to a nonlinear transformation. The nonlinear transformation involves point-by-point squaring of the signal samples. This transformation serves to make all the data positive prior to subsequent integration, and also accentuates the higher frequencies in the signal obtained from the differentiation process. These higher frequencies are normally characteristic of the QRS complex.

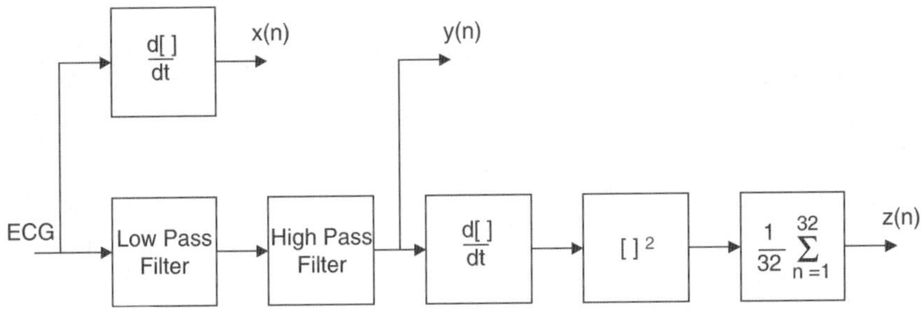

Fig. 2.5. Block diagram of the QRS detector. z(n) is the time averaged signal. y(n) is the band passed ECG. x(n) is the differentiated ECG

The squared waveform passes through a moving window integrator. This integrator sums the area under the squared waveform over a 150 msec interval, advances 1 sample interval and integrates the new 150 msec window. The width of the window was chosen to be long enough to include the time duration of extended abnormal QRS complexes, but short enough so that it does not overlap both the QRS complex and the T wave.

Adaptive amplitude threshold applied to the band pass filtered waveform and to the moving integration waveform are based on continuously updated estimate of the peak signal level and the peak noise. After preliminary detection by the adaptive thresholds, decision processes make the final determination as to whether or not detected event was a QRS complex. A measurement algorithm calculates the QRS duration after the detection of each QRS complex. Thus two waveform features are available for subsequent analysis, RR interval and QRS duration.

Band Pass Integer Filter

The band pass filter for the QRS (heart rate) detection algorithm reduces noise in the ECG signal by matching the spectrum of the average QRS complex. Thus it attenuates T wave interference as well as noise. The pass band that maximizes the QRS energy is approximately in the 5–15 Hz range. The filter implemented in this algorithm is a recursive integer filter in which poles are located to cancel the zeros on the unit circle of the z plane. A low pass filter and a high pass filter are cascaded to form the band pass filter.

Low Pass Integer Filter

The transfer function of the second order low pass filter is

$$\mathrm{H(z)} = \frac{(1 - \mathrm{z}^{-6})^2}{(1 - \mathrm{z}^{-1})^2} \tag{2.2}$$

The difference equation of this filter is

$$y(n\mathrm{T}) = 2y(n\mathrm{T}-\mathrm{T}) - y(n\mathrm{T}-2\mathrm{T}) + x(n\mathrm{T}) - 2x(n\mathrm{T}-6\mathrm{T}) + x(n\mathrm{T}-12\mathrm{T}) \tag{2.3}$$

High Pass Integer Filter

The high pass filter is implemented by subtracting a first order low pass filter from an all pass filter with delay shown in the Fig. 2.6.

The low pass filter is an integer coefficient filter with the transfer function

$$\mathrm{H_{lp}(z)} = \frac{\mathrm{Y(z)}}{\mathrm{X(z)}} = \frac{(1 - \mathrm{z}^{-32})}{(1 - \mathrm{z}^{-1})} \tag{2.4}$$

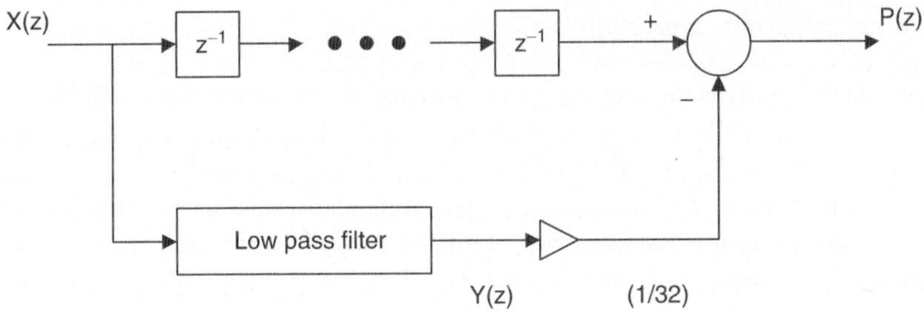

Fig. 2.6. High pass filter

The difference equation of the transfer function is

$$y(n\mathrm{T}) = y(n\mathrm{T} - \mathrm{T}) + x(n\mathrm{T}) - x(n\mathrm{T} - 32\mathrm{T}) \tag{2.5}$$

The high pass filter is obtained by dividing the output of the low pass filter by its dc gain and then subtracting from the original signal. The transfer function of the high pass filter is

$$\mathrm{H_{hp}(z)} = \frac{\mathrm{P(z)}}{\mathrm{X(z)}} = \mathrm{z}^{-16} - \frac{\mathrm{H_{lp}(z)}}{32} \tag{2.6}$$

The difference equation for this filter is

$$p(n\mathrm{T}) = x(n\mathrm{T} - 16\mathrm{T}) - (1/32)[y(n\mathrm{T} - \mathrm{T}) + x(n\mathrm{T}) - x(n\mathrm{T} - 32\mathrm{T})] \tag{2.7}$$

Derivative

After the signal has been filtered it is then differentiated to provide information about the slope of the QRS complex. A five-point derivative has the transfer function

$$\mathrm{H(z)} = 0.1(2 + \mathrm{z}^{-1} - \mathrm{z}^{-3} - 2\mathrm{z}^{-4}) \tag{2.8}$$

This derivative is implemented with the difference equation

$$y(n\mathrm{T}) = (1/8)[2x(n\mathrm{T}) + x(n\mathrm{T} - \mathrm{T}) + x(n\mathrm{T} - 3\mathrm{T}) - 2x(n\mathrm{T} - 4\mathrm{T})] \tag{2.9}$$

The fraction 1/8 is an approximation of the actual gain of 0.1. This derivative approximates the ideal derivative in the dc through 30 Hz frequency range.

Squaring Function

The squaring function that the signal passes through is a nonlinear operation. The equation that implements this operation is

$$y(n\mathrm{T}) = [x(n\mathrm{T})]^2 \tag{2.10}$$

This operation makes all data points in the processed signal positive and it amplifies the output of the derivative process nonlinearly. It emphasizes the higher frequencies in the signal that are mainly due to the QRS complex.

Moving Window Integral

The slope of the R wave alone is not a guaranteed way to detect a QRS event. Many abnormal QRS complexes that have large amplitude and long duration (not very steep slopes) might not be detected using information of the R wave only. Thus we need to extract more information from the signal to detect a QRS event. Moving window integration extracts features in addition to the slope of the R wave. It is implemented with the following difference equation

$$y(n\mathrm{T}) = (1/\mathrm{N})[x(n\mathrm{T} - (\mathrm{N} - 1)\mathrm{T}) + x(n\mathrm{T} - (\mathrm{N} - 2)\mathrm{T}) + \cdots + x(n\mathrm{T})] \quad (2.11)$$

where N is the number of samples in the width of the moving window. The width of the window should be approximately the same as the widest possible QRS complex. If the size of the window is too large the integration waveform will merge the QRS and T complexes together. On other hand, if the size of the window is too small, a QRS complex could produce several peaks at the output of the stage. The width of the window should be chosen experimentally.

QRS Detection Using Adaptive Thresholds

R peak is then detected using an upward and downward threshold. The thresholds are calculated using running estimates of signal peak and noise peak. Signal peaks are defined as those of the QRS complex and the noise peaks are those of the T waves, muscle noise etc. After the ECG has passed through various filter stages, its signal to noise ratio increases. Hence the thresholds can be chosen above the noise peak levels. It increases the overall sensitivity of the detector. Two sets of thresholds are used, each of which has two threshold levels. The set of thresholds that is applied to the waveform from the moving window integrator is given by

SPKI = 0.125 PEAKI + 0.875 SPKI   if PEAKI is the signal peak
NPKI = 0.125 PEAKI + 0.875 NPKI   if PEAKI is the noise peak
THRESHOLD I1 = NPKI + 0.25(SPKI–NPKI)
THRESHOLD I2 = 0.5 THRESHOLD I1

where

PEAKI is the overall peak
SPKI is the running estimate of the signal peak
NPKI is the running estimate of the noise peak
THRESHOLD I1 is the first threshold applied
THRESHOLD I2 is the second threshold applied

A peak is determined when the signal changes direction within a certain time interval. Thus, SPKI is the peak that the algorithm has learned to be that of the QRS complex, while NPKI is any peak that is not related to the signal of interest. It can be seen from the above equations that the new values of thresholds are calculated from the previous ones. In this way, the thresholds are dynamically adjusted to improve detection.

Whenever a new peak is detected it must be categorized as a signal peak or noise peak. If the peak level exceeds the THRESHOLD I1 during the first analysis of the signal, it is detected as a signal peak. A refractory pause of 200 ms is introduced after detection to minimize the detection of the same QRS complex second time.

2.4 Detection of QRS Complex Onset and Offset

The detection of QRS complex generally implies the detection of R peak of QRS complex, which can be used as the reference for extraction of various features on each beat of the ECG signal. To know the QRS complex in its entirety, recognition of the onset and offset is necessary in most types of computer based ECG analysis. QRS onset is generally defined as the beginning of the Q wave or R wave, if no Q wave is present. QRS offset is defined as the end of the S wave [6] if no S wave is visible. QRS delineation is moreover a prerequisite in certain schemes for the classification of different morphologies found. In such schemes, the features, which characterize the QRS complex, are computed with respect to the delineated interval.

Automatic detection of QRS onset and offset points with reasonable accuracy has been a difficult task. The problem is additionally complicated by the presence of power line interference and baseline wander in the original signal. Estimation of QRS onset and offset in multiple lead recordings is based on a delineation function, which is obtained by combining different leads. The function commonly employed is the spatial velocity. The recognition is done by a simple threshold or a template matching technique. Kenneth in 1982 [7] determined the onset as the intersection of the closest fit of a line having a slope of an R wave wall and the base of the wall considered with some offset from the baseline. The slope of the R wave onset and termination walls have been measured by the rule activation, which was used to detect them. This method is time consuming. Leif Sornomo [8] has determined the onset and offset using a maximum likelihood procedure based on the statistical models for the low frequency and high frequency segments of the ECG. A commercially available software product [9] determines the onsets and offsets by an analysis of the simultaneous slopes in all 12 leads. That is, the QRS duration is measured from the earliest onset in any lead to the latest deflection in any lead. Lin *et al* [10] detected onset and offset based on a thresholding technique. This method may be sensitive to noise mainly the baseline noise, which cannot be eliminated without distorting P waves.

In this chapter, we have detected edges as the point with zero slope when there is a sudden change and by a minimum distance method otherwise. The edges are assumed to lie within a window length of 50 samples taken symmetrically around the R peak. 50 samples correspond to the maximum possible edge distance from the peak, which has been computed from the physiological study of ECG and the sampling frequency of the ECG signal. To find whether a beat has a sharp change or not, the signal values were converted to a three-pulse code train after finding the finite differences. If $[y_1, y_2, y_3, \ldots, y_\mathrm{i}, \ldots, y_\mathrm{n}]$ are the filtered signal samples defined at time $t_1, t_2, t_3, \ldots,$ etc. then the finite differences are defined as

$$\mathrm{d}y_i = y_{i+1} - y_i. \tag{2.12}$$

For easy analysis purpose the signal samples were then converted into a train of three-pulse code train,

$$sl = \begin{matrix} 1, & \text{if } \mathrm{d}y_\mathrm{i} > 0 \\ 0, & \text{if } \mathrm{d}y_\mathrm{i} = 0 \\ -1, & \text{if } \mathrm{d}y_\mathrm{i} < 0 \end{matrix} \tag{2.13}$$

Presence of sharp changes was detected as change from −1 and 0 to 1 or vice versa (i.e. change in sign). A sharp change is defined as a point with sudden change to zero or negative slope is considered as an edge. But if no sharp changes were present in the signal, edge is detected as the minimum distance from the projected origin. The projected origin is the point of intersection of vertical projection from the peak on to the time axis and the horizontal projection from the edges of the window on the vertical projection. For onset, the edge of the window is taken as the beginning of the window whereas finding offset is taken as the end of the window. If this point is defined as (p_x, p_y), the distance matrix comprising of distance of all signal points within the window length to the projected origin can be calculated as

$$dis_i = \mathrm{sqrt}((t_i - p_x)^2 + (y_i - p_y)^2). \tag{2.14}$$

The edge is then defined as $[y_i, t_i]$ with minimum dis_i. The validity of the edge as a true edge was checked by using angle criteria. The angle is found between the points at which edge is detected and the fifth sample from that point. The edge was declared as true edge if the angle is less than 8 degrees. In case of false detection the minimum distance method was repeated with this point (point at which angle is found to be greater than 8 degrees) as the end of the window. By detecting the onset and offset of QRS complex, QRS duration is determined.

2.5 ST Segment Analyzer

The isoelectric period (ST segment) following the QRS is the time, at which the entire ventricle is depolarized and roughly corresponds to the plateau phase of the ventricular action potential. The ST segment is important in the diagnosis of ventricular ischemia or hypoxia because under those conditions, the ST segment can become either depressed or elevated.

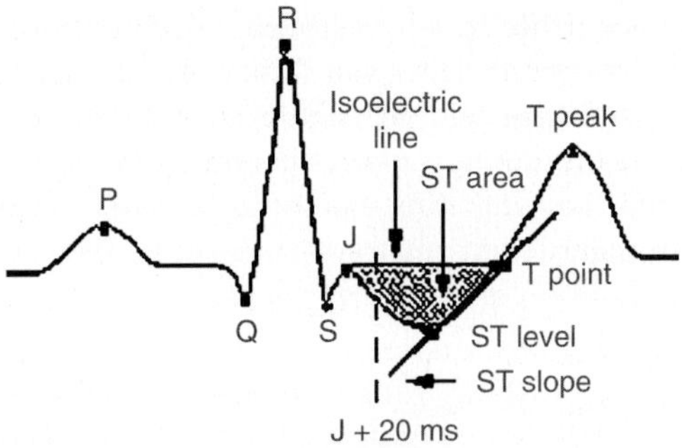

Fig. 2.7. ECG signal indicating ST area

The J point (shown in Fig. 2.7) is the first inflection point after the S point or may be the S point itself in certain ECG waveforms. The T wave peak is the maximal absolute value, relative to the isoelectric line between J + 80 ms and R + 400 ms. The onset of the T wave, the T point, is found by looking for a 35 ms period on the R side of the T wave which has values within one sample unit of each other. The T point is among the most difficult features to identify. If it is not detected, it is assumed to be J + 120 ms.

Having identified various ECG features, ST segment measurements are made using a window search method. Two boundaries, the J + 20 ms and the T point, define the window limits. The point of maximal depression or elevation in the window is then identified.

ST segment length is defined as the distance between the J point and the T point. In addition to the ST segment length, several other parameters are calculated. The ST segment deviation can be expressed as the polarity of change relative to the isoelectric line. The ST slope is defined as the amplitude difference between the ST segment point and the T point divided by the corresponding time interval. The ST area is calculated by summing all sample values between the J and T points after subtracting the isoelectric line value from each point. RT interval is defined as the time between the R peak and the corresponding T peak.

2.6 Complexity Analysis of ECG Signal

Physiological signals such as ECG have a wide variety of forms. To describe them, traditional features discussed previously using amplitude and frequency information is not sufficient. This is because the signals of different bandwidth cannot be compared. In addition such measures do not allow comparison within the same subject groups. The absolute frequency of rhythms may differ within the same subject group due to various reasons. Hence some other features are also determined.

When visually inspecting the ECG signals, one of the first impression they give to the observer is the complexity. Some beats appear at random while others seem to demonstrate a reappearance of certain patterns at various intervals. In medical research, signal variability or system complexity has been correlated with physiological signal conditions. Direct assessment of signal complexity thus offers certain advantages in clinical research. First complexity is an intuitive description and thus eases the interpretation of measurement results. Second, as mentioned, invariant measures allow comparison across different patient population as they are insensitive to absolute measurements of amplitude and frequency.

ECG being nonlinear, non-stationary signal, a number of measures of this complex time series have been developed based on the concepts of nonlinear dynamics like Lypanov exponents, correlation dimension etc, to classify normal/abnormal beats. In this chapter, two measures of clustering the beats in terms of their complexity is used. One of these measures is defined in frequency domain and the other is defined in time domain. Complexity in frequency domain is known as spectral entropy whereas in time domain it is known as temporal complexity.

ECG signal is analyzed in a beat wise manner. Once the boundary of QRS complex is known each beat is analyzed in two groups, one containing the QRS complex and other containing the P & T segments.

2.6.1 Spectral Entropy

Spectral entropy quantifies the spectral complexity of the time series. A variety of spectral transformations exist. Of these the Fourier transformation (FT) is most probably the well-known transformation method from which the power spectral density (PSD) can be obtained. The PSD is a function that represents the distribution of power as function of frequency. Thus normalization of PSD with respect to the total spectral power will yield a probability density function. Application of Shannon's channel entropy gives an estimate of the spectral entropy of the process where entropy is given by

$$\mathrm{H} = \sum_{\mathrm{f}} p_f log\left(\frac{1}{p_f}\right) \tag{2.15}$$

where p_f is the pdf value at frequency f. Heuristically the entropy has been interpreted as a measure of uncertainty about the event at f. Thus entropy H may be used as a measure of system complexity. This spectral entropy H is computed for the two groups of each beat.

2.6.2 Temporal Complexity

Lempel and Ziv define temporal complexity for a string length of width n as

$$c(n) = \frac{h \cdot n}{log_k n} \tag{2.16}$$

where k denotes the number of different characters used to represent the string length and h denotes the normalized source entropy, expressed as

$$h = (-1/log\, n)\Sigma_i\, p_i\, log\, p_i. \tag{2.17}$$

Thus h is determined by the probability p_i for each state i. In finding the temporal complexity for each of the components of the individual beats, the string length is defined with width equal to length of the corresponding components. The string length was defined using binary digits with two characters 1 and 0. For a string length within the boundary of QRS complex originally defined as $x_1, x_2, \ldots, x_\mathrm{n}$ with

$$x_m = (1/n)\Sigma_i x_\mathrm{i}. \tag{2.18}$$

After the binary transformation, it is defined as $s_1, s_2, \ldots, s_\mathrm{n}$, where s_i is obtained as

$$s_i = \begin{matrix} 1, & \text{if } x_\mathrm{i} \geq x_\mathrm{m} \\ 0, & \text{if } x_\mathrm{i} < x_\mathrm{m} \end{matrix} \tag{2.19}$$

The probability pi for each state is defined as

$$\mathrm{p_i} = (1/\mathrm{n})\Sigma_\mathrm{i}\, \mathrm{s_i}, \quad \text{for } \mathrm{s_i} = 1. \tag{2.20}$$

Temporal complexity $c(n)$ was computed in a similar manner for signal length consisting of P and T wave.

2.7 Sample Results and Discussion

The various signal-processing techniques for extracting characteristic features of ECG have been discussed. These characteristic features are used to represent significant characteristics of the signal for detection and diagnosis of cardiac abnormalities. The effectiveness of the techniques is demonstrated in this section. For ease of discussion we have considered three cases case K1024 (Class N), K6528 (Class A) and K9956 (Class C) and the results obtained on applying various techniques of feature extraction such as Noise filtering, QRS complex detection, QRS onset and offset detection, ST segment analyzer are given in this section.

2.7.1 Noise Filtering

The first step in noise filtering involves the elimination of most commonly present noise such as the baseline wander and the AC power noise. Baseline noise falls in the frequency band of 0 Hz and AC power noise is in the region of 50 Hz and its harmonics. According to the recommendations made by American Heart Association for ECG recordings, in the lower frequency region, the frequency components above 0.5 Hz should not be removed. Hence the cutoff frequency is chosen as 0.5 Hz to remove the baseline noise. In ECG the typical high frequency component is the QRS complex which is about 20 Hz. To remove the high frequency noises including the power noise the cut off frequency is chosen in the present study as 21 Hz. The actual and filtered output for Class N, Class A and Class C ECG's are shown below in Figs. 2.8–2.12. As seen from Figs. 2.9, 2.11, and 2.12, we can note that filter removes the baseline wander and attenuates the noise components without any significant distortion in characteristic points or the geometry of the ECG signal. From Fig. 2.12, we can note that even for Class C life-threatening ECG signals, the performance of the filter is good and is able to successfully remove most of the noise components while maintaining the morphology of the signal.

2.7.2 QRS Complex Peak Detection

The main component of ECG signal processing is the QRS complex detection. The implementation results of the QRS detection algorithm are shown in Figs. 2.13 and 2.14.

On implementation of the QRS detection algorithm the R peak is determined. Figures 2.15–2.17 shows the R peaks superimposed on the signal for Class N, Class A and Class C ECG, respectively.

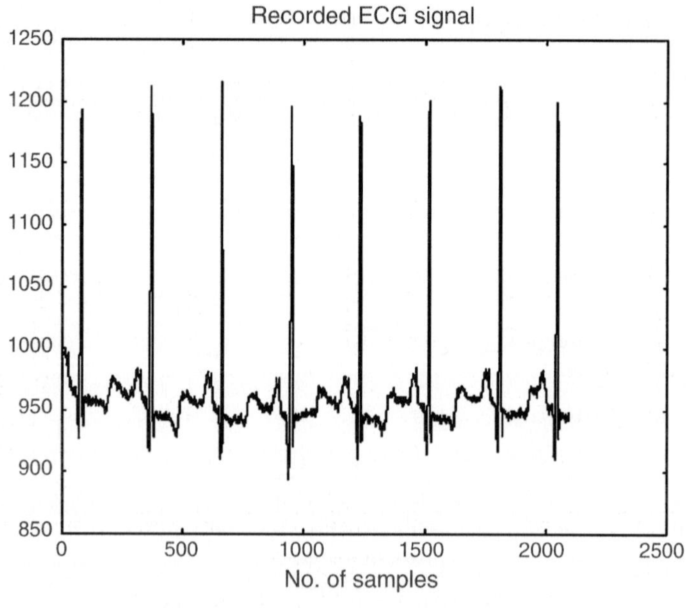

Fig. 2.8. Class N ECG signal (K1024)

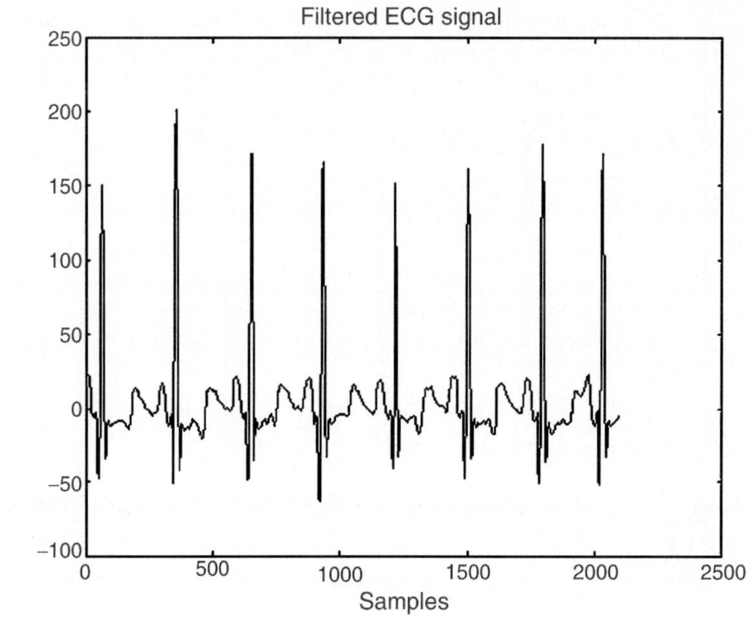

Fig. 2.9. Filtered Class N ECG signal (K1024)

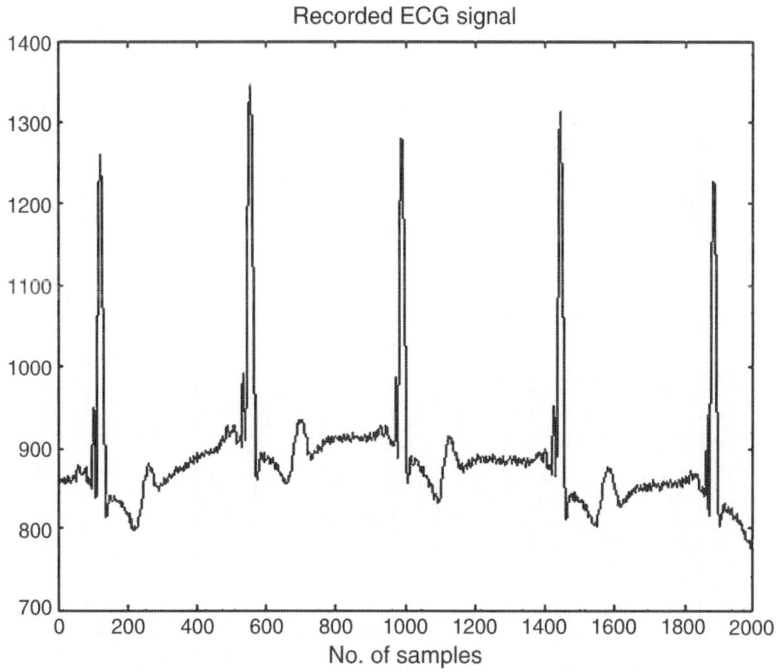

Fig. 2.10. Class A ECG signal (K6528)

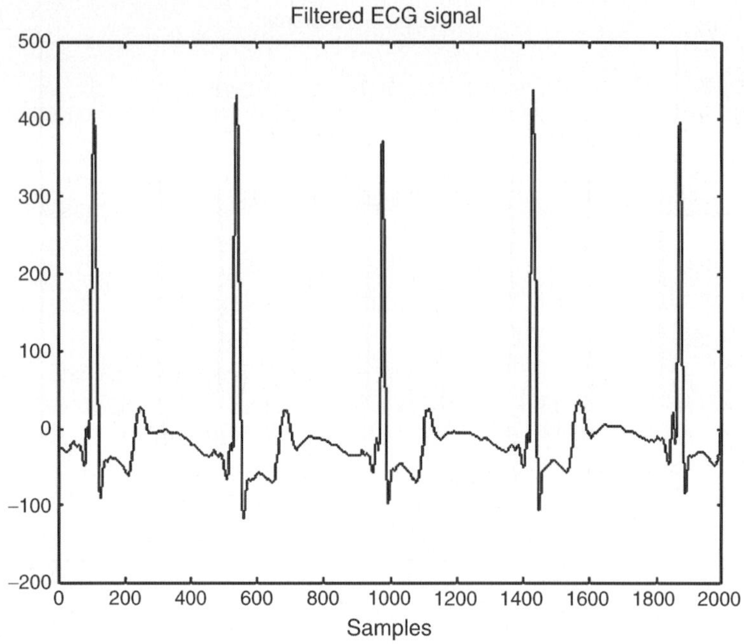

Fig. 2.11. Class C ECG signal (K9956)

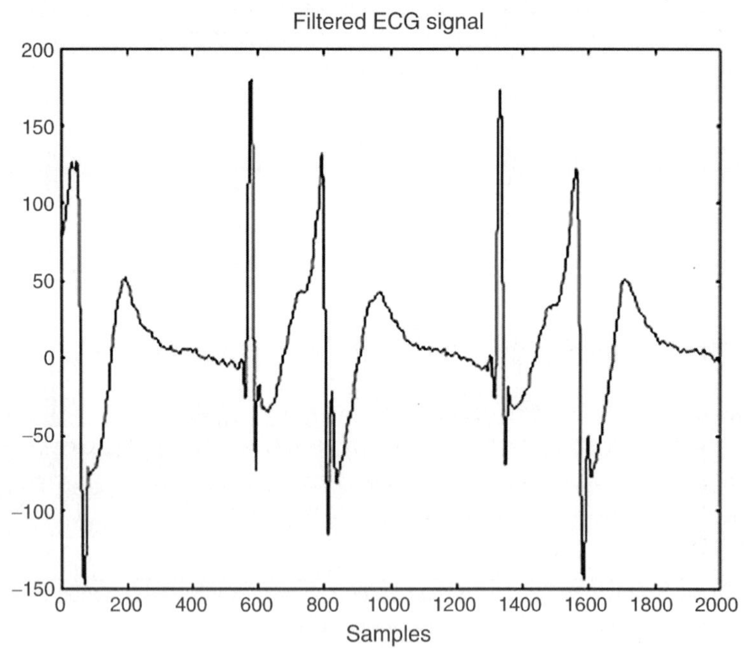

Fig. 2.12. Filtered Class C ECG signal (K9956)

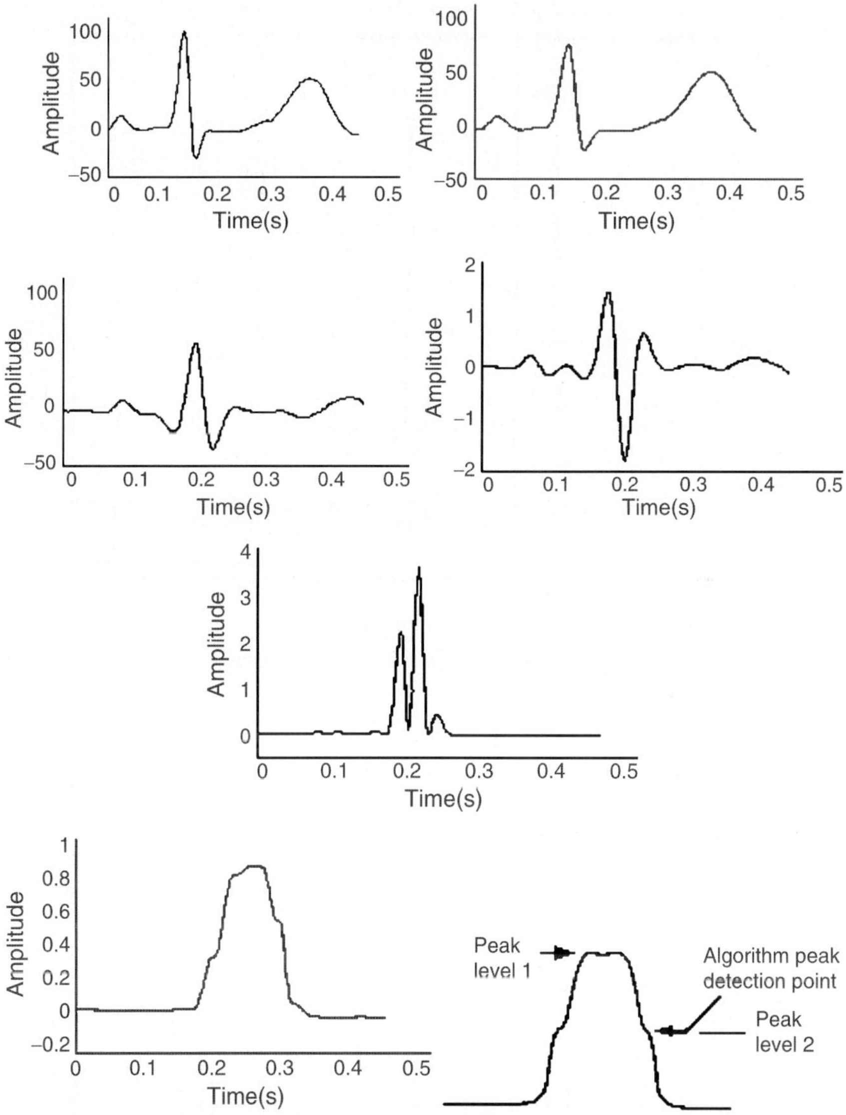

Fig. 2.13. Results of the QRS detection algorithm: (**a**) ECG signal showing one beat, (**b**) low pass filtered ECG, (**c**) band pass filtered ECG, (**d**) ECG after band pass filtering and differentiation, (**e**) signal after squaring function, (**f**) signal after moving window integration, (**g**) output of the moving window integrator with peak detect point

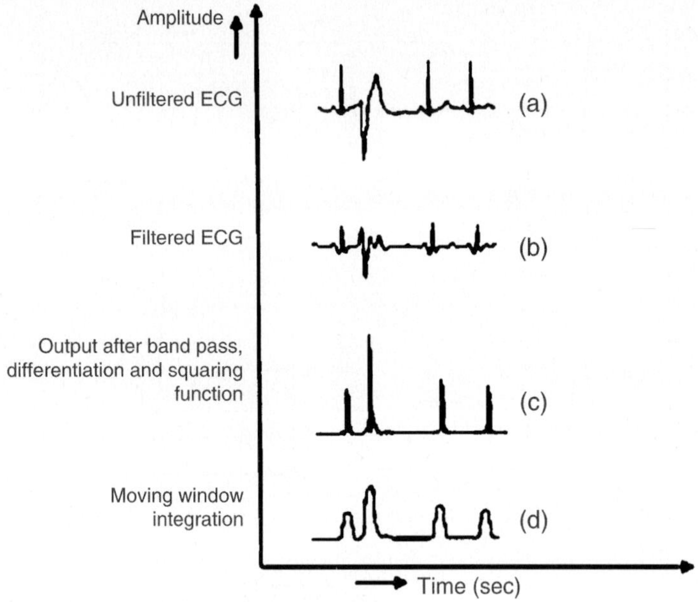

Fig. 2.14. QRS detector signals: (**a**) unfiltered ECG, (**b**) filtered ECG, (**c**) output after band pass, differentiation and squaring function, (**d**) moving window integration

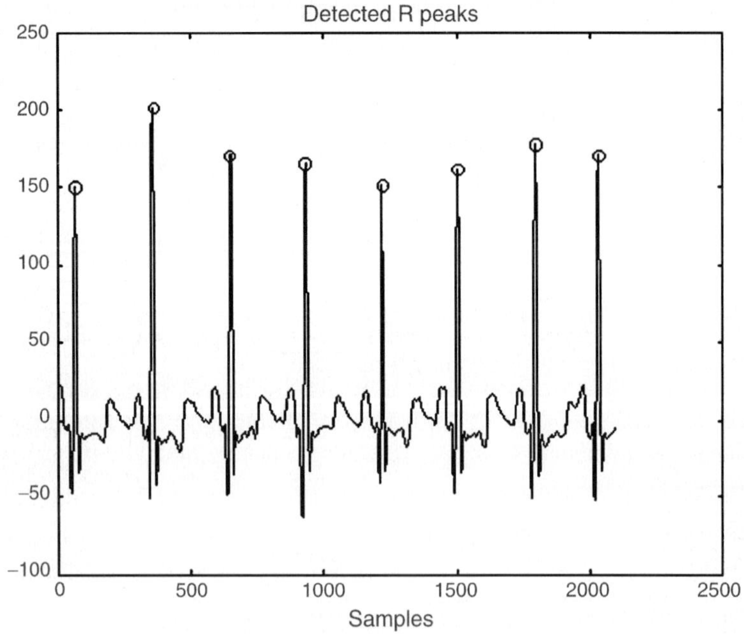

Fig. 2.15. QRS peak detection for Class N ECG signal (K1024)

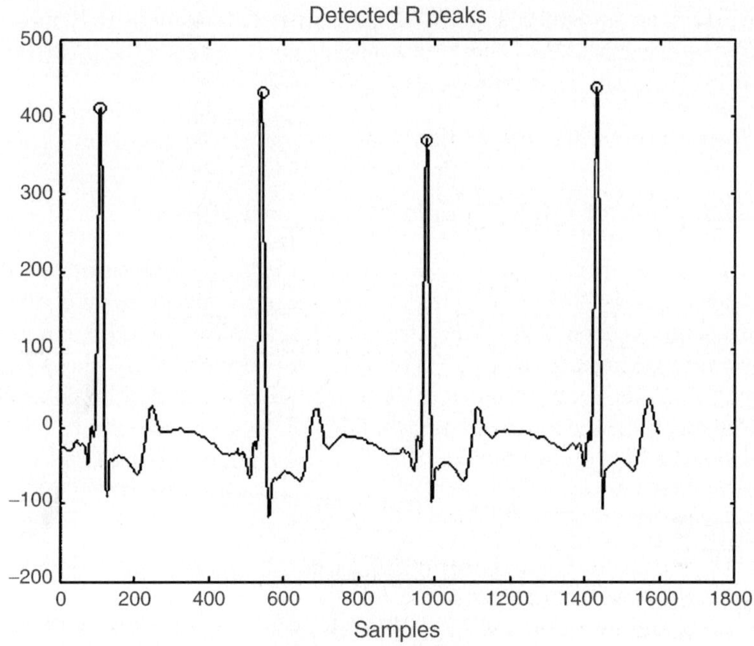

Fig. 2.16. QRS peak detection for Class A ECG signal (K6528)

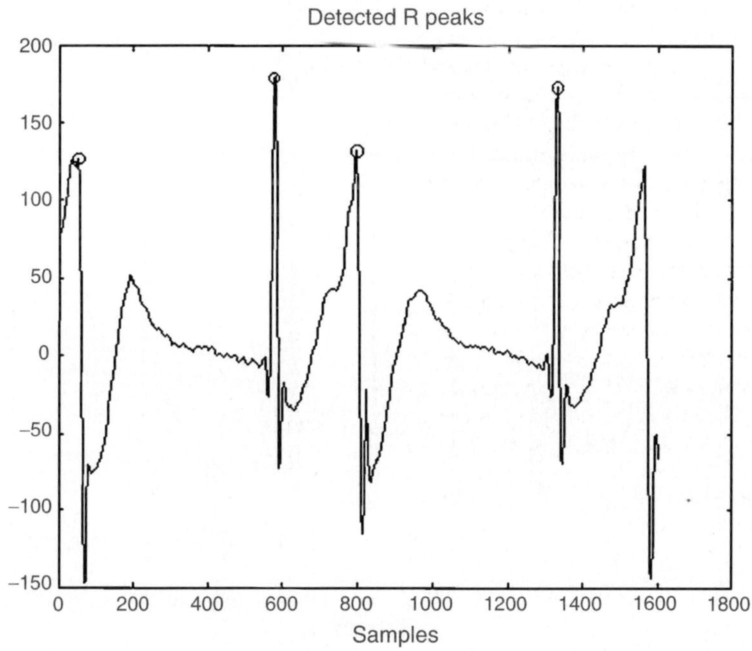

Fig. 2.17. QRS peak detection for Class C ECG signal (K9956)

Once R peak is detected the R-R interval is determined. R-R interval is the time interval between the two consecutive R peaks. The heart rate is determined from the R-R interval as

$$\text{Heart Rate} = (360/\text{R} - \text{R interval in samples})^{*}60\,\text{beats/min} \tag{2.21}$$

2.7.3 Detection of QRS Complex Onset and Offset

Once R peak is known, the next task is to find the QRS complex in its entirety for further analysis. This also helps to determine the irregularity of the QRS complex. Figures 2.18–2.20 shows the onset and offset of QRS complex in Class N, Class A and Class C ECG signals, respectively. Once the onset and offset points of the QRS complex is detected, QRS width can be determined. QRS width is the time interval between QRS onset and QRS offset. The onset and offset points are also used as the boundary points of the QRS complex for complexity analysis.

2.7.4 ST Segment Detection

The ST segment detection process was discussed in Sect. 2.5. This method detects the start and end points of ST segment that provide useful information

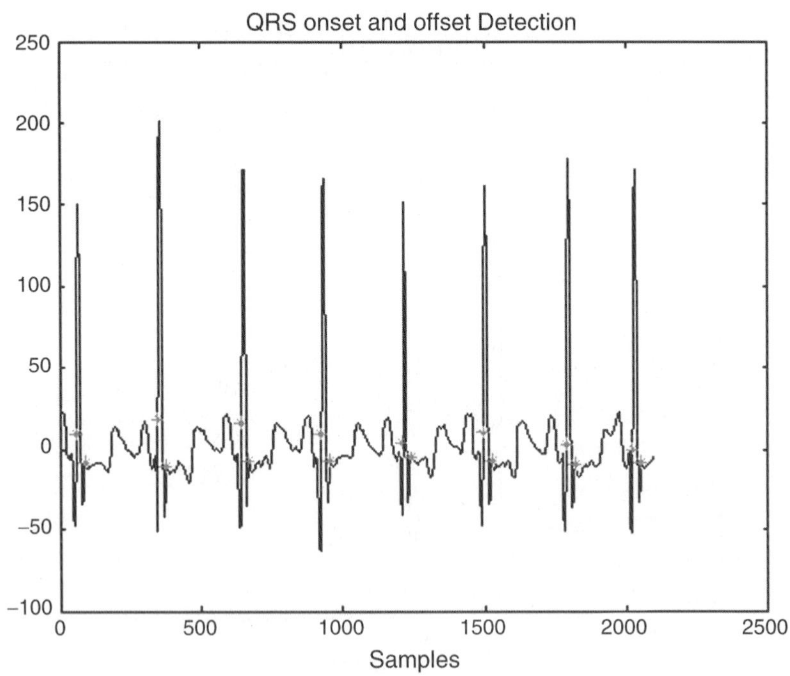

Fig. 2.18. Onset and offset of QRS complex in Class N ECG signal (K1024)

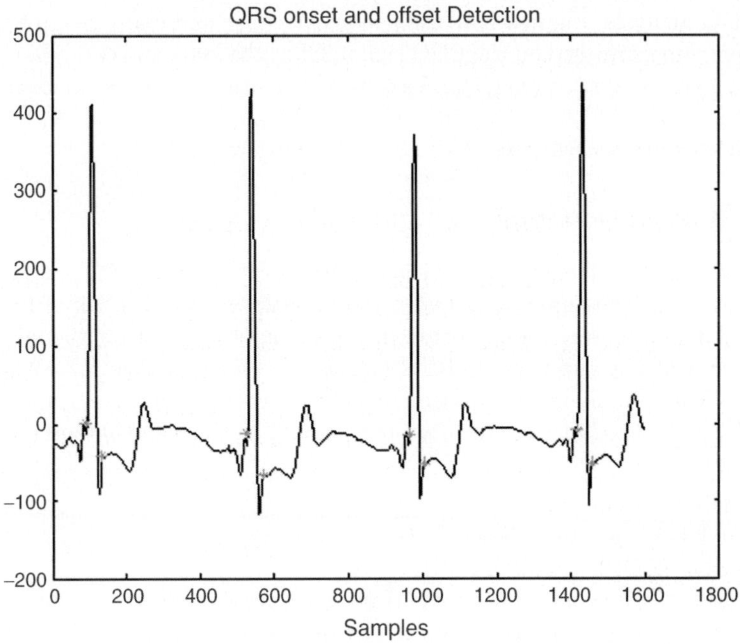

Fig. 2.19. Onset and offset of QRS complex in Class A ECG signal (K6528)

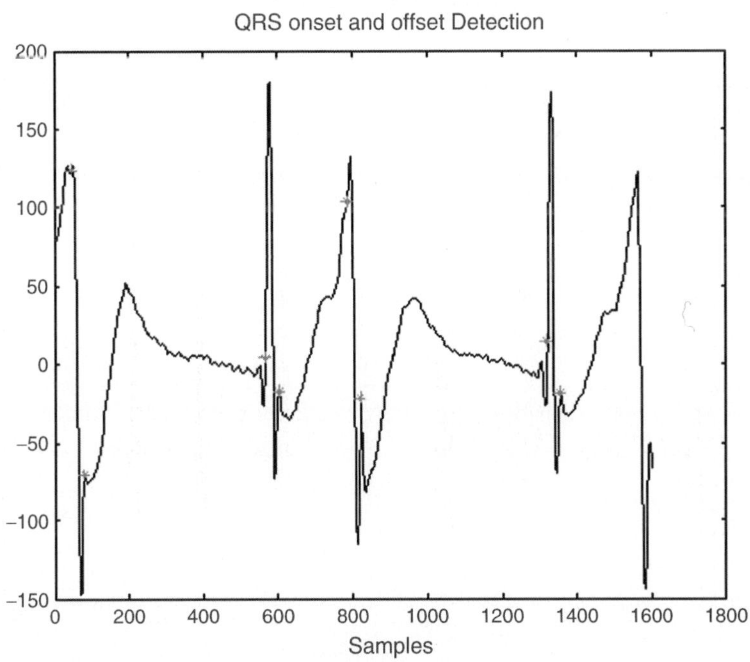

Fig. 2.20. Onset and offset of QRS complex in Class C ECG signal (K9956)

for ECG analysis. Figures 2.21–2.23 show the start and end points of the ST segment superimposed on the ECG signal. Once the start and end points of the ST segment are determined, the characteristic features of the ST segment such as ST segment width, ST segment slope, ST segment deviation, ST segment region area are determined.

2.7.5 T Peak Detection

The peak of the T wave in ECG is determined to find the RT interval. RT interval is the time interval between the R peak and the T peak of the same beat of ECG signal. The detected T peaks superimposed on the signal for Class N, Class A and Class C ECG signals are shown in Figs. 2.24–2.26.

The thirteen characteristic features obtained from each beat of the ECG signal can be used for classification purposes. The thirteen characteristic features are

a. HR – Heart Rate
b. ΔHR – Change in Heart Rate
c. ω_{qrs} – QRS complex width
d. h_{qrs} – Normalized source entropy for QRS complex
e. h_p – Normalized source entropy for ST wave
f. c_{qrs} – Complexity parameter for QRS complex

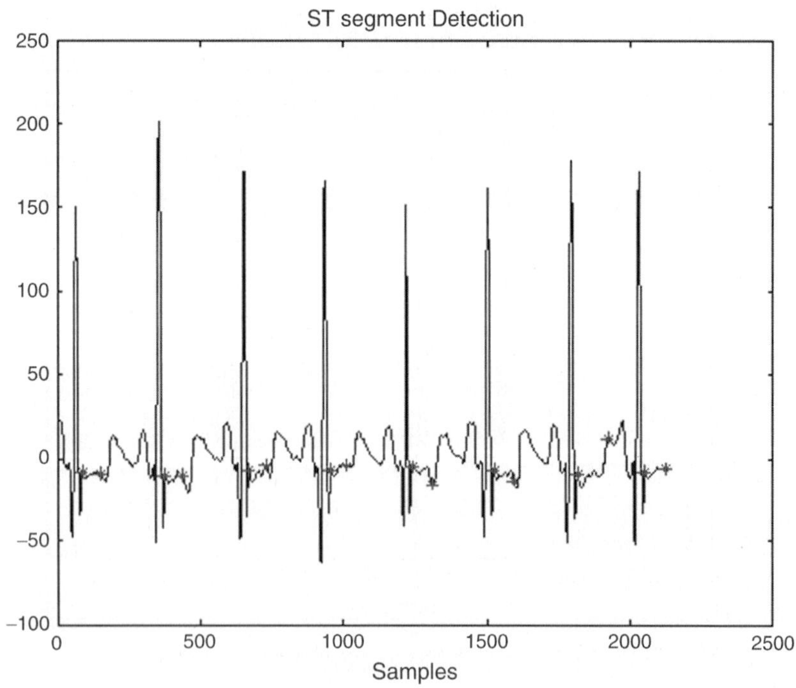

Fig. 2.21. ST segment detection for Class N ECG signal (K1024)

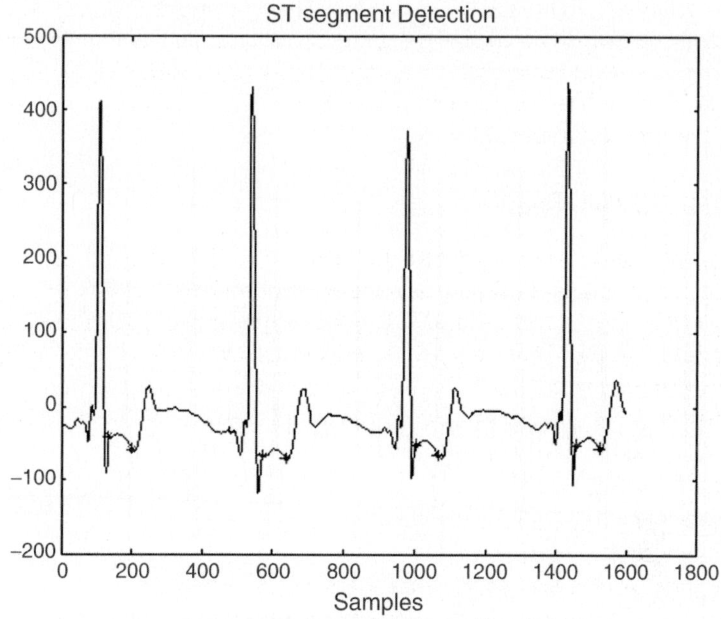

Fig. 2.22. ST segment detection for Class A ECG (K6528)

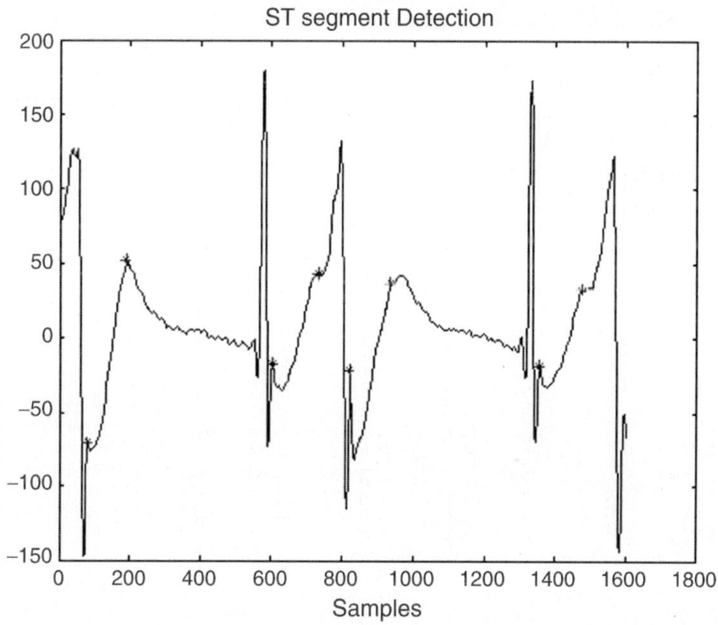

Fig. 2.23. ST segment detection for Class C ECG (K9956)

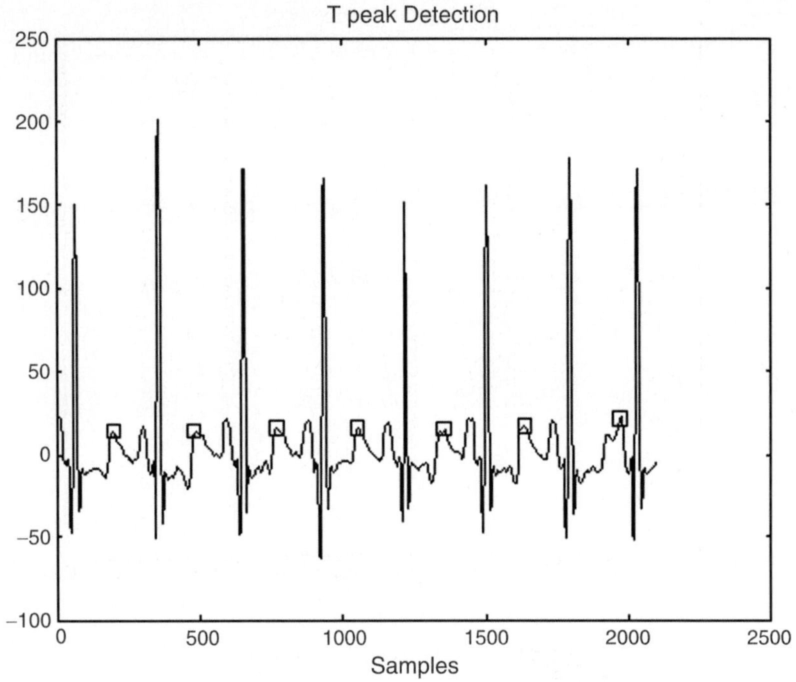

Fig. 2.24. T peak detection for Class N ECG signal (K1024)

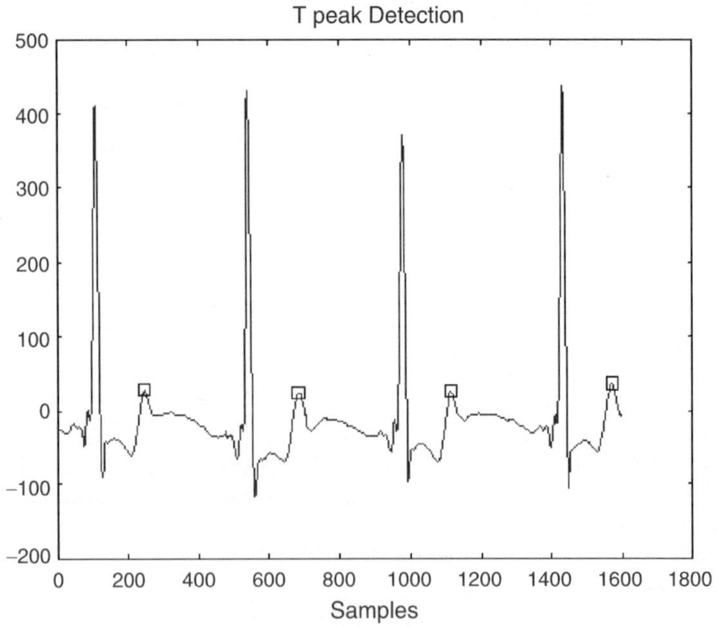

Fig. 2.25. T peak detection for Class A ECG (K6528)

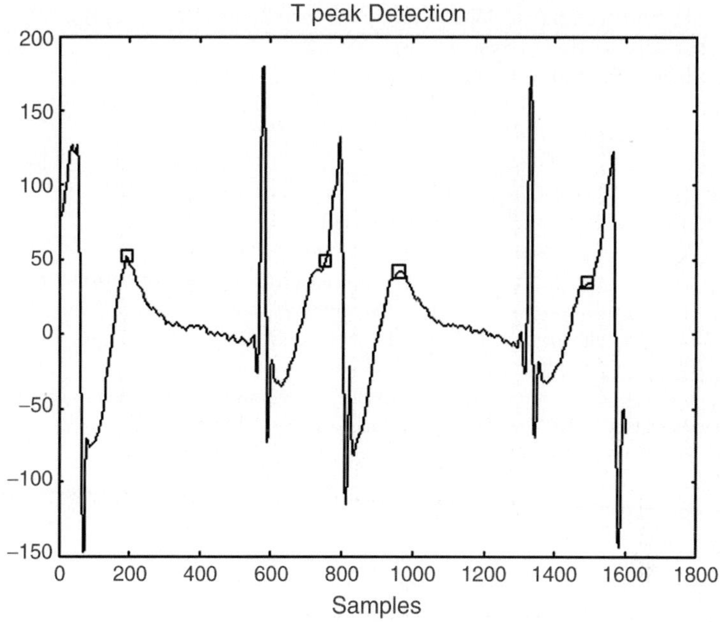

Fig. 2.26. T peak detection for Class C ECG (K9956)

g. c_{st} – Complexity parameter for ST wave
h. H – Spectral entropy
i. τ_{RT} – RT interval
j. ρ_{ST} – ST segment length
k. ϕ – ST segment deviation
l. θ – ST segment angle of deviation and
m. A – ST segment area.

These features can be fed to neural network, fuzzy logic or any other classifier and diagnostic decisions can be made. These features can be extracted from lead II ECG and are very helpful in diagnostics using portable ECG recording instruments.

References

1. Friesen, G.M., Thomas, C.J., Jadallah, M.A., Yates, S.L., Quint, S.R., and Nagle, H.T. (1990): 'A comparison of noise sensitivity of 9 QRS detection algorithms', *IEEE Trans. Biomed. Eng.*, **37**(1), pp. 85–98.
2. Thakor, N.V., Webster, J.G., and Tompkins, W.J. (1984): 'Estimation of QRS complex power spectra for design of a QRS filter', *IEEE Trans. Biomed. Eng.*, **BME-31 (11)**, pp. 702–706.
3. Alste, J.A.V., and Schilder, T.S. (1985): 'Removal of base-line wander and power line interference from the ECG by an efficient FIR filter with a reduced number of taps', *IEEE Trans. Biomed. Eng.*, **32**, pp. 1052–1060.

4. Pan, J., and Tompkins, W.J. (1985): 'A real-time QRS detection algorithm', *IEEE Trans. Biomed. Eng.*, **32**, pp. 230–236.
5. Hamilton, P.S., and Tompkins, W.J. (1986): 'Quantitative investigation of QRS detection rules using the MIT/BIH arrhythmis database', *IEEE Trans. Biomed. Eng.*, **33**, pp. 1157–1165.
6. Mervin, J.G., and Goldschalager, N. (1989): 'Principles of clinical Electrocardiography', *Appleton Lange.*
7. Kenneth, P.B. (1982): 'Rule based learning for more accurate ECG analysis', *IEEE Trans. Pattern analysis and Machine intelligence*, **4**, pp. 369–379.
8. Sornomo, L. (1987): 'A model based approach to QRS delineation', *In Proceedings of Computers and Biomedical Research*, pp. 526–540.
9. Williams, J.L., Benmel, J.H.V., and Zywietz, C. (1986): 'Computer ECG Analysis', *Elseiver Science Publishers.*
10. Lin, P.K., and Chang, W.H. (1997): 'A technique for automated arrhythmia detection of holter ECG', *In Proceedings of Engineering in Medicine and Biology*, **35**, pp. 626–631.

3

Prediction of Cardiac Signals Using Linear and Nonlinear Techniques

N. Kannathal, U. Rajendra Acharya, Lim Choo Min and Jasjit S. Suri

Analysis of variations in the instantaneous heart rate time series using the beat-to-beat RR-intervals (the RR tachogram) is known as heart rate variability (HRV) analysis [1, 2]. HRV analysis has been shown to provide an assessment of cardiovascular disease [3]. The heart rate may be increased by slow acting sympathetic activity or decreased by fast acting parasympathetic (vagal) activity. The balance between the effects of the sympathetic and parasympathetic systems, the two opposite acting branches of the autonomic nervous system, is referred to as the sympathovagal balance and is believed to be reflected in the beat-to-beat changes of the cardiac cycle [1]. The heart rate is given by the reciprocal of the RR-interval in units of beats per minute. Spectral analysis of the RR tachogram is typically used to estimate the effect of the sympathetic and parasympathetic modulation of the RR-intervals. The two main frequency bands of interest are referred to as the LF band (0.04–0.15 Hz) and the HF band (0.15–0.4 Hz) [2]. Sympathetic tone is believed to influence the LF component, whereas both sympathetic and parasympathetic activities have an effect on the HF component [1]. The ratio of the power contained in the LF and HF components has been used as a measure of the sympathovagal balance [1, 2].

The spectral analysis of beat-to-beat HRV is a well-established non-invasive way to investigate the autonomic control of the cardiovascular system [4, 5]. Previous studies in humans [6–8] showed the power spectrum to be characterized by three main components. These are high frequency component (HF), centered at the respiration rate; a low or middle frequency component (LF or MF) that is related to the vasomotor activity regulating arterial blood pressure (ABP); and a very-low frequency component (VLF) most likely related to thermoregulation. Either non-parametric methods, based on the fast Fourier transform algorithm (FFT) [6, 8], parametric methods, based on autoregressive models (AR) [7], moving averaging (MA) and autoregressive moving average models (ARMA) have been used in the previous studies. Methods based on FFT have some technical limitations such as (a) the use of deterministic algorithms that, in principle, are applicable only to periodical phenomena, (b) the need of windowing the data, and (c) uncertainty in defining the relative power of the various spectral components [7, 9, 10]. The AR-based methods avoid some of these limitations because they do not require windowing or filtering the data, that are applicable to non-periodic phenomena and allow autonomic computation of central frequency and power of the principal spectral components [7, 9]. Nihal *et al* have compared the FFT and AR based sonogram outputs and have explained the advantages of the 20 MHz pulsed Doppler data in real time [11]. Natalucci *et al* have used the parametric analysis technique to study the heart rate variability in anaesthetized rats [12]. Recently, Dingfei *et al* have used autoregressive modeling (AR) technique to classify normal sinus rhythm (NSR) and various cardiac arrhythmias including atrial premature contraction (APC), premature ventricular contraction (PVC), supraventricular tachycardia (SVT), ventricular tachycardia (VT) and ventricular fibrillation (VF) from the ECG signal [13]. Inan Guler *et al* have used AR methods for the determination of Behcet disease [14]. Elif Derya *et al* have studied the spectral analysis of internal carotid arterial Doppler signals using FFT, AR, MA and ARMA methods [15].

Recently, a new approach for the discrimination among VF, VT and SVT has been developed using a total least squares-based Prony modeling algorithm [16]. Two features, energy fractional factor (EFF) and predominant frequency (PF) were derived from the total least squares based Prony model. AR modeling has been used extensively to model heart rate variability (HRV) and for power spectrum estimation of ECG and HRV signals [17–21]. Amplitude modulated sinusoidal signal model, which is a special case of the time-dependent AR model have been applied to modeling ECG signals [17]. Adaptive AR modeling with Kalman filtering has also been used [20]. Parameters extracted from AR modeling have been used for arrhythmia classification in conjunction with other features [22]. It was suggested that increasing the model order would not reduce the prediction error, implying that a linear predictor order of two is sufficient for fast cardiac arrhythmia detection [23]. Acharya *et al* have explained all the different types of linear, frequency and non-linear techniques, available for the analysis of heart rate signals [24].

Nonlinear additive autoregressive model-based data analysis was used in the diagnostic analysis of short-term heart rate variability [25]. For this purpose, a nonlinear regression approach, namely, the maximal correlation method is applied to the data of 37 patients with dilated cardiomyopathy as well as of 37 age- and sex-matched healthy subjects. They found that, this approach is a powerful tool in discriminating both groups and promising for further model-based analyses. The R-R interval series, specific to the pre-ictal period, was sought by applying an unsupervised fuzzy clustering algorithm to the N-dimensional phase space of N consecutive interval durations or the absolute value of duration differences [26]. Forecasting success was about 86 and 82%, respectively, at times ranging from 10 min to 30 s prior to seizure onset.

HRV Signal Model

Signal modeling is a basic method in signal processing. Once the model of a signal is identified, the characteristics of signal can be easily controlled by changing the parameters of this model. The simulated signal can then be used to validate and compare various signal processing algorithms. In addition, if the model does faithfully reflect the physiological process of the signal, then it can be used to study the physiological mechanism of this signal.

Extracting useful clinical information from the real (noisy) ECG requires reliable signal processing techniques. These include R-peak detection, QT-interval detection, and the derivation of heart rate and respiration rate from the ECG. The variability of these RR-intervals reveals important information about the physiological state of the subject. At present, new biomedical signal processing algorithms are usually evaluated by applying them to ECGs acquired from real patients. Usually it will be of short duration not sufficiently enough for the evaluator to decide on the accuracy and reliability of a given algorithm. To facilitate this evaluation, it is required to generate longer duration signals from these short duration signals while preserving the time domain and frequency domain characteristics of the signal.

A realistic artificial biomedical signal generator that is able to encompass the range of signals observed for both normal and abnormal subjects is therefore a useful tool. Furthermore, the ability to rapidly create a regeneratable time series, enables a researcher to quickly prototype applications and test theories on both normal and abnormal signals.

This chapter deals with the linear and nonlinear models for generating a synthetic HRV signal with realistic and prescribed heart rate dynamics. The model is to provide a standard realistic HRV signal with known characteristics. The main characteristics of an HRV signal are, in the time domain, the signal is neither periodic nor completely random and in the frequency domain, the signal consists mainly of three spectral peaks, i.e., a high frequency (HF) peak around 0.20 Hz, a low frequency (LF) peak around 0.10 Hz, and a very low frequency (VLF) peak, which is also called the 1/f component because its spectral magnitude increases with the decrease of frequency. Thus, the simulated HRV signal must at least be able to reveal the following characteristic parameters: the HF component frequency, the LF component frequency, and the parameters governing the 1/f spectrum of the VLF component.

The commonly used frequency domain measure for HRV signal is the low frequency/high-frequency (LF/HF) ratio, defined as the ratio of power between 0.015–0.15 Hz and 0.15–0.4 Hz in the RR tachogram. HRV analysis has been shown to provide an assessment of cardiovascular disease. The heart rate may be increased by slow acting sympathetic activity or decreased by fast acting parasympathetic (vagal) activity. The balance between the effects of the sympathetic and parasympathetic systems, the two opposite acting branches of the autonomic nervous system, is referred to as the sympathovagal balance and is believed to be reflected in the beat-to-beat changes of the cardiac cycle. The heart rate is given by the reciprocal of the RR-interval in units of beats per minute. Spectral analysis of the RR tachogram is typically used to estimate the effect of the sympathetic and parasympathetic modulation of the RR-intervals. The ratio of the power contained in the LF and HF components has been used as a measure of the sympathovagal balance.

Generating a long duration HRV signal from the given short duration signal facilitates a comparison of different signal processing techniques. The HRV signal generated with the prescribed time domain and frequency domain characteristics can be used for diagnostic purposes by predicting the nature of the HRV signals. The model can also be used for numerous applications such as (1) the synthetic HRV could be used to assess the effectiveness of different techniques for noise and artifact removal. These could be evaluated by adding noise and/or artifact onto the synthetic signal and then comparing the original with the processed signal. (2) Abnormal morphological changes could be introduced to the lead signal and the long term changes could be observed. (3) Abnormal beats can be predicted on a long run and used for diagnostic purposes.

In this chapter, a detailed discussion is carried out on the prediction of heart rate signals using auto regressive (AR) (Burg's method), and recursive neural network (RNN) (Elman's method). The stimulated signal is validated using the frequency domain measures of LF and HF components. The performance of the prediction methods is evaluated using the Normalized Root Mean Square Error (NRMSE) and the signal to noise ratio (SNR).

3.1 Data Acquisition and Preprocessing

ECG data for the analysis and classification was obtained from the MIT-BIH arrhythmia database, the MIT-BIH Ventricular Arrhythmia database and the MIT-BIH supraventricular arrhythmia database [27]. Various ECG segments were selected from the databases for modeling and classification. The data set included around 200 segments each of Normal Sinus Rhythm (NSR) ECGs, Preventricular Contraction (PVC), Complete Heart Block (CHB), Sick Sinus Syndrome (SSS), Left Bundle Branch Block (LBBB), Ischemic/Dilated Cardiomyopathy, Atrial Fibrillation (AF), Atrial Premature Contraction (APC) and Ventricular Fibrillation (VF). The sampling frequency of the data from the MIT-BIH Arrhythmia database was 360 Hz, the sampling frequency of the data from the MIT-BIH ventricular arrhythmia database was 250 Hz and the sampling frequency of the data from the MIT-BIH supraventricular arrhythmia was 128 Hz. The data from the MIT-BIH arrhythmia and supraventricular arrhythmia databases were re-sampled so that all the data used in the analysis had a sampling frequency of 250 Hz. The ECG signals were preprocessed to remove noise due to power line interference, respiration, muscles tremors, spikes, etc., and to detect the R peaks in the ECG signals. The R peaks of

Table 3.1. Number of datasets in each category

Class	**No. of datasets**
Left bundle branch block	40
Normal sinus rhythm	90
Preventricular contraction	70
Atrial fibrillation	50
Ventricular fibrillation	45
Complete heart block	45
Isc./dil. cardiomyopathy	45
Sick sinus syndrome	45

ECG were detected using Tompkins's algorithm [28–30]. Table 3.1 shows the number of data set in each cardiac class. The interval between two successive R peaks is defined as the RR interval (t_{r-r} seconds) and the heart rate (beats per minute) is given as:

$$HR = 60/t_{r-r} \tag{3.1}$$

The heart signal is generated from ECG signal by using the instantaneous heart rate values.

For the purpose of this study, the cardiac disorders are classified into eight categories namely:

(1) Normal sinus rhythm (NSR)
(2) Preventricular contraction (PVC)
(3) Complete heart block (CHB)
(4) Sick sinus syndrome (SSS)
(5) Atrial fibrillation (AF)
(6) Ventricular fibrillation (VF)
(7) Ischemic/dilated cardiomyopathy
(8) Left bundle branch block (LBBB)

The number of datasets used for the analysis in each category is given in Table 3.1. These cardiac abnormalities are discussed in detail in Chap. 2.

3.2 Modeling Techniques

Signal modeling deals with the representation of signals in an efficient manner. In general, there are two steps in the modeling process. The first is to choose an appropriate parametric form for the model. In this chapter, the model used is one that represents a signal as the output of a causal linear shift-invariant filter that has a rational system function. Once the form of the model has been selected, the next step is to find the model parameters that provide the *best* approximation to the given signal. There are, however, many different ways to define, what is meant by the best approximation and, depending upon the definition that is used, there will be different solutions to the modeling problem along with different techniques for finding the model parameters. Therefore, in developing an approach to signal modeling, it will be important not only to find a model that is useful, i.e., works well, but one that has a computationally efficient procedure for deriving the model parameters from the given data.

3.2.1 Linear Method

Linear modeling techniques are based on the estimation of a linear time-invariant model that has white noise as input and the signal to be analyzed as output. There are power spectrum estimate methods that use models without zeros (AR) and models without poles (MA). Auto-regressive models lead to power spectrum with sharp peaks, and better suited for electric motors fault detection. Moreover the linear equations, to find the coefficients of AR models, are simpler to be solved. The AR methods tested are the Yule-Walker, Burg, covariance, and modified covariance methods. The Yule-Walker and covariance methods solve the set of linear equations by minimizing the forward prediction error in the least squares sense. The Burg and modified covariance methods solve the set of linear equations by minimizing the forward and backward prediction errors in the least squares sense. The Yule-Walker and Burg approaches always guarantee a stable model. Unfortunately, the performance of the Yule-Walker approach degrades when the number of samples decreases. The covariance-based approaches perform well, when the order of the filter chosen is smaller than the number of sinusoids actually present in the analyzed signal. The Burg's approach yields a more stable and robust AR model parameters.

AR Method

The autoregressive (AR) model [9,31] is one of the linear prediction formulas that attempt to predict an output y_n of a system based on the previous inputs $(x_n, x_{n-1}, x_{n-2}, \ldots)$. It is also known in the filter design industry as an infinite impulse response filter (IIR) or an all pole filter, and is sometimes known as a maximum entropy model in physics applications. The definition used here is as follows

$$y(t) = \sum_{i=1}^{m} a(i) \cdot x(t-i) + \varepsilon(t) \tag{3.2}$$

where $a(i)$ is the autoregression coefficient, $x(i)$ is the series under investigation, $\varepsilon(t)$ is the output of uncorrelated errors and m is the order (length) of the filter which is generally less than the length of the series. The noise term or residue, epsilon in the above, is almost always assumed to be Gaussian white noise. The current term of the series can be estimated by a linear weighted sum of previous terms in the series. The weights are the autoregression coefficients. The problem in AR analysis is to derive the "best" values for $a(i)$ given a series $x(i)$. The majority of methods assume the series $x(i)$ is linear and stationary. By convention the series $x(i)$ is assumed to be zero mean, if not this is simply another term $a(0)$ in front of the summation in the equation above.

The power spectrum of a mth order autoregressive process is

$$P_{xx}^{BU}(f) = \frac{\overset{\frown}{\vec{E}}_m}{\left|1 + \sum\limits_{k=1}^{m} \widehat{a}_m(k)\, e^{-j2\pi fk}\right|^2} \tag{3.3}$$

where $\overset{\frown}{\vec{E}}_m$ is total least square error. The Burg method results in high resolution and yields a stable AR model.

It is essential to choose the appropriate model order. The order of the AR model has a major effect on the spectral estimate for the time series. Too low order will result in a smoothed spectrum and too high order will increase the resolution of the spectrum and introduce spurious peaks. The estimate for the power associated with the single component is also dependent on the order that is selected. The orders p = 15-20 are often satisfactory for heart rate signal prediction. Several penalty function methods for model order selection exist that utilize the prediction error variance such as FPE (final prediction error) and AIC (Akaike information criteria) [32].

The order of the model was chosen as the one that minimizes the Akaike information criterion (AIC) figure of merit [33]:

$$AIC(m) = N.In\left(\overset{\wedge}{\lambda}^2\right) + 2m \tag{3.4}$$

where N is the number of data samples and $\overset{\wedge}{\lambda}^2$ is the estimated white noise variance. In this work the order of the AR model is taken as: p = 16 [34].

3.2.2 Nonlinear Method

In conventional modeling, it is assumed that the signal is the output of a linear system driven by random noise. In other words, signals are treated as realizations of some random process and the underlying systems are modeled as linear [9,35]. After the discovery of chaos, some deterministic systems with few degrees of freedom which without any random input can produce signals that exhibit uncertainty and poses noise like spectra. These systems are named chaotic dynamical systems [36]. A chaotic system is a nonlinear dynamical system and the uncertainty existing in its output is originated from the system dynamics instead of an external driving force. Therefore, it is appropriate to apply nonlinear methods to model the underlying dynamics of the chaotic signal such as the heart rate signal. Since the heart rate signal is a chaotic signal, a nonlinear model can be expected to out perform the nonlinear model.

Artificial Neural Networks (ANN) is a powerful nonlinear modeling tool. An artificial neural network can be regarded as a dynamical system [37]. The relaxation of the neural networks can exhibit a rich variety of dynamical behavior [38, 39]. This property is highly desirable in dynamic modeling to preserve the dynamics of the original system. The advantage of ANN is their ability to generalize what they learn during training to new situations. If the signal to be modeled is noisy and has finite length, it is desirable that a model is able to interpolate and extrapolate the mapping from the training examples in a sensible way. Due to their plasticity, function approximation capability, wide spectrum of possible dynamics and generalization capability, ANNs are often used as a nonlinear modeling tool.

There are several architectures available on the neural networks in literature. Recurrent neural networks (RNN) involving dynamic elements and internal feedback connections have been considered to be more suitable for nonlinear modeling purposes [40]. In the last few years, various works have been presented showing that recurrent neural network are quite effective in modeling nonlinear dynamical systems [41,42]. The critical issue in the application of RNN is the choice of network architecture and the suitable training algorithm. For the application of modeling heart rate signals, a recurrent Elman network using backpropagation algorithm is chosen.

Recurrent Neural Network (Elman) Method

Feedforward neural networks have been successfully used to solve problems that require the computation of a static function, i.e., a function, whose output depends only upon the current input, and not on any previous inputs. In the real world however, one encounters many problems which cannot be solved by learning a static function because the function being computed changes with each input received. In such cases, system needs to predict the outputs with some knowledge of how the past inputs affect the processing of the present input, as well as a way of storing the past inputs. In other words such a system must have a memory of the past input and a way to use that memory to process the current input. It should be clear from the architecture of feedforward neural networks, that past inputs have no way of influencing the processing of future inputs. This situation can be rectified by the introduction of feedback connections in the network. Now network activation produced by past inputs can cycle back and affect the processing of future inputs. The class of neural networks which contain feedback connections are called recurrent neural networks (RNNs). While the set of topologies of a feedforward networks is fairly constrained, but, RNN can take on any arbitrary topology, as any node in the network may be linked with any other node (including itself). The only requirement, we make is that the network have clearly defined input and output nodes.

Recurrent networks are the state of the art in nonlinear time series prediction, system identification, and temporal pattern classification. As the output of the network at time t is used along with a new input to compute the output of the network at time $t+1$, the response of the network is dynamic. There are few RNN architectures proposed by Frasconi, Gori-Soda, Narendra-Parthasarathy, Williams and Zipser, and Elman.

Elman Networks are a form of recurrent neural networks, which have connections from their hidden layer back to a special copy layer. This means that the function learnt by the network can be based on the current inputs plus a record of the previous state(s) and outputs of the network. In other words, the Elman net is a finite state machine that learns what state to remember (i.e., what is relevant). The special copy layer is treated as just another set of inputs and so standard backpropagation learning techniques can be used (something which isn't generally possible with recurrent networks).

Architecture of a Simple Elman Network

The architecture of the Elman network is shown in Fig. 3.1. Each ellipse in the figure represents a set of nodes. Recurrent connections in this network (the dashed line) are implemented as follows: at each timestep 't' the activations of the nodes in the hidden layer can be thought of as the internal representation of the input so far. The values in those hidden nodes are stored one-to-one in context nodes (the ellipse on the lower right). Each context node is connected in the forward direction with all of the hidden nodes. The presence of this simple loop implies that the activations of the hidden units at time 't' can influence the activations of the hidden units at time t+1. There are, the same number of context units as hidden units and the connections from the latter

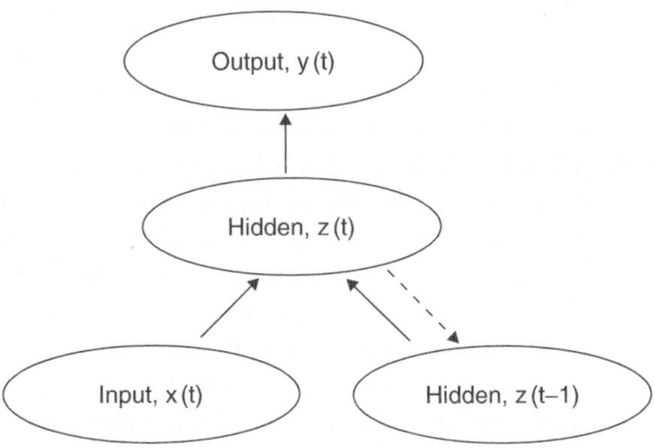

Fig. 3.1. Elman network architecture

to the former are one-to-one and have weights fixed at 1. The context units are connected to the hidden units in a one to many fashion.

Training Elman Networks

At each time step, a copy of the hidden layer units is made to a copy layer. Processing is done as follows:

1. Copy inputs for time t to the input units
2. Compute hidden unit activations using net input from input units and from copy layer
3. Compute output unit activations as usual
4. Copy new hidden unit activations to copy layer

At the end of each timestep $t > 0$, all the activations of the nodes in the input layer of the simple Elman network are known. Furthermore, as all of the trainable weights are in the forward direction, the standard backpropagation algorithm is used to train this network. In the generalized version of the Elman network the activations of hidden units and input units of many previous time steps are stored and a specialized version of the backpropagation algorithm called Backpropagation Through Time (BPTT) operates over all of these units.

The weights from the copy layer to the hidden layer play a special role in error computation. The error signal, they receive comes from the hidden units, and so depends on the error at the hidden units at time t. The activations in the hidden units, however, are just the activation of the hidden units at time $t - 1$. Thus, in training, a gradient of an error function is used, which is determined by the activations at the present and the previous time steps.

A generalization of this approach is to copy the input and hidden unit activations for a number of previous timesteps. The more context (copy layers) maintained, the more history are explicitly included in the gradient computation. This approach is known as Backpropagation Through Time. It can be seen as an approximation to the ideal of computing a gradient which takes into consideration not just the most recent inputs, but all inputs seen so far by the network. Figure 3.2 below illustrates the above process.

The inputs and hidden unit activations at the last three time steps are stored. The solid arrows show how each set of activations is determined from the input and hidden unit activations on the previous time step. A backward pass, illustrated by the dashed arrows, is performed to determine separate values of delta (the error of a unit with respect to its net input) for each unit and each time step separately. Because each earlier layer is a copy of the layer one level up, a new constraint is introduced that the weights at each level be identical. Then the partial derivative of the negative error with respect to $w_{i,j}$ is simply the sum of the partials calculated for the copy of $w_{i,j}$ between each two layers.

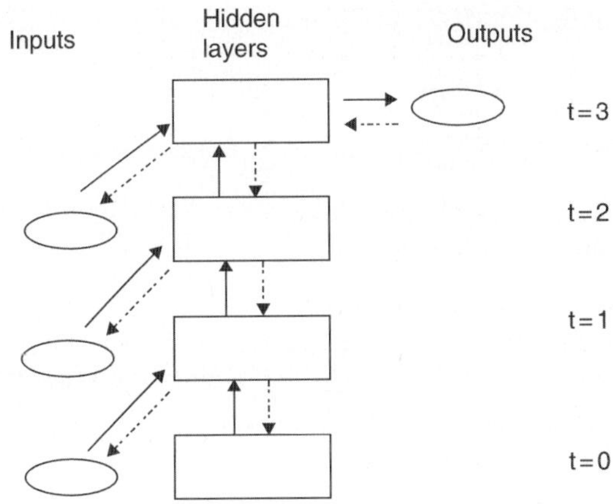

Fig. 3.2. Illustration of backpropagation through time algorithm

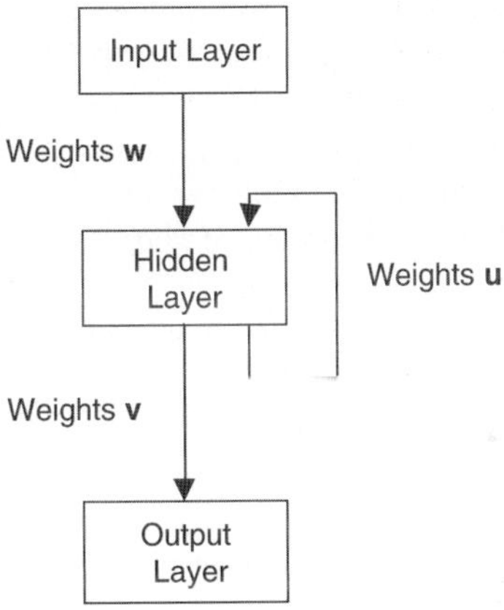

Fig. 3.3. Block diagram of the proposed neural network

Elman networks and their generalization, Backpropagation Through Time, both seek to approximate the computation of a gradient based on all past inputs, while retaining the standard backpropagation algorithm. BPTT has been used in a number of applications. The main task is to produce a particular output sequence in response to specific input sequence.

The block diagram of the recurrent network used is given in Fig. 3.3. It is a two-layer backpropagation network, with the addition of a feedback connection from the output of the hidden layer to its input. This feedback path allows Elman networks to learn to recognize and generate temporal patterns, as well as spatial patterns.

The output of the hidden layer $y_j(t)$ is given by

$$y_j(t) = f(net_j) \tag{3.5}$$

$$net_j(t) = \sum_i w_{ji}x_i(t) + \sum_h u_{jh}y_h(t-1) + \vartheta_j \tag{3.6}$$

The final output $y_k(t)$ is given by

$$y_k(t) = g(net_k) \tag{3.7}$$

$$net_k(t) = \sum_j v_{kj}y_j(t) + \vartheta_k \tag{3.8}$$

3.3 Results

The AR modeling and NN modeling were applied to eight different types of HRV signals discussed in Sect. 3.2. The original NSR, VF, AF, ISCH, CHB, LBBB, PVC and SSS segments, the corresponding AR modeled segments and the error signals are shown in Fig. 3.4. The original NSR, VF, AF, ISCH, CHB, LBBB, PVC and SSS segments, the corresponding NN modeled segments and the error signals are shown in Fig. 3.5. Two main criteria, SNR and NRMSE were used to evaluate the performance of the AR model and the NN model. The SNR was calculated to be from 15 to 35 dB. Figure 3.6 shows the SNR of the predicted HRV signals from the AR model and NN model. The SNR of the predicted signals from the NN model is better than for the signals of the AR model (Table 3.3). For critical abnormalities such as VF, PVC, ISCH, AF, CHB and SSS the SNR is significantly higher in the predicted signals from the NN model. The % NRMSE values is computed for the modeled signals and given in Table 3.2. It can be seen that the error is less for the NN model compared to the AR model.

Validation of the Signal Model

The generated HRV signals are validated using NRMSE, SNR and LF/HF ratio measures. The NRMSE given in Table 3.2 indicates the predicted signal to be a close follower of the actual signal with the NN model performing better than the AR model in the HRV signal types considered. The simulated normal HRV signal from the NN model closely follows the original with the NRMSE less than 0.3. Overall, the NN model generates signal with signal amplitude difference of about 20% and with a higher signal to noise ratio. The modeled signal closely follows the original signal in the time domain. In frequency domain, the results of LF/HF ratio measure indicate the preservance of the frequency domain features in the predicted signal. The % difference of the ratio between the modeled and actual signal is less 10%.

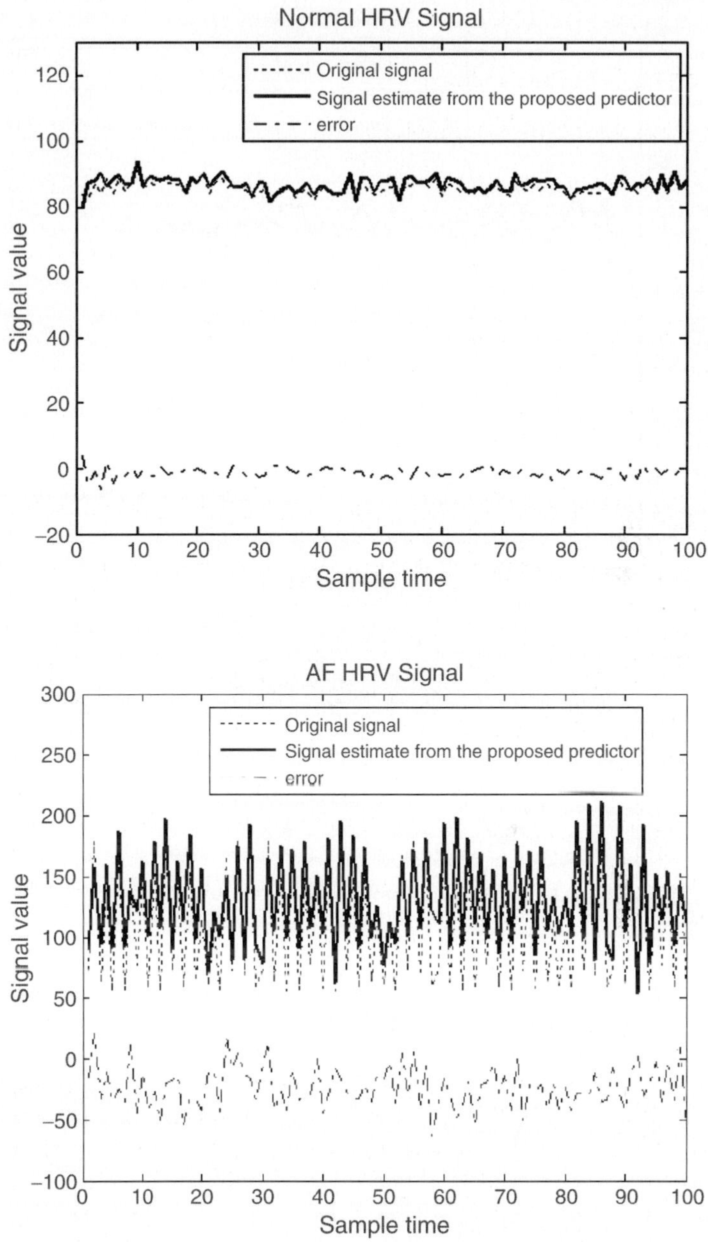

Fig. 3.4. Original, reconstructed and error signals for various heart rate signals using the AR model technique

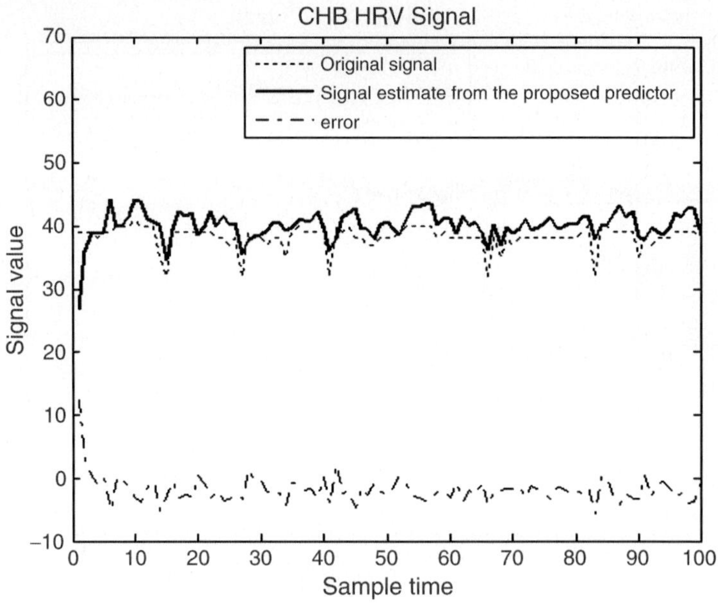

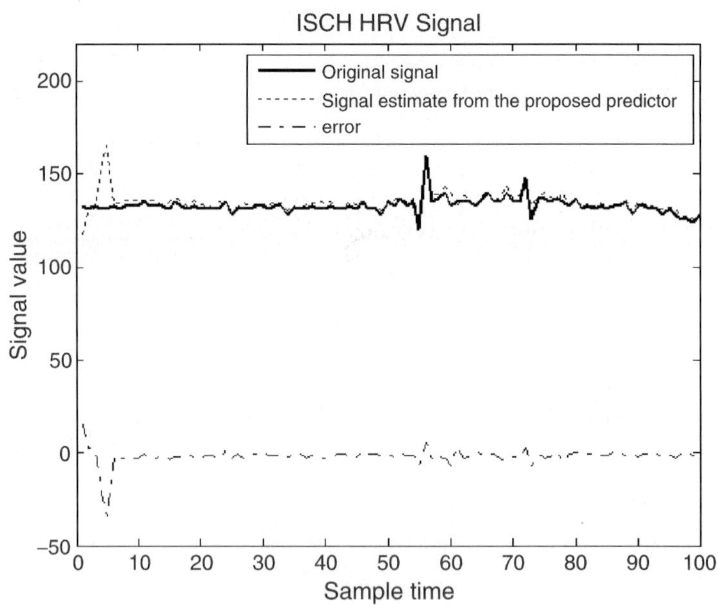

Fig. 3.4. *Continued.*

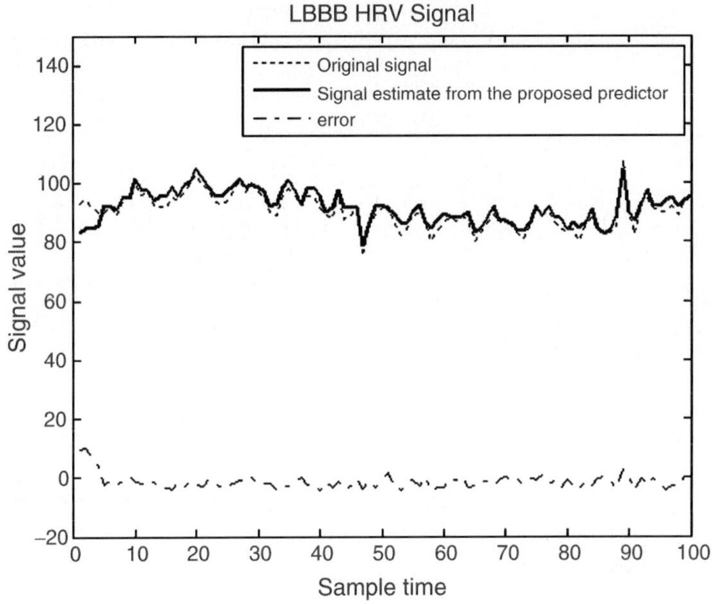

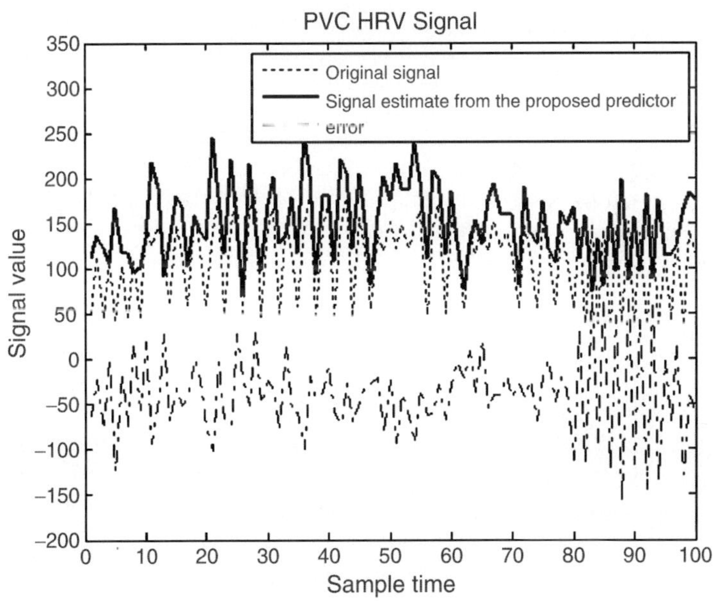

Fig. 3.4. *Continued.*

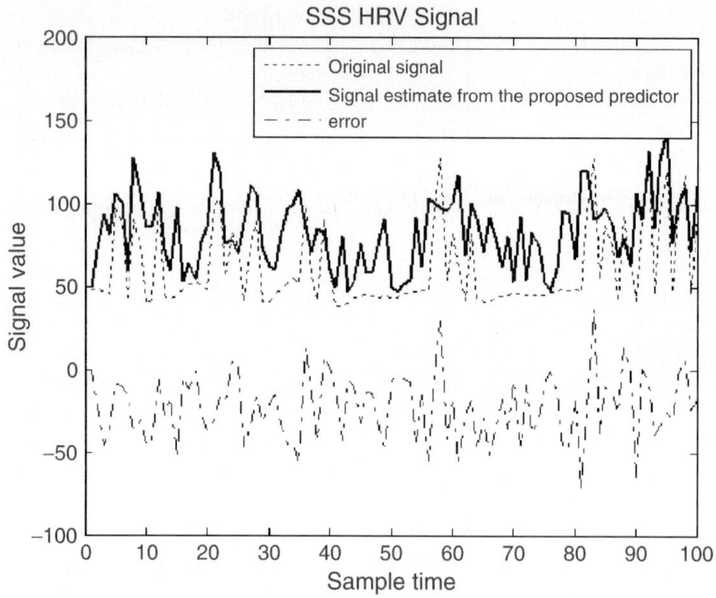

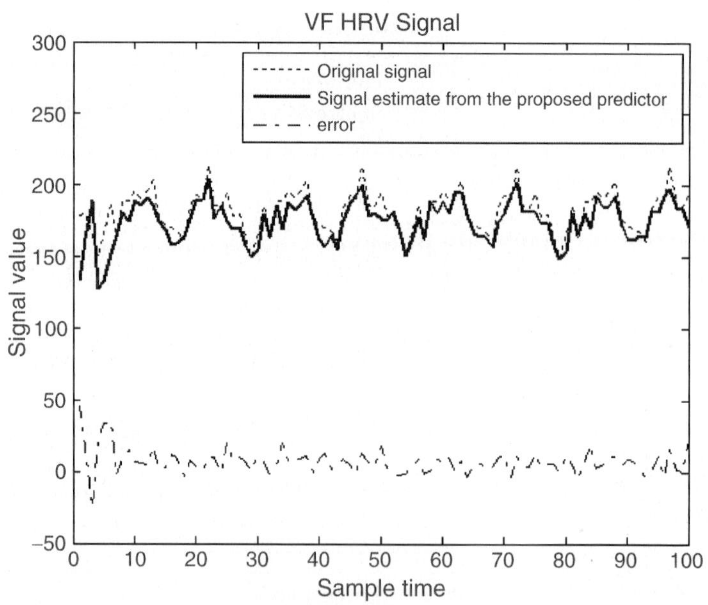

Fig. 3.4. *Continued.*

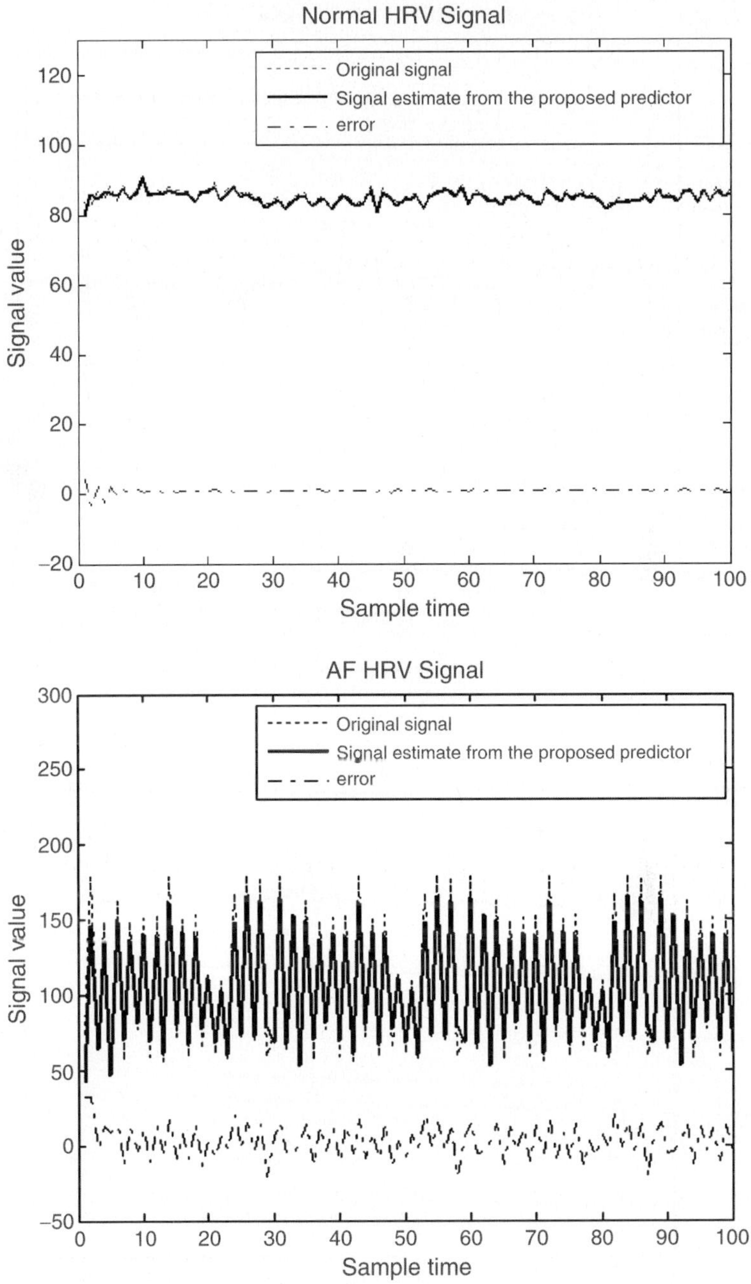

Fig. 3.5. Original, reconstructed and error signals for various heart rate signals using the neural network model

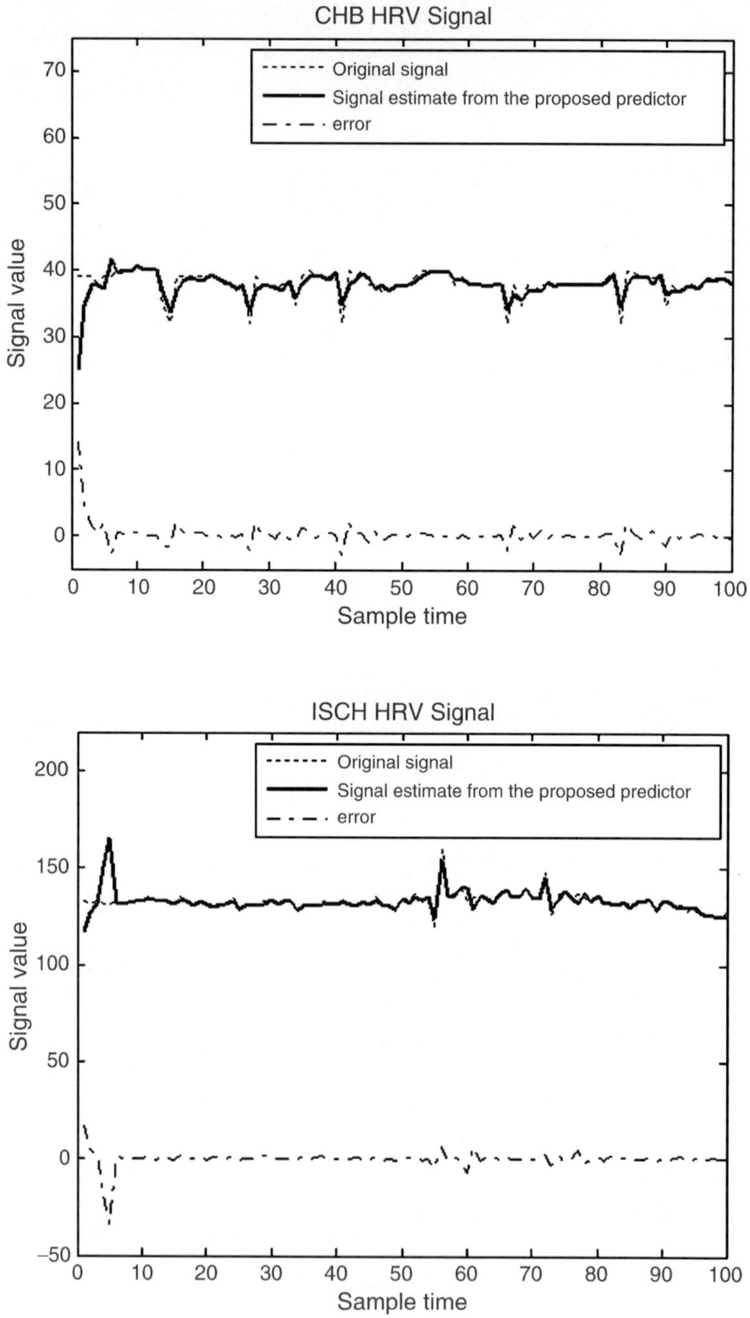

Fig. 3.5. *Continued.*

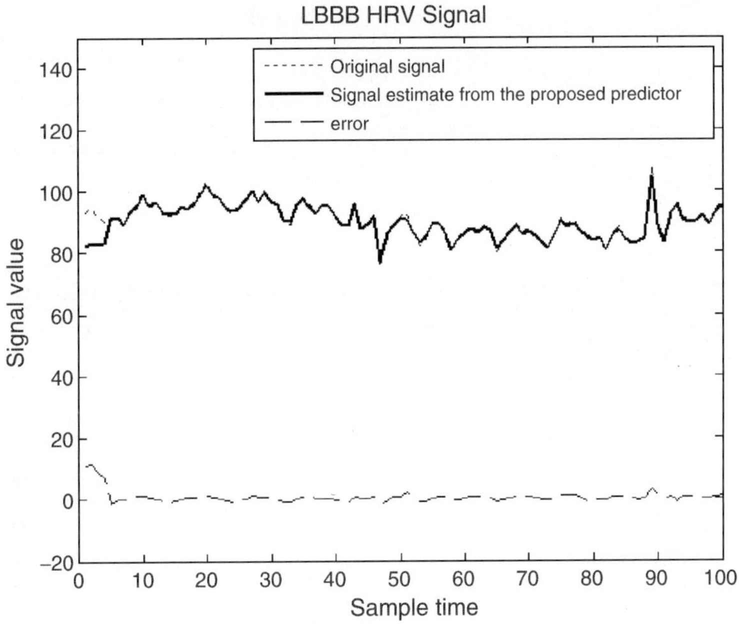

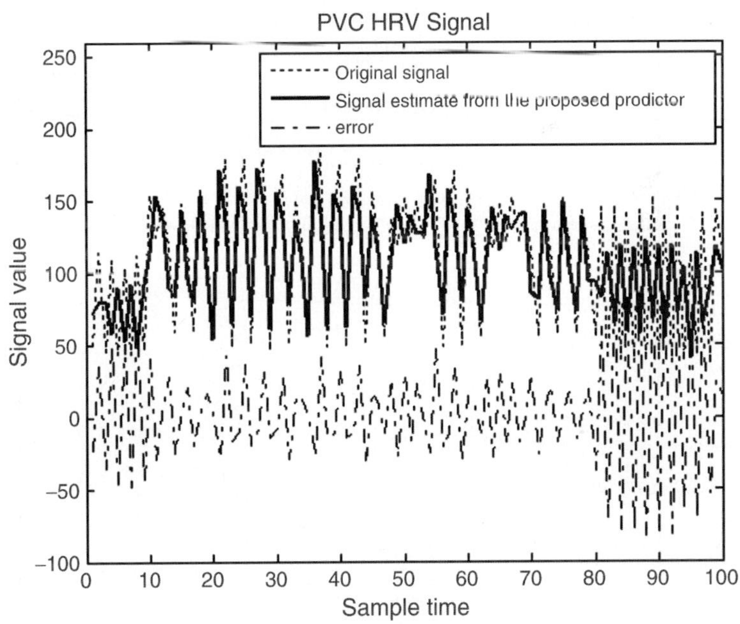

Fig. 3.5. *Continued.*

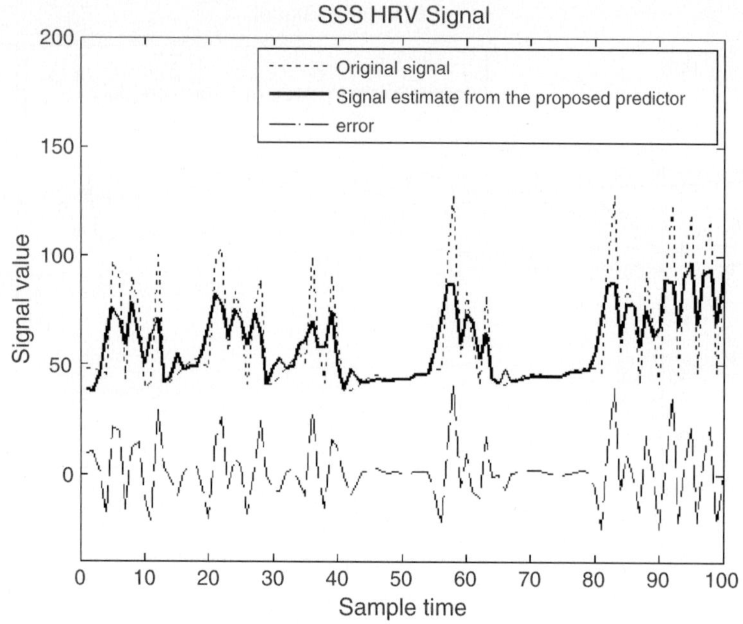

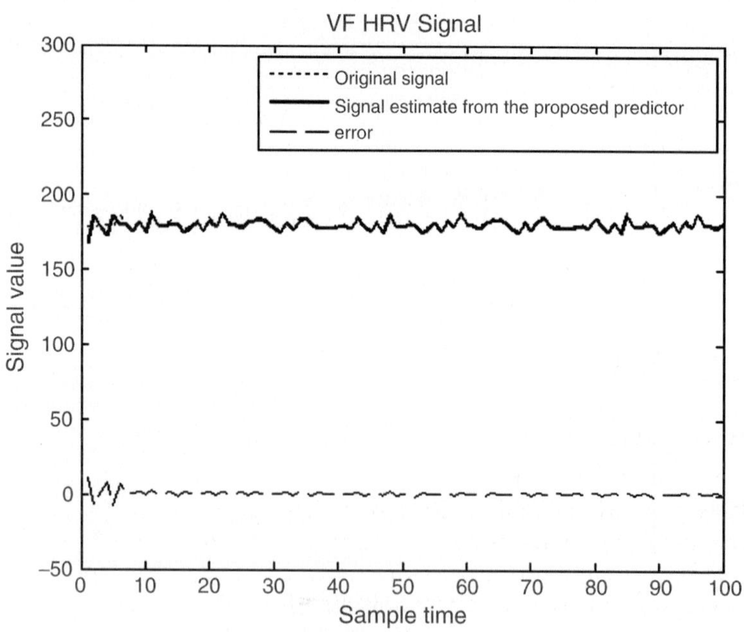

Fig. 3.5. *Continued.*

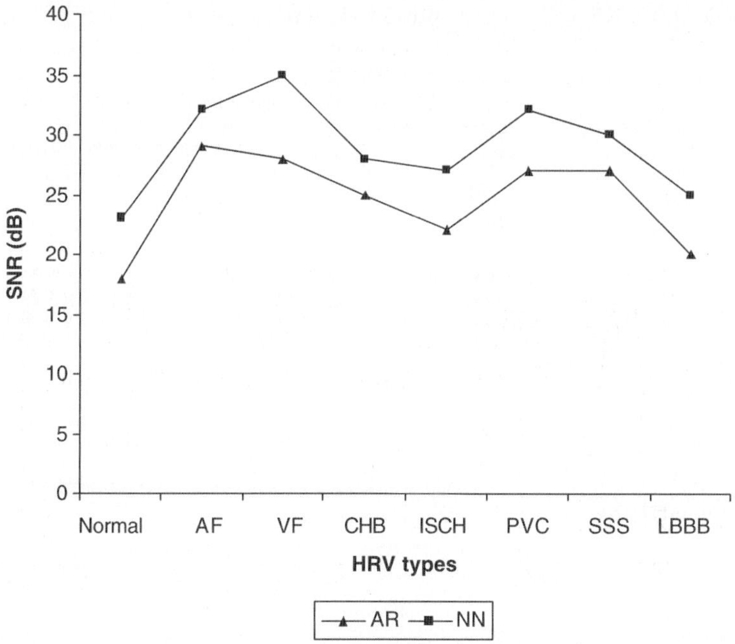

Fig. 3.6. Signal to noise ratio of the predicted signals from the AR and NN model

Table 3.2. NRMSE (%) values of the predicted signals from the AR and NN model

	AR	NN
Normal	0.67 ± 0.21	0.29 ± 0.11
AF	40.35 ± 3.31	10.42 ± 2.13
VF	19.63 ± 1.54	0.93 ± 0.26
CHB	9.37 ± 1.52	5.26 ± 1.04
ISCH	10.71 ± 2.41	9.24 ± 2.11
PVC	49.70 ± 3.86	22.47 ± 3.85
SSS	43.05 ± 3.73	17.69 ± 2.64
LBBB	3.14 ± 0.87	2.24 ± 1.16

3.4 Discussions

The study of heart rate variability (HRV) has become increasingly popular because information on the state of the autonomic nervous system can be non-invasively inferred by the use of relatively basic signal processing techniques. In this chapter, we have discussed various mathematical models that account for the physiological behavior of HRV, giving not only a representation of the autonomic nervous influence on the heart rate, but also providing a tool which helps in indicating which of the HRV representations exhibit better performance in terms of modeling the underlying physiology. Of the two techniques

Table 3.3. SNR values of the predicted signals from the AR and NN model

	AR	NN
Normal	18	23
AF	29	32
VF	28	35
CHB	25	28
ISCH	22	27
PVC	27	32
SSS	27	30
LBBB	20	25

Table 3.4. Comparison of LF/HF ratio of the predicted signals with the original signal

HRV Signal Types		LF/HF ratio			
		AR model		NN model	
	Original Signal	Predicted Signal	% difference	Predicted Signal	% difference
NSR	0.8635	0.8861	2.6141	0.8794	1.8382
LBBB	0.2441	0.2642	8.2516	0.2591	6.1620
PVC	1.3453	1.2122	9.8938	1.3369	0.6245
AF	0.5498	0.5581	1.5010	0.5613	2.0830
VF	0.2853	0.3011	5.5316	0.2912	2.0618
CHB	1.1532	1.2529	8.6417	1.2154	5.3899
Ischemic	2.9948	3.2674	9.1041	3.0329	1.2737
SSS	0.4185	0.4378	4.6202	0.4467	6.7471

discussed, it can be seen that the NN model can generate more reliable and accurate HRV signals. The reconstructed signals from the NN model exhibit higher signal to noise ratio and less error. The reproducibility ability of the NN model is better than that of the AR model. This is because the HRV signal is inherently chaotic and nonlinear. The NN model can model the nonlinear aspects of the underlying system better than the AR model.

The true power and advantage of neural networks lies in their ability to represent both linear and non-linear relationships and in their ability to learn these relationships directly from the data being modeled. Traditional linear models are inadequate, when it comes to modeling data that contains non-linear characteristics.

3.5 Conclusion

In this chapter, linear and nonlinear models that generate the long term HRV signals from the short term recordings is discussed. The simulated results show that, this model can create a simulated HRV signal that approximates the real HRV signal, both in time and frequency domains. The models are useful in the prediction of the heart rate signals and subsequently help in the analysis and diagnosis of cardiac abnormalities. The nonlinear model out perform the linear in terms of NRMSE, SNR and LF/HF ratio. This is due to the nonlinear model's inherent ability to model the underlying nonlinearity in the system.

References

1. Malik, M., Camm, A.J. (1995): 'Heart Rate Variability', *Armonk, NY: Futura.*
2. Task Force of the European Society of Cardiology, The North American Society of Pacing, and Electrophysiology. (1996): 'Heart rate variability: Standards of measurement, physiological interpretation, and clinical use', *Circulation*, **93**, pp. 1043–1065.
3. Crawford, M.H., Bernstein, S., and Deedwania, P. (1999): 'ACC/AHA guidelines for ambulatory electrocardiography', *Circulation*, **100**, pp. 886–893.
4. Akselrod, S., Gordon, D., Shannon, D.C., Barger, A.C., and Cohen, R.J. (1981): 'Power spectrum analysis of heart rate fluctuation: a quantitative probe of beat to beat cardiovascular control', *Science*, **213**, pp. 220–223.
5. Sayers, B, MCA. (1973): 'Analysis of heart rate variability', *Ergonomics*, **16**, pp. 17–32.
6. De Boer, R.W., Karemaker, J.M., and Strackee, J. (1985): 'Relationships between short-term blood pressure fluctuations and heart rate variability in resting subjects', *Medical & Biological Engineering & Computing*, **23**, pp. 352–364.
7. Pagani, M., Lombardi, F., Guzzetti, S., Rimoldi, O., Furlan, R., Pizzinelli, P., Sandrone, G., Malfatto, G., Dell'Orto, S., Piccaluga, E., Turiel, M., Baselli, G., Cerruti, S., and Malliani, A. (1986): 'Power spectral analysis of heart rate and arterial pressure variabilities as a marker of sympatho-vagal interaction in man and conscious dog', *Circulation Research*, **59**, 178–193.
8. Akselrod, S., Eliash, S., Oz, O., and Cohen, S. (1987): 'Hemodynamic regulation in SHR: investigation by spectral analysis', *American Journal of Physiology*, **253**, pp. H176–H183.
9. Marple, L. (1987): 'Digital Spectral Analysis with Applications', *Prentice Hall, Inc.*, London.
10. Owis, M.I., Abou-Zied, A.H., Youssef, A.B., and Kadah, Y.M. (2002): 'Study of features based on nonlinear dynamical modeling in ECG arrhythmia detection and classification', *IEEE Trans Biomed Eng.*, **49**(7), 733–736.
11. Guler, N.F., Kemal, M., and Inan Guler. (1995): "Comparison of FFT and AR based sonogram ouputs of 20 MHz pulsed Doppler data in real time", *Computers in Biology and Medicine*, **25**(4), 383–391.
12. Natalucci, S., Ruggeri, P., Cogo C.E., and Burattini, R. (1999): "Parametric Analysis of Heart Rate Variability in Anaesthetised SHR and WKY Rats", *Computers in Cardiology*, **26**, 1999, 591–594.
13. Dingfei, G., Srinivasan, N., and Krishnan, S.M. (2002) "Cardiac arrhythmia classification using autoregressive modeling", *BioMedical Engineering OnLine*, **1**(1); 5, 1–12.

14. Guler, I., Hardalac, F., and Ubeyli, E.D. (2002): 'Determination of Behcet disease with the application of FFT and AR methods', *Computers in Biology and Medicine*, **32**, pp. 419–434.
15. Ubeyli, E.D., and Guler I. (2003): 'Spectral analysis of internal carotid arterial Doppler signals using FFT, AR, MA, and ARMA methods', *Computers in Biology and Medicine*, 2003, pp. 1–13.
16. Chen, S.W. (2000): "Two-stage discrimination of cardiac arrhythmias using a total least squares-based prony modeling algorithm", *IEEE Trans Biomed Eng.*, **47**, pp. 1317–1326.
17. Mukhopadhyay, S., and Sircar, P. (1996): 'Parametric modelling of ECG signal', *Med Biol Eng Comp.*, **34**, pp. 171–173.
18. Pinna, G.D., Maestri, R., and Cesare, A.D. (1996): 'Application of time series spectral analysis theory: analysis of cardiovascular variability signals', *Med Biol Eng Comp.*, **34**, pp. 142–148.
19. Bennett, F.M., Chrisstini, D.J., Ahmed, H., and Lutchen, K. (1993): 'Time series modeling of heart rate dynamics', *Proc Comp Cardiol.*, pp. 273–276.
20. Arnold, M., Miltner, W.H.R., and Witte, H. (1998): 'Adaptive AR modeling of nonstationary time series by means of Kalman filtering', *IEEE Trans Biomed Eng.*, **45**, pp. 553–562.
21. Mainardi, L.T., Bianchi, A.M., Baselli, G., and Cerutti, S. (1995): 'Pole-tracking algorithms for the extraction of time-variant heart rate variability spectral parameters', *IEEE Trans Biomed Eng.*, **42**, pp. 250–258.
22. Ham, F.M., and Han, S. (1996): 'Classification of cardiac arrhythmias using fuzzy ARTMAP', *IEEE Trans Biomed Eng.*, **43**, pp. 425–430.
23. Lin, K.P., and Chang, W.H. (1989): 'QRS feature extraction using linear prediction', *IEEE Trans Biomed Eng.*, **36**, pp. 1050–1055.
24. Acharya, U.R., Joseph, P.K., Kannathal, N., Lim, C.M., and Suri, J.S. (2006): 'Heart rate variability: A Review', IFMBE *Journal of Medical & Biological Engineering & Computing Journal* (in press).
25. Wessel, N., Malberg, H., Bauernschmitt, R., Schirdewan, A., and Kurths, J. (2006): 'Nonlinear additive autoregressive model-based analysis of short-term heart rate variability', *Med Biol Eng Comput.*, **44**(4), pp. 321–330.
26. Kerem., D.H., Geva, A.B. (2005): 'Forecasting epilepsy from the heart rate signal', *Med Biol Eng Comput.*, 43(2), pp. 230–239.
27. The Mit-Bih Arrhythmia Database, 3rd ed. (1997): 'Harvard-MIT Division of Health Science Technology', *Biomedical Health Center*, Cambridge, MA.
28. Pan, J., and Tompkins, W.J. (1985): 'Real Time QRS Detector algorithm', *IEEE Transactions on Biomedical Engineering*, **32**(3), pp. 230–236.
29. Wariar, R., and Eswaran, C., 'Integer coefficient bandpass filter for the simultaneous removal of baseline wander, 50 and 100 Hz interference from the ECG', *Medical & Biological Engineering & Computing*, **29**, pp. 333–336.
30. Tompkins, W. (1993): 'Biomedical Digital Signal Processing', *Prentice Hall, Englewood Cliffs*, New Jersey. 1993.
31. Kay, S.M. (1998): 'Modern Spectral Estimation: Theory and Application', *Englewood Cliffs, NJ: Prentice-Hall.*
32. Choi, B. (1992): 'ARMA Model Identification', *Springer-Verlag*, New York, 1992.
33. Akaike, H. (1969): 'Fitting autoregressive models for prediction', *Annals of the Institute of Statistical Mathematics*, **21**, pp. 243–247.

34. Broadman, A., Schlindwein, F.S, Rocha, A.P and Leite, A. (2002): 'A study on the optimum order of autoregressive models for heart rate variability', *Physiological Measurement*, **23**, pp. 324–336.
35. Papoulis, A. (1991): 'Probability, Random Variables, and Stochastic Processes', *3rd ed., McGraw-Hill*, New York.
36. Eckmann, J.P., and Ruelle, D. (1985): 'Ergodic theory of chaos and strange attractors', *Reviews of Modern Physics*, **57**(3), part 1, pp. 617–656.
37. Pineda, F.J. (1988): 'Dynamics and architecture for neural computation', *Journal of Complexity*, **4**, pp. 216–245.
38. Chapeau-Blondeau, F., and Chauvet, G. (1992): 'Stable, oscillatory, and chaotic regimes in the dynamics of small neural networks with delay', Neural Networks, **5**, pp. 735–743.
39. van der, M.H.L.J., Verschure, P.F.M.J., and Molenaar, P.C.M. (1990): 'A note on chaotic behavior in simple neural networks', *Neural Networks*, **3**, pp. 119–122.
40. Linkens, D., and Nyongesa, Y. (1996): 'Learning systems in intelligent control: an appraisal of fuzzy, neural and genetic algorithm control applications', *IEE Proc. Control Theory App.*, **134**(4), pp. 367–385.
41. Parlos, A., Chong, K., Atiya, A. (1994): 'Application of the recurrent multilayer perceptron in modelling complex process dynamics', *IEEE Trans. on Neural Networks*, **5**, 1994, 255–266.
42. Draye, J., Pavisic, D., and Libert, G. (1996): 'Dynamic recurrent neural networks: a dynamical analysis', *IEEE Trans. On Systems Man and Cybernetics*, Part B, **26**, pp. 692–706.

4

Visualization of Cardiac Health Using Electrocardiograms

U. Rajendra Acharya, P. Subbanna Bhat, U.C. Niranjan, N. Kannathal, Lim Choo Min and Jasjit S. Suri

The chapter discusses an efficient and novel method to assist the physician to visualize voluminous cardiac data acquired over several hours. The system uses different colors to identify different types of cardiogram signals. In the display strategy each ECG beat is represented by a grid. The visualization strategy is hierarchical; that is, it provides for viewing of data from different level of abstraction, and the physician can have a top down approach to narrow down the time interval and signal details. This display strategy is extended to sector graph, with a menu driven hierarchical display strategy, which progressively unfolds greater details for chosen intervals. Provision is made for changing the parameters of classification, and thus the physician has the option for fine tuning the classification.

4.1 Introduction

Electrocardiogram (ECG) is a representative signal containing important information about the condition of the heart. The shape and size of the P-QRS-T wave, the interval between the various peaks and the changing heart rate etc., contain indications that are very useful in diagnosis of heart ailments. The ECG being a non-stationary signal, the disease indicators may occur at random in the time scale. Therefore, the patient may have to be kept under observation for long intervals for accurate diagnosis.

The ECG signal is usually recorded in magnetic tapes (holter recording), a digitized version of which can be stored in computers. However, the above details are not clearly perceptible for casual observation of the bare signal on the screen. Therefore, it is convenient to analyze and classify the signal using computer, and display the results for clinical observation.

Visual media is a most effective tool for communication, especially when the data is voluminous with subtle variations. Normally the data size in biomedicine is enormously large. For example, the symptoms of disease may show up only during certain period of the day, and that too may occur at random in the time scale. Therefore the physician may have to monitor and study the data over several hours for diagnostic purposes [1]. The tedium of reading voluminous data can be considerably reduced by using the computer for identifying and displaying the abnormalities occurring over a specified time interval [2]. In the present work, a hierarchical visualization technique is developed to aid the physician to identify the problem spots in the data recorded over long intervals. The vast data for the entire period under study is compressed and displayed on the computer screen using a color code to indicate different (ECG and) heart rates. The user can readily identify any deviation from normalcy by scanning the pattern. The user-friendly visualization tool allows the user to progressively expand the selected portion of the pattern for a closer study. Provision is made for the user to define or alter the thresholds of classification.

The signal can be viewed at different levels of hierarchy, which will progressively unfold greater details of the ECG signal and heart rate. At the highest level of abstraction, the data for 24 hours can be viewed in a single frame, and at the other extreme end, the individual ECG patterns can be displayed on the computer screen. Provision is made for the user to define or alter the thresholds of classification. Such techniques of 'software visualization' have been used in the past, to display programs, program artifacts and program behavior [2].

Proper recognition of beats is impeded by power-line interference; electromyogram noise and baseline drift which are often present in the ECG signal. In long-term monitoring, electrode impedance can increase considerably resulting in very low signal-to-noise ratio, which can make detection practically impossible in a single lead. Therefore, usually two or three leads are used for monitoring [4].

Several studies to classify various cardiac arrhythmias have been reported [5–7]. A number of techniques have been used for identification of arrhythmias including correction waveform analysis [8], time-frequency analysis [9], complexity measures [10], and a total least squares-based Prony modeling algorithm [5]. Different features are extracted from the ECG for classification of ventricular arrhythmias including QRS and ST segment based values, heart rate, spectral features, AR coefficients, complexity measures and nonlinear measures [5–7].

In the past, Udupa *et al* have proposed a set of algorithms for interactive visualization, manipulation, and measurement of large 3-D objects (e.g., heart) on general-purpose workstations [11]. They have addressed the use of these algorithms for visualizing medical data. Morikawa *et al* have investigated the feasibility of four-dimensional electrocardiography (4-D ECG), a new display in which the vector loop was rotated and scanned along a timed axis to overcome the shortcomings of vector cardiography (VCG) [12]. The P wave delineation score, signifying good agreement with the intraobserver and interobserver variability, was significantly higher in 4-D ECG than those in the orthogonal leads or those on the transverse and frontal projections. The techniques of 'software visualization' have been used in the past, to display programs, program artifacts and program behavior [2]. Paul *et al* have proposed scalogram plots to scrutinize the co-ordinated atrial activity during ventricular fibrillation [13]. He *et al* have proposed the three-dimensional (3-D) image of cardiac bioelectric source distribution from body-surface electrocardiograms [14]. Cardiac electrical sources were modeled by a current dipole distribution throughout the entire myocardium, and estimated by using the Laplacian weighted minimum norm (LWMN) algorithm from body-surface potentials. The layout of this chapter is as follows: Sect. 4.2 presents the preprocessing of the raw signals. Section 4.3 deals with the arrhythmia detection algorithm. Sections 4.4–4.6 of the chapter discusses the data handling and display, hierarchical display and sector graph display strategies, respectively. The discussions on our results is presented in Sect. 4.7. Finally the chapter concludes in Sect. 4.8.

4.2 Preprocessing

- Re-sample digitized signals to a sampling rate of 512 Hz. Original signal had a sampling rate of 1000 Hz.
- Pass the data through a low pass filter with a cut-off frequency of 15 Hz to remove unwanted high frequencies present in ECG signal.
- Apply a high pass filter with cut-off frequency 0.3 Hz to suppress baseline wander present in ECG signal.
- As the low pass filter is of second order and has a lower dB, a band-reject filter of cut-off frequencies 50 & 60 Hz to suppress the power-line interference noise.
- Similarly, the high pass filter is also of second order and has lower dB. A median filter was used to extract baseline wander of the processed ECG signal, and then subtract it from the processed ECG signal to effectively remove all baseline wander.

4.3 Arrhythmia Detection

An arrhythmia (deviation from normal rhythm) detection algorithm is developed based on three parameters namely, the R-R interval, QRS width and P-R interval [15–18]. The typical range of these parameters for various classes of disease are listed in Table 4.1. The parameters are finalized by trial and error method, after a number of iterations. However, the tool provides the option for changing the threshold levels and comparing it with the previous frame of display.

Table 4.1. Disease classification based on heart rate and ECG

Types	Heart rate (bpm)	P-R interval (sec)	QRS width (sec)
Normal	60–90	0.12–0.21	0.06–0.08
Sinus bradycardia	40–60	0.12–0.21	0.06–0.08
Premature ventricular contraction	90–140	P wave absent	0.12–0.21
Sinus tachycardia	90 >	0.12–0.21	0.06–0.08
Bundle branch block	60–140	0.12–0.2	0.12 >
A–V Block	60–140	0.21 >	> 0.2
Atrial premature contraction	60–140	0.12–0.21	> 0.2 or < 0.12
Supra ventricular tachycardia	> 150	< 0.12	0.12 <
Ventricular flutter	180–250	P wave absent	> 0.16
Ventricular tachycardia	100–200	< 0.12	0.12 ≤

In our work, we have considered the following types of cardiac arrhymias [1,19,20]. The heart rate (beats per minute) is given by the expression

$$\mathrm{HR} = 60/t_{r-r} \qquad \text{(bpm)}$$

It may be noted that three diseases are not listed in Table 4.1 (but are included for general classification), because they are impulsive deviations from normalcy; they are:

- *Bigeminy*: normal–VPC–normal–VPC.
- *Trigeminy*: normal–normal–VPC–normal–normal–VPC.
- *Quadrigeminy*: normal–normal–normal–VPC–normal–normal–normal–VPC.

These different cardiac arrhythmias are discussed in Chap. 1.

4.4 Data Handling and Display

The analog ECG signal – the 'P QRS T' wave – sampled at a suitable rate (512 samples/sec) to retain all relevant details of peaks, troughs and frequency, is codified into a specific color depending on the disease it represents. At the first level, a rectangular grid comprising of 15 × 39 segments – each segment representing a single ECG beat – is displayed on the screen. The color of the segment indicates the nature of deviation from normalcy (Table 4.2). In the present work, 13 classes are identified; for example, GREEN indicates a *normal* ECG, and RED indicates *Ventricular tachycardia*, etc. Provision is made for displaying the analog waveform of any chosen segment. The entire frame corresponds to about 10 minutes of data.

The raw data is sampled and stored in a random access file. The type of an ECG beat and its first and last sample addresses for each ECG are stored

Table 4.2. ECG and heart rate visualization color code

Heart rate (bpm)	Color	Arrhythmia
<40	Blue	Sinus bradycardia
40 to 60	Green	Normal
60 to 80	Cyan	Premature-ventricular contraction
80 to 100	Red	Ventricular tachycardia
100 to 120	Magenta	Sinus tachycardia
120 to 140	Brown	Supra ventricular tachycardia
140 to 160	Light gray	Bigeminy
160 to 180	Light green	Trigeminy
180 to 200	Light cyan	Quadrigeminy
200 to 220	Light red	Bundle branch block
220 to 240	Light magenta	A-V block
240 to 260	Yellow	Ventricular flutter
260 to 280	White	Atrial premature contraction

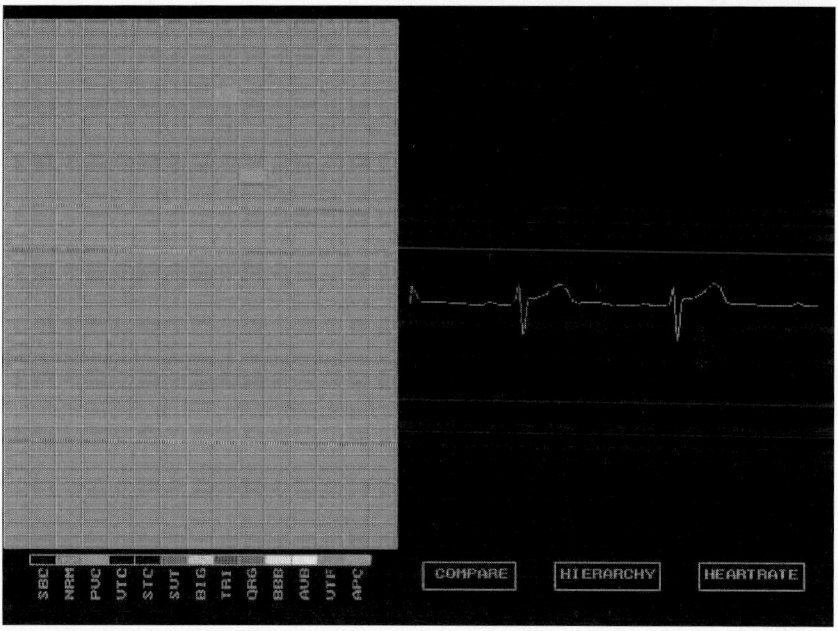

Fig. 4.1. ECG visualization for sample data

in another file named 'information (class) file'. This facilitates easy manipulation of data corresponding to a grid. The algorithm yields the type (class) of each ECG beat in the data. Figure 4.1 represents the result of the ECG visualization for a sample data (with the chosen range for 'normalcy' as :: heart rate: 60–90 bpm; QRS width: 0.06–0.08 sec; PR interval: 0.12–0.2 sec). Here,

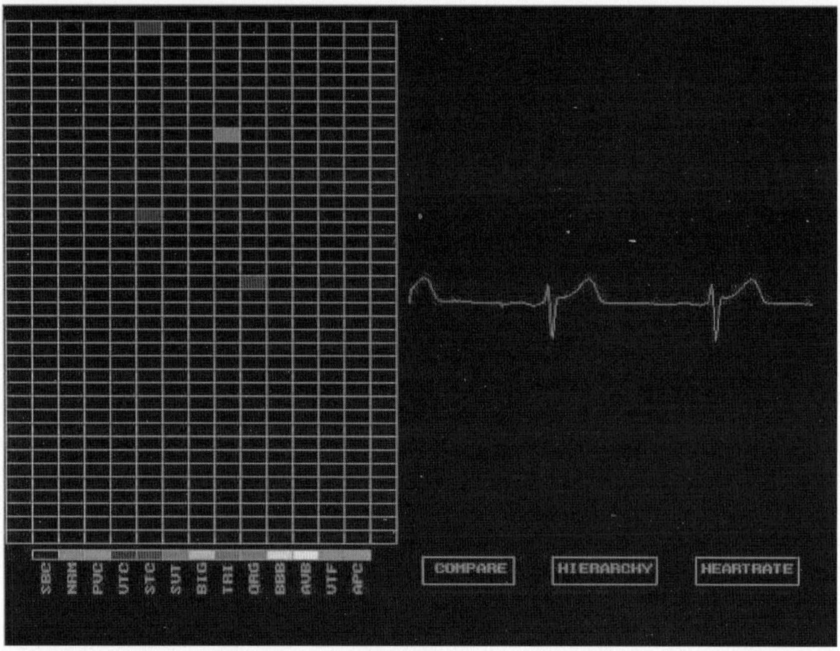

Fig. 4.2. Compare frame ECG visualization

the color CYAN represents *Premature ventricular contraction*, and BLUE represents *Sinus bradycardia*. This figure displays predominantly normal ECG beats.

To study the effect of changing parameter thresholds, the following facility is provided: Two 'information (class) files' containing the ECG types before and after the change of thresholds are created. The display subroutine reads both the 'information (class) files' and updates only those grid segments whose classes have changed (compared to the previous frame), while the unchanged portion of the grid is blanked out. Figure 4.2 displays the 'comparison frame' corresponding to Fig. 4.1, with the range of 'normalcy' changed to :: heart rate: 65–90 bpm; QRS width: 0.06–0.08 sec; and PR interval: 0.12–0.2 sec.

Apart from displaying the ECG classification, the tool has provision to display the heart rate classification with a color code (Table 4.2) [6]. The heart rate is obtained by evaluating the inverse of R-R interval (Fig. 4.4).

4.5 Hierarchical Display

At the next higher level of abstraction, each grid segment represents a group of 10 ECG beats, and its color represents the *most dominant* type of deviation during the group interval. Figure 4.3 represents the higher level abstraction of Fig. 4.1. The less dominant deviations if any, are ignored in this level of display.

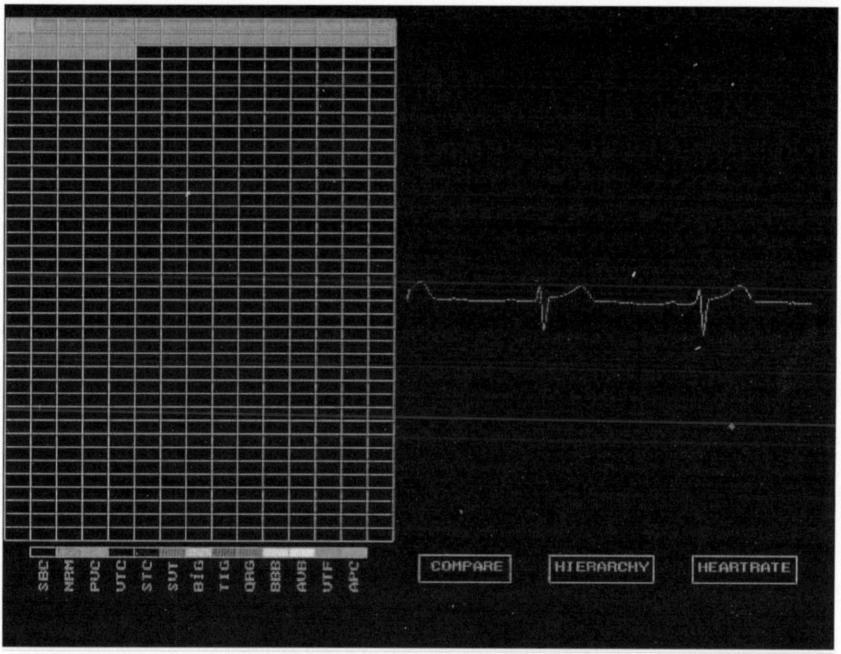

Fig. 4.3. Hierarchical ECG visualization

In the present case, the display pattern of a single frame can represent about, one and half hours of data. Thus an overall view of the patient's state of health is available in a single reference frame. Provision is made for going back to the lower level of display for the chosen portion of the display.

4.6 Sector Graph Display

The sector graph approach is used to provide a comprehensive view of the state of health over 24 hours. The circle is divided into 24 sectors, corresponding to the hours of the day. Within the sector, the color code is employed to indicate the percentage of the type of deviation (from normalcy), during that hour. Figure 4.5 is obtained for the range of thresholds defined as heart rate: 60–90 bpm; QRS width: 0.06–0.08 sec; PR interval: 0.12–0.2 sec. This example displays indications of Premature *ventricular contraction* and *Sinus bradycardia* in certain sectors, in addition to normal data.

In the next lower level of abstraction, the entire circle corresponds to one hour interval, and is divided into 12 segments of 5 minutes each (Fig. 4.6 corresponds to 1.00–2.00 AM interval of Fig. 4.5). At the next lower level, details of 1 minute interval are displayed (Fig. 4.7).

Figure 4.8 provides the heart rate display for 24 hours.

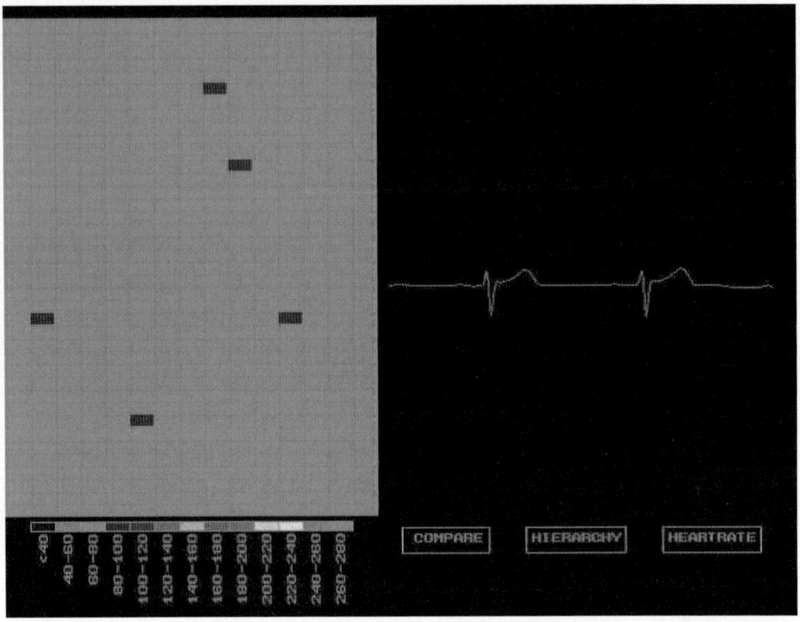

Fig. 4.4. Heart rate visualization

#[Figures 4.1–4.4 are reprinted with permission from Rajendra Acharya U., Subbanna Bhat P., Niranjan U.C, "A display platform for electrocardiograms", Innovations and Technology in Biology and Medicine (ITBM-RBM), France, October 1999, vol. 20, pp 303–306]

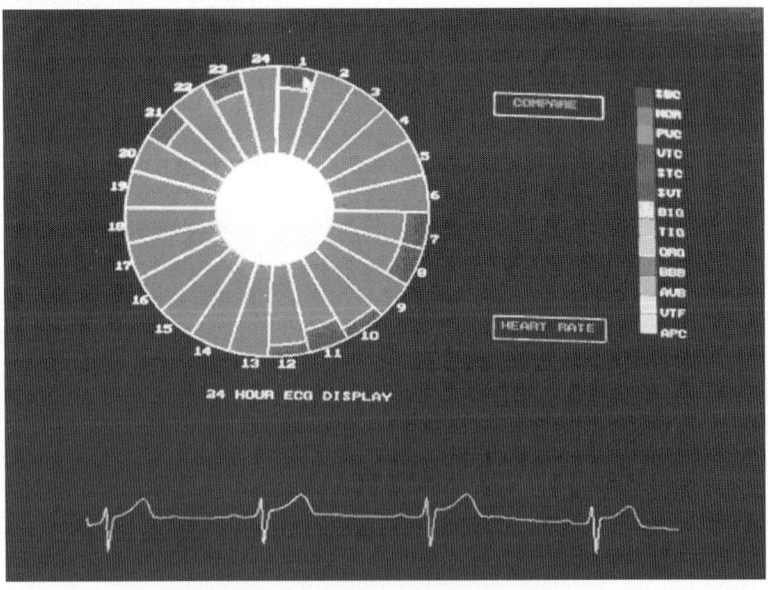

Fig. 4.5. 24 hour ECG visualization

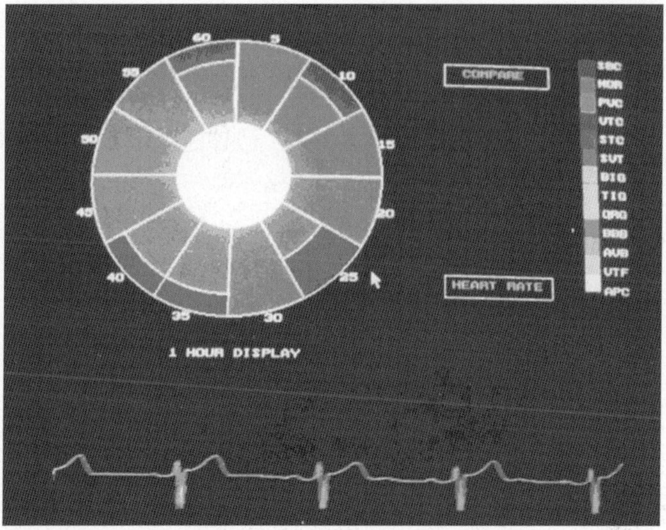

Fig. 4.6. 1 hour ECG visualization

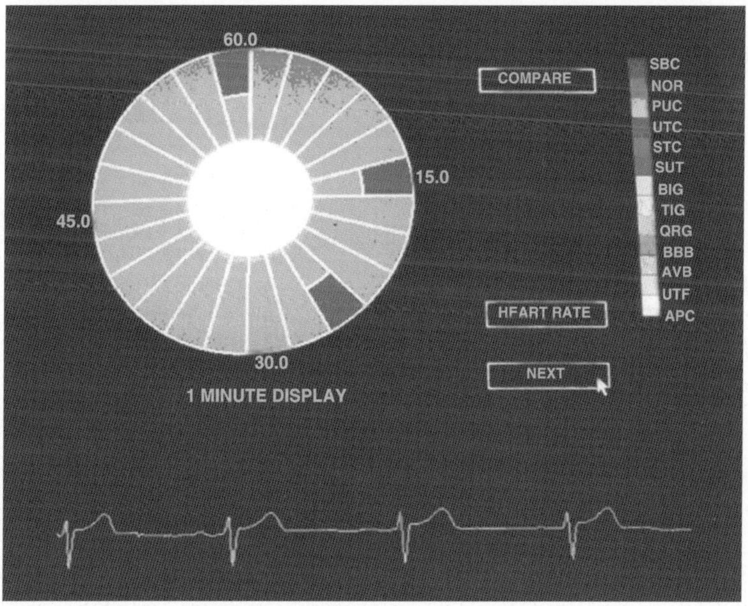

Fig. 4.7. 1 minute ECG visualization

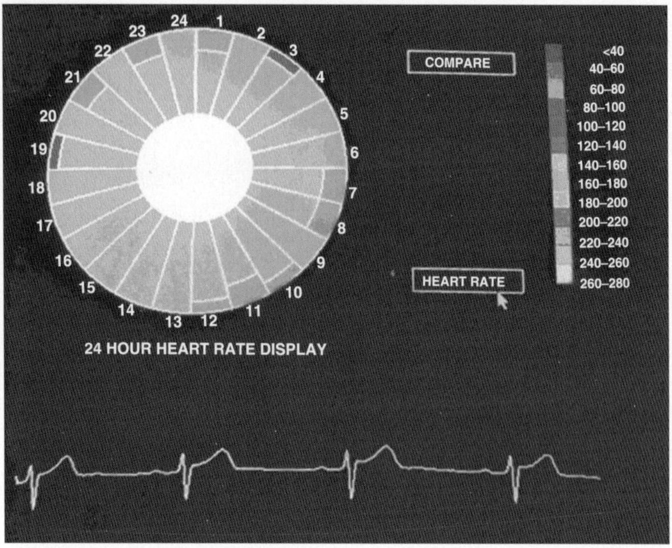

Fig. 4.8. 24 hour heart rate visualization

#[Figures 4.5–4.8 are reprinted with permission from Rajendra Acharya U., Subbanna Bhat P., Niranjan U.C, "Comprehensive visualization of cardiac health", Computers in Biology and Medicine Journal, USA, January 2002, vol. 32/1, pp 49–54]

4.7 Discussion

Disease symptoms are highly subjective signals that may occur at random in the time scale. Therefore, the physician may have to observe the patient for long intervals to diagnose the affliction. The ECG may contain important information about the nature of disease, both in its shape and duration between its salient peaks and troughs. The actual values and variation of these parameters are not obvious for a casual survey of the waveform upon the screen. However, the problem can be effectively handled by using computer for analysis and display of salient features of the signal. In this chapter, features are extracted from the ECG for arrhythmia detection and different colors for various cardiac abnormalities. In this chapter, we have proposed the visualization of cardiac disorders using colored grids and sector graphs. In our work, we have achieved the detection of each abnormal beat immediately with less strain using the color grid method. Also, during the analysis of huge data (24 hours), sector graph method will be very useful. In this technique, one can visualize the whole day data, one hour and even five minutes of ECG data at one glance. In this scheme, normal and twelve other types of abnormal beats can easily be visualized.

In the hierarchical visualization method, we can visualize small duration of the ECG data. But however, by assigning one grid for ten or hundred similar class of ECG data, one can visualize more data at one glance. In the sector graph method of visualization, one can see the abnormality at any instant of the day, just by clicking the mouse at that instant of the day (Fig. 4.5). This method of visualization is very effective and helps in detecting the instant at which the abnormal beat has occurred. In both methods, one can visualize the heart rate information also.

This work can be used to visualize more abnormalities by using robust arrhythmia detection software. ST segment width, ST segment amplitude, P wave amplitude, etc. can be used as additional parameters in addition to HR, QRS width, PR interval in designing the arrhythmia detection. Also, this work can be extended to visualize the real time cardiac health by using the parallel processing concepts.

4.8 Conclusion

A visual display platform for the study of voluminous ECG data collected over several hours is developed. Hierarchical display technique is useful for identification as well as time localization of abnormalities, which would considerably reduce the tedium involved in the manual scrutiny of data. The menu driven system provides for altering the parameter thresholds and study the changes in classification. This technique does not need extensive computations and is simple to operate. The platform is expected to be of significant help to study the effects of drugs, treatment and therapy. The scheme developed can be easily adapted to other biomedical signals.

References

1. Sokolow, M., McLlroy, M.B., and Chiethin, M.D. (1990): 'Clinical cardiology', *VLANGE medical book.*
2. Ball, T., and Eick, S.G. (1996): 'Software Visualization in the Large', *IEEE Computer*, pp. 33–43.
3. Schroeder, W., Martin, K., and Lorensen, B. (1996): 'The Visualization Toolkit', *Prentice Hall.*
4. Moreas, J.C.T.B., Seixas, M.O., Vilani, F.N., and Costa, E.V. (2002): 'A QRS complex detection algorithm using electrocardiogram leads', *Computers in Cardiology*, **29**, pp. 205–208.
5. Chen, S.W. (2000): 'Two-stage discrimination of cardiac arrhythmias using a total least squares-based prony modeling algorithm', *IEEE Trans Biomed Eng.*, **47**, pp. 1317–1326.
6. Melo, S.L., Caloba, L.P., and Nadal, J. (2000): 'Arrhythmia analysis using artificial neural network and decimated electrocardiographic data', *Computers in Cardiology*, **27**, pp. 73–76.

7. Small, M., Yu., D.J., Grubb, N., Simonotto, J., Fox, K.A.A., and Harrison, R.G. (2000): 'Automatic identification and recording of cardiac arrhythmia', *Computers in Cardiology*, **27**, pp. 355–358.
8. Caswell, S.A., Kluge, K.S., and Chiang, C.M.J. (1993): 'Pattern recognition of cardiac arrhythmias using two intracardiac channels', Proceedings of Computers in Cardiology, pp. 181–184.
9. Afonoso, V.X. (1995): 'Tompkins WJ, Detecting ventricular fibrillation: Selecting the appropriate time-frequency analysis tool for the application', *IEEE Eng Med Biol Mag.*, **14**, pp. 152–159.
10. Xu, S.Z., Yi, S.Z., Thakor, N.V., and Wang, Z.Z. (1999): 'Detecting ventricular tachycardia and fibrillation by complexity measure', *IEEE Trans Biomed Eng.*, **46**, pp. 548–555.
11. Udupa, J.K., and Odhner, D. (1991): 'Fast visualization, manipulation, and analysis of binary volumetric objects', *IEEE Transactions on Graphics and Applications*, **11**(6), pp. 53–62.
12. Morikawa, J., Habuchi, Y., Nishio, M., Yamamoto, T., Tanaka, H., and Kashima, K. (1996): 'Delineation of premature P waves on four-dimensional electrocardiography, a new display of electrical forces by computer techniques', *Angiology*, **47**(11), pp. 1101–1106.
13. Paul, S.A., James, N.W., Gareth, R.C., Petter, A.S., and Colin, E.R. (2002): 'Finding coordinated atrial fibrillation during ventricular fibrillation using wavelet decomposition', *IEEE-EMBS Magazine*, **21**(1), pp. 58–65.
14. He, B., and Wu, D. (2001): 'Imaging and visualization of 3-D cardiac electric activity', *IEEE Transactions on Information Technology in Biomedicine*, **5**(3), pp. 181–186.
15. Lynn, P.A. (1977): 'Online digital filter for biological signals: some fast designs for a smallcomputer' *Medical & Biological Engg. & Computing*, **15**, pp. 534–540.
16. Pan, J., and Tompkins, W.J. (1985): 'A real time QRS detection algorithm', *IEEE Trans. on Biomedical Engineering*, **32**(3), pp. 230–236.
17. Tompkins, W.J. (1993): 'Biomedical Signal Processing', Prentice Hall Publications.
18. Goldschlager, N., and Mervin J.G. (1989): 'Principles of Clinical Electrocardiography', Thirteenth Edition, pp. 58–238.
19. Wariar, R., and Eswaran, C. (1991): 'Integer coefficient bandpass filter for the simultaneous removal of baseline wander, 50 and 100 Hz interference from the ECG', *Medical & Biological Engineering & Computing*, **29**, pp. 333–336.
20. Robert, J.H. (2002): 'Basic Dysrhythmias', *Mosby*, Third edition, Interpretation & Management, pp. 110–379.

5

Heart Rate Variability

Rajendra Acharya U, Paul Joseph K, Kannathal N, Lim Choo Min
and Jasjit Suri S

During normal sinus rhythm, the heart rate (HR) varies from beat to beat. Heart rate variability (HRV) results from the dynamic interplay between the multiple physiologic mechanisms that regulate the instantaneous HR. It is believed that Heart Rate Variability (HRV) will become as common as pulse, blood pressure or temperature in patient charts in the near future. In the last ten years more than 2000 published articles have been written about HRV. HRV has been used as a screening tool in many disease processes. Various medical disciplines are looking at HRV. In diabetes and heart disease it has been proven to be predictive of the likelihood of future events.

5.1 Physiological Phenomenon of HRV

The origin of heartbeat is located in a sino-atrial (SA) node of the heart, where a group of specialized cells continuously generates an electrical impulse spreading all over the heart muscle through specialized pathways and creating process of heart muscle contraction well synchronized between both atriums and ventricles. The SA node generates such impulses about 100–120 times per minute at rest. However in healthy individual resting heart rate (HR) would never be that high. This is due to continuous control of the autonomic nervous system (ANS) over the output of SA node activity. Its net regulatory effect gives real HR. In healthy subject at rest it is ranging between 50 and 70 beats per minute.

The autonomic nervous system is a part of the nervous system that non-voluntarily controls all organs and systems of the body. As the other part of nervous system ANS has its central (nuclei located in brain stem) and peripheral components (afferent and efferent fibers and peripheral ganglia) accessing all internal organs. There are two branches of the autonomic nervous system – sympathetic and parasympathetic (vagal) nervous systems that always work as antagonists in their effect on target organs.

For most organs including heart the sympathetic nervous system stimulates organ's functioning. An increase in sympathetic stimulation causes increase in HR, stroke volume, systemic vasoconstriction, etc. The heart response time to sympathetic stimulation is relatively slow.

In contrast, the parasympathetic nervous system inhibits functioning of those organs. An increase in parasympathetic stimulation causes decrease in HR, stroke volume, systemic vasodilatation, etc. The heart's response time to parasympathetic stimulation is almost instantaneous.

At rest both sympathetic and parasympathetic systems are active with parasympathetic dominance. The actual balance between them is constantly changing in an attempt to achieve optimum considering all internal and external stimuli.

There are various factors affecting autonomic regulation of the heart, including but not limited to respiration, thermoregulation, humoral regulation (renin-angiotensin system), blood pressure, cardiac output, diabetes, sleep, age, alcoholism, nervous system, drugs, smoking, renal failure etc. This review focuses on different factors affecting the heart rate, methodology and interpretation of HRV measures.

5.2 Introduction

Heart rate variability (HRV), the variation over time of the period between consecutive heartbeats, is predominantly dependent on the extrinsic regulation of the heart rate. HRV is thought to reflect the heart's ability to adapt to changing circumstances by detecting and quickly responding to unpredictable stimuli. HRV analysis is the ability to assess overall cardiac health and the state of the autonomic nervous system (ANS) responsible for regulating cardiac activity.

HRV is a useful signal for understanding the status of the autonomic nervous system (ANS). HRV refers to the variations in the beat intervals or correspondingly in the instantaneous heart rate (HR). The normal variability in HR is due to autonomic neural regulation of the heart and the circulatory system [1]. The balancing action of the sympathetic nervous system (SNS) and parasympathetic nervous system (PNS) branches of the autonomic nervous system (ANS) controls the HR. Increased SNS or diminished PNS activity results in cardio-acceleration. Conversely, a low SNS activity or a high PNS activity causes cardio-deceleration. The degree of variability in the HR provides information about the functioning of the nervous control on the HR and the heart's ability to respond.

Past 20 years have witnessed the recognition of the significant relationship between autonomic nervous system and cardiovascular mortality including sudden death due to cardiac arrest [2–4]. Numerous papers in connection with HRV related cardiological issues [5–7] reiterated the significance of HRV in assessing the cardiac health. The interest in the analysis of heart rate variability (HRV), (that is, the fluctuations of the heart beating in time), is not new. Furthermore, much progress was achieved in this field with the advent of low cost computers with massive computational power, which paved the way for many recent advances.

Spectral analysis of beat-to-beat variations of heart rate (HR) and blood pressure (BP) was applied in order to obtain non-invasive indices of sympathetic and parasympathetic regulation [8]. HR, diastolic BP, mid-frequency band power (0.07–0.14 Hz) of HR and systolic BP, and plasma adrenaline and noradrenaline concentrations showed significant increases when changing from supine to sitting to standing posture, whereas high-frequency band power (0.15–0.50 Hz) of HR decreased in a posture-dependent fashion. Viktor *et al* have studied the variation of heart rate spectrogram and breathing rates in lateral and supine body positions [9]. Recently, new dynamic methods of HRV quantification have been used to uncover nonlinear fluctuations in heart rate, that are not otherwise apparent. Several methods have been proposed: Lyapunov exponents [10], 1/f slope [7], approximate entropy ($ApEn$) [11] and detrended fluctuation analysis [12].

Heart rate variability, that is, the amount of heart rate fluctuations around the mean heart rate, can be used as a mirror of the cardiorespiratory control system. It is a valuable tool to investigate the sympathetic and parasympathetic function of the autonomic nervous system. The most important application of heart rate variability analysis is in the surveillance of postinfarction and diabetic patients. Heart rate variability gives information about the sympathetic-parasympathetic autonomic balance and thus about the risk for sudden cardiac death in these patients. Heart rate variability measurements are easy to perform, noninvasive, and have good reproducibility, if used under standardized conditions [13, 14].

Boris *et al* have introduced the sample asymmetry analysis (SAA) and illustrated its utility for assessment of heart rate characteristics occurring early in the course of neonatal sepsis and systemic inflammatory response syndrome (SIRS) [15]. Compared with healthy infants, infants who experienced sepsis had similar sample asymmetry in health, and elevated values before sepsis and SIRS ($p = 0.002$). Cysarz *et al* have demonstrated that the binary symbolization of R-R interval dynamics, which at first glance seems to be an enormous waste of information, gives an important key to a better understanding of normal heart rate regularity [16]. Furthermore, differential binary symbolization still enables the identification of nonlinear dynamical properties.

Recently, Verlinde *et al* have compared the heart rate variability of aerobic athletes with the controls and showed that the aerobic athletes have an increased power in all frequency bands [17]. These results are in accordance with values obtained by spectral analysis using the Fourier transform, suggesting that wavelet analysis could be an appropriate tool to evaluate oscillating components in HRV. But, in addition to classic methods, it also gives a time resolution. Time-dependent spectral analysis of HRV using the wavelet transform was found to be valuable for explaining the patterns of cardiac rate control during reperfusion. In addition, examination of the entire record revealed epochs of markedly diminished HRV in two patients, which attribute to vagal saturation [18]. A method for analyzing HRV signals using the wavelet transform was applied to obtain a time-scale representation for very low-frequency (VLF), low-frequency (LF) and high-frequency (HF) bands using the orthogonal multiresolution pyramidal algorithm [19]. Results suggest that wavelet analysis provides useful information for the assessment of dynamic changes and patterns of HRV during myocardial ischemia. Time-frequency parameters calculated using wavelet transform and extracted from the nocturnal heart period analysis appeared as powerful tools for obstructive sleep apnoea syndrome diagnosis [20]. Time-frequency domain analysis of the nocturnal heart rate variability using wavelet decomposition could represent an efficient marker of obstructive sleep apnoea syndrome [20]. Schumacher *et al* have explained the use of linear and non-linear analysis in the analysis of the heart rate signals [21]. The effect of autonomic nervous system (ANS), blood pressure, myocardial infarction (MI), nervous system, age, gender, drugs, diabetes, renal failure, smoking, alcohol, sleep on the heart rate variability are discussed in detail in the following sections.

5.2.1 The Autonomic Nervous System (ANS)

The ANS has sympathetic and parasympathetic components. Sympathetic stimulation, occurring in response to stress, exercise and heart disease, causes an increase in heart rate by increasing the firing rate of pacemaker cells in the heart's sino-atrial node. Parasympathetic activity, primarily resulting from the function of internal organs, trauma, allergic reactions and the inhalation of irritants, decreases the firing rate of pacemaker cells and the heart rate, providing a regulatory balance in physiological autonomic function. The separate rhythmic contributions from sympathetic and parasympathetic autonomic activity modulate the heart rate (RR) intervals of the QRS complex in the electrocardiogram (ECG), at distinct frequencies. Sympathetic activity is associated with the low frequency range (0.04–0.15 Hz) while parasympathetic activity is associated with the higher frequency range (0.15–0.4 Hz) of modulation frequencies of the heart rate. This difference in frequency ranges allows HRV analysis to separate sympathetic and parasympathetic contributions. This should enable preventive intervention at an early stage when it is most beneficial.

5.2.2 HRV and Blood Pressure

Some studies using non-pharmacological approaches to hypertension have shown that a reduction in respiration rate is associated with decreased blood pressure. Decrease in respiration rate to less than 10 breaths per minute with elongated expiration have been achieved with musical tone breath control and demonstrated a reduction in the blood pressure [22, 23]. Various breathing exercises including yoga breathing have demonstrated reduction in breath rate associated with a reduction in blood pressure in hypertensives [24, 25]. Guyton (1992) points out that body fluid volumes that are regulated by the kidney determine long-term control of BP [26]. This in turn is significantly influenced by salt and its excretion.

A method to describe relationships between short-term blood pressure fluctuations and heart-rate variability in resting subjects was analyzed in the frequency domain [27]. The European Society of Hypertension working group on baroreflex and cardiovascular variability has produced a comprehensive database for testing and comparison of methods [28]. Westerhof *et al* have proposed a cross-correlation baro-reflex sensitivity (xBRS) technique for the computation of time-domain baroreflex sensitivity on spontaneous blood pressure and heart rate variability using EUROBAVAR data set [29]. They proved that, the xBRS method may be considered for experimental and clinical use, because the values yielded were correlated strongly with and was close to the EUROBAVAR averages.

5.2.3 HRV and Myocardial Infarction

A predominance of sympathetic activity and reduction in parasympathetic cardiac control has been found in patients with acute myocardial infarction (MI) [30]. It was shown that, the HRV decreases with the recent myocardial infarction [31, 32]. Despite the beneficial effects on clinical variables, exercise training did not markedly alter HRV indexes in subjects after MI [33]. A significant decrease in SDRR and high-frequency power in the control group suggested an ongoing process of sympathovagal imbalance in favor of sympathetic dominance in untrained patients after MI with new-onset left ventricular dysfunction. Although previous studies demonstrated an association between depressive symptoms and cardiac mortality after acute myocardial infarction (AMI) little is known about the possible mechanisms of this association. Patients with post-AMI depression have a cardiac autonomic dysfunction as reflected by decreased HRV and increased HR. This autonomic dysfunction seems not to be an independent mediator of the increased mortality observed in depressed patients during a 5-year follow-up [34]. Decreased vagal activity after myocardial infarction results in reduced heart-rate variability and increased risk of death. Impaired heart rate deceleration capacity is a powerful predictor of mortality after myocardial infarction and is more accurate than ventricular ejection fraction (LVEF) and the conventional measures of heart-rate variability [35]. Several studies showed that thrombolysis reduces ventricular arrhythmias and improves heart rate variability (HRV) in patients with acute myocardial infarction (AMI). Larosa *et al* have failed to show any significant benefit of primary percutaneous coronary intervention (PCI) compared to thrombolysis on ventricular arrhythmias and HRV in patients with ST-segment elevation AMI [36].

5.2.4 HRV and Nervous System

Disorders of the central and peripheral nervous system have effects on heart rate variability. The vagally and sympathetically mediated fluctuations in heart rate may be independently affected by some disorders. All normal cyclic changes in heart rate are reduced in the presence of severe brain damage [37] and depression [31, 32]. The instantaneous heart rate pattern was studied in 102 patients admitted to a neurosurgical intensive care unit. Short-term (STV) and long-term (LTV) heart rate variability were compared to the Glasgow coma scale as a method for patient assessment. LTV seems to be the most useful heart rate parameter in the clinical setting, and both STV and LTV performed better in the serial evaluation of patients [38]. In serial determinations, the rate of return of normal heart rate variability may reflect the subsequent state of neuronal function.

The significance of HRV analysis in psychiatric disorders arises from the fact that one can easily detect a sympathovagal imbalance (relative cholinergic and adrenergic modulation of HRV), if it exists in such pathologies. There are conflicting reports about HRV and major depression. It is proved that, in physically healthy depressed adults the HRV does not vary from healthy subjects [39].

5.2.5 HRV and Cardiac Arrhythmia

A complex system like cardiovascular system cannot be linear in nature and by considering it as a nonlinear system, can lead to better understanding of the system dynamics. Recent studies have also stressed the importance of nonlinear techniques to study HRV in issues related to both health and disease. The progress made in the field using measures of chaos has attracted the scientific community to apply these tools in studying physiological systems, and HRV is no exception. There have been several methods of estimating invariants from nonlinear dynamical systems being reported in the literature. Fell *et al* and Radhakrishna *et al* have tried the nonlinear analysis of ECG and HRV signals, respectively [40, 41]. Also, Paul *et al* showed that coordinated mechanical activity in the heart during ventricular fibrillation may be made visible in the surface ECG using wavelet transform [42]. Mohamed *et al* [43] have used nonlinear dynamical modeling in ECG arrhythmia detection and classification. Acharya *et al* have classified the HRV signals using non-linear techniques, and artificial intelligence into different groups [44–46]. Dingfie *et al* have classified cardiac arrhythmia into six classes using autoregressive modeling [47].

5.2.6 HRV in Diabetes

Effects of hypoglycemia on cardiac autonomic regulation may contribute to the occurrence of adverse cardiac events. Koivikko *et al* have concluded that hypoglycemia results in the reduction of cardiac vagal outflow in both diabetic and nondiabetic subjects [48]. Altered autonomic regulation may contribute to the occurrence of cardiac events during hypoglycemia. The spectral components of short-term HRV calculated by using the FFT and AR methods were not interchangeable and FFT analysis was preferred in diabetic patients [49]. In frequency domain, the analysis of sympathetic (LF) and parasympathetic (HF) component evidenced an association between the offspring of type 2 diabetic subjects and a sympathetic over activity [50]. A global reduction and alteration of circadian rhythm of autonomic activity are present in offspring of type 2 diabetic patients with and without insulin resistance. Diabetes patients had lower values for time-domain and frequency-domain parameters than controls [51]. Most heart rate variability parameters were lower in diabetes patients with chronic complications than in those without chronic complications. Type 2 diabetic patients with microalbuminuria have diminished heart rate variability in response to deep breathing, change of position and the Valsalva maneuver, but they preserve BP response to postural change [52]. Therefore, microalbuminuria seems to be associated with early diabetic autonomic neuropathy (DAN), but not with advanced DAN.

It was concluded that cardiac (parasympathetic) autonomic activity was diminished in diabetic patients before clinical symptoms of neuropathy become evident [53–55].

5.2.7 HRV and Respiration

HRV has been shown to increase with decreasing respiration frequency [56,57]. Even though respiration is known to greatly affect the heart rate variability, it is often not measured when assessing heart rate variability [25, 58–60]. The respiration has a variable phase relationship with the cardiac cycle. The different influences of the respiratory inspiration and expiration cycle phases on heart rate are usually not considered. Voluntary cardiovascular respiratory synchronization (VCRS) uses a signal to guide an individual to inspire and expire phase locked with a certain pattern of heart beats. Not only is the phase of inspiration recorded, this phase locking allows for an analysis of the influence of respiration on heart rate variability. An advantage of VCRS is the ability to know the respiration rate, when respiration is occurring, and the respiration phase in relation to the beat-by-beat heart rate and how it may influence it.

Baselli *et al* (1995) noted that respiration effects on cardiovascular variability are not exogenous and emphasize a model where there is 'a common central drive that modulates both breathing and cardiovascular control, such as during periodic breathing and synchronous slow Mayer waves [61]. Sleight *et al* showed that there is no major difference in autonomic control as suggested by comparisons of spontaneous free breathing and controlled breathing at the same rate and depth [62]. It was found that the timing of inspiration as well as the timing of expiration in relationship to subsequent heart beats affected heart rate variability [63, 64].

5.2.8 HRV and Renal Failure

In patients with renal failure, autonomic function tests have been done [65], followed by heart rate variability indices [66] and spectral analysis of heart rate [67]. Although autonomic function tests revealed predominant impairment of the parasympathetic nervous system [65], spectral analysis exhibited a strong reduction in the heart rate power spectrum at all frequency ranges, both sympathetically and parasympathetically [67]. The relationship between heart rate variability (HRV) parameters and electrolyte ion concentrations in both pre- and post-dialysis were studied [68]. 5-minute HRV of 20 chronic renal failure patients were analyzed. Results revealed that calcium is negatively correlated to the mean of RR intervals and normalized high-frequency (HF) power after hemodialysis. A model of baroreflex control of blood pressure (BP) was proposed in terms of a delay differential equation and was used to predict the adaptation of short-term cardiovascular control in chronic renal failure (CRF) patients [69]. They showed that in CRF patients, the mean power in the LF band was higher and lower in the HF bands than the corresponding values in the healthy subjects.

5.2.9 HRV and Gender, Age

It is proved that, the heart rate variability depends on the age and sex also. The heart rate variability was more in the physically active young and old women [70, 71]. It was proved by Emese *et al* that the alert new borns have lower heart rate variation in the boys than in the case of girls [72]. The Heart rate variation for healthy subjects from 20–70 yrs was studied by Hendrik *et al* and found that the HRV decreases with age and variation is more in the case of female than male [73].

Previous studies have assessed gender and age-related differences in time and frequency domain indices [74] and some nonlinear component of HRV. There also seemed to be a significant difference between day and night hours when studying HRV indices using spectral and time domain methods [74, 75].

The amount of heart rate variability is influenced by physiologic and maturational factors. Maturation of the sympathetic and vagal divisions of the autonomic nervous system results in an increase in heart rate variability with gestational age [76] and during early postnatal life [76]. Heart rate variability decreases with age [47]. This decline starts in childhood [77]. Infants have a high sympathetic activity that decreases quickly between ages 5 and 10 years [78]. The influence of provocation on heart rate variability (that is, standing and fixed breathing) is more pronounced at younger ages [77]. In adults, an attenuation of respiratory sinus arrhythmia with advancing age usually predominates [79, 80]. It was shown that compared to men, women are at lower risk of coronary heart disease [81].

5.2.10 HRV and Drugs

Heart rate variability can be significantly influenced by various groups of drugs. The influence of medication should be considered, while interpreting heart rate variability. On the other hand, heart rate variability can be used to quantify the effects of certain drugs on the autonomic nervous system.

The effects of beta-blockers and calcium channel blockers on the heart rate variability have been studied in postinfarction and hypertensive patients [82–84]. With spectral analysis it is possible to unravel the sympathetic and parasympathetic activities of these drugs and thus explain their protective effects in cardiac diseases. In normotensive adults, the beta-adrenergic blocker atenolol appears to augment vagally mediated fast fluctuations in heart rate [85]. Guzzetti and colleagues [83] studied the effect of atenolol in patients with essential hypertension. They found not only an increase in high-frequency fluctuations, but also a decrease in the sympathetically mediated low-frequency oscillations. This decrease in sympathetic activity was also noticed in postinfarction patients using metoprolol [82] and in patients with heart failure using acebutolol [84]. Thus beta-blockers are able to restore the sympathetic-parasympathetic balance in cardiovascular disease. Effect of Omacor on HRV parameters in patients with recent uncomplicated myocardial infarction was studied [86]. And the study, quantified the improvement in time domain HRV indices and can assess the safety of administering Omacor to optimally treated post-infarction patients. Eryonucu *et al* have investigated the effects of β_2-adrenergic agonist therapy on heart rate variability (HRV) in adult asthmatic patients by using frequency domain measures of HRV [87]. The LF and LF/HF ratio increased and total power (TP) decreased at 5, 10, 15 and 20 min after the salbutamol and the terbutaline inhalation, HF will not change significantly after the salbutamol and terbutaline inhalation.

5.2.11 HRV and Smoking

Studies have shown that smokers have increased sympathetic and reduced vagal activity as measured by HRV analysis. Smoking reduces the heart rate variability. One of the mechanisms by which smoking impairs the cardiovascular function is its effect on autonomic nervous system (ANS) control [88–90]. Altered cardiac autonomic function, assessed by decrements in HRV, is associated with acute exposure to environmental tobacco smoke (ETS) and may be part of the pathophysiologic mechanisms linking ETS exposure and increased cardiac vulnerability [91]. Philip *et al* have shown that cigarette exposed fetuses have lower HRV and disrupted temporal organization of autonomic regulation before effects of parturition, postnatal adaptation, and possible nicotine withdrawal contributed to differences in infant neurobehavioral function [92]. Also, it was proved that, the vagal modulation of the heart had blunted in heavy smokers, particularly during a parasympathetic maneuver. Blunted autonomic control of the heart may partly be associated with adverse event attributed to cigarette smoking [93].

5.2.12 HRV and Alcohol

HRV reduces with the acute ingestion of alcohol, suggesting sympathetic activation and/or parasympathetic withdrawal. Malpas *et al* have demonstrated vagal neuropathy in men with chronic alcohol dependence using 24 hour HRV analysis [94].

Ryan *et al* have previously reported a strong positive association between average day time and night time heart rate measured during 24 hour ambulatory blood pressure monitoring and usual alcohol intake [95]. ECG indices of vagal activity have been reported to have significantly lower indices of cardiac vagal nerve activity than normal volunteers, in acute alcoholic subjects [94, 96, 97].

5.2.13 HRV and Sleep

The results from Fumiharu *et al* suggest that mechanisms involving electroencephalographic desynchronization and/or conscious states of the brain are reflected in the fractal component of HRV [98]. Compared to stage 2 and stage 4 non-REM sleep, the total spectrum power was significantly higher in REM sleep and its value gradually increased in the course of each REM cycle [99]. The value of the VLF component (reflects slow regulatory mechanisms, e.g. the renin-angiotensin system, thermoregulation) was significantly higher in REM sleep than in stage 2 and stage 4 of non-REM sleep. The LF spectral component (linked to the sympathetic modulation) was significantly higher in REM sleep than in stage 2 and stage 4 non-REM sleep. Patients with sleep apnoea tend to have a spectral peak lying between 0.01 and 0.05 cycles/beat, with the width of the peak indicating variability in the recurrence rate of the apnoea. In most of the subjects, the frequency spectrum immediately below the apnoea peak was relatively flat. The first visual analysis of the single computed spectrum from each subject led to a correct classification score of 28/30 (93%) [100]. Gregory *et al* suggested that long-lasting alterations existed in autonomic function in snoring subjects [101].

5.2.14 HRV and Fatigue

Night shift work has often been associated with increasing degree and frequency of various psychologic complaints. Munakata *et al* have showed that the, psychologic disturbances after night work were associated with altered cardiovascular and endocrine responses in healthy nurses [102]. Some of the psychologic complaints may be attributable to lower waking blood pressure. Spectral analysis could be a means of demonstrating impairment of autonomic balance for the purpose of detecting a state of fatigue that could result in overtraining with a decrease in sympathetic vasomotor control (−18%) and a reduction in diastolic pressure (−3.2%) [103]. Smoking and overwork such as frequent business trips may amplify the autonomic dysfunction in relation to

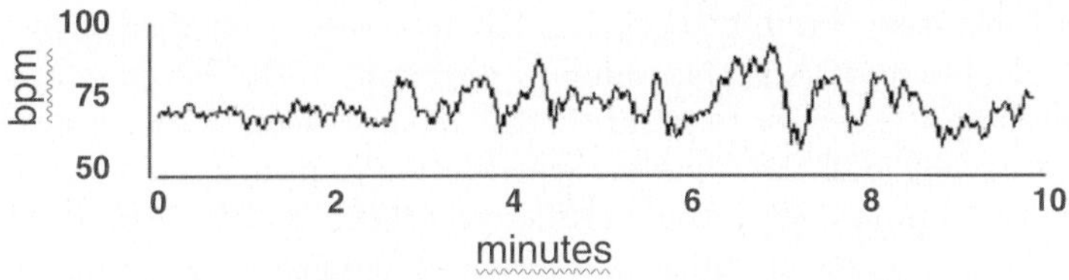

Fig. 5.1. Heart rate variation of a normal subject

vital exhaustion (VE) among workers with a pronounced feeling of VE [104]. It was found that a modulating effect of magnitopuncture on sympathetic and parasympathetic nerve activities in healthy subjects was associated with the acupuncture points. The findings represent physiological evidence that magnitopuncture may reduce mental fatigue in healthy drivers [105]. Jouanin *et al* have studied the effects of prolonged physical activities on resting heart rate variability (HRV) during a training session attended by 23 cadets of the French military academy [106]. These results as a whole suggest that parasympathetic nervous system activity increases with fatigue. It was shown that the modulating effect of acupuncture on heart rate variability not only depended on the points of stimulation such as acupuncture or non-acupuncture points but also on the functional state of the subject, namely whether the subjects are in a state of fatigue or not [107]. Alcohol dependence compromises vagal output measured before sleep onset, which correlates with loss of delta sleep and with morning reports of sleep impairments. Testing of interventions that target sympathovagal balance might identify new strategies for partial amelioration of the sleep disturbances and impairments in daytime functioning observed in persons with alcohol dependence [108].

Heart rate variability (HRV) is a measure of variations in the heart rate. Figure 5.1 shows the variation of the heart rate of a normal subject. It is usually calculated by analyzing the time series of beat-to-beat intervals from ECG or arterial pressure tracings. Various measures of heart rate variability have been proposed, which can roughly be subdivided into time domain, frequency domain and non-linear domain measures.

5.3 Methods

5.3.1 Time Domain Analysis

Two types of heart rate variability indices are distinguished in time domain analysis. Beat-to-beat or short-term variability (STV) indices represent fast changes in heart rate. Long-term variability (LTV) indices are slower fluctuations (fewer than 6 per minute). Both types of indices are calculated from the R-R intervals occurring in a chosen time window (usually between

0.5 and 5 minutes). From the original RR intervals, a number of parameters can be calculated: SDNN, the standard deviation of the NN intervals, SENN is the Standard Error, or Standard Error of the Mean, is an estimate of the standard deviation of the sampling distribution of means, based on the data, SDSD is the standard deviation of differences between adjacent NN intervals. RMSSD, the root mean square successive difference of intervals, pNN50%, the number of successive difference of intervals which differ by more than 50 msec expressed as a percentage of the total number of ECG cycles analyzed. The statistical parameters SDNN, SENN, SDSD, RMSSD, NN50(%), and pNN50% [4] can be used as time domain parameters.

5.3.2 Analysis by Geometrical Method

Geometrical methods present RR intervals in geometric patterns and various approaches have been used to derive measures of heart rate variability from them. The triangular index is a measure, where the length of RR intervals serves as the x-axis of the plot and the number of each RR interval length serves as the y-axis. The length of the base of the triangle is used and approximated by the main peak of the RR interval frequency distribution diagram. The triangular interpolation of NN interval histogram (TINN) is the baseline width of the distribution measured as a base of a triangle, approximating the NN interval distribution (the minimum of HRV). Triangular interpolation approximates the RR interval distribution by a linear function and the baseline width of this approximation triangle is used as a measure of the heart rate variability index [109, 110]. This triangular index had a high correlation with the standard deviation of all RR intervals. But it is highly insensitive to artifacts and ectopic beats, because they are left outside the triangle. This reduces the need for preprocessing of the recorded data [109]. The major advantage of geometric methods lies in their relative insensitivity to the analytical quality of the series of NN intervals.

5.3.3 Poincare Geometry

The Poincare plot (or Return Map), a technique taken from nonlinear dynamics, portrays the nature of R-R interval fluctuations. It is a graph of each R-R interval plotted against the next interval. Poincare plot analysis is an emerging quantitative-visual technique whereby the shape of the plot is categorized into functional classes that indicate the degree of the heart failure in a subject [111]. The plot provides summary information as well as detailed beat-to-beat information on the behavior of the heart [112].

The geometry of the Poincare plot is essential. A common way to describe the geometry is to fit an ellipse to the graph [113]. The ellipse is fitted onto the so called line-of-identity at 45^0 to the normal axis. The standard deviation of the points perpendicular to the line-of-identity denoted by SD1 describes short-term variability which is mainly caused by respiratory sinus arrhythmia

(RSA). The standard deviation along the line-of-identity denoted by SD2 describes long-term variability.

Statistically, the plot displays the correlation between consecutive intervals in a graphical manner. Nonlinear dynamics considers the Poincare plot as the two dimensional (2-D) reconstructed R-R interval phase-space, which is a projection of the reconstructed attractor describing the dynamics of the cardiac system [114]. The R-R interval Poincare plot typically appears as an elongated cloud of points oriented along the line-of-identity. The dispersion of points perpendicular to the line-of-identity reflects the level of short term variability. The dispersion of points along the line-of-identity is thought to indicate the level of long-term variability.

To characterize the shape of the plot mathematically, most researchers have adopted the technique of fitting an ellipse to the plot. A set of axis oriented with the line-of-identity is defined [115]. The axis of the Poincare plot is related to the new set of axis by rotation of $\theta = \pi/4\,\text{rad}$.

$$\begin{bmatrix} x_1 \\ x_2 \end{bmatrix} = \begin{bmatrix} \cos\theta & -\sin\theta \\ \sin\theta & \cos\theta \end{bmatrix} \begin{bmatrix} RR_n \\ RR_{n+1} \end{bmatrix} \tag{5.1}$$

In the reference system of the new axis, the dispersion of the points around the x_1 axis is measured by the standard deviation denoted by $SD1$. The quantity measures the width of the Poincare cloud and, length of the cloud and therefore indicates the level of short term HRV [116], Hausdorff *et al.*, 1996. The length of the cloud along the line-of-identity measures the long-term HRV and is measured by $SD2$ which is the standard deviation around the x_2 axis [Kamen *et al.*, 1996]. These measures are related to the standard HRV measures in the following manner:

$$SD1^2 = Var\,(x_1) = Var\left(\frac{1}{\sqrt{2}}RR_n - \frac{1}{\sqrt{2}}RR_{n+1}\right) \tag{5.2}$$

$$SD1^2 = \frac{1}{2}Var\,(RR_n - RR_{n+1}) = \frac{1}{2}SDSD^2 \tag{5.3}$$

Thus, the $SD1$ measures of Poincare width is equivalent to the standard deviation of the successive intervals, except that is scaled by $\frac{1}{\sqrt{2}}$. This means that one can relate $SD1$ and $SD2$ to the autocovariance function

$$SD1^2 = \phi_{RR}\,(0) - \phi_{RR}\,(1) \tag{5.4}$$

$$SD2^2 = \phi_{RR}\,(0) + \phi_{RR}\,(1) \tag{5.5}$$

By adding the above two equations (5.4) and (5.5), we get

$$SD1^2 + SD2^2 = 2SDRR^2 \tag{5.6}$$

Finally,

$$SD2^2 = 2SDRR^2 - \frac{1}{\sqrt{2}}SDRR^2 \tag{5.7}$$

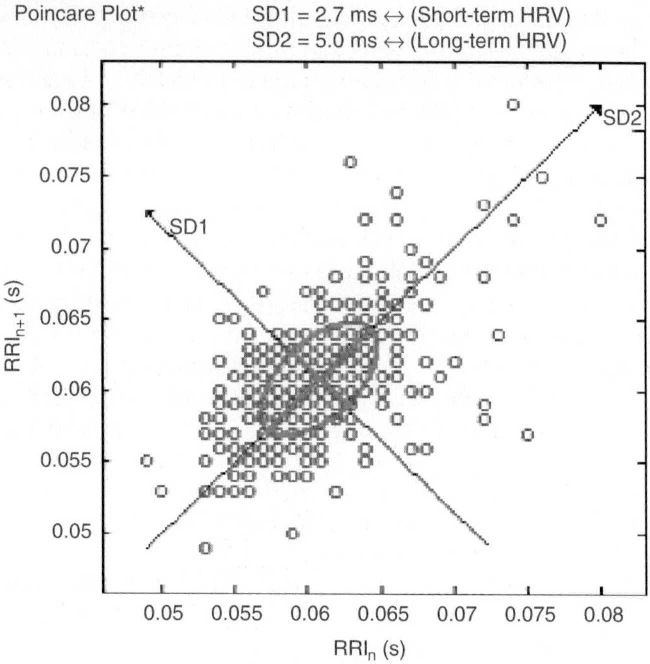

Fig. 5.2. Poincare plot of a normal subject

Where $SDRR = \sqrt{E\left[RR_n^2\right] - \overline{RR}}$ is the square root of the variance of the RR intervals. The standard deviation of the successive differences of the RR intervals, denoted by $SDSD$.

$$SDSD = \sqrt{E\left[\left(RR_n - RR_{n+1}\right)^2\right]}. \tag{5.8}$$

The plots of the HRV changes can be envisioned as values distributed over an area defined by four quadrants. If the changes from one heart rate value to the next occur randomly and independently, then all the four quadrants will have equal number of points. Also, if the heart rate tends to increase quickly and decrease slowly, then there would be more number of points in the 1st and 3rd quadrants. The opposite would be true, if the heart rate is decreased more quickly than it increased. Figure 5.2 shows the Poincare plot of normal heart rate.

5.4 Frequency Domain Analysis

Frequency-domain measures pertain to HR variability at certain frequency ranges associated with specific physiological processes. Before frequency-domain analysis is performed, all abnormal heartbeats and artifacts must be detected and removed, then cardiotachogram (sequence of RR intervals) must be resampled to make it as if it is a regularly sampled signal. A standard spectral analysis routine is applied to such modified recording and the following parameters evaluated on 5-min time interval: Total Power (TP), High Frequency (HF), Low Frequency (LF) and Very Low Frequency (VLF). When long-term data is evaluated an additional frequency band is derived – Ultra Low Frequency.

The HF power spectrum is evaluated in the range from 0.15 to 0.4 Hz. This band reflects parasympathetic (vagal) tone and fluctuations caused by spontaneous respiration known as respiratory sinus arrhythmia.

The LF power spectrum is evaluated in the range from 0.04 to 0.15 Hz. This band can reflect both sympathetic and parasympathetic tone.

The VLF power spectrum is evaluated in the range from 0.0033 to 0.04 Hz. The physiological meaning of this band is most disputable. With longer recordings it is considered representing sympathetic tone as well as slower humoral and thermoregulatory effects. There are some findings that in shorter recordings VLF has fair representation of various negative emotions, worries, rumination, etc.

The TP is a net effect of all possible physiological mechanisms contributing in HR variability that can be detected in 5-min recordings, however sympathetic tone is considered as a primary contributor.

The LF/HF ratio is used to indicate balance between sympathetic and parasympathetic tone. A decrease in this score might indicate either increase in parasympathetic or decrease in sympathetic tone. It must be considered together with absolute values of both LF and HF to determine what factor contributes in autonomic imbalance.

The frequency domain analysis is traditionally performed by means of Fast Fourier Transformation (FFT). This method is simple in calculation but for fair representation of all frequency-domain HRV scores at least 5-min data should be collected. FFT assumes that time series represents a steady-state process. Because of that all data recordings should be conducted at highly stable standardized conditions, when no other factors other than current autonomic tone contributes in HRV. One of the most serious disadvantages is, its insensitivity to rapid transitory processes, which often possess very valuable information about how physiology or certain pathological processes behave dynamically.

Some most recent studies implemented an alternative way to estimate power spectrum of HRV. It is based on autoregression methods. One of its major advantages is that it doesn't require to have analyzed data series to be in steady state. Thus any HRV data can be analyzed and fair HRV information still derived. Such analysis can be also performed at relatively shorter time intervals (less than 5 minutes) without missing meaningful HRV information. Finally this method is sensitive to rapid changes in HR properly showing tiny changes in autonomic balance. The drawback of this approach is a necessity to perform massive calculations to find best order of autoregression model.

In AR method, the estimation of AR parameters can be done easily by solving linear equations. In AR method, data can be modeled as output of a causal, all pole, discrete filter whose input is white noise. AR method of order p is expressed by the following equation:

$$x(n) = -\sum_{k=1}^{p} a(k)x(n-k) + w(n) \tag{5.9}$$

where $a(k)$ are AR coefficients and $w(n)$ is white noise of variance equal to σ^2. AR (p) model can be characterized by AR parameters $\{a[1], a[2], \ldots, a[p], \sigma^2\}$.

The important aspect of the use of AR method, is the selection of the order p. Much work has been done by various researchers on this problem and many experimental results have been given in literature such as the papers presented by Akaike [117–119]. The order of the AR model $p = 16$ can be taken [119].

An autoregressive process, x(n), may be represented as the output of an all-pole filter that is driven by unit variance white noise. The Burg method is used to get the AR model parameter. The power spectrum of a pth order autoregressive process is

$$P_{xx}^{BU}(f) = \frac{\widehat{E}_p}{\left|1 + \sum_{k=1}^{p} \widehat{a}_p(k) e^{-j2\pi fk}\right|^2} \tag{5.10}$$

where $\widehat{E}$ is total least square error. The Burg method results in high resolution and yields a stable AR model.

Spectral analysis of heart rate variability can be a powerful tool to assess autonomic nervous system function. It is not only useful when studying the pathophysiologic processes in certain diseases but also may be used in daily clinical practice. Figure 5.3 indicates the AR spectrum of a normal subject.

5.4.1 Limitations of Fourier Analysis

Conventional signal methods of analysis based on Fourier transform technique are not very suitable for analyzing non-stationary signals. The Fourier transform technique resolves the time domain signal into complex exponential functions, along with information of their phase shift measured with respect to a specific reference instant. Here the frequency components extend from $-\infty$ to $+\infty$ in the time scale. That is, even finite length signals are expressed as the sum of frequency components of infinite duration. Besides, the phase angle being a modular measure, it fails to provide the exact location of an 'event' along the time scale. This is a major limitation of the Fourier transform approach.

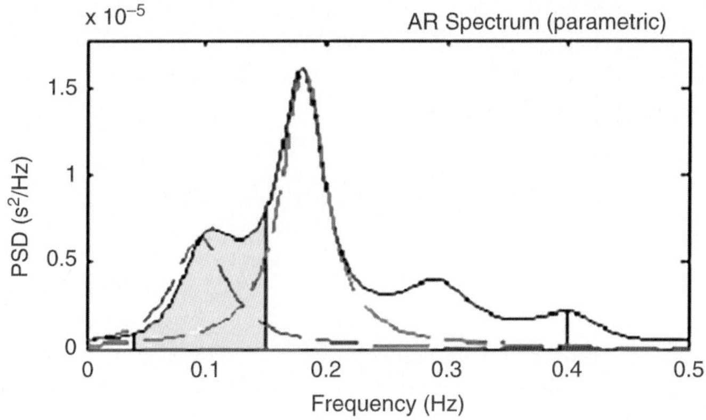

Fig. 5.3. AR spectrum of normal subject

5.4.2 Higher Order Spectra (HOS)

The HRV signal can be analyzed using different higher order spectra (also known as polyspectra) that are spectral representations of higher order moments or cumulants of a signal. The Bispectrum is the Fourier transform of the third order correlation of the signal and is given by

$$\mathrm{B(f_1, f_2) = E[X(f_1)X(f_2)X^*(f_1 + f_2)]} \tag{5.11}$$

where X(f) is the Fourier transform of the signal x(nT) and E[.] stands for the expectation operation. In practice, the expectation operation is replaced by an estimate that is an average over an ensemble of realizations of a random signal. For deterministic signals, the relationship holds without an expectation operation with the third order correlation being a time-average. For deterministic sampled signals, X(f) is the discrete-time Fourier transform and in practice is computed as the discrete Fourier transform (DFT) at frequency samples using the FFT algorithm. The frequency f may be normalized by the Nyquist frequency to be between 0 and 1.

The bispectrum may be normalized (by power spectra at component frequencies) such that it has a value between 0 and 1, and indicates the degree of phase coupling between frequency components [120]. The normalized bispectrum or bicoherence is given by

$$B_{co}(f_1, f_2) = \frac{E[(X(f_1)X(f_2)X^*(f_1 + f_2))]}{\sqrt{P(f_1)P(f_2)P(f_1 + f_2)}} \tag{5.12}$$

where $P(f)$ is the power spectrum.

Higher Order Spectral Features

One set of features are based on the phases of the integrated bispectrum derived by Chandran *et al* [121] and is described briefly below:

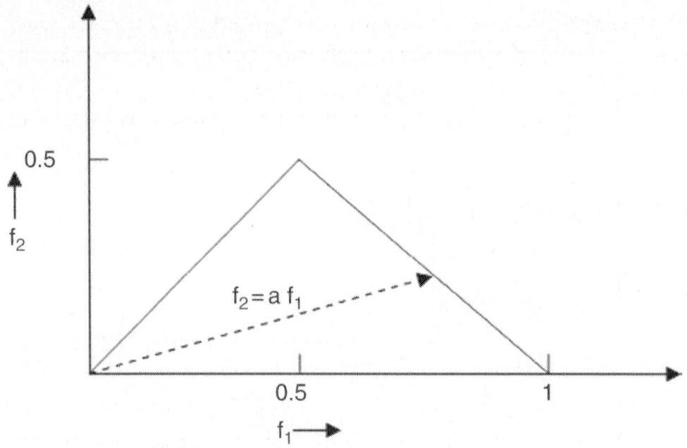

Fig. 5.4. Non-redundant region of computation of the bispectrum for real signals. Features are calculated by integrating the bispectrum along the dashed line with slope = a. Frequencies are shown normalized by the Nyquist frequency

Assuming that there is no bispectral aliasing, the bispectrum of a real signal is uniquely defined with the triangle $0 \leq f_2 \leq f_1 \leq f_1 + f_2 \leq 1$. Parameters are obtained by integrating along the straight lines passing through the origin in bifrequency space. The region of computation and the line of integration are depicted in Fig. 5.4. The bispectral invariant, $P(a)$, is the phase of the integrated bispectrum along the radial line with the slope equal to a. This is defined by

$$P(a) = \arctan\left(\frac{I_i(a)}{I_r(a)}\right) \tag{5.13}$$

where

$$\begin{aligned} I(a) &= I_r(a) + jI_i(a) \\ &= \int_{f_1=0^+}^{\frac{1}{1+a}} B(f_1, af_1)df_1 \end{aligned} \tag{5.14}$$

for $0 < a \leq 1$, and $j = \sqrt{-1}$. The variables I_r and I_i refer to the real and imaginary part of the integrated bispectrum, respectively.

These bispectral invariants contain information about the shape of the waveform within the window and are invariant to shift and amplification and robust to time-scale changes. They are particularly sensitive to changes in the left-right asymmetry of the waveform. For windowed segments of a white Gaussian random process, these features will tend to be distributed symmetrically and uniformly about zero in the interval $[-\pi, +\pi]$. If the process is chaotic and exhibits a colored spectrum with third order time-correlations or phase coupling between Fourier components, the mean value and the distribution of the invariant feature may be used to identify the process.

Another set of features are based on the work of Ng *et al* [122]. These features are the mean magnitude and the phase entropy. However, unlike their work, we calculated these features within the region defined in Fig. 5.4. The formulae of these features are

$$\text{Mean of Magnitude} : \mathrm{M_{ave}} = \frac{1}{L}\sum_{\Omega} |b(f1, f2)| \tag{5.15}$$

$$\text{Phase Entropy} : \mathrm{P_e} = \sum\nolimits_n p(\psi_n)\log p(\psi_n) \tag{5.16}$$

where

$$\Omega = \{(\mathrm{f_1}, \mathrm{f_2})|\mathrm{f_1}, \mathrm{f_2} \text{ in the region in Fig. 5.4}\}$$
$$\psi_n = \{\phi| -\pi + 2\pi n/N \leq \phi < -\pi + 2\pi(n+1)/N, \quad n = 0, 1, \ldots, N-1$$

$$p(\psi_n) = \frac{1}{L}\sum\nolimits_{\Omega} 1(\phi(b(f1, f2)) \in \psi_n), 1(.) = \mathit{indicator\ function}$$

L is the no. of point within the region in Fig. 5.4,
ϕ refers to the phase angle of the bispectrum.

The formulae for these bispectral entropies are given as:

$$\text{Normalized Bispectral Entropy (BE1)} : \mathrm{P_1} = -\sum\nolimits_n p_n \log p_n \tag{5.17}$$

where $p_n = \frac{|B(f_1,f_2)|}{\sum_{\Omega}|B(f_1,f_2)|}$, $\Omega =$ the region as in Fig. 5.4.

$$\text{Normalized Bispectral Squared Entropy (BE2)} : \mathrm{P_2} = -\sum\nolimits_n p_n \log p_n \tag{5.18}$$

where $p_n = \frac{|B(f_1,f_2)|^2}{\sum_{\Omega}|B(f_1,f_2)|^2}$, $\Omega =$ the region as in Fig. 5.4.

Bispectrum and bicoherence plots, mean amplitude of bispectrum, bispectral entropies, invariant features can be used to find different cardiac abnormalities.

5.4.3 Short Time Fourier Transform (STFT)

In order to locate a particular event along the time scale, a finite length window is used at that point. This window may, then be moved along the signal in time producing a succession of estimates of the spectral components of the signal. This works well for signals composed of stationary components and for slowly varying signals. However, for the signal containing, both slowly varying components and rapidly changing transient events, STFT fails. If we use a window of infinite length, we get the FT, which gives perfect frequency resolution, but no time information. Furthermore, in order to obtain the stationarity, we have to have a short enough window, in which the signal is stationary. The narrower we make the window, the better the time resolution, and better the assumption of stationarity, but poorer the frequency resolution.

Wavelet transform [123] overcomes this problem. It uses small windows at the high frequency and longer windows at low frequency. The wavelet can be classified into two types: (1) Discrete Wavelet transform (DWT), (2) Continuous Wavelet transform (CWT).

5.4.4 Continuous Time Wavelet Transform (CWT) Analysis

A *'wavelet'* implies a small wave of finite duration and finite energy, which is correlated with the signal to obtain the wavelet coefficients [124]. The reference wavelet is known as the *mother wavelet*, and the coefficients are evaluated for the entire range of dilation and translation factors [123]. Initially the *mother wavelet* is shifted (translated) continually along the time scale for evaluating the set of coefficients at all instants of time. In the next phase, the wavelet is dilated for a different width – also normalized to contain the same amount of energy as the mother wavelet – and the process is repeated for the entire signal. The wavelet coefficients are real numbers usually shown by the intensity of a chosen color, against a two dimensional plane with y-axis representing the dilation (scaling factor) of the wavelet, and the x-axis, its translation (shift along the time axis). Thus the wavelet transform plot (*scalogram*) can be seen as a color pattern against a two dimensional plane. In the CWT the wavelet coefficients are evaluated for infinitesimally small shifts of translation as well as scale factors. That is, the color intensity distribution in the *scalogram* pattern contains information about the location of the 'event' occurring in the time domain [125–127]. Thus the color patterns in the scalogram can be useful in highlighting the abnormalities and is specific to different types of diseases.

For a given wavelet $\psi_{a,b}(t)$, the coefficients are evaluated using Eq. given below.

$$W(a,b) \equiv \int_{-\infty}^{\infty} f(t)\frac{1}{\sqrt{|a|}}\psi^*\left(\frac{t-b}{a}\right)dt \qquad (5.19)$$

where $\psi^*\left(\frac{t-b}{a}\right) = \psi^*_{a,b}(t)$; a $\rightarrow$ scale factor; b $\rightarrow$ translation factor

The *scalogram* patterns thus obtained also depend on the wavelet chosen for analysis. Bio-signals usually exhibit self similarity patterns in their distribution, and a wavelet which is akin to its fractal shape would yield the best results in terms of clarity and distinction of patterns.

5.5 Nonlinear Methods of Analysis

Recent developments in the theory of nonlinear dynamics have paved the way for analyzing signals generated from nonlinear living systems [128, 129]. It is now generally recognized that these nonlinear techniques are able to describe the processes generated by biological systems in a more effective way. The technique has been extended here to study various cardiac arrhythmias. The parameters like correlation dimension (CD), largest Lyapunov exponent (LLE), SD1/SD2 of Poincare plot, approximate entropy ($ApEn$), Hurst exponent, fractal dimension, α slope of detrended fluctuation analysis and recurrence plots.

5.5.1 Capacity Dimension

The simplest method for measuring the dimension of a data set is to measure its capacity dimension or box counting dimension [128]. In this method, the set is covered with smaller elements of size e. The number of elements required to cover the segment is inversely proportionate to the size of the element. Thus for one-dimensional objects, $N(e) = \frac{k}{e}$, where e is the size of the square, $N(e)$ is the number of squares of that size required to cover the set, and k is a constant. For an arbitrary set it is given by, $N(e) = \frac{k}{e^D}$. where D is the dimension of the set. We can solve the formula for D, by taking the limit as $e \to 0$. This is the capacity method of estimating D. Then D will be given by

$$D_{cap} = \lim_{e \to 0} \frac{\log (N(e))}{\log (1/e)} \tag{5.20}$$

5.5.2 Correlation Dimension

The phase space plot is a plot, in which, X-axis represents the heart-rate X[n] and the Y-axis represents the heart-rate after a delay X[n+delay]. The choice of an appropriate delay is calculated using the minimal mutual information technique [130, 131]. Figure 5.6 shows the typical phase space plot of a normal subject. The phase space plot shows unique spread for various cardiac disorders [46].

Correlation Dimension is one of the most widely used measures of Fractal Dimension. Here we adapt the algorithm proposed by Grassberger and Procaccia [132]. The idea is to construct a function $C(r)$ that is the probability that two arbitrary points on the orbit are closer together than r. This is done by calculating the separation between every pair of N data points and sorting them into bins of width dr proportionate to r. A correlation dimension can be calculated using the distances between each pair of points in the set of N number of points, $s(i,j) = |X_i - X_j|$

A correlation function, $C(r)$, is then calculated using, $C(r) = \frac{1}{N^2} \times$ (Number of pairs of (i,j) with $s(i,j)\langle r)$. $C(r)$ has been found to follow a power law similar to the one seen in the capacity dimension: $C(r) = kr^D$. Therefore, we can find D_{corr} with estimation techniques derived from the formula:

$$D_{corr} = \lim_{r \to 0} \frac{\log (C(r))}{\log (r)} \tag{5.21}$$

5.5.3 Lyapunov Exponent

To discriminate between chaotic dynamics and periodic signals Lyapunov exponent (λ) are often used. It is a measure of the rate at which the trajectories separate one from other [129,133]. The trajectories of chaotic signals in phase space follow typical patterns. Closely spaced trajectories converge and diverge exponentially, relative to each other. For dynamical systems, sensitivity to initial conditions is quantified by the Lyapunov exponent (λ). They characterize the average rate of divergence of these neighboring trajectories. A negative exponent implies that the orbits approach a common fixed point. A zero exponent means the orbits maintain their relative positions; they are on a stable attractor. Finally, a positive exponent implies the orbits are on a chaotic attractor.

For two points in a space $\mathrm{X_0}$ and $\mathrm{X_0} + \Delta\mathrm{x_0}$, that are function of time and each of which will generate an orbit in that space using some equations or system of equations, then the separation between the two orbits $\Delta\mathrm{x}$ will also be a function of time. This separation is also a function of the location of the initial value and has the form $\Delta\mathrm{x}(\mathrm{X_0}, \mathrm{t})$. For chaotic data set, the function $\Delta\mathrm{x}(\mathrm{X_0}, \mathrm{t})$ will behave erratically. The mean exponential rate of divergence of two initially close orbits is characterized by:

$$\lambda = \lim_{t \to \alpha} \frac{1}{t} \ln \frac{|\Delta x(X_0, t)|}{|\Delta X_0|} \tag{5.22}$$

The Lyapunov exponent "λ" is useful for distinguishing various orbits.

Largest Lyapunov Exponent (LLE) quantifies sensitivity of the system to initial conditions and gives a measure of predictability. Presence of positive Lyapunov exponent indicates chaos. Even though a m dimensional system has m Lyapunov exponents, in most applications it is sufficient to compute only the largest Lyapunov exponent (LLE). For heart rate variability analysis, the method proposed by Rosenstien *et al* [10], can be used, which is robust with data length. This method looks for nearest neighbor of each point in phase-space and tracks their separation over certain time evolution. The LLE is estimated using a least squares fit to "average" line and is defined by:

$$y(n) = \frac{1}{\Delta t} \langle \ln (d_i (n)) \rangle \tag{5.23}$$

where $d_i(n)$ is the distance between ith phase-space point and its nearest neighbor at nth time step, and $\langle . \rangle$ denotes the average overall phase space points. This last averaging step is the main feature that allows an accurate evaluation of LLE even when we have a short and noisy data.

5.5.4 Hurst Exponent

The Hurst exponent is a measure that has been widely used to evaluate the self-similarity and correlation properties of fractional Brownian noise, the time series produced by a fractional (fractal) Gaussian process. Hurst exponent is used to evaluate the presence or absence of long-range dependence and its degree in a time series. However, local trends (nonstationarities) are often present in physiological data and may compromise the ability of some methods to measure self-similarity. Hurst Exponent is the measure of the smoothness of a fractal time series based on the asymptotic behavior of the rescaled range of the process. In time series analysis of EEG, Hurst Exponent H is used by Dangel *et al* [134] to characterize the non-stationary behavior of the sleep EEG episodes. The Hurst exponent H is defined as,

$$H = \log(R/S)/\log(T), \tag{5.24}$$

where T is the duration of the sample of data and R/S the corresponding value of rescaled range. The above expression is obtained from the Hurst's generalized equation of time series that is also valid for Brownian motion. If $H = 0.5$, the behavior of the time-series is similar to a random walk. If $H < 0.5$, the time-series cover less "distance" than a random walk. But if $H > 0.5$, the time-series covers more "distance" than a random walk. H is related to the dimension D_2 given by,

$$H = E + 1 - D_2 \tag{5.25}$$

Here, E is the Euclidean dimension.

5.5.5 Detrended Fluctuation Analysis

The concept of a fractal is most associated with geometrical objects satisfying two criteria: self-similarity and fractal dimensionality. Self-similarity means that an object is composed of sub-units and sub-sub-units on multiple levels that statistically resemble the structure of the whole object. The second criteria for fractal object is that it has a fractional dimension, also called fractal, that can be defined to be any curve or surface that is independent of scale. This concept of fractal structure can be extended to the analysis of heart rate signals.

The Detrended Fluctuation Analysis (DFA) is used to quantify the fractal scaling properties of short interval RR signals. This technique is a modification of root-mean-square analysis of random walks applied to nonstationary signals [135]. The root-mean-square fluctuation of an integrated and detrended time series is measured at different observation windows and plotted against the size of the observation window on a log-log scale.

First, the R-R time series (of total length N) is integrated using the equation:

$$y(k) = \sum_{i=1}^{k} [RR(i) - RR_{avg}] \tag{5.26}$$

where $y(k)$ is the k th value of the integrated series, $RR(i)$ is the ith inter beat interval, and the RR_{avg} is the average inter beat interval over the entire series.

Then, the integrated time series is divided into windows of equal length, n. In each window of length n, a least-squares line is fitted to the RR interval data (representing the trend in that window). The y coordinate of the straight line segments are denoted by $y_n(k)$. Next, we detrend the integrated time series, $y_n(k)$, in each window. The root-mean-square fluctuation of this integrated and detrended series is calculated using the equation:

$$F(n) = \sqrt{1/N \sum_{k=1}^{N} [y(k) - y_n(k)]^2} \tag{5.27}$$

This computation is repeated over all time scales (window sizes) to obtain the relationship between $F(n)$ and the window size n (i.e., the number of beats in a window that is the size of the window of observation). In this study, the box size is ranged from 4 to $\sim$ 300 beats. A box size larger than 300 beats would give a less accurate fluctuation value because of finite length effects of data [136].

Typically, $F(n)$ will increase with window size. The fluctuation in small windows are characterized by a scaling exponent (self-similarity factor), $\propto$, representing the slope of the line relating $\log F(n)$ to $\log(n)$. Figure 5.5 shows the DFA plot for the normal heart rate signal. In this method, a fractal like signal results in a scaling exponent value of 1 ($\propto = 1$). This value for white Gaussian noise (totally random signal) will be 0.5, and a Brownian noise signal with spectrum rapidly decreasing power in the higher frequencies results in an exponent value of 1.5 [135]. The $\propto$ can be viewed as an indicator of the "roughness" of the original time series: the larger the value of the $\propto$ the smoother the time series. A good linear fit of the $\log F(n)$ to $\log(n)$ plot (DFA plot) indicates $F(n)$ is proportional to n^α, where α is the single exponent

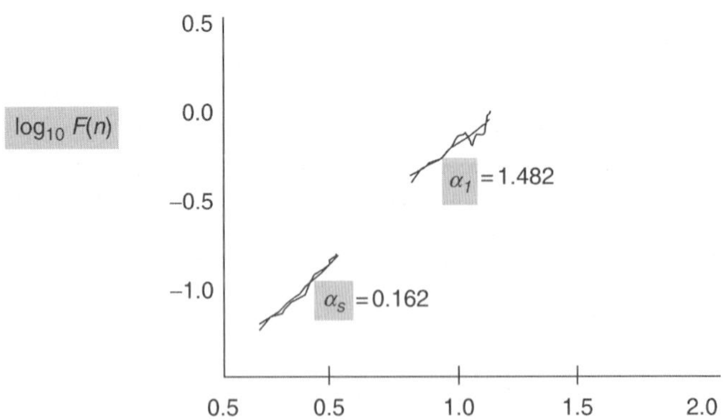

Fig. 5.5. $F(n)$ plotted against several box sizes, n, on a log-log scale

describing the correlation properties of the entire range of heart rate data. However in some cases, we found that the DFA plot was not strictly linear but rather consisted of two distinct regions of different slopes separated at a break point n_{bp} [133]. This observation suggests there is a short range scaling exponent, α_s, over periods of 3 to n_{bp} beats, and a long-range exponent, α_l, over long periods [133].

5.5.6 Entropies

Entropy is a thermodynamic quantity describing the amount of disorder in the system. From an information theory perspective, the above concept of entropy is generalized as the amount of information stored in a more general probability distribution. First Shannon applied the concept of information or logical entropy to the science of information theory and data communications. Recently a number of different entropy estimators [137] have been applied to quantify the complexity of the signal. Entropy estimators are broadly classified into two categories-spectral entropies and embedding entropies. The spectral entropies use the amplitude components of the power spectrum of the signal as the probabilities in entropy calculations. In this topic the spectral entropies – Shannon entropy, Renyi's entropy are discussed. The embedding entropies use the time series directly to estimate the entropy. Kolmogorov-Sinai entropy, approximate entropy and sample entropy are the embedding entropies discussed here.

Spectral Entropy (SEN)

Spectral entropy (SEN) [138, 139] is the normalized form of Shannon's entropy. It quantifies the spectral complexity of the time series. A variety of spectral transformations exist. Of these the Fourier transformation (FT) is most probably the well-known transformation method from which the power spectral density (PSD) can be obtained. The PSD is a function that represents the distribution of power as function of frequency. For each frequency, the power level P_f obtained from Fourier Transform is summed and the total power, $\sum P_f$ is calculated. Normalization of PSD with respect to the total spectral power will yield a probability density function. Each frequency's power level is divided by the total power $[p_f = \frac{P_f}{P_T}; P_T = Total\ Power]$, yielding in the end the total; $\sum p_f = 1$. Entropy is computed by multiplying the power in each frequency by the logarithm of the same power, $p_f \times \log(p_f)$ and multiplying the result by -1. Total entropy is the sum of entropy computed over entire frequency range. Thus the spectral entropy is given by

$$SEN = \sum_f p_f \log\left(\frac{1}{p_f}\right) \tag{5.28}$$

Heuristically the entropy has been interpreted as a measure of uncertainty about the event at f. Thus entropy H may be used as a measure of system complexity. It measures the spread of data. Data with broad, flat probability distribution have high entropy. Data with narrow, peaked distribution will have low entropy. SEN is also a special case of a series of entropies termed Renyi entropies $REN(\alpha)$.

Renyi's Entropy

Renyi's entropy [137, 140] is given by

$$REN(\alpha) = -\frac{\alpha}{1-\alpha} \sum \log p_k^{\alpha} \quad (\alpha \neq 1) \tag{5.29}$$

Embedding Entropies

The embedding entropies use the time series directly to estimate the entropy. Kolmogorov-Sinai entropy, approximate entropy and sample entropy are the embedding entropies discussed below.

Kolmogorov Sinai Entropy (K)

Entropy is determined from the embedded time series data by finding points on the trajectory that are close together in phase space but which occurred at different times (i.e., are not time correlated). These two points are then followed into the future to observe how rapidly they move apart from one another. The time it takes for point pairs to move apart is related to the so-called Kolmogorov entropy [141], K, by

$$\langle t_{div} \rangle = 2^{-Kt} \tag{5.30}$$

where $\langle t_{div} \rangle$ is the average time for the pair to diverge apart and K is expressed in bits per second.

The calculation of K from a time series typically starts from reconstructing the system's trajectory in an embedding space. According to Grassberger and Procaccia [132], K can be determined from the correlation function. $C_m(r, N_m)$ as

$$K = \lim_{r \to 0} \lim_{m \to \infty} \frac{1}{\tau} \frac{C_m(r, N_m)}{C_{m+1}(r, N_{m+1})} \tag{5.31}$$

The correlation function $C_m(r, N_m)$ indicates the probability that two arbitrary points on the orbit are closer together than r. This is done by calculating the separation between every pair of N data points and sorting them into bins of width dr proportionate to r.

A correlation function, $C(r)$, for the embedding dimension m is then calculated using,

$$C_m(r, N_m) = \frac{2}{N_m(N_m - 1)} \sum_{i=1}^{N_m} \sum_{\substack{j=1 \\ j \neq i}}^{N_m} \Theta\left(r - \|\mathbf{x}_i - \mathbf{x}_j\|\right) \tag{5.32}$$

where $\mathbf{x}_i$ and $\mathbf{x}_j$ are the points of the trajectory in the phase space, r is the radial distance around each reference point $\mathbf{x}_i$, $\Theta \rightarrow$ is the Heaviside function and $N_m = N-(m-1)\tau$ is the number of points in the multidimensional state space.

τ is called the delay time and m is the embedding dimension.

Entropy reflects how well one can predict the behavior of each respective part of the trajectory from the other. Higher entropy indicates less predictability and a closer approach to stochasticity.

Approximate Entropy (ApEn)

K-S entropy measure diverges to a value of infinity when the signal is contaminated by the slightest noise. Pincus [11] proposed Approximate Entropy (*ApEn*) as a solution to these problems and successfully applied it to relatively short and noisy data. *ApEn* is scale invariant and model independent and discriminates time series for which clear future recognition is difficult. *ApEn* detects the changes in underlying episodic behavior not reflected in peak occurrences or amplitudes [142]. *ApEn* assigns a nonnegative number to a time series, with larger values corresponding to more complexity or irregularity in the data [11]. For N data points $x(1), x(2), \ldots, x(N)$, with an embedding space of $\Re^m$, the *ApEn* measure is given by

$$\begin{aligned} ApEn(m,r,N) = &\frac{1}{N-m+1}\sum\nolimits_{i=1}^{N-m+1} \log C_i^m(r) \\ &-\frac{1}{N-m}\sum\nolimits_{i=1}^{N-m} \log C_i^{m+1}(r) \end{aligned} \tag{5.33}$$

where $C_i^m(r) = \dfrac{1}{N-m+1}\sum\limits_{j=1}^{N-m+1} \Theta\left(r - \|\mathbf{x}_i - \mathbf{x}_j\|\right)$ is the correlation integral. The values of m and r may be chosen based on the results of previous studies by Pincus indicating good statistical validity for *ApEn* [143].

Sample Entropy (SampEn)

SampEn agreed with theory much more closely than *ApEn* over a broad range of conditions. The improved accuracy of SampEn statistics make them useful in the study of experimental clinical cardiovascular and other biological time series [144]. Abnormal heart rate characteristics of reduced variability and transient decelerations are present early in the course of neonatal sepsis. To investigate the dynamics, SampEn, which is of less biased measure than the popular *ApEn* was calculated [145]. They proposed more information on the selection of parameters of the SampEn.

The basic idea of the SampEn is very similar to the *ApEn*, but there is a small computational difference [145]. For the calculation of SampEn we first take the original time series $x[i], i = 1, \ldots, N$, and construct vector sequences of size m, $\mathbf{u}[1]$ through $\mathbf{u}[N-m+1]$, defined by $\mathbf{u}[i] = \{x[i], \ldots, x[i+m-1]\}$.

The vectors length m, is known as the embedded dimension. The constructed vectors represent m consecutive x values commencing with the ith point. The distance $d(\mathbf{u}[i], \mathbf{u}[j])$ between vectors $\mathbf{u}[i]$ and $\mathbf{u}[j]$ is defined as $d(\mathbf{u}[i], \mathbf{u}[j]) = \max\{|\mathbf{u}(i+k) - \mathbf{u}(j+k)|, 0 \leq k \leq m-1\}$ where k accounts for the vector component index. The probability of finding another vector within distance r from the template vector $\mathbf{u}[i]$ is estimated by

$$C_i'^m(r) = \{the\ number\ of\ j, j \neq i, j \leq N - m + 1, such\ that\ d\left[u(i), u(j)\right] \leq r\} / (N - m + 1)$$

Now we can determine

$$\phi'^m(r) = (N - m + 1)^{-1} \sum_{i=1}^{N-m+1} C_i'^m(r) \tag{5.34}$$

and

$$SampEn(m, r, N) = -\ln\left[\phi'^m(r)/\phi'^{m+1}(r)\right] \tag{5.35}$$

SampEn measures complexity of the signal in the same manner as *ApEn*. However, the dependence on the parameters N and r is different. SampEn decreases monotonically when r increases. In theory, SampEn does not depend on N. In analyzing time series including < 200 data points, however, the confidence interval of the results is unacceptably large. When r and N are large, SampEn and *ApEn* give the same results.

5.5.7 Fractal Dimension (FD)

The term "fractal" was first introduced by Mandelbrot in 1983 [146]. A fractal is a set of points that when looked at smaller scales, resembles the whole set. The concept of fractal dimension (FD) that refers to a non-integer or fractional dimension originates from fractal geometry. In traditional geometry, the topological or Euclidean dimension of an object is known as the number of directions each differential of the object occupies in space. This definition of dimension works well for geometrical objects whose level of detail, complexity or "space-filling" is the same. However, when considering two fractals of the same topological dimension, their level of "space-filling" is different, and that information is not given by the topological dimension. The FD emerges to provide a measure of how much space an object occupies between Euclidean dimensions. The FD of a waveform represents a powerful tool for transient detection. This feature has been used in the analysis of ECG and EEG to identify and distinguish specific states of physiologic function [147]. Many algorithms are available to determine the FD of the waveform. In this work, algorithms proposed by Higuchi and Katz are implemented for analysis of ECG and EEG signals. FD can be calculated by using (1) Higuchi's Algorithm, (2) Katz's algorithm.

Higuchi's Algorithm

Let us consider $x(1), x(2), \ldots, x(N)$ the time sequence to be analyzed. We construct k new time series $x_m^k as : x_m^k = \{x(m), x(m+k), x(m+2k), \ldots, x\left(m + \left[\frac{N-m}{k}\right] k\right), \}$ for $m = 1, 2, \ldots, k$, where m indicates the initial time value, and k indicates the discrete time interval between points, and $\lfloor a \rfloor$ means the integer part of a. For each of the k time series or curves x_m^k, the length $L_m(k)$ is computed by,

$$L_m(k) = \frac{\sum_{i=1}^{\lfloor a \rfloor} |x(m+ik) - x(m+(i-1)k)| (N-1)}{\lfloor a \rfloor k} \tag{5.36}$$

where N is the total length of the data sequence x, $(N-1)/\lfloor a \rfloor k$ is a normalization factor and $a = \frac{N-m}{k}$. An average length is computed as the mean of the k lengths $L_m(k)$ for $m = 1, 2, \ldots, k$. This procedure is repeated for each k ranging from 1 to $k_{\max}$, obtaining an average length for each k. In the curve of $\ln(L_m(k))$ versus $\ln(1/k)$, the slope of the least-squares linear best fit is the estimate of the fractal dimension ($D^{Higuchi}$) [148].

Katz's Algorithm

Using Katz's method [149] the FD of a curve can be defined as,

$$D^{Katz} = \frac{\log_{10}(L)}{\log_{10}(d)} \tag{5.37}$$

where L is the total length of the curve or sum of distances between successive points, and d is the diameter estimated as the distance between the first point of the sequence and the point of the sequence that provides the farthest distance. Mathematically, d can be expressed as $d = \max\left(\|x(1), x(i)\|\right)$.

Considering the distance between each point of the sequence and the first, point i is the one that maximizes the distance with respect to the first point. The FD compares the actual number of units that compose a curve with the minimum number of units required to reproduce a pattern of the same spatial extent. FDs computed in this fashion depend upon the measurement units used. If the units are different, then so are the FDs. Katz's approach solves this problem by creating a general unit or yardstick: the average step or average distance between successive points, a. Normalizing the distances D^{Katz} is then given by,

$$D^{Katz} = \frac{\log_{10}(L/a)}{\log_{10}(d/a)} \tag{5.38}$$

5.5.8 Recurrence Plots (RP)

Recurrence plots are graphical devices specially suited to detect hidden dynamical patterns and nonlinearities in data. It is a visualization technique to detect the recurrence or correlations in the data. It is a graph which shows all those times at which a state of the dynamical system recurs. In other words, the RP reveals all the times when the phase space trajectory visits roughly the same area in the phase space.

In practice, one chooses r_i such that the ball of radius r_i centered at X_i in $\Re^d$ contains reasonable number of points X_j of the trajectory. One dots a point at each (i, j) for which X_j is in the ball of radius r_i centered at X_i. This plot is called the recurrence plot.

RP is usually symmetric because the distance measure is symmetric. But complete symmetry is obtained because $r_i \neq r_j$. The RP exhibits characteristic large-scale and small-scale patterns.

As mentioned above, RPs provide a visual impression of the trajectory of a dynamical system in phase space. Suppose that the time series $\{X_i\}_{i=1}^{N}$ representing the trajectory of a system in phase space is given, with $X_i \in \Re^d$. The RP is based on the following matrix

$$\Re_{i,j} = \Theta\left(\varepsilon - \|X_i - X_j\|\right), i, j = 1, \ldots, N \tag{5.39}$$

where $\Theta(.)$ is the Heaviside function, $\|.\|$ denotes a norm and ε is a predefined threshold. We will use the maximum norm throughout this work. We obtain a 2-dimensional NxN matrix, which is symmetric with respect to the main diagonal $i = j$.

Structures in Recurrence Plots

The initial purpose of RPs is the visual inspection of higher dimensional phase space trajectories. The view on RPs gives hints about the time evolution of these trajectories. The advantage of RPs is that they can also be applied to rather short and even nonstationary data.

The RPs exhibit characteristic large scale and small scale patterns. The first patterns were denoted by Eckmann *et al* [150] as *typology* and the latter as *texture*. The typology offers a global impression which can be characterized as *homogeneous, periodic, drift* and *disrupted.*

Homogeneous RPs are typical of stationary and autonomous systems in which relaxation times are short in comparison with the time spanned by the RP. An example of such an RP is that of a random time series.

Oscillating systems have RPs with diagonal oriented, *periodic* recurrent structures (diagonal lines, checkerboard structures). However, even for those oscillating systems whose oscillations are not easily recognizable, the RPs can be used in order to find their oscillations. The *drift* is caused by systems with slowly varying parameters. Such slow (adiabatic) change brightens the RP's upper-left and lower-right corners. Abrupt changes in the dynamics as well as extreme events cause *white areas or bands* in the RP. RPs offer an easy possibility to find and to assess extreme and rare events by using the frequency of their recurrences.

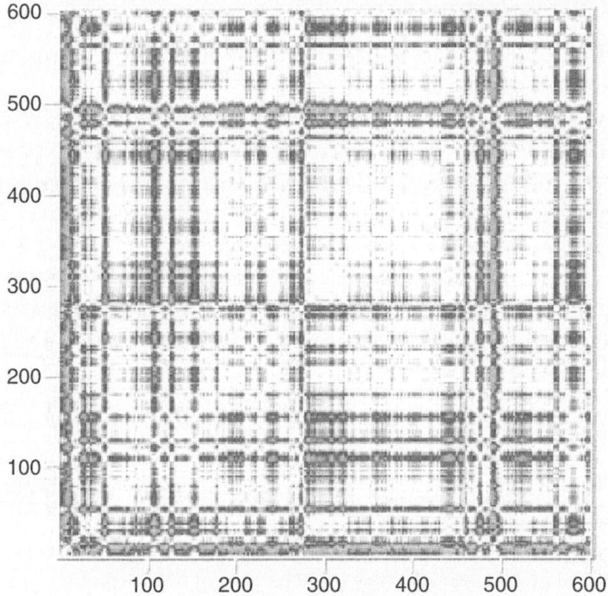

Fig. 5.6. Recurrence plot of normal heart rate

The recurrence plot of normal heart rate is given in Fig. 5.6. For *normal* cases, the RP has diagonal line and less squares indicating more variation in the heart rate. Abnormalities like *Complete Heart Block* (CHB) and in Ischemic/dilated cardiomyopathy cases, show more squares in the plot indicating the inherent periodicity and the lower heart rate variation [151].

5.6 Requirements for Non-Linear Analysis

Specifics of the biological systems require modifications of standard nonlinear dynamics algorithms. The main problems of the nonlinear analysis when applying it to biological signals can be summarized as follows: (a) high level of random noise in the biological data. The applied nonlinear dynamics methods should be robust to the noise influence; (b) short experimental data sets due to the low frequencies of the biological signals. Short realizations cause large error bars in the estimation of the chaos parameters; (c) nonstationarity of the biological systems, i.e., ECG have different modulations influenced by various external factors with different characteristic times; (d) spatially extended character of the system.

5.6.1 Surrogate Data

It is necessary to check the data for the nonlinearity. One of the tests for nonlinearity is the surrogate data test.

The method of using surrogate in nonlinear time series analysis was introduced by Theiler *et al* [152] in 1992. Surrogate signal is produced by phase randomizing the original data. It has similar spectral properties as of the given data. The surrogate data sequence has the same mean, the same variance, the same autocorrelation function and therefore the same power spectrum as the original sequence, but phase relations are destroyed. In the case of data shuffling the histograms of the surrogate sequence and the reference sequence are identical. The random phase spectrum is generated by using any of the three methods described below.

1. Random phase: here the complex phase values of the Fourier transformed input signal are chosen randomly.
2. Phase shuffle: here the phase values of the original spectrum are used in random order.
3. Data shuffle: here the phase values of the original spectrum are used in random order and the sorted values of the surrogate sequence are substituted by the corresponding sorted values of the reference sequence additionally.

The measured topological properties of the experimental time series are then compared with that of the measured topological properties of the surrogate data sets. If both the experimental data and the surrogate data results differ more than 50%, then it rejects the null hypothesis and indicates that the experimental data contain nonlinear features.

5.7 Discussion

Statistical measures of variability are easy to compute and provide valuable prognostic information about patients. Time domain measures are susceptible to bias secondary to nonstationary signals. A potential confounding factor in characterizing variability with standard deviation is the increase in baseline heart rate that may accompany diminished HRV indices. The clinical significance of this distinction is unclear, because the prognostic significance of altered SDNN remains clinically useful. Another limitation of time domain measures is that they do not reliably distinguish between distinct biological signals. There are many potential examples of data series with identical means and standard deviations but with very different underlying rhythms [153]. Therefore, additional, more sophisticated methods of variability analysis are necessary to characterize and differentiate physiological signals. It is nonetheless encouraging that, using rather crude statistical measures of variability, it is possible to derive clinically useful information.

In order to derive a valid and meaningful analysis using a fast Fourier transform and frequency domain analysis, the assumptions of stationarity and periodicity are to be fulfilled. The signal must be periodic, namely it is a signal that is comprised of oscillations repeating in time, with positive and negative alterations [154]. In the interpretation of experimental data, periodic behavior may or may not exist when evaluating alterations in spectral power in response to intervention. The assumption of stationarity may also be violated with prolonged signal recording. Changes in posture, level of activity and sleep patterns will alter the LF and HF components of spectral analysis [155]. Spectral analysis is more sensitive to the presence of artifact and/or ectopy than time domain statistical methods. In addition, given that different types of Holter monitors may yield altered LF signals [156], it is essential to ensure that the sampling frequency of the monitor used to read QRS complexes does not contribute to error in the variability analysis [157]. Thus, the performance and interpretation of spectral analysis must incorporate these limitations. Recommendations based upon the stationarity assumption include the following: short-term and long-term spectral analyses must be distinguished; long-term spectral analyses are felt to represent averages of the alterations present in shorter term recordings and may hide information; traditional statistical tests should be used to test for stationarity when performing spectral analysis; and physiological mechanisms that are known to influence HRV throughout the period of recording must be controlled.

A nonlinear deterministic approach appears to be more appropriate to describe more complex phenomena, showing that apparently erratic behavior can be generated even by a simple deterministic system with nonlinear structure [158, 159]. It is therefore not surprising that a specific subtype of nonlinear dynamics, chaos theory and fractals, has recently been applied to the study of HRV signal.

Approximate entropy is a measure and parameter that quantifies the regularity or predictability of time series data. This shows higher values for the normal heart signals and will have smaller values for the cardiac abnormal signals [46]. Detrended fluctuation analysis (DFA) quantifies fractal-like correlation properties of the data [160]. The root-mean square fluctuation of the integrated and detrended data are measured in observation box of various sizes and then plotted against the size of the box [161]. The scaling exponent represents the slope of this line, which relates log(F(n)-fluctuation) to log(n-box size). The short-term (F-fast) and long-term (S-slow) scaling exponents are also calculated [162].

The CD is a measure of the complexity of the process being investigated. It is calculated from the phase space plots and shows different range of values for different cardiac diseases [46]. Another nonlinear dynamical parameter of great importance is the Lyapunov exponent (LE), which quantifies the average growth of infinitesimally small errors in initial points. Chaotic processes are characterized by one or more positive LEs, which means that the neighboring points of trajectory in the phase space diverge. Converging processes are characterized by a Lyapunov spectrum of negative exponents. The LE is a measure of predictability of the process, which quantifies the exponential divergence of initially close state-space trajectories.

The Hurst scaling exponent (H) characterizes the shape of self-similar signals and ranges from 0 to 1. A self-similar signal with $H \approx 0$ resembles white noise with spiky oscillations. A signal with $H \approx 0.5$ shows brownian noise-like oscillations, whereas signals with $H \approx 1$ exhibit smooth oscillations. Similarly, Fractal Dimension allows us to measure the degree of complexity by evaluating how fast our measurements increase or decrease as our scale becomes larger or smaller. These values will be higher for the normal subjects and small for the abnormal subjects due to reduced or rhythmic variation.

Several investigators have stressed the importance of non-linear techniques such as fractal dimension (FD) and approximate entropy ($ApEn$) to analyze HR time series, as these series are essentially non-linear in nature [163, 164]. There are several ways to determine FD, which measures the space-filling propensity and complexity of the time series [165]. Acharya *et al* have explained all the different types of linear and non-linear techniques, available for the analysis of heart rate signals [166].

Slowly varying heart rate diseases like, CHB, Ischemic/dilated cardiomyopathy have more number of squares in the RP. This is due to the inherent periodicity of the time series. The RP show more patches of colors in cardiac diseases, where the heart rate signal is varying rapidly [151]. Censi *et al* performed a quantitative study of coupling patterns between respiration and spontaneous rhythms of heart rate and blood pressure variability signals by using the Recurrence Quantification Analysis (RQA) [167]. They applied RQA to both simulated and experimental data obtained in control breathing at three different frequencies (0.25, 0.20, and 0.13 Hz) from ten normal subjects. RP concept was used to detect the life threatening arrhythmias like ventricular tachycardias [168].

Jamsek *et al* [169] used bispectral analysis to study the coupling between cardiac and respiratory activity. Witte *et al* [170] too studied the coupling between cardiac and respiratory activity but this research was on neonatal subjects. Pinhas *et al* [171] have used the bispectrum to analyze the coupling between blood pressure (BP) and HRV in heart transplant patients.

5.8 Conclusion

The science of analyzing biological signals has undergone tremendous growth over the past decade, with the development of advanced computational methods that characterize the variation, oscillation, complexity and regularity of signals. These methods were developed in response to theoretical limitations of the others; however, all appear to have clinical significance. There is no consensus that any single technique is the single best means of characterizing and differentiating biological signals; rather, investigators agree that multiple techniques should be performed simultaneously to facilitate comparison between methods, techniques and studies. Variability analysis represents a novel means to evaluate and treat individual patients, suggesting a shift from epidemiological analytical investigation to continuous individualized variability analysis. The existing literatures show that, the nonlinear parameters are more effective in analyzing the cardiac health of the subjects.

References

1. Saul, J.P. (1990): 'Beat-to-beat variations of heart rate reflect modulation of cardiac autonomic outflow', *News Physiological Sciences*, **5**, pp. 32–37.
2. Schwartz, P.J., and Priori, S.G. (1990): 'Sympathetic nervous system and cardiac arrythmias', In: Zipes, D.P., and Jalife, J. eds. *Cardiac Electrophysiology*, From Cell to Bedside. Philadelphia: Saunders, W.B. pp. 330–343.
3. Levy, M.N., and Schwartz, P.J. (1994): 'Vagal control of the heart: Experimental basis and clinical implications', *Armonk: Future.*
4. Task Force of the European Society of Cardiology and North American Society of Pacing and Electrophysiology. (1996): 'Heart Rate Variability: Standards of measurement, physiological interpretation and clinical use', *European Heart Journal*, **17**, pp. 354–381.
5. Berger, R.D., Akselrod, S., Gordon, D., and Cohen, R.J. (1986): 'An efficient algorithm for spectral analysis of heart rate variability', *IEEE Transactions on Biomedical Engineering*, **33**, pp. 900–904.
6. Kamath, M.V., and Fallen, E.L. (1995): 'Correction of the heart rate variability signal for ectopics and missing beats', In: Malik, M., and Camm, A.J. eds. *Heart rate variability*, Armonk: Futura, pp. 75–85.
7. Kobayashi, M., and Musha, T. (1982): '1/f fluctuation of heart beat period', *IEEE transactions on Biomedical Engineering*, **29**, pp. 456–457.
8. Boomsma, F.T., and Manintveld. (1999): 'Cardiovascular control and plasma catecholamines during restand mental stress: effects of posture', *Clinical Science*, **96**, pp. 567–576.
9. Viktor, A., Jurij-Matija, K., Roman, T., and Borut, G. (2003): 'Breathing rates and heart rate spectrograms regarding body position in normal subjects', *Computers in Biology and Medicine*, **33**, pp. 259–266.
10. Rosenstien, M., Colins, J.J., and De Luca, C.J. (1993): 'A practical method for calculating largest Lyapunov exponents from small data sets', *Physica D*, **65**, pp. 117–134.
11. Pincus, S.M. (1991): 'Approximate entropy as a measure of system complexity', *Proceedings of National Acadamic Science, USA*, **88**, pp. 2297–2301.
12. Peng, C.K., Havlin, S., Hausdorf, J.M., Mietus, J.E., Stanley, H.E., and Goldberger, A.L. (1996): 'Fractal mechanisms and heart rate dynamics', *Journal on Electrocardiology*, **28** (suppl), pp. 59–64.
13. Grossman, P., Karemaker, J., and Wieling, W. (1991): 'Prediction of tonic parasympathetic cardiac control using respiratory sinus arrhythmia: the need for respiratory control', *Psychophysiology*, **28**, pp. 201–216.

14. Kleiger, R.E., Bigger, J.T., Bosner, M.S., Chung, M.K., and Cook, J.R., Rolnitzky, L.M., Steinman, R., and Fleiss, J.L. (1991): 'Stability over time of variables measuring heart rate variability in normal subjects', *Am J Cardiol.*, **68**, pp. 626–630.
15. Kovatchev, B.P., Farhy, L.S., Cao, H., Griffin, M.P., Lake, D.E., and Moorman, J.R. (2003): 'Sample Asymmetry Analysis of Heart Rate Characteristics with Application to Neonatal Sepsis and Systemic Inflammatory Response Syndrome', *Pediatric Research*, **54**, pp. 892–898.
16. Cysarz, D., Bettermann, H., and Van Leeuwen, P. (2000): 'Entropies of short binary sequences in heart period dynamics', *Am J Physiol Heart Circ Physiol*, **278**, pp. H2163–H2172.
17. Verlinde, D., Beckers, F., Ramaekers, D., and Aubert, A.E. (2001): 'Wavelet decomposition analysis of heart rate variability in aerobic athletes', *Auton Neurosci.*, **90**(1–2), pp. 138–141.
18. Toledo, E., Gurevitz, O., Hod, H., Eldar, M., and Akselrod, S. (2003): 'Wavelet analysis of instantaneous heart rate: a study of autonomic control during thrombolysis', *Am J Physiol Regul Integr Comp Physiol.*, **284**(4), pp. R1079–R1091.
19. Gamero, L.G., Vila, J., and Palacios, F. (2002): 'Wavelet transform analysis of heart rate variability during myocardial ischaemia', *Medical & Biological Engineering & Computing*, **40**, pp. 72–78.
20. Roche, F., Pichot, V., Sforza, E., Court-Fortune, I., Duverney, D., Costes, F., Garet, M., and Barthélémy, J.C. (2003): 'Predicting sleep apnoea syndrome from heart period: a time-frequency wavelet analysis', *Eur Respir J*, **22**, pp. 937–942.
21. Schumacher, A. (2004): 'Linear and nonlinear approaches to the analysis of R-R interval variability', *Biol Res Nurs.*, **5**(3), pp. 211–221.
22. Schein, M., Gavish, B., Herz, M., Rosner-Kahana, D., Naveh, P., Knishkowy, B., Zlotnikov, E., Ben-Zvi, N., and Melmed, R.N. (2001): 'Treating hypertension with a device that slows and regularizes breathing: a randomized double-blind controlled study', *J Human Hyperten*, **15**(4), pp. 271–278.
23. Grossman, E., GRossman, A., Schein, M.H., Zimlichman, R., and Gavish, B. (2001): 'Breathing-control lowers blood pressure', *J Human Hyperten*, **15**(4), pp. 263–269.
24. Spicuzza, L., Gabutti, A., Porta, C., Montano, N., and Bernardi, L. (2000): 'Yoga and chemoreflex response to hypoxia and hypercapnia', *Lancet*, **356**(9240), pp. 1495–1496.
25. Bernardi, L., Sleight, P., Bandinelli, G., Cencetti, S., Fattorini, L., Wdowczyc-Szulc, J., and Lagi, A. (2001): 'Effect of rosary prayer and yoga mantras on autonomic cardiovascular rhythms: comparative study', *Br. Med. J.*, 32322–**29**, pp. 1446–1449.
26. Guyton, A.C. (1992): 'Kidneys and fluids in pressure regulation. Small volume but large pressure changes', *Hypertension*, **191**, pp. I2–I8.
27. De Boer, R.W., Karemaker, J.M., and Strackee, J. (1985): 'Relationships between short-term blood-pressure fluctuations and heart-rate variability in resting subjects. I: A spectral analysis approach', *Med Biol Eng Comput.*, **23**(4), pp. 352–358.
28. Laude, D., Elghozi, J.L., Girald, A., Bellard, E., Bouhaddi, M., Castiglioni, P., Catherine, C., Andrei, C., Marco, D.I.R., Jacques-Olivier, F., Ben, J., Karemaker John M., Georges, L., Gianfranco, P., Pontus, P.B., Alberto,

P., Luc, Q., Jacques, R., Heinz, R., and Harald, M.S. (2004): 'Comparison of various techniques used to estimate spontaneous baroreflex sensitivity (the EUROVAR study)', *Am J Physiol Regul Integr Comp Physiol*, **286**, pp. R226–R231.
29. Westerhof, B.E., Gisolf, J., Stok, W.J., Wesseling, K.H., and Karemaker, J.M. (2004): 'Time-domain cross-correlation baroreflex sensitivity: performance on the Eurobavar data set', *J Hypertens.*, **22**(7), pp. 1259–1263.
30. Rothschild, M., Rothschild, A., and Pfeifer, M. (1988): 'Temporary decrease in cardiac parasympathetic tone after acute myocardial infarction', *Am J Cardiol.*, **18**, pp. 637–639.
31. Carney, R.M., Blumenthal, J.A., Stein, P.K., Watkins, L., Catellier, D., Berkman, L.F., Czajkowski, S.M., O'Connor, C., Stone, P.H., and Freedland, K.E. (2001): 'Depression, Heart Rate Variability, and Acute Myocardial Infarction', *Circulation*, **104**, pp. 2024.
32. Carney, R.M., Blumenthal, J.A., Freedland, K.E., Stein, P.K., Howells, W.B., Berkman, L.F., Watkins, L.L., Czajkowski, S.M., Hayano, J., Domitrovich, P.P., and Jaffe, A.S. (2005): 'Low Heart Rate Variability and the Effect of Depression on Post-Myocardial Infarction Mortality', *Arch Intern Med.*, **165**, pp. 1486–1491.
33. Duru, F., Candinas, R., Dziekan, G., Goebbels, U., Myers, J., Dubach, P., and Chur, K. (2000): 'Effect of Exercise Training on Heart Rate Variability in Patients With New-Onset Left Ventricular Dysfunction After Myocardial Infarction', *Am Heart J*, **140**(1), pp. 157–161.
34. Stefano, D., Serena, B., Anselmino, M., Paolo, V.G., Barbara, C., Luigi, P., Giuseppe, A., and Paolo, T.G. (2006): 'Depression in patients with acute myocardial infarction: Influence on autonomic nervous system and prognostic role. Results of a five-year follow-up study', *Int J Cardiol.*, **67**(7), pp. 1172–1177.
35. Bauer, A., Kantelhardt, J.W., Barthel, P., Schneider, R., Makikallio, T., Ulm, K., Hnatkova, K., Schomig, A., Huikuri, H., Bunde, A., Malik, M., and Schmidt, G. (2006): "Deceleration capacity of heart rate as a predictor of mortality after myocardial infarction: cohort Study", **367**(9523), pp. 1674–1681.
36. Larosa, C., Infusino, F., Sgueglia, G.A., Aurigemma, C., Sestito, A., Lombardo, A., Niccoli, G., Crea, F., and Lanza, G.A. (2005): 'Effect of primary percutaneous coronary intervention versus thrombolysis on ventricular arrhythmias and heart rate variability in acute myocardial infarction', *Ital Heart J.*, **6**(8), pp. 629–633.
37. Lowensohn, R.I., Weiss, M., and Hon, E.H. (1977): 'Heart-rate variability in brain-damaged adults', *Lancet*, **1**, pp. 626–628.
38. Leipzig, T.J., and Lowensohn, R.I. (1986): 'Heart rate variability in neurosurgical patients', *Neurosurgery*, **19**, pp. 356–362.
39. Sayar, K., Güleç, H., Gökçe, M., and Ismail, AK. (2002): 'Heart Rate Variability in Depressed Patients', Bulletin of Clinical Psychopharmacology, **12**(3), pp. 130–133.
40. Fell, J., Mann, K., Roschke, J., and Gopinathan, M.S. (2000): 'Nonlinear analysis of continuous ECG during sleep I. Reconstruction', *Biological Cybernatics*, **82**, pp. 477–83.
41. Radhakrishna, R.K.A., Vikram, K.Y., Narayana, D.D., and Vedavathy, T.S. (2001): 'Characterizing Chaos in heart rate variability time series of panic

disorder patients', *Proceedings of Proceedings of ICBME*, Biovision Bangalore India, pp. 163–167.

42. Paul, S.A., James, N.W., Gareth, R.C., Petter, A.S., and Colin, E.R. (2002): 'Finding Coordinated Atrial Activity During Ventricular Fibrillation Using Wavelet Decomposition', *IEEE Engineering In Medicine and Biology Magazine*, **21**(1), pp. 58–61.
43. Dingfei Ge, Srinivasan, N., Krishnan, and S.M. (2002): 'Cardiac arrhythmia classification using autoregressive modeling', *BioMedical Engineering OnLine*, **1**(1): 5.
44. Owens, M.I., Ahmed, H., Abou-Zied, Abou-Bakr, M., Youssef, and Yasser, M.K.(2002): 'Study of features on nonlinear dynamical modeling in ECG arrhythmia detection and classification', *IEEE transactions on Biomedical Engineering*, **49**(7), pp. 733–736.
45. Acharya, U.R., Bhat, P.S., Iyengar, S.S., Rao, A., and Dua, S. (2003): 'Classification of heart rate using artificial neural network and fuzzy equivalence relation', *Pattern Recognition*, **36**, pp. 61–68.
46. Acharya, U.R., Kannathal, N., and Krishnan, S.M. (2004): 'Comprehensive analysis of cardiac health using heart rate signals', *Physiological Measurement Journal*, **25**, pp. 1130–1151.
47. Acharya, U.R., Kannathal, N., Seng, O.W., Ping, L.Y., and Chua, T.L. (2004): 'Heart rate analysis in normal subjects of various age groups', *Biomedical Online Journal*, **3**, pp. 24.
48. Koivikko, M.L., Salmela, P.I., Airaksinen, K.E., Tapanainen, J.S., Ruokonen, A., Makikallio, T.H., and Huikuri, H.V. (2005): 'Effects of sustained insulin-induced hypoglycemia on cardiovascular autonomic regulation in type 1 diabetes', *Diabetes*, **54**(3), pp. 744–750.
49. Chemla, D., Young, J., Badilini, F., Maison-Blanche, P., Affres, H., Lecarpentier, Y., and Chanson, P., 'Comparison of fast Fourier transform and autoregressive spectral analysis for the study of heart rate variability in diabetic patients', *Int J Cardiol.*, **104**(3), pp. 307–313.
50. Fiorentini, A., Perciaccante, A., Paris, A., Serra, P., and Tubani, L. (2005): 'Circadian rhythm of autonomic activity in non diabetic offsprings of type 2 diabetic patients', *Cardiovasc Diabetol.*, **1**; 4:15.
51. Kudat, H., Akkaya, V., Sozen, A.B., Salman, S., Demirel, S., Ozcan, M., Atilgan, D., Yilmaz, M.T., and Guven, O. (2006): 'Heart rate variability in diabetes patients, *J Int Med Res.*, **34**(3), pp. 291–296.
52. Chen, H.S., Wu, T.E., Jap, T.S., Lee, S.H., Wang, M.L., Lu, R.A., Chen, R.L., and Lin, H.D. (2006): 'Decrease heart rate variability but preserve postural blood pressure change in type 2 diabetes with microalbuminuria', *J Chin Med Assoc.*, **69**(6), pp. 254–258.
53. Pfeifer, M.A., Cook, D., Brodsky, J., Tice, D., Reenan, A., and Swedine, S., et al. (1982): 'Quantitative evaluation of cardiac parasympathetic activity in normal and diabetic man', *Diabetes*, **3**, pp. 339–345.
54. Villareal, R.P., Liu, B.C., and Massumi, A. (2002): 'Heart rate variability and cardiovascular mortality', *Curr Atheroscler Rep.*, **4**(2), pp. 120–127.
55. Singh, J.P., Larson, M.G., O'Donell, C.J., Wilson, P.F., Tsuji, H., Lyod-Jones, D.M., and Levy, D. (2000): 'Association of hyperglycemia with reduced heart rate variability : The Framingham Heart Study', *American Journal of Cardiology*, **86**, pp. 309–312.

56. Hainsworth, R.(1995): 'The control and physiological importance of heart rate Heart Rate Variability', In: Malik, M., and Camm, A.J. eds. (*Armouk, NY: Futura*), pp. 3–19.
57. Patterson, R., and Kaiser, D. (1997): 'Heart rate change as a function of age, tidal volume and body position when breathing using voluntary cardiorespiratory synchronization', *Physiol. Meas.*, **18**, pp. 183–189.
58. Bernardi, L., Wdowczyc-Szulc, J., Valenti, C., Castoldi, S., Passino, C., Spadacini, G., and Sleight, P. (2000): 'Effects of controlled breathing, mental activity and mental stress with or without verbalisation on heart rate variability', *J. A m. College Cardiol.*, **35**, pp. 1462–1469.
59. Bernardi, L., Gabutti, A., Porta, C., and Spicuzza, L. (2001): 'Slow breathing reduces chemoreflex response to hypoxia and hypercapnia, and increases baroreflex sensitivity', *J. Hypertens.*, **1912**, pp. 2221–2229.
60. Bernardi, L., Porta, C., Gabutti, A., Spicuzza, L., and Sleight, P. (2001): 'Modulatory effects of respiration', *Review Auton*, **90**(1–2), pp. 47–56.
61. Baselli, G., Porta, A., and Ferrari, G. (1995): 'Models of the analysis of cardiovascular variability signals Heart Rate Variability', In: Malik, M., and Camm, A.J. eds. (*Armouk, NY: Futura*), pp. 135–145.
62. Sleight, P., and Casadei, B. (1995): 'Relationship between heart rate respiration and blood pressure variability Heart Rate Variability', In: Malik, M., and Camm, A.J. eds. (*Armouk, NY: Futura*), pp. 311–327.
63. Jennings, J.R., and Van Der Molen, M.W. (2002): 'Cardiac timing and the central regulation of action', *Psychol. Res.*, **66**, pp. 337–349.
64. Jennings, J.R., Mcknight, J.D., and Van Der Molen, M. (1996): 'Phase sensitive interaction of cardiac and respiratory timing in humans', *Psychophysiology*, **33**, pp. 514–521
65. Zoccali, C., Ciccarelli, M., and Maggiore, Q. (1982): 'Defective reflex control of heart rate in dialysis patients: Evidence for an afferent autonomic lesion', *Clin Sci.*, **63**, pp. 285–292.
66. Forsstrom, J., Forsstrom, J., Heinonen, E., Valimaki, I., and Antila, K. (1986): 'Effects of haemodialysis on heart rate variability in chronic renal failure', *Scand J Clin Lab Invest.*, **46**, pp. 665–670.
67. Axelrod, S., Lishner, M., Oz, O., Bernheim, J., and Ravid, M. (1987): 'Spectral analysis of fluctuations in heart rate: an objective evaluation of autonomic nervous control in chronic renal failure', *Nephron*, **45**, pp. 202–206.
68. Tsai, A.C., and Chiu, H.W. (2002): 'Relationship between heart rate variability and electrolyte concentration in chronic renal failure patients under hemodialysis' *International Journal of Bioelectromagnetism*, **4**(2), pp. 307–308.
69. Lerma, C., Minzoni, A., Infante, O., and José, M.V. (2004): 'A Mathematical Analysis for the Cardiovascular Control Adaptations in Chronic Renal Failure', *Artificial Organs*, **28**(4), pp. 398–409.
70. Ryan, S.M., Goldberger, A.L., Pincus, S.M., Mietus, J., and Lipsitz, L.A. (1994): 'Gender-and age-related differences in heart rate: are women more complex than men', *Journal of American College of Cardiology*, **24**(7), pp. 1700–1707.
71. Kevin, P.D., Christoper, A.D., Pamela, P.J., and Seals, D.R. (1998): 'Elevated heart rate variability in physically active young and older adult women', *Clinical Science*, **94**, pp. 579–584.

72. Emese, N., Hajnalka, O., Bardos, G., and Molnar, P. (2000): 'Gender-related heart rate differences in human neonates', *Pediatric Research*, **47**(6), pp. 778–780.
73. Hendrik, B., Uwe, K.H.W., Axel, B., Kluge, N., Hugo, A.K., Richardt, G., and Jurgen, P. (2003): 'Circadian profile of cardiac autonomic nervous modulation in healthy subjects: Differing Effects of aging and Gender on heart rate variability', *Journal of Cardiovascular Electrophysiology*, **14**(8), pp. 791–799.
74. Ramaekers, D., Ector, H., Aubert, A.E., Rubens, A., and Van De, W.F. (1993): 'Heart rate variability and heart rate in healthy volunteers: Is the female autonomous nervous system cardioprotective?', *European Heart Journal*, **19**, pp. 1334–1341.
75. Yamasaki, Y., Kodama, M., and Matsuhisa, M. (1996): 'Diurnal heart rate variability in healthy subjects: effects of aging and sex differences', *American Journal of Physiology*, **271**, pp. 303–310.
76. Van Ravenswaaij, C.M., Hopman, J.C., Kollee, L.A., Van Amen, J.P., Stoelinga, G.B., and Van Geijn, H.P. (1991): 'Influences on heart rate variability in spontaneously breathing preterm infants', *Early Hum Dev.*, **27**, pp. 187–205.
77. Schwartz, J.B., Gibb, W.J., and Tran, T. (1991): 'Aging effects on heart rate variation', *J Gerontol.*, **46**, pp. M99–M106.
78. Finley, J.P., Nungent, S.T., and Hellenbrand, W. (1987): 'Heart rate variability in children, spectral analysis of developmental changes between 5 and 25 years', *Can J Physiol Pharmacol*, **65**, pp. 2048–2052.
79. Weise, F., Heydenreich, F., Kropf, S., and Krell, D. (1990): 'Intercorrelation analyses among age, spectral parameters of heart rate variability and respiration in human volunteers', *J Interdiscipl Cycle Res.*, **21**, pp. 17–24.
80. Lipsitz, L.A., Mietus, J., Moody, G.B., and Goldberger, A.L. (1990): 'Spectral characteristics of heart rate variability before and during postural tilt: relations to aging and risk of syncope', *Circulation*, **81**, pp. 1803–1810.
81. Wilson, P.W., and Evans, J.C. (1993): 'Coronary artery prediction', *American Journal of Hypertens*, **6**, pp. 309S–313S.
82. Bekheit, S., Tangella, M., El-Sakr, A., Rasheed, Q., Craelius, W., and El-Sherif, N. (1990): 'Use of heart rate spectral analysis to study the effects of calcium channel blockers on sympathetic activity after myocardial infarction', *Am Heart J.*, **119**, pp. 79–85.
83. Guzzetti, S., Piccaluga, E., Casati, R., Cerutti, S., Lombardi, F., and Pagani, M., et al. (1988): 'Sympathetic predominance in essential hypertension: a study employing spectral analysis of heart rate variability', *J Hypertens.*, **6**, pp. 711–717.
84. Coumel, P., Hermida, J.S., Wennerblom, B., Leenhardt, A., Maison-Blanche, P., and Cauchemez, B. (1991): 'Heart rate variability in left ventricular hypertrophy and heart failure, and the effects of beta-blockade', *Eur Heart J.*, **12**, pp. 412–22.
85. Muller, J.E., Morrison, J., Stone, P.H., Rude, R.E., Rosner, B., and Roberts, R., Pearle, D.L., Turi, Z.G., Schneider, J.F., and Serfas, D.H. (1984): 'Nifedipine therapy for patients with threatened and acute myocardial infarction: a randomized, double-blind, placebo-controlled comparison', *Circulation*, **69**, pp. 740–747.
86. Cornel, P., Compagnone, D., Luszick, J., and Cees-Nico, V. (2003): 'Effect of Omacor on HRV parameters in patients with recent uncomplicated myocardial

infarction – A randomized, parallel group, double-blind, placebo-controlled trial: study design', *Current Controlled Trials in Cardiovascular Medicine*, **4**, pp. 2.
87. Eryonucu, B., Uzun, K., Güler, N., and Bilge, M. (2001): 'Comparison of the acute effects of salbutamol and terbutaline on heart rate variability in adult asthmatic patients', *Eur Respir J*, **17**, pp. 863–867.
88. Hayano, J., Yamada, M., and Sakakibara, Y., et al. (1990): 'Short, Longterm, effects of cigarette smoking on Heart Rate Variability', *Am J Cardiol*, **65**, pp. 84–88.
89. Luchini, D., Bertocchi, F., and Maliani, A., et al. (1996): 'A controlled study of the autonomic changes produced by habitual cigarette smoking in healthy subjects', *Cardiovascular Research*, **31**, pp. 633–639.
90. Niedermaier, O., Smith, M., Beightol, L., Zukowska-Grojec, Z., Goldstein, D.S., and Eckberg, D.L. (1993): 'Influence of Cigarette Smoking on Human Autonomic Function', *Circulation*, **88**, pp. 562–571.
91. Pope, C.A., Eatough, D.J., Gold, D.R., Pang, Y., Nielsen, K.R., Nath, P., Verrier, R.L., and Kanner, R.E. (2001): 'Acute exposure to environmental tobacco smoke and heart rate variability', *Environmental Health Perspectives*, **109**(7), pp. 711–716.
92. Philip, S.Z., and Jeannine, L.G. (2006): 'Maternal Cigarette-Smoking During Pregnancy Disrupts Rhythms in Fetal Heart Rate', *Journal of Pediatric Psychology*, **31**(1), pp. 5–14.
93. Irfan, B., Metin, A.E., Dayimi, K., Muhsin, T., Osman, K., Mehmet, M., Ozlem, B.E., and Yelda, B. (2005): 'Cigarette Smoking and Heart Rate Variability: Dynamic Influence of Parasympathetic and Sympathetic Maneuvers', *The Annals of Noninvasive Electrocardiology*, **10**(3), pp. 324–329.
94. Malpas, S.C., Whiteside, E.A., and Maling, T.J. (1991): 'Heart rate variability and cardiac autonomic function in men with chronic alcohol dependence', *Br Heart J*, **65**, pp. 84–88.
95. Ryan, J.M., and Howes, L.G. (2002): 'Relations between alcohol consumption, heart rate, and heart rate variability in men', *Heart*, **88**, pp. 641–642.
96. Rossinin, J., Vitasalo, M., and Partanen, J., et al. (1997): 'Effects of acute alcohol ingestion on heart rate variability in patients with documented coronary artery disease and stable angina pectoris', *Am J Cardiol.*, **79**, pp. 487–491.
97. Pellizzer, A.M., Kamen, P.W., and Esler, M.D., et al. (2001): 'Comparative effects of mibefradil and nifedipine gastrointestinal transport system on autonomic function in patients with mild to moderate essential hypertension', *J Hypertens*, **19**, pp. 279–285.
98. Fumiharu, T., and Yoshiharu, Y. (2000): 'Decreased fractal component of human heart rate variability during non-REM sleep', *Am J Physiol Heart Circ Physiol*, **280**, pp. H17–H21.
99. Bušek, P., Vaòková, J., Opavský, J., Salinger, J., and Nevšímalová, S. (2005): 'Spectral Analysis of Heart Rate Variability in Sleep', *Physiol. Res.*, **54**, pp. 369–376.
100. Drinnan, M.J., Allen, J., Langley, P., and Murray, A. (2000): 'Detection of sleep apnoea from frequency analysis of heart rate variability', *Computers in Cardiology*, **27**, pp. 259–262.
101. Gregory, J.G., Susan, E.M., and Jason, H.M. (2005): 'Heart rate variability in non-apneic snorers and controls before and after continuous positive airway pressure', *BMC Pulm Med.*, **5**, pp. 9.

102. Munakata, M., Ichi, S., Nunokawa, T., Saito, Y., Ito, N., Fukudo, S., and Yoshinaga, K. (2001): 'Influence of night shift work on psychologic state and cardiovascular and neuroendocrine responses in healthy nurses', *Hypertens Res.*, **24**(1), pp. 25–31.
103. Portier, H., Louisy, F., Laude, D., Berthelot, M., and Guezennec, C.Y. (2001): 'Intense endurance training on heart rate and blood pressure variability in runners', *Med Sci Sports Exerc.* **33**(7), pp. 1120–1125.
104. Watanabe, T., Sugiyama, Y., Sumi, Y., Watanabe, M., Takeuchi, K., Kobayashi, F., and Kono, K. (2002): 'Effects of vital exhaustion on cardiac autononomic nervous functions assessed by heart rate variability at rest in middle-aged male workers', *Int J Behav Med.*, **9**(1), pp. 68–75.
105. Li, Z., Jiao, K., Chen, M., and Wang, C. (2003): 'Effect of magnitopuncture on sympathetic and parasympathetic nerve activities in healthy drivers – assessment by power spectrum analysis of heart rate variability', *Eur J Appl Physiol.*, **88**(4–5), pp. 404–410.
106. Jouanin, J.C., Dussault, C., Peres, M., Satabin, P., Pierard, C., and Guezennec, C.Y. (2004): 'Analysis of heart rate variability after a ranger training course', *Military Med.*, **169**(8), pp. 583–587.
107. Li, Z., Wang, C., Mak, A.F., and Chow, D.H. (2005): 'Effects of acupuncture on heart rate variability in normal subjects under fatigue and non-fatigue state', *Eur J Appl Physiol.*, **94**(5–6), pp. 633–640.
108. Irwin, M.R., Valladares, E.M., Motivala, S., Thayer, J.F., and Ehlers, C.L. (2006): 'Association between nocturnal vagal tone and sleep depth, sleep quality, and fatigue in alcohol dependence', *Psychosom Med.*, **68**(1), pp. 159–166.
109. Malik, M., Farrell, T., Cripps, T.R., and Camm, A.J. (1989): 'Heart rate variability in relation to prognosis after myocardial infarction: selection of optimal processing techniques', *Eur Heart J*, **10**, pp. 1060–1074.
110. Farrell, T.G., Bashir, Y., Cripps, T., Malik, M., Poloniecki, J., Bennett, E.D., Ward, D.E., and Camm, A.J. (1991): 'Risk stratification for arrhythmic events in postinfarction patients based on heart rate variability, ambulatory electrocardiographic variables and the signal-averaged electrocardiogram', *J Am Coll Cardiol*, **18**, pp. 687–697.
111. Woo, M.A., Stevenson, W.G., Moser, D.K., Trelease, R.B., and Harper, R.H. (1992): 'Patterns of beat-to-beat heart rate variability in advanced heart failure', *American Heart Journal*, **123**, pp. 704–710.
112. Kamen, P.W., Krum, H., and Tonkin, A.M. (1996): 'Poincare plot of heart rate variability allows quantitative display of parasympathetic nervous activity', *Clinical science*, **91**, pp. 201–208.
113. Brennan, M., Palaniswami, and Kamen, P. (2001): 'Do existing measures of Poincare plot geometry reflect nonlinear features of heart rate variability?', *IEEE transactions on Biomedical Engineering*, **48**(14), pp. 1342–1347.
114. Takens, F. (1981): 'Detecting strange attractors in turbulence', in Dynamical Systems and Turbulence.ser. Lecture notes in Mathematics. Rand, D.A., and Young, L.S. eds. *Berlin, Germany: Springer-Verlag*, **898**, pp. 366–381.
115. Tulppo, M.P., Makikallio, T.H., Takala, T.E.S., Seppanen', T., and Huikuri, H.V. (1996): 'Quantitative beat-to-beat analysis of heart rate dynamics during exercise', *American Journal of Physiology*, **271**, pp. H244–H252.
116. Hausdorff, J.M., and Peng, C.K. (1996): 'Multi-scaled randomness: a possible source of $1/f$ noise in biology', *Phys Rev E*, **54**, pp. 2154–2157.

117. Akaike, H. (1969): 'Fitting autoregressive models for prediction', *Annals of the Institute of Statistical Mathematics*, **21**, pp. 243–247.
118. Akaike, H. (1974): 'A new look at statistical model identification', *IEEE transaction on Automatics Control*, **19**, pp. 716–723.
119. Broadman, A., Schlindwein, F.S., Rocha A.P., and Leite, A. (2002): 'A study on the optimum order of autoregressive models for heart rate variability', *Physiological Measurements*, **23**, pp. 324–336.
120. Nikias, C.L., and Rughuveer, M.R. (1987): 'Bispectrum estimation: A digital signal processing framework', *Proc. IEEE*, **75**, pp. 869–890.
121. Chandran, V., and Elgar, S. (1993): 'Pattern Recognition Using Invariants Defined from Higher Order Spectra – One-Dimensional Inputs', *IEEE Trans. on Signal Processing*, **41**(1), pp. 205–212.
122. Ng, T.T., Chang, S.F., and Sun, Q. (2004): 'Blind Detection of Photomontage Using Higher Order Statistics', *IEEE International Symposium on Circuits and Systems (ISCAS)*, Vancouver, Canada.
123. Rao R.M., and Bopardikar, A.S. (1998): 'Wavelet Transforms Introduction to theory and applications', *Addison Wesley Longman, Inc.*
124. Vetterli, M. (1992): 'Wavelet and filter banks: Theory and design,' *IEEE Trans. Signal Proc.*, **40**(9), pp. 2207–2232.
125. Vetterli, M., and Kovacevic, J. (1995): 'Wavelets and Subband Coding', Englewood Cliffs, NJ: Prentice-Hall.
126. Chui, C.K. (1992): 'Wavelet Analysis and its Applications', *Boston, MA: Acadamic Press.*
127. Grossmann, A., Kronland-Martinet, R., and Morlet, J. (1990): 'Reading and understanding continuous wavelet transforms' In Wavelets: Time-Frequency Methods and Phase Space. Combes, J.M., Grossmann, A., and Tchamitchian, Ph. eds. *New York, NY: Springer-Verlag*, pp. 2–20.
128. Metin, A. (2001): 'Nonlinear Biomedical Signal Processing – Dynamic Analysis and Modeling', *IEEE Press.*
129. Bruce, J.W. (2000): 'Fractal Physiology And Medicine: Studies of Nonlinear Phenomena in Life', Science, 1, *World Scientific Publishing Co. Pte. Ltd*, Singapore.
130. Fraser, A.M., and Swinney, H.L. (1986): 'Independent coordinates for strange attractors from mutual information', *Physical Review A*, **33**, pp. 1134–1140.
131. Fraser, A.M. (1989): 'Information and entropy in strange attractors', *IEEE transactions on Information Theory*, **35**, pp. 245–262.
132. Grassberger, P., and Procaccia, I. (1983): 'Measuring the strangeness of strange attractors', *Physica D*, **9**, pp. 189–208.
133. Wolf, A., Swift, J.B., Swinney, H.L., and Vastano, J.A. (1985): 'Determining Lyapunov exponents from a time series', *Physica D*, **16**, pp. 285–317.
134. Dangel, S., Meier, P.F., Moser, H.R., Plibersek, S., and Shen, Y. (1999): 'Time series analysis of sleep EEG', *Computer assisted Physics*, pp. 93–95.
135. Huikuri, H.V., Makikallio, T.H., Peng, C.K., Goldberger, A.L., Hintze, U., and Moller, M. (2000): 'Fractal correlation properties of R-R interval dynamics and mortality in patients with depressed left ventricular function after an acute myocardial infarction', *Circulation*, **101**, pp. 47–53.
136. Iyengar, N., Peng, C.K., Morin, R., Goldberger, A.L., and Lipsitz, L.A. (1996): 'Age-related alterations in the fractal scaling of cardiac interbeat interval dynamics', *American Journal of Physiology*, pp. 1078–1084.

137. Grassberger, P., Schrieber, T., and Schaffrath, C. (1991): 'Nonlinear time sequence analysis', *Int J Bifurcation and Chaos*, **1**(3), pp. 512–547.
138. Shannon, C.E. (1948): 'A mathematical theory of communication', *Bell Syst. Tech. Journal*, **27**, pp. 379–423.
139. Fell, J., Roschke, J., Mann, K., and Schaffner, C. (1996): 'Discrimination of sleep stages: a comparison between spectral and nonlinear EEG measures', *Electroencephalogr Clin Neurophysiol*, **98**(5), pp. 401–410.
140. Waheed, K., and Salam, F.M. (2002): 'A Data-Derived Quadratic nobreak Independence Measure for Adaptive Blind Source Recovery in Practical Applications', *In 45th IEEE International Midwest Symposium on Circuits and Systems*, Tulsa, Oklahoma, pp. 473–476.
141. Kantz, H., and Schreiber, T. (1997): 'Nonlinear Time Series Analysis', *Cambridge Univ. Press*: U.K.
142. Pincus, S.M., and Viscarello, R.R. (1992): 'Approximate entropy: a regularity measure for heart rate analysis', *Obstetrics Gynecology*, **79**, pp. 249–55.
143. Pincus, S.M., and Goldberger, A.L. (1994): 'Physiological time-series analysis: what does regularity quantify?', *American Journal of Physiology*, **266**, pp. 1643–56.
144. Richman, J.S., and Moorman, J.R. (2000): 'Physiological time series analysis using approximate entropy and sample entropy', *Am J Physiol*, **278**(6), pp. H2039–H2049.
145. Lake, D.E., Richman, J.S., Griffin, M.P., and Moorman, J.R. (2002): 'Sample entropy analysis of neonatal heart rate variability', *Am J Physiol*, **283**(3), pp. R789–R797.
146. Mandelbrot, B.B. (1983): 'Geometry of Nature', *Freeman Sanfrancisco.*
147. Vikram, K.Y., Sobolewski, E., Jampala, V.C., and Kay, J. (1998): 'Fractal dimension and approximate entropy of heart period and heart rate: awake versus sleep differences and methodological issues', *Clinical Science*, **95**, pp. 295–301.
148. Higuchi, T. (1988): 'Aproach to an irregular time series on the basis of the fractal theory', *Physica D*, **31**, pp. 277–283.
149. Katz, M.J. (1988): 'Fractals and analysis of waveforms', *Computers in Biology and Medicine*, **18**, pp. 145–156.
150. Eckmann, J.P., Kamphorst, S.O., and Ruelle, D. (1987): 'Recurrence Plots of Dynamical Systems', *Europhys. Lett.* **4**, pp. 973–977.
151. Chua, K.C., Chandran, V., Acharya, U.R., and Min, L.C. (2006): 'Computer-based analysis of cardiac state using entropies, recurrence plots and Poincare geometry', *Journal of Medical & Engineering Technology* (in press).
152. Theiler, J., Eubank, S., Longtin, A., Galdrikian, B., and Farmer, J.D. (1992): 'Testing for nonlinearity in time series: the method of surrogate data', *Physica D*, **58**, pp. 77–94.
153. Stanley, H.E., Amaral, L.A., Goldberger, A.L., Havlin, S., Ivanov, P., and Peng, C.K. (1999): 'Statistical physics and physiology: monofractal and multifractal approaches', *Physica A*, **270**, pp. 309–324.
154. Mansier, P., Clairambault, J., Charlotte, N., Medigue, C., Vermeiren, C., Lepape, G., Carre, F., Gounaropoulou, A., and Swynghedauw, B. (1996): 'Linear and non-linear analyses of heart rate variability: a minireview', *Cardiovasc Res.*, **31**, pp. 371–379.

155. Furlan, R., Guzzetti, S., Crivellaro, W., Dassi, S., Tinelli, M., Baselli, G., Cerutti, S., Lombardi, F., Pagani, M., and Malliani, A. (1990): 'Continuous 24-hour assessment of the neural regulation of systemic arterial pressure and RR variabilities in ambulant subjects', *Circulation*, **81**, pp. 537–547.
156. Pinna, G.D., Maestri, R., Di Cesare, A., Colombo, R., and Minuco, G. (1994): 'The accuracy of power-spectrum analysis of heart-rate variability from annotated RR lists generated by Holter systems', *Physiol Meas.*, **15**, pp. 163–179.
157. Merri, M., Farden, D.C., Mottley, J.G., and Titlebaum, E.L. (1990): 'Sampling frequency of the electrocardiogram for spectral analysis of the heart rate variability', *IEEE Trans Biomed Eng.*, **37**, pp. 99–106.
158. Denton, T.A., Diamond, G.A., Helfant, R.H., Khan, S., and Karagueuzian, H. (1990): 'Fascinating rhythm: a primer on chaos theory and its application to cardiology', *Am Heart J.*, **120**, pp. 1419–1440.
159. Goldberger, A.L. (1996): 'Non-linear dynamics for clinicians: chaos theory, fractals, and complexity at the bedside', *Lancet*, **347**, pp. 1312–1314.
160. Peng, C.K., Havlin, S., Stanley, H.E., and Goldberger, A.L. (1995): 'Quantification of scaling exponents and crossover phenomena in nonstationary heartbeat time series', *Chaos*, **5**, pp. 82–87.
161. `http://reylab.bidmc.harvard.edu/tutorial/DFA/node5.html`.
162. Pikkujamsa, S.M., Makikallio, T.H., Sourander, L.B., Raiha, I.J, Puukka, P., Skytta, J. Peng, C.K., Goldberger, A.L., and Huikuri, H.V. (1999): 'Cardiac interbeat interval dynamics form childhood to senescence', *Circulation*, **100**(4), pp. 393–399.
163. Goldberger, A. L., and West, B.J. (1987): 'Fractals in physiology and medicine', Yale J. Biol. Med., **60**, pp. 421–435.
164. Pincus, S.M., Gladstone, I.M., and Ehrenkranz, R.A. (1991): 'Aregularity statistic for medical data analysis', *J. Clin. Monit.* **7**, pp. 335–345.
165. Glenny, R.W., Robertson, H.T., Yamashiro, S., and Bassingthwaighte, J.B. (1991): 'Applications of fractal analysis to physiology', *J. Appl. Physiol.*, **70**, pp. 2351–2367.
166. Acharya, U.R., Joseph, P.K., Kannathal, N., Lim, C.M., and Suri, J.S. (2006): 'Heart rate variability: A Review', IFMBE *Journal of Medical & Biological Engineering & Computing Journal* (in press).
167. Censi, F., Calcagnini, G., and Cerutti, S. (2002): 'Coupling patterns between spontaneous rhythms and respiration in cardiovascular variability signals', *Computer Methods and Programs in Biomedicine*, **68**(1), pp. 37–47.
168. Marwan, N., Wessel, N., Meyerfeldt, U., Schirdewan, A., and Kurths, J. (2002): 'Recurrence-plot-based measures of complexity and their application to heart-rate-variability data', *Physical Review E*, **66**(2), pp. 1–8.
169. Jamsek, J., Stefanovska, A., and Mcclintock, P.V.E. (2004): 'Nonlinear cardio-respiratory interactions revealed by time-phase bispectral analysis", *Physics in Medicine and Biology*, **49**, pp. 4407–4425.
170. Witte, H., Putsche, P., Eiselt, M., Arnold, M., and Schmidt, K. (2001): 'Technique for the quantification of transient quadratic phase couplings between heart rate components', *Biomed. Tech. (Berl)*, **46**, pp. 42–49.
171. Pinhas, I., Toledo, E., Aravot, D., and Akselrod, S. (2004): 'Bicoherence Analysis of New Cardiovasculr Spectral Components Observed in Heart-Transplant Patients: Statistical Approach for Bicoherence Thresholding', *IEEE Trans on Biomedical Engeering*, **51**, pp. 1774–1783.

6

Data Fusion of Multimodal Cardiovascular Signals

Kannathal N, Rajendra Acharya U, E.Y.K. Ng, Lim Choo Min, Jasjit S Suri, Jos A E Spaan

Computer technology has an important role in structuring biological systems. The explosive growth on high performance computing techniques in recent years with regard to the development of good and accurate models of biological systems has contributed significantly to the new approaches on the modeling transient behavior of biological system. Data Fusion is the process of combining data from several sources, inputs from sensors with information from other sensors, information processing blocks, data bases or knowledge bases into unified representational format [1, 2]. A data fusion system must identify when data represents different views of the same object, when data is redundant, and when mismatch occurs between data items. Data fusion deals with the synergistic combination of information made available by different measurement sensors, information sources and decision makers. Thus, sensor fusion is concerned with distributed detection, sensor registration, data association, state estimation, target identification, decision fusion, user interface and database management [3]. Various techniques involved in fusion are least square method, Bayesian method, fuzzy logic, neural network and so on, but they lack information on how they are applied [4,5]. Attempts have been made to relate these fusion techniques with fusion tasks in the fusion architecture framework. Data fusion [3, 6] architecture has gone through various developmental phases and gradually has evolved into two techniques, the rule-based decision-making and fuzzy logic decision-making [7].

Multiple sensor systems were originally motivated by their applications in military surveillance but are now being employed in a wide range of applications [8–11]. Location of a moving object (such as an aircraft) using radar can be taken as an example [12]. Even tough data fusion methods were developed primarily for military applications, many non-military applications including in the area of biomedical engineering are emerging. They include applications to condition monitoring, monitor of machines, robotics and medicine [13–20]. A typical application in medicine is the detection of patient status based on the data obtained from the recording of multi channel electrocardiogram (ECG), arterial blood pressure (ABP) and respiration. Using of multimodal data can improve disease detection in various ways. In the past, multisensor fusion for arterial and ventricular activity detection in coronary care monitoring was carried out. Alfredo *et al* [21] have presented multisensor and multisource data fusion skills to improve atrial and ventricular activity detection in critical care environments. Fracisco *et al* [22] proposed a framework for fusion of structured and unstructured data based on case based reasoning concept. A novel approach for robust cardiac rhythm tracking based on data fusion has been described by Thoraval *et al* [13]. They have reported that their approach gives better detection of abnormal ventricular contractions. Hence, one can expect better results with regard to diagnosis by fusion of biological signals from various sources.

6.1 Approaches for Fusion

Patient monitoring systems are used in critical-care units (CCU) to detect, characterize, and automatically generate alarms for each potential life-threatening event. Data acquired about the patient consists of one or more measurements from different types of data gathering devices, such as electrocardiogram, blood pressure meters, transthoracic impedance and plethysmograph. After processing, this raw data is turned into information streams containing multiple measurements of heart rate, respiratory rate, systolic and diastolic blood pressure and SpO_2 [23–29]. These measurements can be fused to yield more accurate estimates of the actual patient parameters and status information such as the detection of sensor failures [13]. This can aid in the elimination of false-positive cases [4]. Fusion of multimodal data can be modelled as multi-dimensional process.

$$\mathrm{Y(k) = [E(k)R(k)B(k)P(k)]} \tag{6.1}$$

where k denotes the discrete time index, while E(k), R(k), B(k), P(k) refer, respectively to ECG, Respiratory, ABP, and PLETH channels in Eq. (6.1).

$$\mathrm{E(k) = (e(k), e(k+1), e(k+2), \ldots\ldots\ldots)} \tag{6.2}$$
$$\mathrm{R(k) = (r(k), r(k+1), r(k+2), \ldots\ldots\ldots)} \tag{6.3}$$
$$\mathrm{B(k) = (b(k), b(k+1), b(k+2), \ldots\ldots\ldots)} \tag{6.4}$$
$$\mathrm{P(k) = (p(k), p(k+1), p(k+2), \ldots\ldots\ldots)} \tag{6.5}$$

In Eq. (6.2) e(k) refers to ECG data, at $(k)^{th}$ instant of time, r(k) refers to respiratory data, b(k) refers to blood pressure data and p(k) refers to plethysmograph data at k^{th} instant of time respectively in Eqs. (6.3,6.4,6.5).

$$\mathrm{E(k) = [(e_1(k), e_1(k{+}1), e_1(k{+}2), \ldots\ldots\ldots), e_2(k), e_2(k{+}1), e_2(k{+}2), \ldots\ldots\ldots)]} \tag{6.6}$$

where $e_1(k), e_2(k)$ are two parameters heart rate and change in heart rate extracted from ECG signal E(k) at k^{th} instant given in Eq. (6.6).

$$R(k) = [(r_1(k), r_1(k{+}1), r_1(k{+}2), \ldots\ldots\ldots), (r_2(k), r_2(k{+}1), r_2(k{+}2), \ldots\ldots\ldots)] \quad (6.7)$$

where $r_1(k), r_2(k)$ are two parameters respiratory rate and change in respiratory rate extracted from respiratory signal R(k) at k^{th} instant given in Eq. (6.7).

$$\begin{aligned} B(k) = \; & [(b_1(k), b_1(k+1), b_1(k+2), \ldots\ldots\ldots), \\ & (b_2(k), b_2(k+1), b_2(k+2), \ldots\ldots\ldots) \\ & (b_3(k), b_3(k+1), b_3(k+2), \ldots\ldots\ldots)] \end{aligned} \quad (6.8)$$

where $b_1(k), b_2(k), b_3(k)$ are three parameters systolic pressure, diastolic and mean pressures extracted from blood pressure signal B(k) at k^{th} instant given in Eq. (6.8).

$$P(k) = [(p_1(k), p_1(k+1), p_1(k+2), \ldots\ldots\ldots)] \quad (6.9)$$

where $p_1(k)$ is the parameter oxygen saturation extracted from plethysmograph signal P(k) at k^{th} instant and is given in Eq. (6.9)

Multiple measurements of the same data are considered competitive data. For example, three measurements of heart rate must be fused to yield one estimate of the actual heart rate of the patient. If one sensor fails or is erratic, while the other two are very close, then the average of those two should be used as the correct estimate. Thus competitive integration yields two outputs. The first is the integrated data, in this case the heart rate. The second is status information, such as information about an erratic or failed sensor.

When multi-modal data is fused, it is considered complementary integration, which is defined as the integration of overlapping (partial) data. The data is partial because it only covers a certain aspect of the patient state. However, it is overlapping because the different types of data change together, as patient state changes. Complementary integration does not produce better estimates of the patient parameters as competitive data does. However, it does yield status information. The most obvious example is, if one type of sensor fails, the others continue to function normally.

In the model developed (Fig. 6.1), there are two heart rate measurements, one respiratory rate, blood pressure systolic, diastolic and mean pressures and one SpO_2 measurement. The relation between the two heart rate measurements is competitive, and can thus be used to yield a more accurate measurement and status information about the two sensors. The relationship between the heart rate, respiratory rate and respiratory volume measurements are complementary. Each of them partially cover the patient state, and can be fused to yield status information about the sensors. For example, if heart rate from lead 1, indicates zero but heart rate from lead 2 indicates some valid

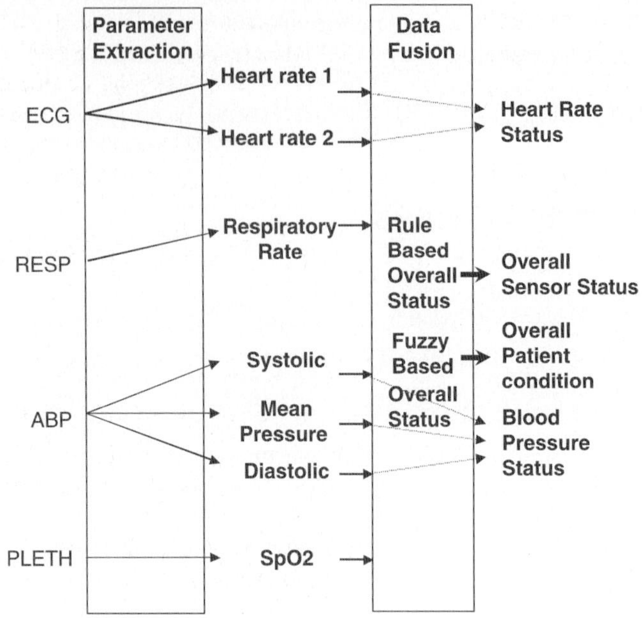

Fig. 6.1. Data Fusion Model of multi-modal signals

value, the ECG lead1 sensor has most likely failed. Similarly, the heart rate may indicate that the patient's heart has stopped beating, but the respiratory rate may show normal breathing.

It is unlikely that the patient's heart has stopped, and the heart rate sensor has failed [13]. These are the ways in which data fusion can provide better information to eliminate false-positive detections.

The rules used to create the heart rate status are as follows. If there is a difference of more than 5% between the two heart rate measurements, or if the status of the heart rate sensors differs, a "Heart rate discrepancy" is flagged. Otherwise, the heart rate is set to the average of the two values. To create the overall status, a tally is taken of the number of sensors that report a status of OK. If they do not indicate OK, they may indicate SENSOR_ERROR, which indicates a hardware failure, NO_DATA, which indicates that not enough data has yet been acquired to calculate patient parameters, or STALE_DATA which indicates no new data has been received within a certain time period. All of these results can be indicative of patient deterioration or sensor failures. However, if the problems are consistent between sensors, they are more likely to flag a problem with the patient rather than the sensors. The table on sensor discrepancy and the sensors reporting OK is shown in the Table 6.1. Parameters extracted from the multi-modal data are combined using rule based approach and fuzzy reasoning methods.

Table 6.1. Sensor discrepancies

Sensors reporting OK	Message Returned
Four (all sensors OK)	No sensor abnormalities detected
Three	Sensor discrepancy (one sensor)
Two	Sensor discrepancy (two sensors)
One	Possible acute patient deterioration; Sensor discrepancy (three sensor)
Zero (no sensors OK)	Highly likely acute patient deterioration

6.2 Rule Based Approach

Set theory has been found to be at the center of universe for the modern computing world. Every element in the world either belongs or doesn't belong to a set; either member or not a member of a set; either true or false. New rules can be derived from existing knowledge by using the true or false statements. Utilizing the information obtained from analyzing multimodal data, new rules can be formulated for detecting critical conditions of the patient. Rules can be developed such as: "If heart rate exceed 90 bpm" or "If respiratory rate exceeds 20" or "If mean Pressure has dropped by 20 bpm" and "Spo2 has decreased below 95%" then patient is diagnosed with left ventricular failure. A rule-based decision making system employs a series of Boolean result parameter tests, combined together with a series of Boolean operators such as AND, OR to indicate whether a particular condition is present or not.

There are many different categories of problems that could be detected by the rule-based approach, such as:

- A drop in heart rate over time is indicative of cardiac problem
- A drop in blood pressure combined with a rise in heart rate indicates that the heart is not pumping forcefully or low preload. The rise in heart rate occurs as the body attempts to increase blood flow.
- A drop in SpO_2 % indicates that oxygen content is reduced.
- Ventricular failures could be detected by a sudden change in heart rate to a very high rate.

The percentage mentioned above are just approximate test values and the clinician has choice of providing the accurate conditions and limits to the Table 6.3 based upon his clinical experience and clinical literature.

Some life-threatening episodes can be detected by using the rules documented in Table 6.2. These rules are not easily available in literature and are assigned based on the discussions with experienced cardiac surgeons. ECG, respiratory, blood pressure and plethysmograph signals were considered for fusion in Table 6.2, however if more parameters such as pressure signals from individual chamber of heart such as left atrium, pulmonary artery, right atrium if fused along with oxygen saturation, a more specific diagnosis can be made. Table 6.3 gives a general idea for fuzzification of parameters.

Table 6.2. Fusion Rules for Cardiovascular Condition Diagnosis

Condition	Constraints	Typical Test Values
Left Ventricular Failure	High Heart Rate High Respiratory Rate Drop in Blood Pressure Low Oxygen Saturation	HR > 90 BPM and RR > 20 RPM/minute SBP < 80 mm/HG SpO_2 < 95%
Right Ventricular Failure	High Heart Rate Very high Respiratory Rate Less drop in Blood Pressure Very Low Oxygen Saturation	HR > 95 BPM and RR > 25 RPM/minute SBP < 110 mm/HG SpO_2 < 90%
Pulmonary Oedema	Very high Heart Rate Very high Respiratory Rate Large drop in Blood Pressure Very Low Oxygen Saturation	HR > 120 BPM and RR > 25 RPM/minute SBP < 80 mm/HG SpO_2 < 90%
Tachycardia	Very high Heart Rate	HR > 120 BPM
Bradycardia	Very low Heart Rate	HR < 60 BPM

Table 6.3. Probabilistic rules assigned by the physician

(HR)		(RR)	(SBP)	(SpO_2)	Failure
		Single Constraint			
1–1	HR > 150	∈ all RR	∈ all SBP	∈ all SpO_2	80%
1–2	110 < HR < 130	∈ all RR	∈ all SBP	∈ all SpO_2	50%
1–3	90 < HR < 110	∈ all RR	∈ all SBP	∈ all SpO_2	30%
1–4	80 < HR < 90	∈ all RR	∈ all SBP	∈ all SpO_2	10%
		Double Constraints			
2–1	110 < HR < 130	RR > 20	∈ all SBP	∈ all SpO_2	80%
2–2	90 < HR < 110	15 < RR < 20	∈ all SBP	∈ all SpO_2	50%
2–3	80 < HR < 90	10 < RR < 15	∈ all SBP	∈ all SpO_2	15%
...	...	...	...	...	...
		Triple Constraints			
3–1	110 < HR < 130	RR > 20	SBP < 70	∈ all SpO_2	80%
3–2	90 < HR < 110	15 < RR < 20	70 < SBP < 80	∈ all SpO_2	60%
3–3	80 < HR < 90	10 < RR < 15	80 < SBP < 100	∈ all SpO_2	20%
...	...	...	...	...	...
		Four Constraints			
4–1	110 < HR < 130	RR > 20	SBP < 70	SpO_2 < 90%	85%
4–2	90 < HR < 110	15 < RR < 20	70 < SBP < 80	95% < SpO_2 < 90%	70%
4–3	80 < HR < 90	10 < RR < 15	80 < SBP < 100	90% < SpO_2 < 99%	25%
...	...	...	...	...	...

6.3 Introduction to Fuzzy Based Decision Making

The world around us is very uncertain and unpredictable. No bivalent logic (also called Boolean or binary logic or law of excluded middle) can possibly solve real world problems without oversimplification. Many approximate

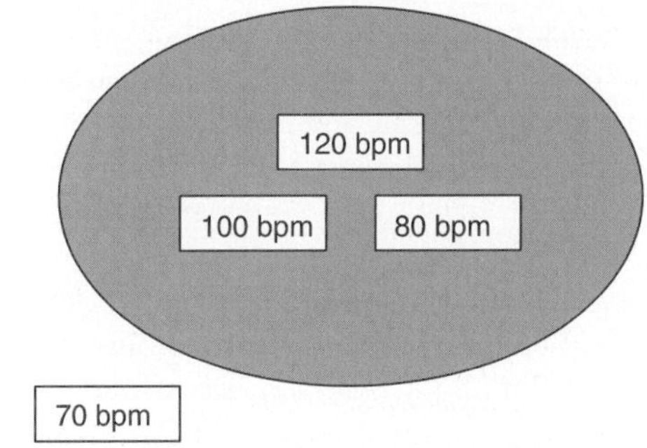

Fig. 6.2. Heart rates defined exactly by 72 bpm

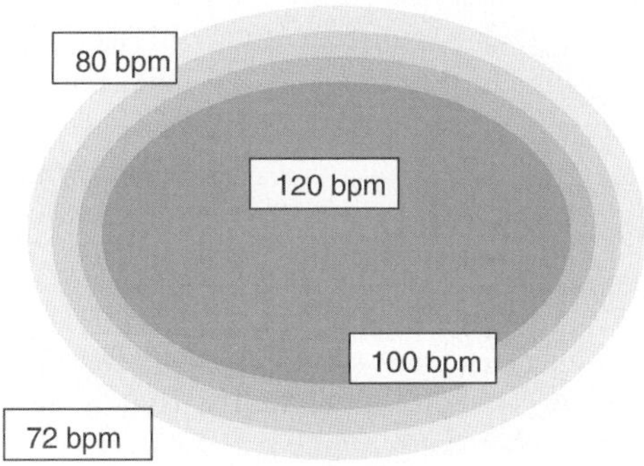

Fig. 6.3. Fuzzy Set of patients with high heart rates

reasoning (also called multi-valued, or continuous or fuzzy logic) methods have been proposed and applied to real world problems [4]. Fuzzy logic allows the representation of human decision and evaluation processes. In reality a crisp rule for certain case cannot be defined. These rules are discrete points in the continuum of possible cases and approximation is done between them. The full scope of human thinking, creativity cannot be mimicked by fuzzy logic. Solution can be derived for a given case by applying fuzzy logic techniques to the rules that have been defined for similar cases.

Most of the medical books describe that normal heart rate is 72 beats per minute, which doesn't mean that a patient with 71 bpm has abnormally low heart rate and a patient with 73 bpm has abnormally high heart rate. Figure 6.2 gives an example of the 'patients with high heart rate' (dark area), where the indicator function defines high heart rate as rates higher than 72 bpm. Figure 6.3 gives an example of a set, where certain elements can also be "more-or-less" members. The shades of grey indicate the degree to which the heart rate belongs to the set of high heart rate. This shades of grey which makes the dark grey area in Fig. 6.2 look "fuzzy" and gave Fuzzy Logic its name.

In Fig. 6.3, each heart rate is associated with a certain degree to which it matches the prototype for high heart rate. This degree is called the "degree of membership" $\mu_{\mathrm{HR}}(\mathrm{x})$ of the element $\mathrm{x} \in \mathrm{X}$ to the set high heart rate, where X is set of all high heart rates and heart rate is called a base variable 'x' with universe X. Different classes of events are specified, along with a membership function.

Fuzzy set is an extension of regular set in which for each element there is also a degree of membership associated with it. The degree of membership can be any value between 0 and 1. An element, whose degree of membership in a set is 0, doesn't belong to that set at all. An element whose degree of membership in a set is 1, belongs one hundred percent to that set. An element whose degree of membership in a set is 0.8, belongs eighty percent to that set and so on. Also an element can belong to more than one set to various degrees of membership. This provides a powerful scheme for representation of uncertainity. Thus this continuous or fuzzy logic includes conventional or binary logic as a special case and extends beyond that.

6.4 Fuzzy Logic Approach

In a rule-based system, a Boolean response is assigned to a condition. Thus a patient either has or does not have a certain condition. In contrast, a fuzzy-logic system attempts to assign a probability that a patient has a certain condition. Thus, a fuzzy-logic system may produce the response that a patient is 70% likely to have a Left Ventricular Failure, but only 8% likely to be having Pulmonary edema. This type of system has some advantages. Firstly, it rids the system of rigid thresholds, such as HR $>$ 90 bpm. A patient with high respiratory rate may be having respiratory troubles without crossing this threshold. In the rule-based system, this case would not be triggered. In contrast, the fuzzy-logic system would assign it a probability almost as high as for a patient with RR $>$ 20. It is for this reason the system is denoted fuzzy, as the boundaries become fuzzy rather than rigid. A second advantage of fuzzy-logic is it can be used to prioritize tasks. If one patient has a high probability of a serious state, while another patient has a lower probability of a less serious state, limited resources can tend to the highest risk group first. Fuzzy-logic is implemented using a decision function, as modeled in Fig. 6.4.

The inputs to the function are a set of patient parameters. The outputs are the probabilities that different conditions are occurring. For example, condition 1, Left Ventricular Failure may have a severity or probability of 70% while condition 2, Pulmonary edema, may have a probability of 5%. The challenge in fuzzy-logic is to create the decision function, as it may be a complex mathematical function. To partially automate the task, the method used is, for each condition, to create a closed object in n-space based on the Boolean rules for that condition, to fuzzify its boundaries, and finally to create a mathematical function to correspond to the fuzzified object.

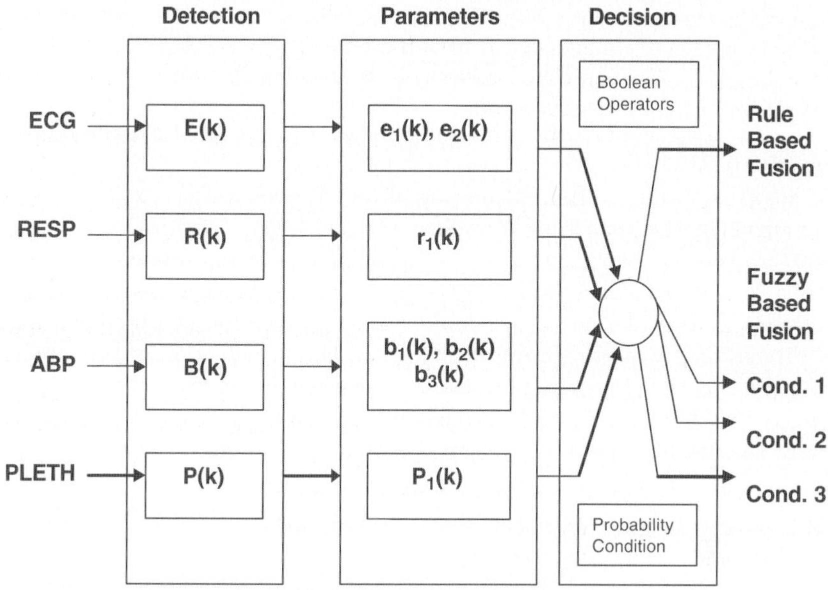

Fig. 6.4. Fuzzy-Logic Decision Function Structure

6.4.1 Fuzzy-Logic Decision Function Created by Fuzzifying Boolean Rules

This method performs a transformation of the Boolean rules to smoothen their boundaries, creating a probabilistic region. The algorithm consists of the following steps. (i) Quantize the patient parameters. For example, heart rate is broken down into 18 steps of 10 BPM/step, blood pressure is broken down into 10 steps of 10 mm/hg /step, oxygen saturation is broken down into 10 steps of 5% per step and respiratory rate is broken down into 10 steps of 10/step. (ii) A 4-dimensional matrix is created with one quantized patient parameter on each dimension. Each element in matrix is assigned a probability P_r of either 100% or 0% based on the Boolean rules. (iii)A new 4-dimensional matrix is created with the same axis as the first. Each element of matrix P_r is fuzzified over window of size 'w' and assigned a probability P_f using the Eq. 6.10

$$P_f[i][j][k][l] = \sum_{p=-w}^{w} \sum_{q=-w}^{w} \sum_{r=-w}^{w} \sum_{s=-w}^{w} P_r[p][q][r][s]/d \qquad (6.10)$$

where d is given by Eq. 6.11, and p, q, r, s indicate the four patient parameters and i, j, k, l indicate the dimension of first, second, third and fourth parameters respectively.

$$d = [1 + \sqrt{(p^2 + q^2 + r^2 + s^2)}] \qquad (6.11)$$

$P_f(\max)$ is the maximum value in Eq. (6.12)

$$P_{f^I}[i][j][k][l] = (P_f[i][j][k][l]/P_f(\max))x100\% \tag{6.12}$$

Eq. 6.12 is used to normalize the probabilities P_f so that the highest probability is 100%.

P_f (max) is obtained after calculating all of the probabilities, P_f.

Currently, the algorithm is implemented in 4-space with four patient parameters and the fuzzification is based on the distances between parameters over a four dimensional window of size 'w'. It can be extended to more dimensions by considering 'n' parameters and fuzzification can be done over a 'n' dimensional window. In implementation, the fuzzification of probabilities is not done over entire matrix but only over elements in the neighborhood of window 'w'. This is done to reduce computational time. Also, by reducing the size of the neighborhood, the fuzzified boundaries become sharper.

6.4.2 Fuzzy-Logic Patient Deterioration Index

Life threatening events like Left Ventricular Failure, Pulmonary edema, and Right Ventricular Failure are assigned with weights based on the risk factor. Patient Deterioration Index is modeled as shown in the Fig. 6.5. Weights assigned to risk factors are given in Table 6.4. Patient deterioration index is formulated to assess the criticality of the condition of the patient. It is based on the fuzzy logic probabilities of three different critical conditions of the heart. Patient Deterioration Index (μ_{di}) is given by the Eq. 6.13.

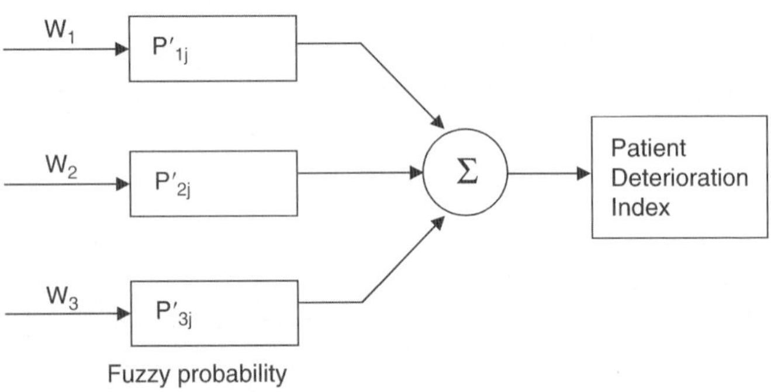

Fig. 6.5. Patient Deterioration Index Function Structure

Table 6.4. Weights assigned to Cardiovascular Problems

Condition	Risk Factor(W_k)
Left Ventricular Failure	0.40
Right Ventricular Failure	0.10
Pulmonary edema	0.50

$$\mu_{di}(j) = \sum_k W_k P'_{kj} \tag{6.13}$$

Where k refers to the different life threatening problems (left ventricular failure, right ventricular failure and pulmonary edema) and P'_{kj} is the corresponding fuzzy logic probability. Patient deterioration index ranges between 0 and 1 where 0 indicates that patient has no deterioration and 1 indicates maximum deterioration.

The weights (W_k) are assigned depending on the seriousness of the disease. Among the four cardiac abnormalities, Pulmonary edema is considered to be most critical and hence assigned a higher weight. Patient Deterioration Index (μ_{di}) value depends on the weights assigned in Table 6.4. Normal subjects do not have any risk factors, so the weight assigned is 0. Fuzzy probability (P'_{kj}) will yield the percentage of the disorders in the subject considered. The weighted sum of these probabilities will yield a single index, which indicates the cardiac health state.

6.5 Patient States Diagnosis System Implementing Data Fusion

Overview of the patient state diagnosis system using data fusion is shown in Fig. 6.6. Data from specific coronary (CCU) events is required. Acquired data undergoes signal processing and parameters are extracted. These parameters are fused and patient's condition is diagnosed. Patient data is acquired from the Physiobank's MIMIC database. Figure 6.7 shows the multi-modal data from MIMIC Database.

The acquired data undergoes preliminary signal processing to extract patient parameters. Tompkins algorithm is used to detect ECG QRS complexes. The signal is digitally bandpass filtered using cascaded integer high-pass and low-pass filters. Differentiation is done to detect the slope of the ECG and to exaggerate the QRS-complex. Then differentiated signal is squared to make all data points positive and non-linear amplification of the output of the derivative to emphasize the higher frequencies. QRS complexes are detected using an upward and downward threshold called adaptive threshold. These are calculated using running estimates of signal peak and noise peak. Thus the thresholds are dynamically adjusted to improve detection.

Function used to calculate respiratory rate detects global peak, global trough, local peak and local trough. The global peak and trough are defined as the largest and smallest values over the range of the entire data. The local peak and trough are the values of the largest and smallest pieces of data until the respective local maximum or minimum is left. The respiratory rate is extracted based on the number of respirations divided by time period between local peaks in which the breaths occur. The respiratory volume is calculated based on the assumption that the patient's vital capacity (VC), that is, the

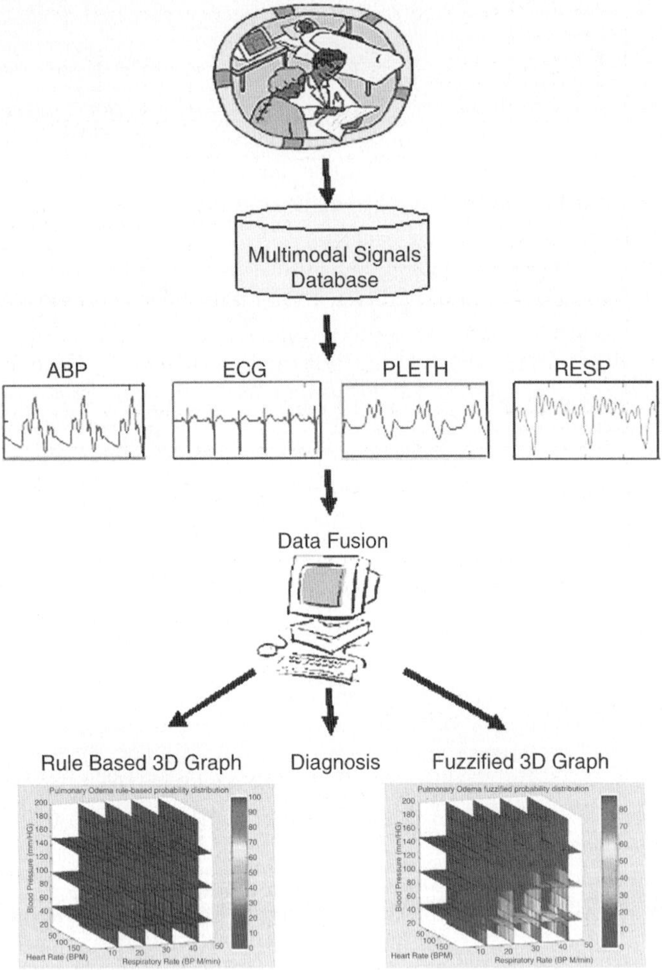

Fig. 6.6. Overview of the Patient state diagnosis system using data fusion

difference between their total lung capacity and residual volume, is 5 litres. It is further assumed that the difference between the global peak and trough corresponds to the vital capacity.

ABP Peaks and troughs are detected based on the local maxima and local minima. Lowest value is stored in the local trough and it is compared with the next data. Minimum value of the data before a peak arrives gives the diastolic and maximum value of the data before a trough arrives gives systolic pressures. Systolic and diastolic pressures are calculated based on the calibrations given in the header file of the data file. Plethysmograph signal is not calibrated and cannot be used in isolation to determine O_2 saturation. The text file found in the same data directory contains the SpO_2 measurements provided by the pleth module.

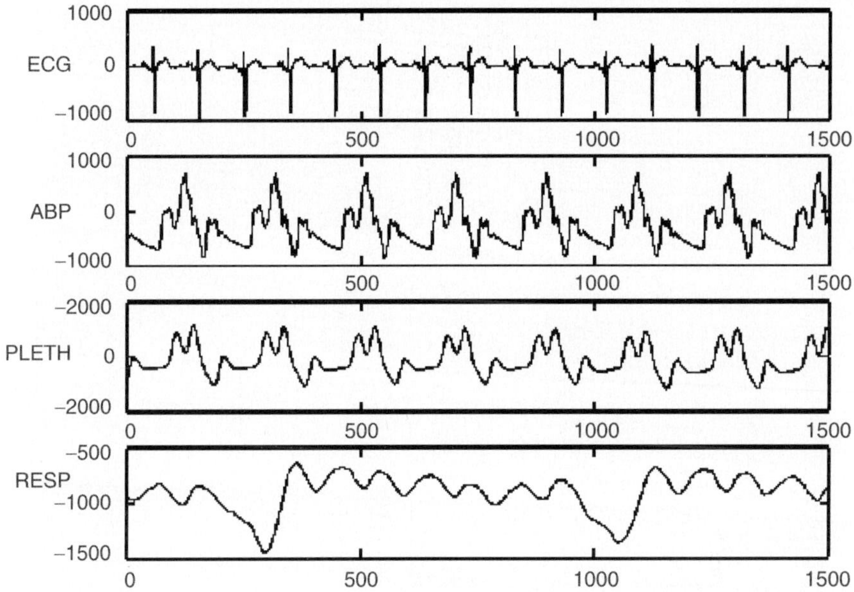

Fig. 6.7. Multi-Modal Data from MIMIC Database

6.6 Results and Discussion

The parameters extracted from the multi-modal signals are used for decision-making using data fusion techniques. The signal information is fused to yield more accurate estimates of the actual patient status information such as the detection of sensor failures, if it has occurred. Status information is based on the fact that "if heart rate indicates the patient's heart has stopped beating but blood pressure signal shows the pressure signal, it is unlikely that the patient's heart has stopped, and more likely that the heart rate sensor (ECG) has failed." This can aid in the elimination of false-positive cases. Status information can be found in lowest part of the Figs. 6.8 and 6.9. Figure. 6.8 indicates the sensor abnormalities as respiratory data is not available and Fig. 6.9 shows that there are no sensor abnormalities detected.

Fuzzy-logic decision function created by fuzzifying Boolean rules is applied to ECG, ABP, PLETH and respiratory signals derived from MIT-MIMIC database. Parameters extracted from the signals are, heart rate, respiratory rate, systolic blood pressure, diastolic blood pressure, mean blood pressure and oxygen saturation. Four-dimensional matrix is formed based on the rule based decision function using the parameters extracted from the signals for each of the three abnormalities discussed below. Elements of the matrix are fuzzified based on the fuzzy-logic decision function.

A program was written in Matlab to graph the rule-based and fuzzy-logic distribution files. It reads the three-dimensional probabilities and graphs

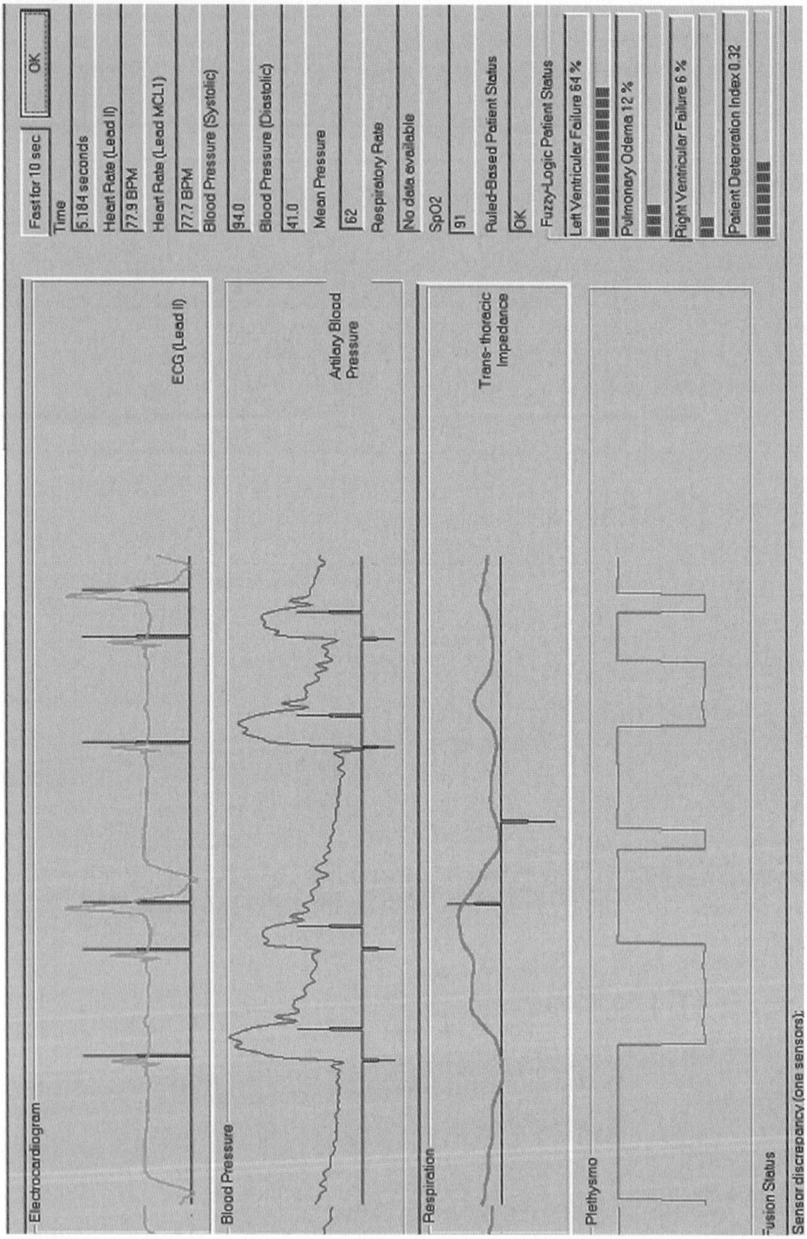

Fig. 6.8. Snapshot of the system showing the sensor discrepancy

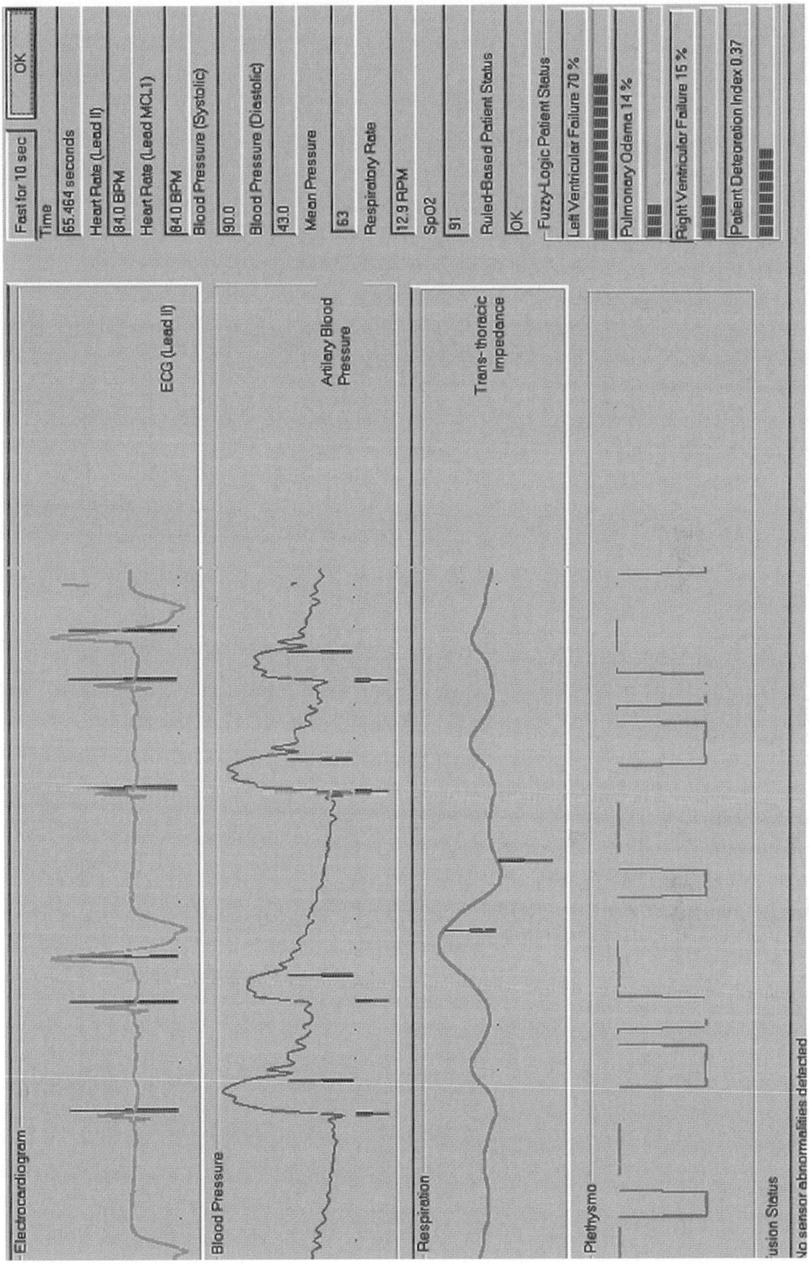

Fig. 6.9. Snapshot of the system showing the patient's deteriorated state

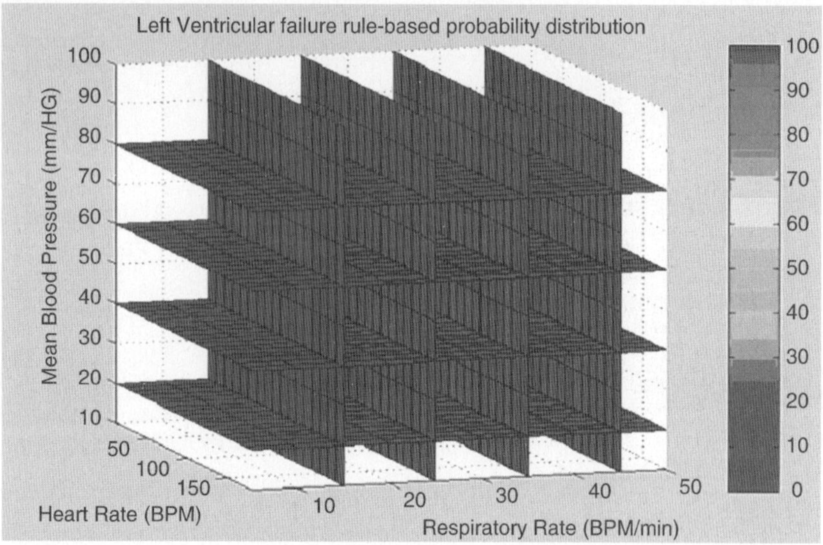

Fig. 6.10. Rule based probablility distribution for Left Ventricular Failure# (color image)

them in slices of a three-dimensional Cartesian graph. It is customized for Left Ventricular Failure, Pulmonary edema and Right Ventricular Failure. Based on the parameter values given in Table 6.1, three dimensional graphs for rule-based and fuzzyfied distributions are shown in Fig. 6.10 and Fig. 6.11 respectively.

Figure 6.10 shows three dimensional graph for left ventricular failure. Three main parameters mean pressure, respiratory rate and heart rate are used to form a three dimensional matrix. Three dimensional graph is formed based on the values of the rule-based matrix. Respiratory axis is divided into 10 slices of 5 respirations per minute each, heart rate axis is divided into 18 slices of 10 beats per minute and mean blood pressure axis is divided into 10 slices of 10 mm/hg. Red colored region indicates that patient has deteriorated based on the rules given in Table 6.1. Patient having heart rate higher than 90 beats per minute and respiratory rate higher than 20 respirations per minute and mean blood pressure lower than 80 mm/hg is diagnosed with left ventricular failure. Figure 6.10 clearly shows the above three conditions and patient's state.

In a rule-based system, a Boolean response is assigned to a condition. Thus a patient either has or does not have a certain condition. In contrast, a fuzzy-logic system attempts to assign a probability that a patient has a certain condition. Diagnosis made by physicians is not based only on crisp rules. Patient having heart rate 89 beats per minute and respiratory rate higher than 19 respirations per minute and mean blood pressure 85 mm/hg is not considered as healthy condition by physician.

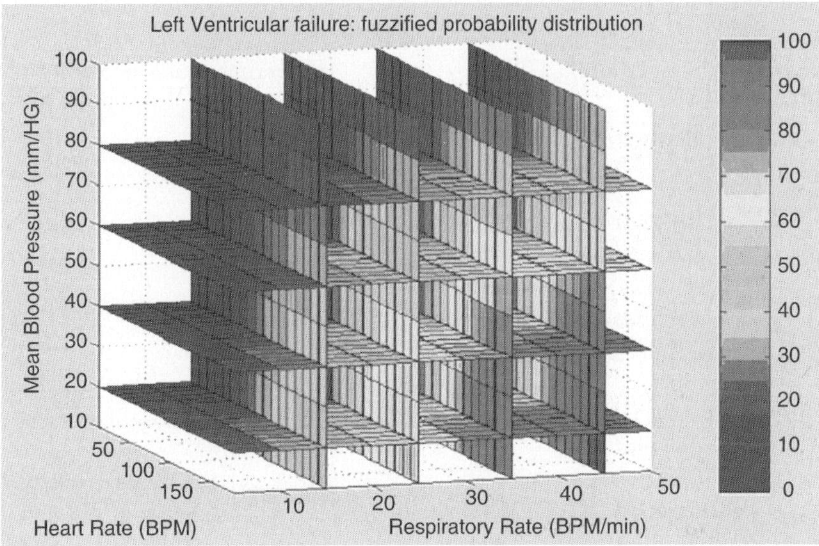

Fig. 6.11. Fuzzy based probability distribution for Left Ventricular Failure# (color image)
[Reprinted with permission from Kannathal N, Rajendra Acharya U, Ng, E. Y. K, Lim Choo Min, Swamy Laxminarayan, "Cardiac health diagnosis using data fusion of Cardiovascular and haemodynamic signals", Computer methods and Programs in Biomedicine, Sweden, 82(2), 2006, 87–96]

The proposed system performance in recognition and classification is evaluated by means of three performance indices viz. classification accuracy, sensitivity, and specificity. The results are given in Table 6.5 and Table 6.6. From the Table 6.6, it can be seen that the proposed system produces promising results with more than 93% diagnostic accuracy. The system is evaluated to have a sensitivity of more than 90% and specificity of more than 91%. The proposed system is intended to aid physicians in ICU environment, where multimodal signals are monitored for long hours. It is not possible for the physician to be attending the patient when the patients are subjected to long term continuous monitoring. The proposed system can give preliminary diagnostics in monitoring the patient.

It has been shown that by combining data sources, better results can be obtained with reduction in the number of false-positive cases. More important, the introduction of fuzzy logic based decision-making improves the likelihood of catching false negative cases that are close to the boundaries of the rule based reasoning. An index called patient deterioration index has been calculated. Testing with limited data has been done and the system is found to perform satisfactorily.

Table 6.5. Results of correct classifications for different clinical classes

Record No	Clinical Class	Correct Classifications		
		Total	TP	TN
211	Respiratory Failure	1705	539	1166
212	CHF / Pulmonary edema	3065	831	2234
213	CHF / Pulmonary edema	4541	2048	2493
214	CHF / Pulmonary edema	2415	1351	1064
215	CHF / Pulmonary edema	2382	1466	916
216	Respiratory Failure	3017	1495	1522
218	Respiratory Failure	1723	782	941
219	Respiratory Failure	1942	797	1145

Table 6.6. Results of accuracy, sensitivity and specificity for different clinical classes

Record No	Accuracy	Sensitivity	Specificity
211	94.41	93.74	94.72
212	93.33	94.22	93.01
213	95.22	90.74	99.24
214	93.42	91.35	96.20
215	94.11	92.26	97.24
216	93.38	93.50	93.26
218	93.19	91.14	94.95
219	95.57	91.71	98.45

6.7 Conclusion

This chapter presents a novel fusion system involving heterogeneous electrophysiological and haemodynamic data for detection of patient states in CCU. Accurate diagnosis of cardiac health using ECG alone is difficult. Hence we have shown that by combining data sources, better results can be obtained with reduction in the number of false-positive cases. And also, to evaluate the severity of the cardiac abnormality, a parameter called patient deterioration index has been proposed. Testing with limited data has been done and the system is found to perform satisfactorily with a diagnostic accuracy of more than 93%.

References

1. Mckerrow, P.J., and Volk, S.J. (1996): 'A systems approach to data fusion', *Data Fusion Symposium*, ADFS '96., First Australian, pp. 217–222.
2. Hall, D.L., and Linas, J. L. (1997): 'An introduction to multisensor data fusion', *Proc. IEEE*, **85**, pp. 6–23.

3. Kokar, M., and Kim, K. (1993): 'Review of Multisensor Data Fusion Architectures and Techniques', *International Symposium on Intelligent Control*, pp. 261–266.
4. Constantin, V.A. (1998): 'Fuzzy Logic and Neuro Fuzzy Applications Explained', *Prentice Hall.*
5. Taur, J.S., and Kung, S.Y. (1993): 'Fuzzy decision neural networks and application to data fusion', *Proceedings of the IEEE-SP Workshop on Neural Networks for Signal Processing.*
6. Sheng, T.(2000): 'Data Fusion of Multisensor Data', *Fourth International conference on Knowledge-Based Intelligent Engineering Systems*, pp. 537–540.
7. Palacharla, P., and Nelson, P. (1994): 'Understanding relations between fuzzy logic and evidential reasoning methods', IEEE World Congress on Computational Intelligence., *Proceedings of the Third IEEE Conference on Fuzzy Systems*, **3**, pp. 1933–1938.
8. Chair, Z., and Varshney, P.K. (1986): 'Optimal data fusion in multiple sensor detection systems', *IEEE Transactions on Aerospace Electronic Systems*, **22**, pp. 98–101.
9. Parhami, B. (1996): 'Multi-sensor data fusion and reliable multi-channel computation: unifying concepts and techniques', *Signals, Systems and Computers*, **1**, pp. 745–749.
10. Lopez, M., and Jimenez, R. (1995): 'Fuzzy reasoning for multisensor management', *IEEE International Conference on Systems, Man and Cybernetics*, **2**, pp. 1398–1403.
11. Toh, K.A., and Yau, W.Y. (2004): 'Combination of Hyperbolic Functions for Multimodal Biometrics Data Fusion', *IEEE Transactions on Systems, Man and Cybernetics*, **34** (2), pp. 1196–209.
12. Llinas, J., and Waltz, E. (1990): 'Multisensor data fusion', *Artech house*, Boston, MA.
13. Thoraval, L., Carrault, G., Schleigh, J.M., Summers, R., Van De Velde, M., and Diaz, J. (1997): 'Data Fusion of Electrophysiological and Haemodynamic Signals for Ventricular Rhythm Tracking', *IEEE Engineering in Medicine and Biology*, pp. 48–55.
14. Azuaje, F., Dubitzky, W., Lopes, P., Black, N., Adamson, K., Wu, X., and White, J. (1999): 'Predicting coronary disease risk based on short-term RR intervals measurements: A neural network approach,' *Artificial Intelligence in Medicine*, **15**, pp. 275–297.
15. Cerutti, S., and Saranummi, N. (1997): 'Improving Control of Patient Status in Critical Care', *IEEE Engineering in Medicine and Biology*, **16** (6), 19–20.
16. Keller, P.E., Kangas, L.J., Hashem, S., Kouzes, R.T., and Allen, P.A. (1995): 'Novel approach to modeling and diagnosing the cardiovascular system', *National Technical Information Services Database* (NTIS).
17. Held, C.M., and Roy, R.J. (2000): 'Hemodynamic management of congestive heart failure by means of a multiple mode rule-based control system using fuzzy logic', *IEEE Transactions on Biomedical Engineering*, **47**, pp. 115–123.
18. Iyengar, S.S., Sastry, S., and Balakrishnan, N. (2003): 'Foundations of data fusion for automation', *IEEE Instrumentation & Measurement Magazine*, **6**, pp. 35–41.
19. Su, S.F., and Chen, K.Y. (2004): 'Fuzzy Hierarchical Data Fusion Networks for Terrain Location Identification Problems', *IEEE Transactions on Systems, Man and Cybernetics*, **34**, pp. 731–739.

20. Perez, P., Vermaak, J., and Blake, A. (2004): 'Data Fusion for Visual Tracking With Particles', *Proceedings of the IEEE,* **92**, pp. 495–513.
21. Alfredo, I.H., Carrault, G., Fernando, M., Laurent, T., and Gianfranco, P. (1999): ' Multisensor Fusion for Atrial and Ventricular Activity Detection in Coronary Care monitoring', *IEEE Transactions on Biomedical Engineering,* **46** (10), pp. 1186–1190.
22. Azuaje, F., Dubitzky, W., Black, N., and Adamson, K. (1999): 'Improving clinical decision support through case-based data fusion', *IEEE Transactions on Biomedical Engineering,* **46** (10), pp. 1181–1185.
23. Baselli, G., Cerutti, S., Livarghi, M., Meneghin, C.I., Pagani, M., and Rimoldi, O. (1988): 'Casual relationship between heart rate and arterial blood pressure variability signal', *IEEE Transactions on Engineering Medicine and Biology,* **26**, pp. 374–378.
24. De Boer, R.W., Karemaker, J. M., and Strackee, J.(1985): 'Relationships between short-term blood pressure fluctuations and heart-rate variability in resting subjects', *IEEE Transactions on Engineering Medicine and Biology,* **23**, pp. 352–364.
25. Kitney, R.I., Fultron, T., McDonald, A.H., and Linkens, D.A. (1985): 'Transient interactions between blood pressure, respiration and heart rate in man', *IEEE Transactions on Biomedical Engineering,* **7**, pp. 217–224.
26. Jiapu, P., and Tompkins, W.J. (1985): 'A Real-Time QRS Detection Algorithm', *IEEE Trasactions on Biomedical Engineering,* **BME-32** (3), 230–237.
27. Tompkins, W.J. (1993): 'Biomedical Digital Signal Processing : C-Language Examples and Laboratory Experiments for the IBM PC', *Prentice Hall.*
28. Gritzali, F. (1988): 'Towards a generalized scheme for QRS detection in ECG', *Signal Processing,* **15**, pp. 183–192.
29. Daskalov, I.K., Ivan, A.D., and Ivailo, I.C. (1998): 'Developments in ECG Acquisition, Preprocessing, Parameter Measurement, and Recording', *IEEE Engineering in Medicine and Biology,* pp. 50–58.

7

Classification of Cardiac Patient States Using Artificial Neural Networks

N. Kannathal, U. Rajendra Acharya, Lim Choo Min and Jasjit S. Suri

Bio-signals are essentially non-stationary signals; they display a fractal like self-similarity. They may contain indicators of current disease, or even warnings about impending diseases. The indicators may be present at all times or may occur at random–in the time scale. However, to (study and) pinpoint anomalies in voluminous data collected over several hours is strenuous and time consuming. Therefore, computer based analytical tools for in-depth study and classification of data over day long intervals can be very useful in diagnostics.

The electrocardiogram (ECG) belongs to the above category of bio-signals. It displays an apparent periodicity (of about 60 to 80 bpm in a healthy adult), but is not exactly periodic. The heart rate of a healthy individual is not a constant even under serene conditions; it keeps on changing throughout the day, which can be directly monitored from the ECG. Disease and affliction do influence the heart rate, and therefore, the pattern and the range of heart rate variability would contain important information about the robustness of health, type of disease etc. Therefore, classification based on the spread and pattern of this parameter can provide useful insight about the type and intensity of the affliction.

Many researchers have suggested various techniques including non-conventional approaches like engineering diagnostic techniques for determining patient conditions. Literature in this area reveals that experimentation is being done for applying artificial intelligence approaches [1] and neural networks [2, 3] for automatic ECG analysis. The work by Mark *et al* [3], shows that the neural network techniques namely, back propagation (BP) and self organizing map (SOM) are used for classification purposes. The multimodal cardiovascular data was used as input to the neural network. Other approaches like Bayesian and heuristic approaches [4] and Markov models [5] were also experimented for classification purposes. Ischemic episode detection using artificial neural network trained with isolated ST-T segments have been developed by Frenkel *et al* [6].

Mark and Janson [7] have used multimodal cardiovascular signals from IMPROVE data library recorded from 60 patients to classify the ICU patient states using neural networks. The IMPROVE data library contains typical disorders of oxygen delivery which are found as Hypovolaemia, Cardiac Failure (CF), Sepsis (High Blood Flow State-HBFS) and Gas exchange availability. Based on this, using neural network techniques such as back propagation (BP) and self organizing maps (SOM) patient states are identified as Cardiac Failure (CF), High Blood Flow State (HBFS), neither HBFS nor CF. Adaptive Resonant Theory (ART) has been applied to a variety of domains ranging from medical applications, such as classifying ECG patterns [8], to associative memory and semantic data processing.

Barro *et al*, [9] have used multichannel adaptive resonance theory (MART) to classify the ECG patterns. Implementation results show that this classifier can classify normal and ventricular beats with an accuracy of more than 90%. Olmez [10] on his work on classification of ECG waveforms has detected 4 types such as normal beat, left bundle branch block beat, premature ventricular contraction and paced beat using Restricted Coulomb Energy (RCE) neural networks and genetic algorithms and obtained a classification accuracy of more than 94%.

Several studies have presented the performance of neural network systems for detection and recognition of abnormal ECG. The use of neural network system in ECG signal analysis offers several advantages over conventional techniques. The neural network can perform the necessary transformation and clustering operations automatically and simultaneously and also the neural network is able to recognize complex and nonlinear groups in the hyperspace. This is a distinct advantage over many conventional techniques. However little work has been devoted to deriving better parameters for reducing the size of the network while maintaining good classification accuracy.

Neural networks, emerged from the field of cognitive engineering, can learn to perform complex tasks. They are especially effective at recognizing patterns, classifying data, and processing noisy signals. Neural networks are loosely modeled on the networks of neurons in biological systems. Artificial neural networks [11,12] are collections of mathematical models that emulate some of the observed properties of biological nervous systems and draw on the analogies of adaptive biological learning. An artificial neural network (ANN) is an information-processing paradigm inspired by the way the densely interconnected, parallel structure of the mammalian brain processes information. The key element of the ANN paradigm is the novel structure of the information processing system. It is composed of a large number of highly interconnected processing elements that are analogous to neurons and are tied together with weighted connections that are analogous to synapses.

Today ANNs are being applied to an increasing number of real world problems of considerable complexity. They are good pattern recognition engines and robust classifiers, with the ability to generalize in making decisions about imprecise input data. They offer ideal solutions to a variety of classification problems such as speech, character and signal recognition, as well as functional prediction and system modeling, where the physical processes are not understood or are highly complex. The advantage of ANNs lies in their resilience against distortions in the input data and their capability of learning. They are often good at solving problems that are too complex for conventional technologies (For example, problems that do not have an algorithmic solution or for which an algorithmic solution is too complex to be found). ANNs are often well suited to problems that people are good at solving, but for which traditional methods are not.

Neural networks are most suitable for situations that have no clear set of rules or relationships. Because neural nets learn from experience, they can perform tasks that are difficult to describe precisely. Neural nets can classify complex data. Applications range from medical diagnostics to financial forecasting to process control [1–3, 13–19]. There has been significant interest recently for developing ANN application to solve biomedical problems such as assistance in diagnosing dermatologic conditions, discrimination of various leukocytes in blood, real-time drug interaction warning system, monitoring and managing physiologic data during surgery and in ICU, drug infusion during surgery and in ICU, mapping electromyography (EMG)-torque relation in joints, EMG signal processing for prosthesis control and medical image analysis.

For the past few years lot of death cases reported are due to cardio vascular abnormalities. Early detection and subsequent treatment are essential in helping millions of patients around the world. Automatic detection of problems in the intensive care unit can efficiently help in prompt attention and treatment, thus saving lives. Among the various cardiovascular signals monitored and recorded from the intensive care patients, ECG plays the key role in determining the state of the patient [20–25]. The ECG signal is the electrical signal generated by the heart's muscle measured on the skin surface of the body.

Electrocardiography has a basic role in cardiology since it consists of effective simple noninvasive low cost procedures for the diagnosis of cardiac disorders that have high epidemiological incidence and are very relevant for their impact on patient life and social costs [26]. Pathological alterations observable by ECG are cardiac rhythm disturbances (or arrhythmia), Dysfunction of myocardial blood perfusion (or cardiac ischemia), Chronic alteration of the mechanical structure of the heart [27]. Cardiac rhythm disturbances are considered to lead to life threatening conditions. Thus the abnormality detection in intensive care patients is very essential and critical.

The present work deals with the application of ANN by using various significant and relevant characteristic features from ECG for classifying intensive care unit (ICU) patient states. In this work, the input feature set uses morphological information, duration and complexity details of the ECG. The performance of the three neural networks viz., back propagation algorithm (BPA), self organizing map (SOM) and radial basis function (RBF) networks have been studied.

7.1 Neural Networks

An Artificial Neural Network (ANN) is an information processing paradigm that is inspired by the way biological nervous systems, such as the brain, process information. The key element of this paradigm is the novel structure of the information processing system. It is composed of a large number of highly interconnected processing elements (neurons) working in unison to solve specific problems. The characteristics of a artificial neural network are

- Adaptive learning: An ability to learn to do tasks based on the data given for training or initial experience.
- Self-organization: An ANN can create its own organization or representation of the information it receives during learning time.
- Real time operation: ANN computations may be carried out in parallel, and special hardware devices are being designed and manufactured to take advantage of this capability.

In the human brain, a typical neuron (Fig. 7.1) collects signals from others through a host of fine structures called *dendrites*. The neuron sends out spikes of electrical activity through a long, thin stand known as an *axon*, which splits into thousands of branches. At the end of each branch, a structure called a *synapse* converts the activity from the axon into electrical effects that inhibit or excite activity in the connected neurons. When a neuron receives excitatory input that is sufficiently large compared with its inhibitory input, it sends a spike of electrical activity down its axon. Learning occurs by changing the effectiveness of the synapses so that the influence of one neuron on another changes. Artificial neural network is a model (Fig. 7.2) to simulate these features.

The common type of artificial neural network consists of three groups, or layers, of units: a layer of "**input**" units is connected to a layer of "**hidden**" units, which is connected to a layer of "**output**" units. The activity of the input units represents the raw information that is fed into the network. The activity of each hidden unit is determined by the activities of the input units

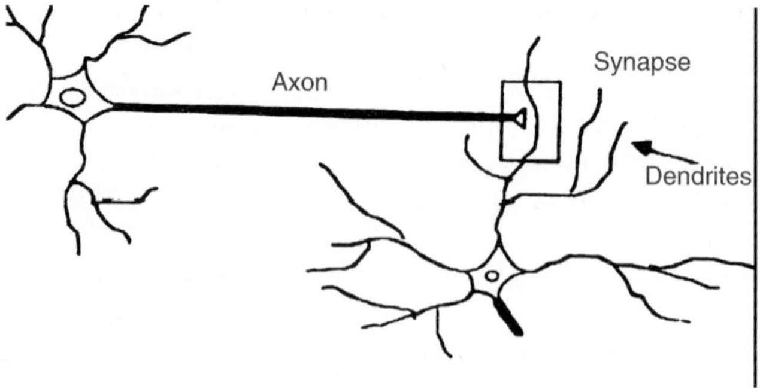

Fig. 7.1. A typical neuron

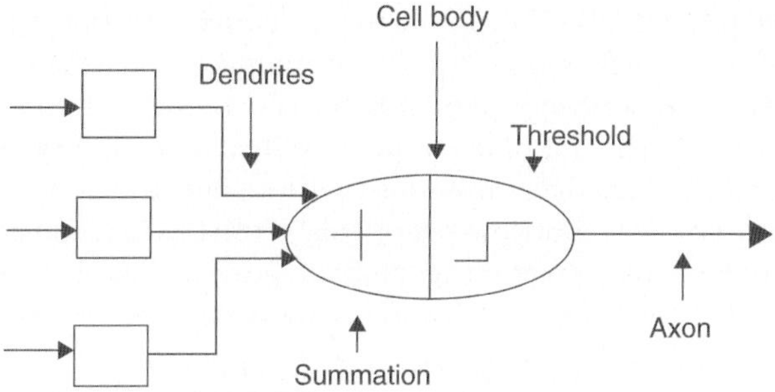

Fig. 7.2. Neuron model

and the weights on the connections between the input and the hidden units. The behavior of the output units depends on the activity of the hidden units and the weights between the hidden and output units. The research on neural networks has led to the development of different types of neural networks to suit the purpose. In this chapter, back propagation networks, self organizing maps networks and radial basis function networks are implemented as classifiers.

7.2 Backpropagation Learning Algorithm

Learning algorithms are of two kinds: supervised and unsupervised. In the former, the system weights are randomly assigned at the beginning; and progressively modified in the light of *desired* outputs for a set of training inputs. The difference between the desired output and the actual output is calculated for every input, and the weights are altered in proportion to the error factor. The process is continued until the system error is reduced to an acceptable limit.

The modified weights correspond to the boundary between various classes, and to draw this boundary accurately, the ANN requires a large training data set which is evenly spread throughout the class domain. For quick and effective training, it is desirable to feed the data from each class in a routine sequence, so that the right message about the class boundaries is communicated to the ANN.

The error *backpropagation* algorithm (BPA), aims at reducing the overall system error to minimum. The weight increment is directed towards the minimum system error and therefore termed as 'gradient descent' algorithm [7]. There is no definite rule to select the step size for weight increment; but the step length certainly has a bearing on the speed of convergence. It has been observed that for good speed, the step size should neither be 'too large', nor 'too small'. In the present case, an near optimum learning constant $\eta = 0.9$ (which controls the step size), is chosen by trial and error. Since weight increment is accomplished in small steps, the algorithm also bears the name '*Delta Rule*'.

The whole process of updating the weight matrix is a slow (in small incremental steps) movement towards a global minima of system error function. Sometimes, there may arise a possibility of the system entering a local minima and unable to come out of it. To remedy such a possibility, the algorithm incorporates a 'momentum term' into its update increment. The term is fraction of increment of its previous step; this term tends to push the present increment in the same direction as that of *previous* step. This term is no guarantee against the algorithm getting stuck at the local minima, but helps to get out of 'small' dips in the path.

In the *backpropagation* algorithm, the modifications are affected starting from the last (output) layer, and progresses towards the input. The schematic of the algorithm is shown in Fig. 7.3 and the weight increments are calculated according to the formula listed in Eqs. (7.1) and (7.2). The BPA requires both the activation function of the neuron and its (first) derivative to be of finite magnitude and single valued. This is a powerful argument in favor of the sigmoid function to be used in ANN of this kind.

The outputs of the hidden layer (${S_j}^h$) and output layer (b_k) are given by the equations

$$ {S_j}^h = f\left(\sum_{i=1}^{n} {w_{ji}}^h S_i - {\theta_j}^h\right) \tag{7.1} $$

$$ b_k = f\left(\sum_{j=1}^{n} {w_{kj}}^o {S_j}^h - {\theta_k}^o\right) \tag{7.2} $$

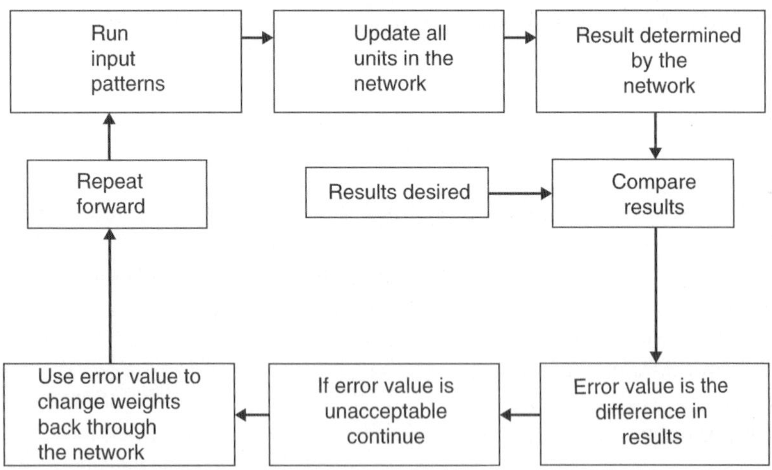

Fig. 7.3. The schematic of the backpropagation algorithm

where ${w_{ji}}^h$ and ${w_{kj}}^o$ represent the weights corresponding to the hidden and output layers respectively and ${\theta_j}^h$ and ${\theta_k}^o$ are the *bias* terms of hidden and output layers respectively.

To calculate the kth component of the output error vector (e_k) and hidden layer error (e_j) vector:

$$e_k = b_k\,(1 - b_k)\,(d_k - b_k) \tag{7.3}$$

$$e_j = {S_j}^h\left(1 - {S_j}^h\right)\sum_{k=1}^{4} w_{kj} e_k \tag{7.4}$$

where d_k is the desired output.

The update equations of the output and hidden layers are given below:

$$w_{kj}(new) = w_{kj} + \eta {s_j}^h e_k \tag{7.5}$$

$$w_{ji}(new) = w_{ji} + \eta s_i e_j \tag{7.6}$$

The thresholds are adapted using the following equations:

$$\theta_k{}^o(new) = \theta_k{}^o + \eta e_k \tag{7.7}$$

$$\theta_j{}^h(new) = \theta_j{}^h + \eta e_j \tag{7.8}$$

It is obvious from Eqs. (7.5) and (7.6), that the updating of the hidden layer is more computation intensive than the output layer. If there are more hidden layers, the computation too progressively increases. For most practical cases a single hidden layer network is adequate. However, a change in the number of neurons in the hidden layer (n) can affect the sophistication of classification. However, an increase in the number of hidden neurons may cause a delay in the convergence of weights. In the present case, a trial and error approach is adopted to fix the number of neurons in the hidden layer.

7.2 Self-Organizing Maps as Classifier

Self-Organizing Maps (SOM) are applied for classification of ECG signals as these networks learn to detect regularities and correlation in their input and adapt their future responses to that input accordingly. The neurons of competitive networks learn to recognize groups of similar input vectors. Self-organizing maps learn to recognize groups of similar input vectors in such a way that neurons physically close together in the neuron layer respond to similar input vectors [28].

Self-organizing maps learn to classify input vectors according to how they are grouped in the input space. The neighboring neurons in the self-organizing map learn to recognize neighboring sections of the input space. Thus self-organizing maps learn both the distribution and topology of the input vectors they are trained on.

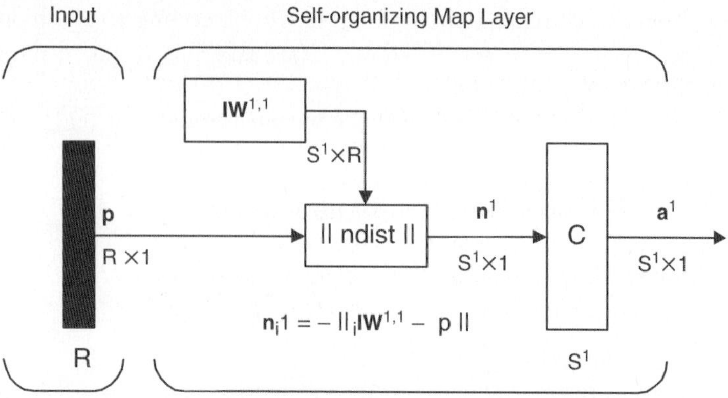

Fig. 7.4. Self-Organizing Map architecture

7.2.1 Architecture

The architecture for the self-organizing maps network [24] is shown in Fig. 7.4.

The ||ndist|| box in this figure accepts the input vector **p** and the input weight matrix $\mathbf{IW}^{1,1}$, and produces a vector having S^1 elements. The elements are the negative of the distances between the input vector and vectors $\mathrm{i}\mathbf{IW}^{1,1}$ formed from the rows of the input weight matrix. The net input $\mathbf{n}^1$ is computed by finding the negative distance between input vector **p** and the weight vectors. The maximum net input a neuron can have is 0. This occurs when the input vector **p** equals that neuron's weight vector. The competitive transfer function accepts a net input vector for a layer and returns neuron outputs of 0 for all neurons except for the winner, the neuron associated with the most positive element of net input $\mathbf{n}^1$. The winner's output is 1. The neuron whose weight vector is closest to the input vector has the least negative net input, and therefore wins the competition to output a 1. Thus a self-organizing feature map network identifies the winning neuron i^*. The weights of the winning neuron (a row of the input weight matrix) and all neurons within a certain neighborhood of the winning neuron are updated using the Kohonen learning rule.

Specifically, weights of all neurons $i \in N_{i^*}(d)$ are adjusted as follows.

$$ {}_{i}\mathbf{w}(q) = {}_{i}\mathbf{w}(q-1) + \alpha(\mathbf{p}(q) - {}_{i}\mathbf{w}(q-1)) \quad (7.9) $$

The neighborhood $N_{i^*}(d)$ contains the indices for all of the neurons that lie within a radius of the winning neuron i^* and is given by

$$ N_i(d) = \{j, dij \leq d\} \quad (7.10) $$

Thus, when a vector **p** is presented, the weights of the winning neuron and its close neighbors will move towards **p**. Consequently, after many presentations, neighboring neurons would have completely learned vectors similar to each other.

7.2.2 Training

In a self-organizing network, learning occurs one vector at a time. First the network identifies the winning neuron. Then the weights of the winning neuron, and the other neurons in its neighborhood, are moved closer to the input vector at each learning step. The winning neuron's weights are altered proportional to the learning rate. The weights of neurons in its neighborhood are altered proportional to half the learning rate. The learning rate and the neighborhood distance are used to determine which neurons are in the winning neuron's neighborhood altered during two training phases (ordering phase and tuning phase).

Ordering Phase of Training

The number of steps for this phase is given by the parameter ordering phase steps. The neighborhood distance starts as the maximum distance between two neurons, and decreases to the tuning neighborhood distance specified. The learning rate starts at the given ordering phase learning rate and decreases until it reaches the given tuning phase learning rate. As the neighborhood distance and learning rate decrease over this phase, the neuron's of the network will typically order themselves in the input space with the same topology which they are ordered physically.

Tuning Phase of Training

Tuning phase lasts for the rest of training period. The neighborhood distance stays at the tuning neighborhood distance. The learning rate continues to decrease from the tuning phase learning rate, but very slowly. The small neighborhood and slowly decreasing learning rate, fine tune the network, while keeping the ordering learnt in the previous phase stable. The number of epochs for the tuning part of training should be much larger than the number of steps in the ordering phase, because the tuning phase usually takes much longer.

The specific values of learning parameters used are

Ordering phase learning rate	– 0.9
Ordering phase steps	– 1000
Tuning phase learning rate	– 0.02
Tuning phase neighborhood distance	– 1

The weight change dw for a given neuron is calculated from the neuron's input p, activation a2, and learning rate μ:

$$\mathrm{dw} = \mu^{*}\mathrm{a2}^{*}(\mathrm{p}' - \mathrm{w}) \tag{7.11}$$

where the activation a2 is found from the layer output a and neuron distances D and the current neighborhood size and a2 is given by,

$$
\begin{aligned}
\mathrm{a2(i,q)} &= 1, && \text{if } \mathrm{a(i,q)} = 1 \\
&= 0.5, && \text{if } \mathrm{a(j,q)} = 1 \text{ and } \mathrm{D(i,j)} <= \mathrm{nd} \\
&= 0, && \text{otherwise.}
\end{aligned}
\tag{7.12}
$$

The learning rate lr and neighborhood size ns are altered through two phases: an ordering phase and a tuning phase.

The ordering phase lasts as many steps as denoted in ordering phase steps. During this phase lr is adjusted from ordering phase learning rate down to tuning phase learning rate and is adjusted from the maximum neuron distance down to 1. It is during this phase that neuron weights are expected to order themselves in the input space consistent with the associated neuron positions.

During the tuning phase μ decreases slowly from tuning phase learning rate and nd is always set to tuning phase neighborhood distance. During this phase the weights are expected to spread out relatively evenly over the input space while retaining their topological order found during the ordering phase.

Thus, the neuron's weight vectors initially take large steps all together toward the area of input space where input vectors are occurring. Then as the neighborhood size decreases to 1, the map tends to order itself topologically over the presented input vectors. Once the neighborhood size is 1, the network should be fairly well ordered and the learning rate is slowly decreased over a longer period to give the neurons time to spread out evenly across the input vectors.

The neurons of a self organizing map will order themselves with approximately equal distances between them, if input vectors appear with even probability throughout a section of the input space. Also, if input vectors occur with varying frequency throughout the input space, the feature map layer will tend to allocate neurons to an area in proportion to the frequency of input vectors there. Thus self-organizing maps, while learning to categorize their input, also learn both the topology and distribution of their input.

7.3 Radial Basis Function Networks as Classifiers

Radial basis function networks may require more neurons than the standard feed-forward networks [7]. They work best when training vectors are more.

7.3.1 Radial Basis Functions

Neuron Model

The net input of a radial basis neuron is different from that of neurons in back propagation and self-organizing networks. Here, the net input to the radial basis transfer function is the vector distance between its weight vector

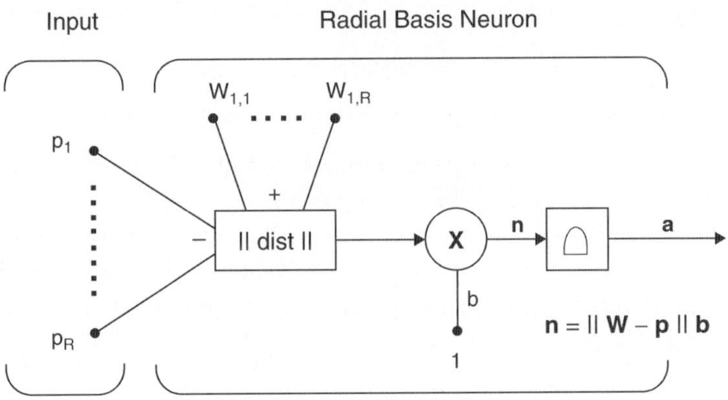

Fig. 7.5. A radial basis network with R inputs

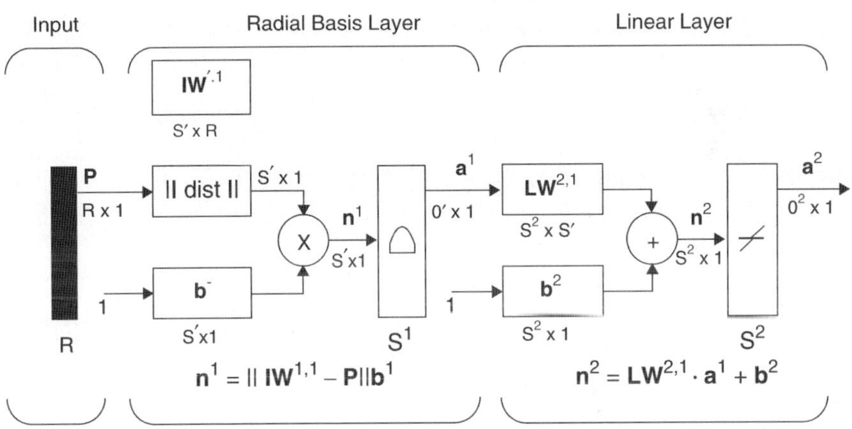

Fig. 7.6. Radial basis network architecture

w and the input vector p, multiplied by the bias b as shown in Fig. 7.5. The architecture of the radial basis network [29] is shown in Fig. 7.6.

The transfer function for a radial basis neuron is:

$$radbas(n) = \mathrm{e}^{-\mathrm{n}^2} \tag{7.13}$$

The plot of the radial basis transfer function is shown in Fig. 7.7.

The radial basis function has a maximum value of 1 when its input is 0. As the distance between w and p decreases, the output increases. Thus a radial basis neuron acts as a detector, which produces 1 whenever the input p is identical to its weight vector p.

The three techniques of neural network back propagation, self-organizing maps and radial basis functions are implemented as classifiers for classifying the cardiac abnormalities. Various characteristics features are extracted

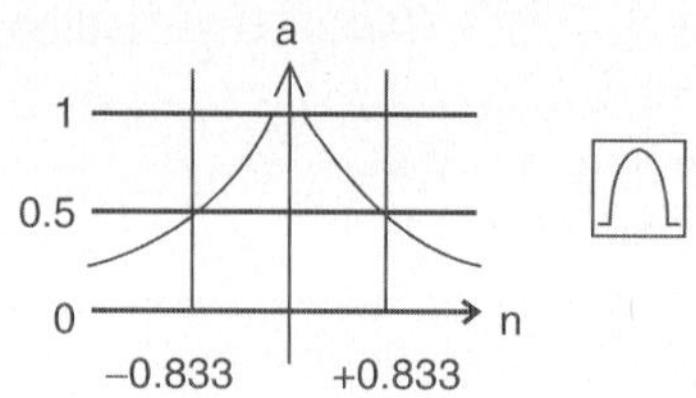

Fig. 7.7. Radial basis Transfer function

from the ECG signal (discussed in Chap. 2) and fed as input to these neural networks for classification.

7.4 Overview of Cardiac Patients States Classification System

The proposed approach for classification of ICU patients (mainly cardiac patients) states involves preprocessing of the ECG signal, extraction of characteristic features and classification using artificial neural network techniques. The overview of the proposed approach is shown in Fig. 7.8. Features set extraction methods are discussed in Chap. 2. Classification of ICU patient states from the extracted feature set is implemented using three different types of neural networks namely back propagation, self organizing maps and radial basis function. A comparative evaluation of the results of the three ANN techniques is performed. Using NN the patient states are classified into ten classes namely normal sinus rhythm (NSR), ventricular bigeminy (VB), atrial bigeminy (AB), pre-ventricular contraction (PVC), ventricular fibrillation (VF), atrial fibrillation (AF), sinus bradycardia (SBR), ventricular tachycardia (VT), supraventricular tachyarrhythmia (SVTA) and paced rhythm (PR).

The data for analysis is obtained from MIT-BIH Arrhythmia database. The training data is formed using 685 datasets and test data with 660 datasets. Details of training and test datasets are given in Table 7.1.

7.4.1 Implementation using Back Propagation

The feed forward neural network using back propagation algorithm and with one input layer, one output layer and three hidden layers is implemented. Input layer consists of thirteen nodes representing the thirteen features. The output layer consists of six nodes to represent the six output classes. The three hidden layers consist of thirteen, thirteen and six nodes respectively. As mentioned earlier, the network is trained with 400 data sets and tested with 400 data sets. The results are tabulated in Table 7.2. All classification statistics are obtained on the mutually exclusive categories of true negatives (TN), true positives (TP), false positives (FP) and false negatives (FN). The four categories are defined as

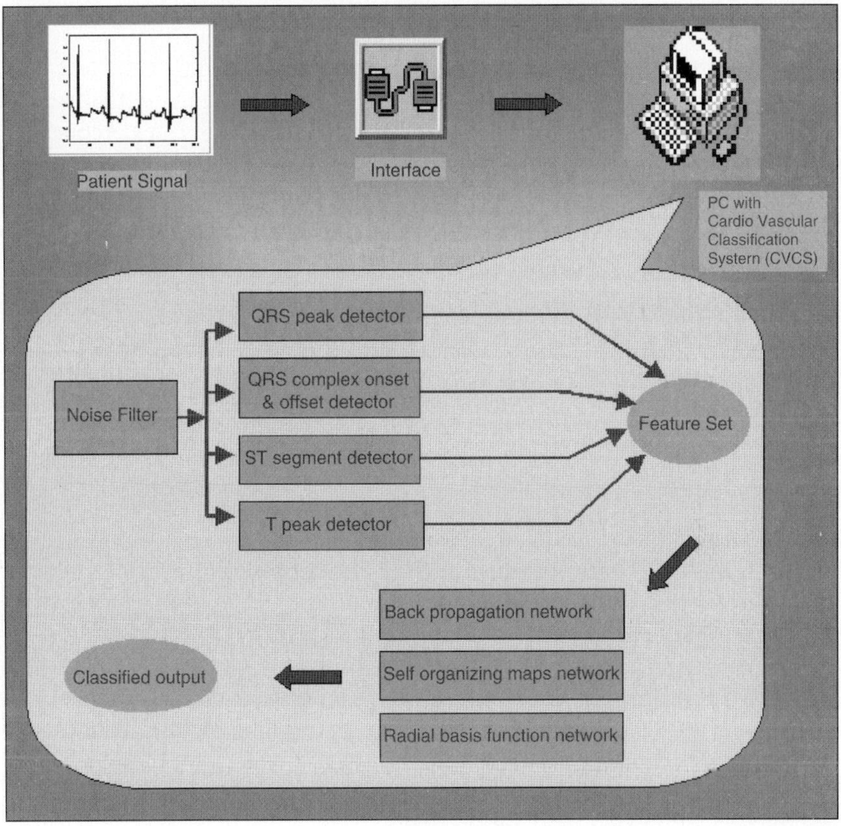

Fig. 7.8. Overview of Cardiac Patients states classification system

Table 7.1. Details of training and test data

Type	Number of training datasets	Number of test datasets
NSR	75	67
VB	60	66
AB	65	67
PVC	70	70
VF	65	67
AF	65	63
SBR	75	70
VT	70	70
SVTA	65	60
PR	75	60
	685	660

Table 7.2. Classification results using BP network

	Test data size	TN	TP	FN	FP	% Classification Accuracy
NSR	67	64	–	–	3	95.52
VB	66	–	62	4	–	93.94
AB	67	–	62	5	–	92.54
PVC	70	–	65	5	–	92.86
VF	67	–	62	5	–	92.54
AF	63	–	60	3	–	95.24
SBR	70		65	5		92.86
VT	70		65	5		92.86
SVTA	60		54	6		90.00
PR	60	–	54	6	–	90.00
	660	64	549	44	3	92.88

TN - Normal beats classified as Normal.
TP - Abnormal beats classified in their respective classes.
FN - Abnormal beats classified as Normal.
FP - Normal beats classified as Abnormal.

Figure 7.9 shows the training graph. The performance characteristics of the network are

Training Time : 2.96 secs
Convergence point : 591 epochs
Convergence point error : 0.1
Mean squared error : 0.01
Classification accuracy : 92.88%

7.4.2 Implementation using Self-Organizing Maps

A SOM is implemented with one input layer, one output layer and three hidden layers. The results obtained are presented in Table 7.3. The performance characteristics are

Training Time : 14.56 secs
Training Epochs : 10000
Classification accuracy : 95.24%

7.4.3 Implementation using Radial Basis Functions

A probabilistic radial basis function network is implemented as a classifier. The network topology is the same as for back propagation network. The results

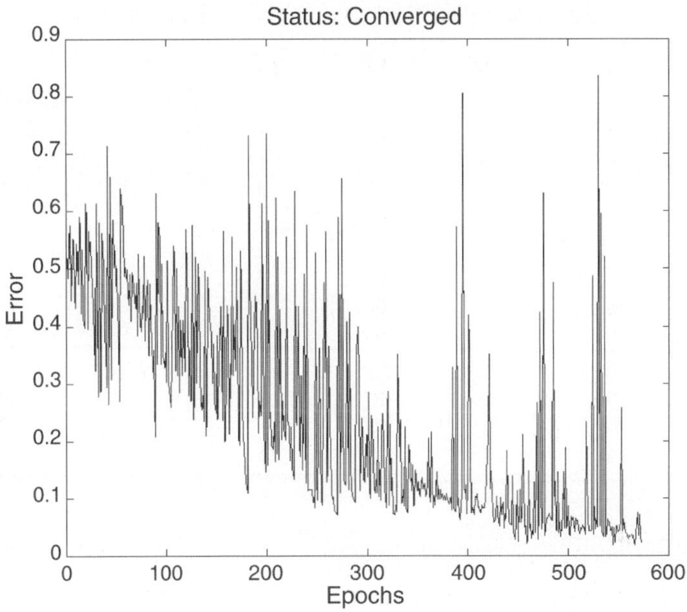

Fig. 7.9. Training graph of Back Propagation network

Table 7.3. Classification results using SOM network

	Test data size	TN	TP	FN	FP	% Classification Accuracy
NSR	67	65	–	–	2	97.01
VB	66	–	62	4	–	93.94
AB	67	–	64	3	–	95.52
PVC	70	–	67	3	–	95.71
VF	67	–	63	4	–	94.03
AF	63	–	60	3	–	95.24
SBR	70		67	3		95.71
VT	70		69	1		98.57
SVTA	60		56	4		93.33
PR	60	–	57	3	–	95.00
	660	65	565	28	2	95.45

obtained are presented in Table 7.4. The performance characteristics of the network are,

Training Time	: 2.64 secs
Classification accuracy	: 96.67%

Table 7.4. Classification results using RBF network

	Test data size	TN	TP	FN	FP	% Classification Accuracy
NSR	67	65	–	–	2	97.01
VB	66	–	64	2	–	96.97
AB	67	–	64	3	–	95.52
PVC	70	–	67	3	–	95.71
VF	67	–	65	2	–	97.01
AF	63	–	62	1	–	98.41
SBR	70		67	3		95.71
VT	70		69	1		98.57
SVTA	60		58	2		96.67
PR	60	–	57	3	–	95.00
	660	65	573	20	2	96.67

Table 7.5. Performance characteristics of Back Propagation network

Number of training data set	Training time	Total number of epochs	Convergence Point	Mean squared error	Classification accuracy
200	3.82	800	748	0.212	82.00%
300	3.61	900	693	0.136	89.00%
400	3.38	800	632	0.023	93.00%
500	3.12	1000	611	0.011	95.50%
600	2.96	600	591	0.010	96.00%

7.5 Effect of Number of Training Data on Network Performance

The performance characteristics results of back propagation network, SOM network and RBF network with thirteen characteristic features as inputs are tabulated in Tables 7.5–7.7 respectively.

From the results the effect of training data size on classification accuracy and training time is studied. Figure 7.10 shows the graph drawn between classification accuracy and training data size.

From the graph it can be seen that for back propagation network the classification accuracy increases rapidly as the training data size is increased. From Table 7.5, we can see that for the current configuration of the network, the input parameters, requires at least 600 epochs (datasets) to train the network and the network performance stabilizes at 600 sets of input training data. From Tables 7.6 and 7.7, it can be seen that the SOM and RBF networks performance saturates at 500 sets of input training data. This emphasizes the

Table 7.6. Performance characteristics of SOM network

Number of training data set	Training time	Total number of epochs	Convergence Point	Mean squared error	Classification accuracy
200	12.62	10000	4000	0.12	90%
300	13.61	10000	5400	0.63	93%
400	13.41	10000	6000	0.014	95%
500	14.58	10000	7500	0.01	97%
600	14.56	10000	8400	0.01	97%

Table 7.7. Performance characteristics of RBF network

Number of training	Training time	Classification accuracy
200	2.32	97.00%
300	2.41	98.00%
400	2.48	98.50%
500	2.56	99.00%
600	2.64	99.00%

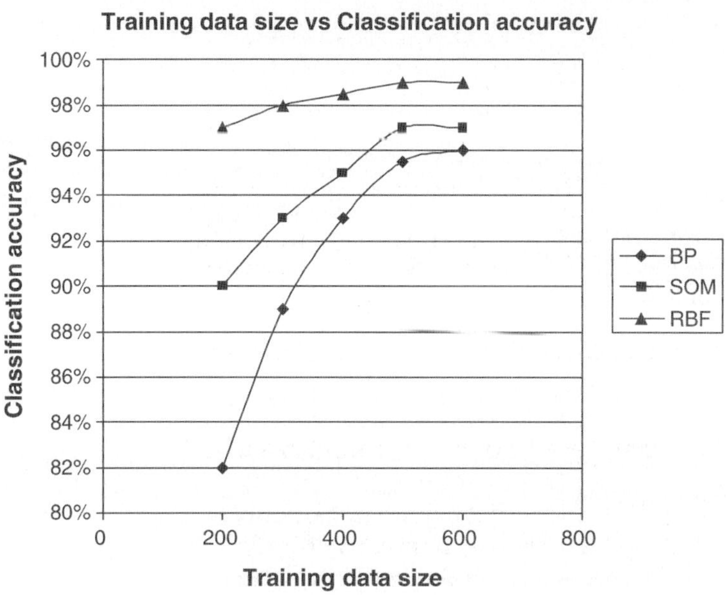

Fig. 7.10. Impact of training data size on classification accuracy

importance of training data size on the network performance. For comparative evaluation of the three techniques of the neural networks, 600 training data is chosen.

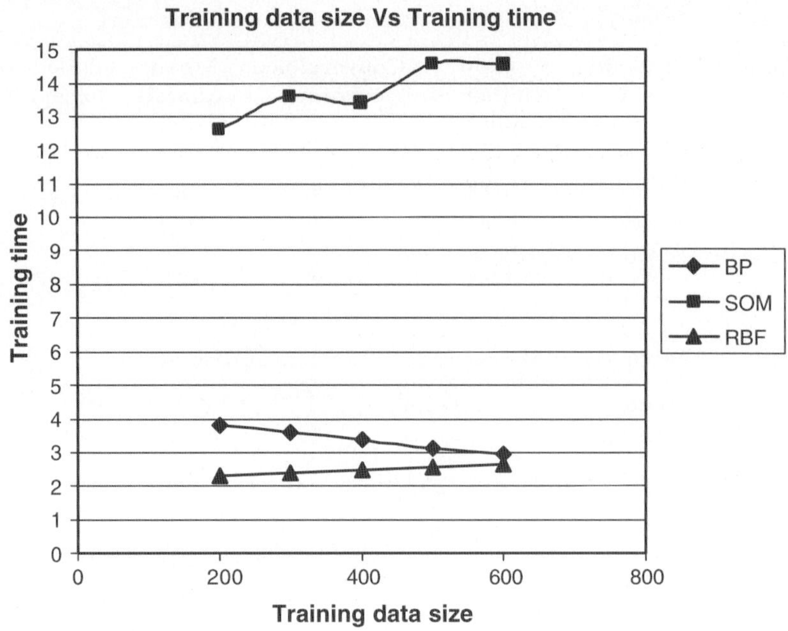

Fig. 7.11. Impact of training data size on training time

The graph drawn in Fig. 7.11 shows the variation of training time with respect to change in training data size. From the graph we can see that the training time required by the SOM is more compared to the other techniques of neural networks. The reason for this behavior is explained before. During initial stages, for back propagation network the training time decreases. This shows that the network learns faster with new input than when iterating with previously used inputs.

7.6 Network Performance Analysis

The neural networks performance in recognition and classification is usually evaluated by means of the following four performance indices. They are

- Classification accuracy,
- Sensitivity,
- Specificity and
- Positive predictive accuracy.

Let,

N_C	number of diagnostic classes
$E_A(i)$	number of classified positive events
$E_B(i)$	Total number of true positive events

$E_{\overline{B}}(i)$ Total number of true negative events
$E_{AB}(i)$ Number of classified true positive events
$E_{\overline{AB}}(i)$ Number of classified true negative events

Then the performance indices are given as follows:

Classification accuracy indicates the accuracy of classification. It is a measure of true positive and true negative events. It is given by

$$Classification\ accuracy = \frac{Number\ of\ correctly\ classified\ events}{Total\ number\ of\ events}$$
$$= \frac{\sum_{i=1}^{N_c} E_{AB}(i) + E_{\overline{AB}}(i)}{\sum_{i=1}^{N_c} E_B(i) + \sum_{i-1}^{N_c} E_{\overline{B}}(i)} \quad (7.14)$$

Sensitivity indicates the rate of true positive events for class i. It is given by

$$sensitivity = \frac{Number\ of\ classified\ true\ positive\ events}{Total\ number\ of\ true\ positive\ events}$$
$$= \frac{\sum_{i=1}^{N_c} E_{AB}(i)}{\sum_{i=1}^{N_c} E_B(i)} \quad (7.15)$$

Specificity is the measure of rate of true negative events for class i. It is given as

$$specificity = \frac{Number\ of\ classified\ true\ negative\ events}{Total\ number\ of\ true\ negative\ events}$$
$$= \frac{\sum_{i=1}^{N_c} E_{\overline{AB}}(i)}{\sum_{i=1}^{N_c} E_{\overline{B}}(i)} \quad (7.16)$$

Positive predictive accuracy is the rate of true positive events among all the classified events in class i.

$$positive\ predictive\ accuracy = \frac{Number\ of\ classified\ true\ positive\ events}{Total\ number\ of\ classified\ positive\ events}$$
$$= \frac{\sum_{i=1}^{N_c} E_{AB}(i)}{\sum_{i=1}^{N_c} E_A(i)} \quad (7.17)$$

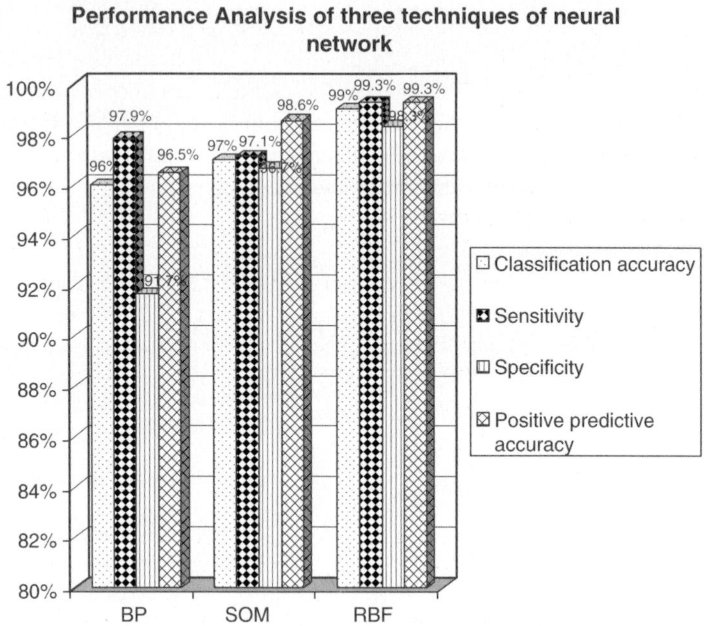

Fig. 7.12. Performance of back propagation, self organizing maps and radial basis function networks

Higher values are desired for all the four indices. The performance parameter values are calculated for the networks. The performance of the three networks is shown in Fig. 7.12.

From the results given in Sects. 7.4–7.6, it can be seen that the radial basis function networks produced results with a classification accuracy of 99%. Classification accuracy of 97% and 96% is achieved when using self-organizing networks and back propagation respectively. The performance of the self-organizing network is moderate with performance indices better than back propagation and lower than radial basis function networks. The sensitivity of the self-organizing networks is lower than the back propagation and radial basis function networks. Radial basis functions exhibited a high sensitivity of about 99.3%. The radial basis functions produced more specific, accurate and sensitive results for classification of cardiac health states compared to back propagation and self-organizing maps.

7.7 Conclusion

The results and performance characteristics of the three neural network techniques are compared and evaluated. From the results, the performance of the radial basis functions is observed to be better compared to self-organizing maps and back propagation techniques and this is because of the inherent ability of the probabilistic radial basis functions to act as classifiers. The radial basis function network requires pre classified teaching signals and the number of the classes has to be specified explicitly. In case of self-organizing maps, the network will group the input training set based on the correlation of the features of the input signal. This is useful when the number of classes is unknown or if the pre classified training set is unavailable. Self-organizing maps requires more computational power during training as it has to group the input set by itself. Back propagation network has simple network architecture and the computations required are reduced. The choice of network to be used is a tradeoff between the performance, computational power and teaching input.

References

1. Gang, Li., Wenyer,Ye., and Ling, Lin. (2000): 'An artificial intelligence approach to ECG analysis', *IEEE EMB Mag.*, **19** (2), pp. 95–100.
2. Rosaria, S., and Carlo, M. (1998): 'Artificial neural networks for automatic ECG analysis', *IEEE Trans. Signal Processing*, **46** (5), pp. 1417–1425.
3. Mark, V.G., and Jansen, H. (1997): 'Using artificial neural network for classifying ICU patient states', *IEEE EMB Mag.*, **16** (6), pp. 41–47.
4. Williams, J.L., and Lesaffre, E. (1987): 'Comparison of multi-group logistic and linear discriminant ECG and VCG classification', *J. Electrocardiol.*, **20**, pp. 83–92.
5. Coast, A.D., Stern, R.M., Casio, G.G., and Britter, S.A. (1990): 'An approach to Cardiac arrhythmia analysis using hidden Markov models', *IEEE Trans. Biomed. Eng.*, **37**, pp. 826–835.
6. Frenkel, D., and Nadal, J. (1999) "Ischemic episode detection using artificial neural network trained with isolated ST-T segments", *Computers in Cardiology*, **26**, pp. 53–56.
7. Gils, M.V., and Jansen, H. (1997): 'Using artificial neural network for classifying ICU patient states', *IEEE EMB Mag.*, **16** (6), pp. 41–47.
8. Silipo, R., and Marchesi, C. (1998): 'Artificial neural networks for automatic ECG analysis', *IEEE Trans. Signal Processing*, **46** (5), pp. 1417–1425.
9. Barro, S., and Fernandez, M. (1998): 'Classifying multi-channel ECG patterns with adaptive neural network', *IEEE EMB Mag.*, **17** (1), pp. 45–55.
10. Ham, F., and Ham, S. (1996): 'Classification of cardiac arrhythmias using fuzzy ARTMAP', *IEEE Trans. Biomed. Eng.*, **41** (4), pp. 425–430.
11. Olmez, T. (1997): 'Classification of ECG waveforms using RCE neural network and genetic algorithm', *Electronics Letters*, **33** (8), pp. 1561–1562.
12. Haykin, S. (1995): 'Neural Networks: A Comprehensive Foundation', *MacMillan College Publishing Company*, New York, USA.
13. Wasserman, P.D. (1993): 'Advanced Methods in Neural Computing', *New York: Van Nostrand Reinhold.*
14. Wu, S., and Chow, T.W.S. (2004): 'Induction Machine Fault Detection Using SOM-Based RBF Neural Networks', *IEEE Trans. Industrial Electronics*, **51**, pp. 183–194.

15. Hu, W., Xie, D., and Tan, T. (2004): 'A Hierarchical Self-Organizing Approach for Learning the Patterns of Motion Trajectories', *IEEE Trans.on Neural Networks.*, **15**, pp. 135–144.
16. Barro, S., and Fernandez, M. (1998): 'Classifying multi-channel ECG patterns with adaptive neural network', *IEEE EMB Mag.*, **17** (1), pp. 45–55.
17. Nicos, M., and Costas, P. (1998): 'An adaptive backpropogation neural networks for realtime ischemia episode detection', *IEEE Trans. Biomed. Eng.*, **45** (7), pp. 805–813.
18. Sun, Y., Chan, K.L., Krishnan, S.M., and Dutt, D. (2000): 'Unsupervised classification of ECG beats using MLVG neural network', *World Congress on Medical Physics and Biomedical Engineering.*
19. Fikret, G:(1999) "Neural network based decision making in diagnostic applications", *IEEE EMB Mag.*, **18** (4), pp. 89–93.
20. Cerutti, S., and Saranummi, N. (1997): "Improving Control of Patient Status in Critical Care," *IEEE EMB Mag.*, **16** (6), pp. 19–20.
21. Fedric, H., and Sowhan, H. (1996): 'Classification of cardiac arrhythmias using fuzzy ARTMAP', *IEEE Trans. Biomed. Eng.*, **41** (4), pp. 425–430.
22. Olmez, T. (1997): 'Classification of ECG waveforms using RCE neural network and genetic algorithm', *Electronics Letters*, **33** (18), pp. 1561–1562.
23. Dokin, Z., Olmez, T., and Yazgan, E. (1999): 'Comparison of discrete WT and FT for ECG beat classification', *Electronics Letters.*
24. Kang Lin., P., and Chang, W.H. (1997): 'A technique for automated arrhythmia detection of holter ECG', *In Proceedings of Engineering in Medicine and Biology*, **35**, pp. 626–631.
25. Williams, J.L., Benmel, J.H.V., and Zywietz, C. (1986): 'Computer ECG Analysis', *Elseiver Science Publishers.*
26. Linh, T.H., Osowski, S., and Stodolski, M. (2003): 'On-line heart beat recognition using Hermite polynomials and neuro-fuzzy network', *IEEE Trans.on Instrumentation and Measurement*, **52**, pp. 1224–1231.
27. Mervin, J., Goldmanand, N., and Goldschalager. (1989): 'Principles of clinical Electrocardiography', *Appleton Lange.*
28. Catherine, Parker, A., and Gary, A.T. (1983): 'Textbook of Anatomy & Physiology', *St. Louis: Mosby.*
29. Kohonen, T. (1987): 'Self-Organization and Associative Memory', Berlin: *Springer-Verlag.*
30. Chen, S., Cowan, C.F.N., Grant, P.M. (1991): 'Orthogonal Least squares Learning Algorithm for Radial Basis Function Networks', *IEEE Trans. Neural Networks*, **2** (3), pp. 302–309.

8

The Application of Autoregressive Modeling in Cardiac Arrhythmia Classification

Dingfei Ge, Narayanan Srinivasan, S M Krishnan

8.1 Introduction

As discussed in previous chapters, automatic detection and classification of rhythms in electrocardiogram (ECG) and heart rate signals play an important role in diagnosis and treatment of critically ill patients. Due to the large number of patients in ICU and the need for continuous observation, numerous methods for cardiac arrhythmias classification have been proposed. In fact, even small improvements to the reliability of such algorithms will save many lives. Ventricular tachycardia (VT) and ventricular fibrillation (VF) are life-threatening cardiac arrhythmias. Irreversible brain damage may occur in two or three minutes if a serious rhythm is not treated [1]. Other arrhythmias including premature ventricular contraction (PVC), atria premature contraction (APC) etc, are not so lethal, but are also important for diagnosing the heart diseases. So quick and reliable automatic detection and classification of these arrhythmias are crucial for physician to diagnose the cardiac diseases.

Various techniques for ECG arrhythmia classification have been proposed such as knowledge-based systems with statistical methods [2], time sequenced adaptive filter analysis [3], total least squares-based Prony modeling algorithm [4], complexity measures [5, 6], Fourier transform [7, 8], modified sequential detection algorithm [9], wavelets based method [10, 11], time domain based algorithm [12], Lyapunov transform [13] and correction waveform analysis [14]. Generally, these techniques only classify two or three arrhythmias or have significantly large processing times, and some are difficult to implement and compute. Thus, there is a need for extending a particular technique for a large number of arrhythmias and easy real-time implementation.

Cardiac arrhythmias using two intracardiac channels can be detected using the correlation waveform analysis (CWA) [14]. CWA was used to detect morphologic changes in the intracardiac electrogram, when compared to electrograms during sinus rhythm. Each electrogram had its respective template and the templates were obtained by signal averaging the waveform from a passage of sinus rhythm. The software trigger was used to align the template with the cycle being tested. A window size of N point was defined in each cycle. The main objective of the direct ECG feature detection was to investigate the number of ventricular conduction defect (VCD) categories that can be formed by advanced cluster analysis methods to reduce the number of classification parameters into a reasonably small set for a meaningful classification. The second objective was to investigate the extent to which, the selected set of repolarization parameters, would help in identification of distinct VCD subgroups. The study of detection of ECG features, show that the key features of ECGs are QRS duration, T axis angle, T amplitude, QRS axis angle and spatial angle. Typically, morphological features related to the P, QRS and ST waves are used for ECG signal analysis [15].

A technique based on averaged threshold crossing intervals was proposed for the detection of VT and VF based on heart rate measurements [16]. A modified sequential detection algorithm was further proposed to improve the accuracy of detecting VT and VF [9]. A Fourier transform based algorithm has been proposed for the detection of supraventricular rhythms from ventricular rhythms [17]. Changes in QRS complexes due to rhythm origination and conduction path were observed with the Fourier transform, and three kinds of rhythms were discriminated by a neural network. High sensitivity and specificity values greater than 98% have been reported for discriminating supraventricular rhythms from ventricular rhythms. However, supraventricular tachycardia (SVT) and VT were grouped together as ventricular rhythms. A new algorithm based on complexity measures was proposed for the detection of normal sinus rhythm (NSR), VT and VF [5, 6]. The algorithm was tested for varying lengths of data and very high accuracy values were achieved for data lengths of 7 seconds for classifying NSR, VT and VF. The algorithm was suggested for real-time implementation in automatic external defibrillators.

A new approach for the discrimination of VF, VT and SVT has been developed using a total least squares-based Prony modeling algorithm [4]. Two features, energy fractional factor (EFF) and predominant frequency (PF) were derived from the total least squares-based Prony model. A two-stage classification method was used, in which the EFF was used for discriminating SVT from VT and VF in the first stage followed by using PF for further separation of VF and VT in the second stage. A classification accuracy of 95.24%, 96.00% and 97.78% were reported for SVT, VF and VT respectively for the Prony modeling algorithm. However, the total least squares-based Prony modeling technique did not consider NSR, APC and PVC for feature extraction and discrimination.

Autoregressive modeling (AR) has been used in various applications, including classification of physiological signals like ECG, EEG, heart rate etc. The advantage of AR modeling is its simplicity and is suitable for real-time classification at the ICU or ambulatory monitoring. AR models are popular due to the linear form of the system, simultaneous equations involving the unknown AR model parameters and the availability of efficient algorithm for computing the solution [18, 19]. AR modeling has been used extensively to model heart rate variability (HRV) and for power spectrum estimation of ECG and HRV signals [20–23]. Amplitude modulated sinusoidal signal model, is a special case of the time-dependent AR model and have been applied to modeling ECG signals [20]. Adaptive AR modeling with Kalman filtering has also been used [23]. The parameters of the estimated time-varying model can be used to calculate instantaneous measures of linear dependence. The usefulness of the procedures in the analysis of physiological signals is discussed in two examples: First, in the analysis of respiratory movement, heart rate fluctuation, and blood pressure, and second, in the analysis of multichannel electroencephalogram (EEG) signals. Parameters extracted from AR modeling have been used for arrhythmia classification in conjunction with other features [24]. Two AR coefficients, along with the mean-square value of the QRS complex segments were utilized as features for classification of normal and abnormal PVC, where the prediction order was only 2 and a fuzzy adaptive resonance theory mapping (ARTMAP) was used for classification. The best result of PVC correct detection was 92% under the ratios of the training data size and testing data size was 2 to 4 [24]. AR modeling was adapted for extracting good features from ECG signals, thus enabling the discrimination of certain ECG arrhythmias [25].

In this chapter, the simple AR modeling technique is proposed to classify six types of cardiac arrhythmias namely, NSR, APC, PVC, SVT, VT and VF. Quadratic discriminant function (QDF) based classification algorithm was performed in various stages. Two hundred sample patterns each from the six classes were selected for classification in this study. A training data set of 60 sample patterns each from the six known classes were used. A QDF classifier was to be trained with the training data set and tested with the remaining data. In addition, the AR modeling technique can be utilized for ECG data compression and can be subsequently used for classification and diagnosis in telemedicine system. Thus, AR modeling is especially suitable for arrhythmia classification and diagnosis in telemedicine system.

8.2 Methods

The methodologies are described in this section. The data used in the analysis was preprocessed to remove the noise before AR modeling. The signal-to-noise ratio, minimum description length principle was used to evaluate the AR model order. AR coefficients and the other two features extracted from amplitude distribution of modeling error were utilized to represent ECGs and classified by QDF-based classification.

8.2.1 Preprocessing

The data in the analysis and classification was obtained from MIT-BIH database. The NSR, PVC and APC from the MIT-BIH arrhythmias database were sampled at 360 Hz, the VT and VF from the MIT-BIH ventricular arrhythmia database were sampled at 250 Hz, and the SVT from MIT-BIH supraventricular arrhythmia database were sampled at 128 Hz. Two hundred segments each from NSR, APC, PVC, VT, VF and SVT were included in the data set. The ECG signals in the current study were re-sampled in order that all the data used in the analysis had a sampling frequency of 250 Hz.

Prior to modeling, the ECG signals used in this study, has been filtered to remove the noise including respiration, baseline drift and wandering etc. A notch filter with cut off frequency of 0.4 Hz with a linear phase characteristic based on the frequency of 250 Hz was used. Thus, the drift caused by respiration at about 0.2 Hz is sufficiently removed. The other noise caused by the motion from the electrode is also minimized. The data was then filtered to eliminate the power line interference using a 50-Hz notch filter.

The R peaks of the ECGs were detected using Tompkin's algorithm [26]. RR intervals in APC are shorter than NSR, the RR intervals in VF and VT are much shorter than normal. In current study, the sample size of the various segments are 1.2 seconds. 0.4 seconds duration before R peak and 0.8 seconds after R peak were picked for modeling. It is adequate to capture most of the information from a particular cardiac cycle.

8.2.2 AR Modeling

The six classes of ECG signals with a sampling frequency of 250 Hz were utilized for AR modeling after filtering. A general AR model of order P can be expressed as [18, 19]

$$HR(n) = \sum_{k=2}^{P} HR(n-k+1)a_i(k) + e(n) \tag{8.1}$$

where $HR(n)$ represents ECG time series, $e(n)$ represents unknown, zero mean white noise, which is called modeling error, a_i represents the AR model coefficients.

It is a crucial issue to determine the model order, which best fits the data when constructing an AR model. The model order P means that P past data samples are needed, in order to predict the present value of the data. The model was estimated from 300 points of data (1.2 seconds) from the ECG segments in this chapter. The model order selection was performed on the six types of ECG signals, namely NSR, APC, PVC, VT, VF and SVT. Various model orders were pre-selected to estimate the model order. Burg's algorithm was used to compute the AR coefficients [18, 19]. In order to evaluate the AR model order, the criteria used in the research were the signal-to-noise *(SNR)* ratio, minimum description length *(MDL)*, and sensitivity of *MDL* (S_P^{MDL}). The signal-to-noise *(SNR)* ratio was computed using [20]

$$SNR = 10 \log \frac{\sum_{i=1}^{N} (HR(i))^2}{\sum_{i=1}^{N} \left(HR(i) - H\tilde{R}(i)\right)^2} \tag{8.2}$$

where $HR(i)$ and $H\tilde{R}(i)$ are the original and simulated signals at the ith instant, and N is the length of the modeled signals.

The minimum description length (MDL) is given by [27,28]

$$MDL = (N - P) \ln \sigma_n^2 + P \ln(N - P) \tag{8.3}$$

where σ_n^2 is the variance of white noise $e(n)$.

The sensitivity of $MDL(S_P^{MDL})$ is represented as

$$S_P^{MDL} = \left| \frac{\Delta MDL/MDL}{\Delta P/P} \right| \tag{8.4}$$

where ΔMDL is the change of MDL corresponding to AR model order change ΔP, ΔP is equal to 1 in this study, and the S_P^{MDL} reflects the sensitivity of MDL with ΔP.

8.2.3 ECG Feature Extraction for Classification

Besides AR coefficients, the other two features n_1 and n_2 obtained from zero mean white noise $e(n)(n = P+1, P+2, \ldots, N)$ were used to classify cardiac arrhythmias in this chapter. The algorithm to generate n_1 and n_2 are described as: (i) select a threshold value *th*, $th = C \times \max(|c(n)|)$, C is a positive value which needs to be determined. (ii) Compute e_1 and e_2: e_1 is the number of the amplitude $|e(n)|$ more than *th* within one signal segment, and e_2 is the one less than *th* within the same signal segment. Thus, the feature vector used as the ECG feature is $x = [a_2, a_3, \ldots, a_{p+1}, n_1, n_2]$ in this chapter.

8.2.4 QDF-based Classification

The ECG features described as above were utilized to classify the cardiac arrhythmias. The various cardiac arrhythmias have been classified by a stage-by-stage QDF-based algorithm. The QDF is given by [29]

$$y_i = \beta_0 + \sum_{k=1}^{d} \beta_k x_k + 2 \sum_{m=1}^{d-1} \sum_{n=m+1}^{d} \beta_{mn} x_m x_n + \varepsilon_i \tag{8.5}$$

and the matrix form is

$$y_i = X_i \beta + \varepsilon_i \tag{8.6}$$

where $x = [x_1, x_2, \ldots, x_d]$ represents an ECG feature vector, y_i is an observed response, ε_i is the QDF error; Xi is a $(d(d+3)/2+1)$-dimensional row vector, ß is a $(d(d+3)/2+1)$-dimensional column vector, that is

$$X_i = [1, x_1, x_2, \ldots, x_d, x_1^2, x_2^2, \ldots, x_d^2, 2x_1x_2, 2x_1x_3 \ldots, 2x_1x_d, 2x_2x_3, 2x_2x_4, \ldots, 2x_2x_d, \ldots, 2x_{d-1}x_d]$$

$$\beta = [\beta_0, \beta_1, \beta_2, \ldots \beta_d, \beta_{11}, \beta_{22}, \ldots, \beta_{dd}, \beta_{12}, \beta_{13}, \ldots, \beta_{1d}, \beta_{23}, \beta_{24}, \ldots, \beta_{2d}, \ldots, \beta_{(d-1)d}]^T$$

The ECG feature vector of a particular ECG segment was mapped to a response (1 or −1). Assume the total number of the ECG segments used for training at a particular stage is D. The following equation can be given

$$\tilde{Y} = A\beta + E \tag{8.7}$$

where $\tilde{Y} = [y_1, y_2, \ldots, y_D]^T$ is a D-dimensional column vector of the observed responses, and made up of "1" and "−1", which correspond to different classes respectively, $A = [X_1, X_2, \ldots, X_D]^T$ is a $D \times (d(d+3)1/2 + 1)$ matrix, $E = [\varepsilon_1, \varepsilon_2, \ldots, \varepsilon_D]^T$ is a D-dimensional column vector of the error.

The least squares estimator is given by

$$\beta = (A^T A)^{-1} A^T \tilde{Y} \tag{8.8}$$

Quadratic discriminant function of the classifier is

$$YI = X_i \beta \tag{8.9}$$

Quadratic discriminant function based classification was performed in stages to differentiate between the normal ECG signals and the various cardiac arrhythmias. The classification algorithm including the groupings, membership and decision-making criteria in every stage are shown in Table 8.1. During the training phase, the estimator β was computed using the training data set from the known classes of ECG segments. The feature vector x's and the previously estimated β were used to compute the correct response at a particular

Table 8.1. The QDF-based classification algorithm

	Stage 1			**Stage 2**			**Stage 3**	
Groups	Membership	Decision-making	Groups	Membership	Decision-making	Groups	Membership	Decision-making
NSR,APC, PVC,SVT	1	$Y1 > 0$	NSR,APC, PVC	1	$Y3 > 0$	APC,NSR	1	$Y4 > 0$
						PVC,NSR	−1	$Y4 < 0$
			SVT	−1	$Y3 < 0$			
			VT	1	$Y3 > 0$			
VT,VF	−1	$Y1 < 0$	**VF**	−1	$Y3 < 0$			
	Stage 4							
Groups	Membership	Decision-making						
NSR	1	$Y5 > 0$						
APC	−1	$Y5 < 0$						
NSR	1	$Y6 > 0$						
PVC	−1	$Y6 < 0$						

stage of classification during the testing phase. To perform the stage-by-stage classification, Euclidean center distance measure between the different classes was used to determine the groupings of classes at each stage. The feature vector $x = [a_2, a_3, \ldots, a_{p+1}, n_1, n_2]$ of a particular ECG segment was mapped to a response (1 or -1) in every stage of classification. In the current chapter, the observation matrix $A = [X_1, X_2, \ldots, X_D]^T$ was constructed according to the above mentioned method using AR coefficients and the features n_1 and n_2 of all the ECG segments are selected for training. The elements of vector $\tilde{Y}$ were assigned values 1 or -1 depending on the member-ship of an ECG segment to a corresponding class or group. The estimator β was computed at each stage of classification based on the selected training sets. During testing, the output response (Y1 in stage 1, Y2 in stage 2, etc) was computed by the equation (8.9) using the features and the previously estimated β at each stage. A threshold value of zero was used to classify the output response at a particular stage. Sixty samples from each class were used for training and the remaining was used for testing in the classification phase. The training sets were randomly selected. The average sensitivity and specificity were computed for all the classes for measuring the performance of the classification. The sensitivity and specificity were defined as follows, respectively [30].

$$Sensitivity = (TE - FN)/TE \tag{8.10}$$

$$Specificity = (TE - FP)/TE \tag{8.11}$$

where *TE* represents the total number of events, *FN* represents false negative, and *FP* represents false positive.

8.3 Results

The six different types of ECG signals from the MIT-BIH database were used for AR modeling. The ECG features were extracted from these ECG signals using AR modeling technique, and the classification was performed using the QDF-based classification algorithm. The AR modeling and classification results is given below.

8.3.1 AR Modeling and Feature Extraction Results

The *SNR*, *MDL* and S_P^{MDL} were used as criteria to evaluate the performance of the AR model with different model orders. The *SNR* was calculated to be from 16.73 dB to 28.63 dB. Figure 8.1 shows a graph of variation of *SNR* as a function of AR model order *P*. It can seen from the graph that, the *SNR* decreases initially with the model order *P*, and remains almost constant for model order greater than or equal to three. Similarly, Fig. 8.2 shows the *MDL* criterion for AR model order selection. In addition, one can also see from Fig. 8.3 that the AR model becomes less sensitive to model order *P*, for

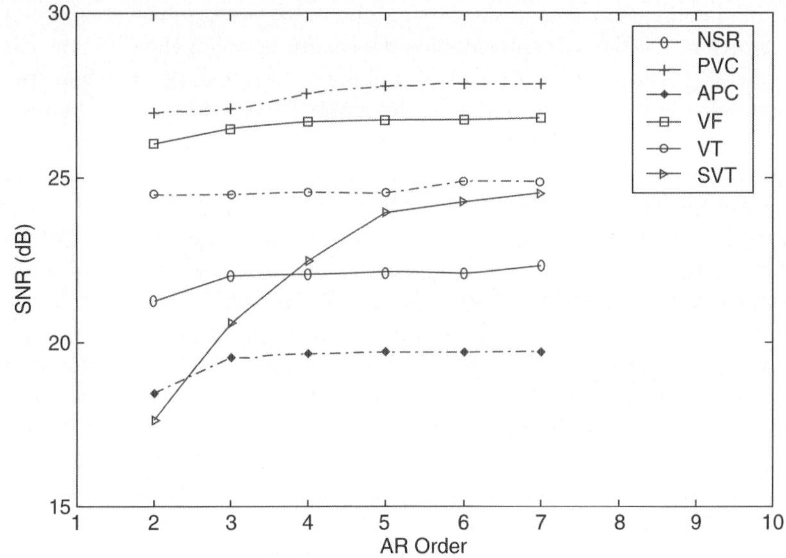

Fig. 8.1. *SNR* for various AR model orders

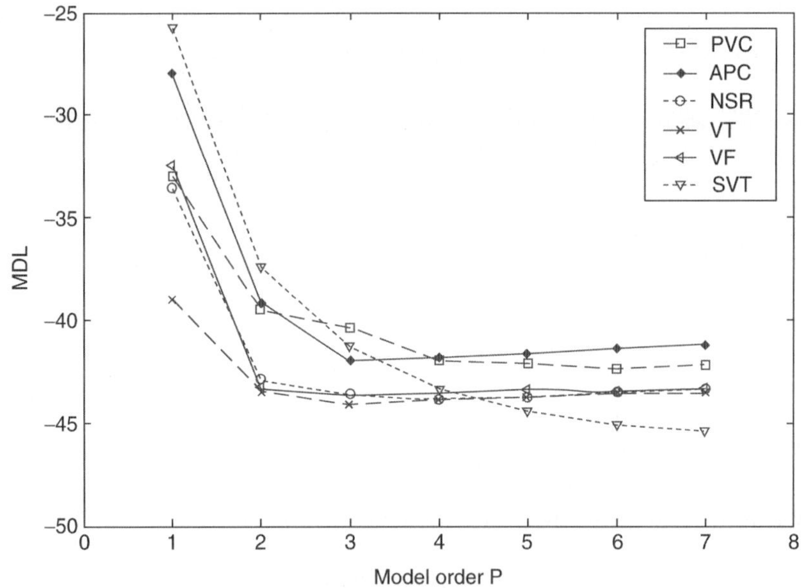

Fig. 8.2. *MDL* for various AR model orders

P more than three. However, the AR model of order four was selected for extracting the features. If the order of the model increases, the number of the AR coefficients and computation will increase. The AR model of order 4 was chosen.

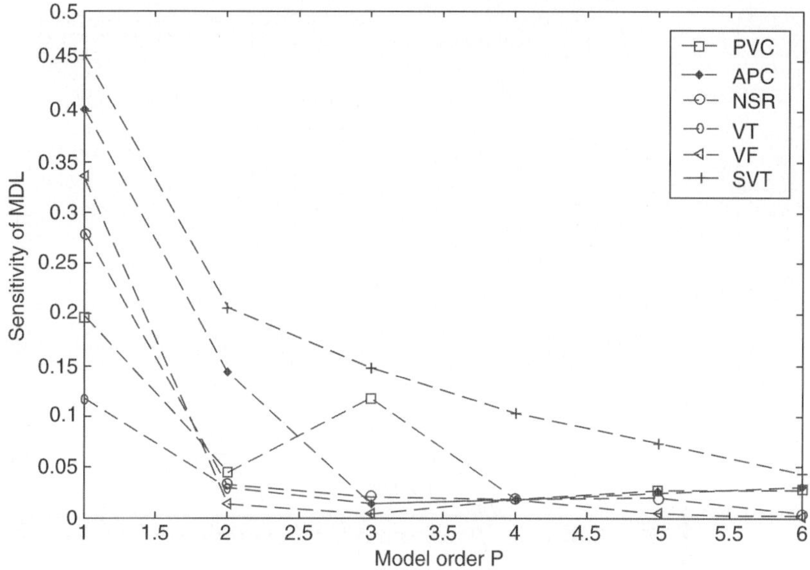

Fig. 8.3. S_P^{MDL} for various AR model orders

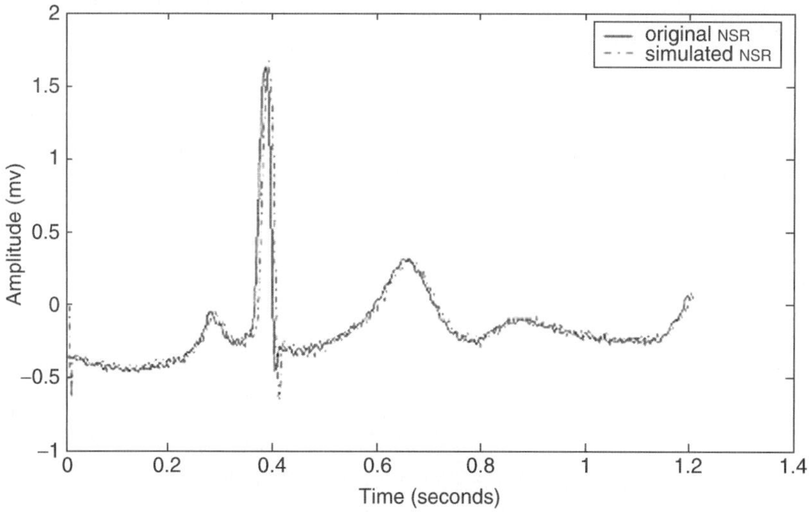

Fig. 8.4. Original and simulated ECG with NSR

The parameters computed using this model order were good enough to achieve a good *SNR* and were found to be sensitive enough to differentiate the six types of ECG signals. The original NSR, APC, PVC, SVT, VT and VF segments as well as the modeled segments are shown in (Figures) Figs. 8.4–8.9.

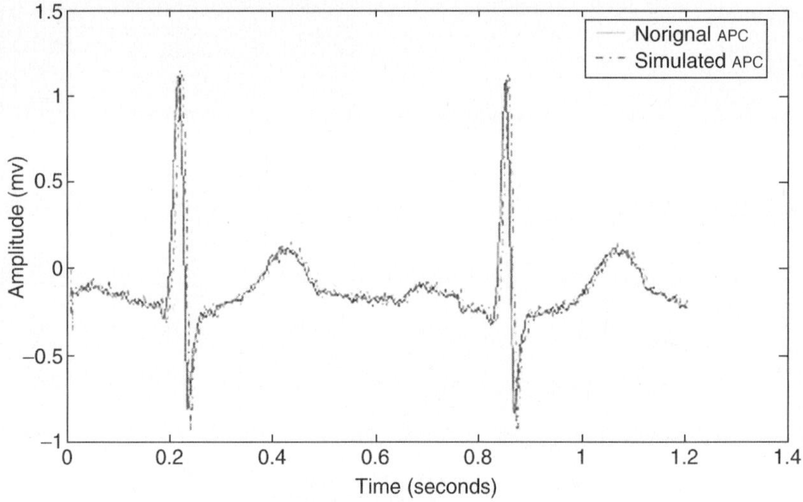

Fig. 8.5. Original and simulated ECG with APC

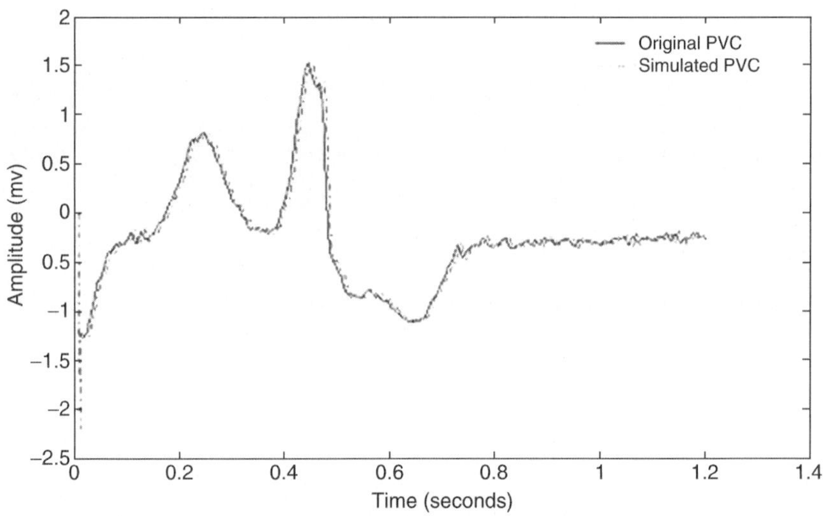

Fig. 8.6. Original and simulated ECG with PVC

The ECG features were extracted by applying AR process of order 4 to the ECG signals. The features n_1 and n_2 were extracted under the AR model of order 4. The distribution of error amplitudes for six different classes is shown in Fig. 8.10. In the current case, the positive value $C = 0.25$ was found to be the best selection to obtain good features n_1 and n_2. Thus, six-dimensional feature vectors comprising of 4 AR coefficients and n_1 and n_2 were used to represent an ECG segment in this chapter.

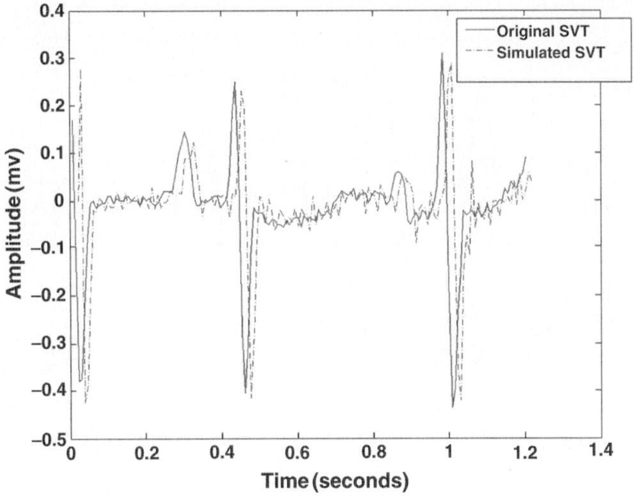

Fig. 8.7. Original and simulated ECG with SVT

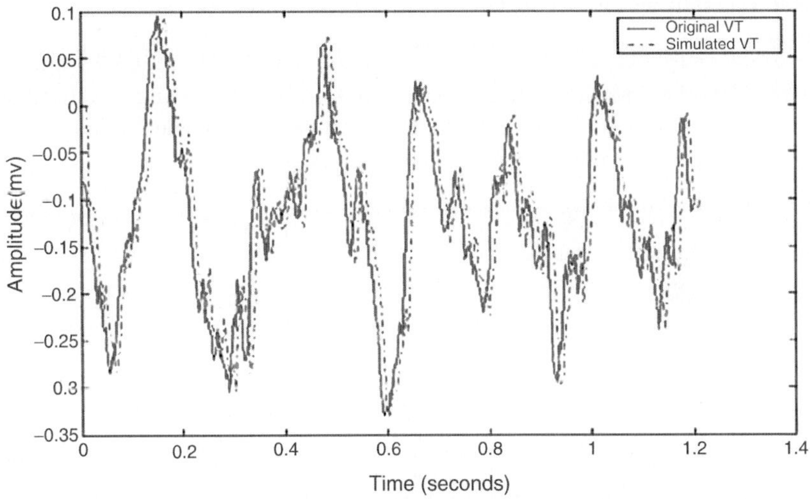

Fig. 8.8. Original and simulated ECG with VT

The results were consistent with other studies on the selection of model order for AR modeling. AR modeling has been used for compression and it has been found that the increase in accuracy by increasing the order of the predictor is negligible, for the predictors of order higher than 3 [25]. The mean feature values for different types of ECG signals, used in the current study are shown in Table 8.2.

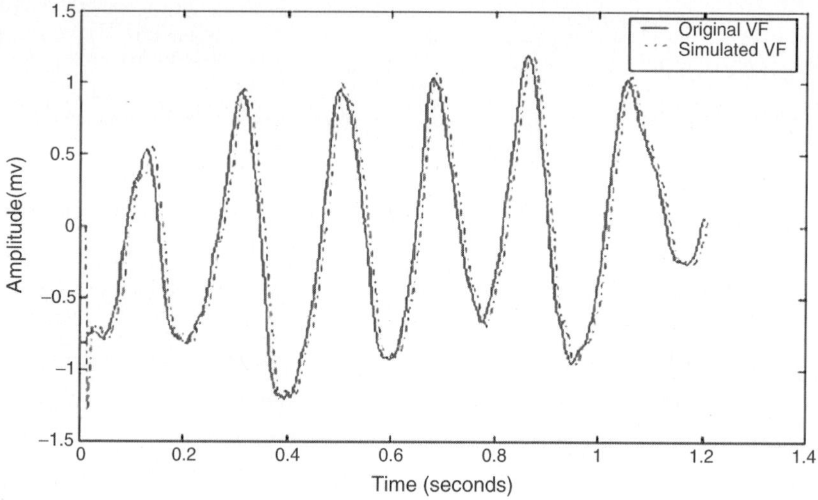

Fig. 8.9. Original and simulated ECG with VF

8.3.2 Classification Results

Six types of ECGs namely, NSR, APC, PVC, SVT, VT, and VF were considered for classification. Classification was performed using a QDF-based classifier, which was applied in various stages. Table 8.1 shows the stage-by-stage QDF-based classification algorithm for classifying various ECG signals. It is a multiple dichotomy of six classes of samples. The Euclidean center distance between these classes were computed for determining the groupings of classes at each stage. Table 8.3 shows the values of the Euclidean center distance. One can see that APC/NSR/PVC, VT/VF and SVT form one group respectively due to small values of the Euclidean center distance within the same group and large values between different groups. Therefore, VT/VF was separated from APC/NSR/PVC and SVT in stage one (Y1). The membership of VT/VF was defined as "−1", and the membership of APC/NSR/PVC and SVT was defined as "+1". The least squares estimator β was computed from equation of (8.8). The output response $Y1$ was computed from equation (8.9). The value of $Y1$ was used to determine the classes. Similarly, in the second stage ($Y2$), VT and VF were differentiated. In the third stage ($Y3$), SVT was distinguished from NSR, APC and PVC. In the later stages ($Y4$, $Y5$ and $Y6$), NSR, APV and PVC were distinguished from each other and classified. Sixty cases each from the six classes were selected randomly, to estimate β in training phase, and the remaining were used for testing in testing phase. The classification results on test data are given in Table 8.4 and 8.5. Table 8.4 shows classification results for a sample training set. Table 8.5 shows the performance of classification for the various classes, which were averaged

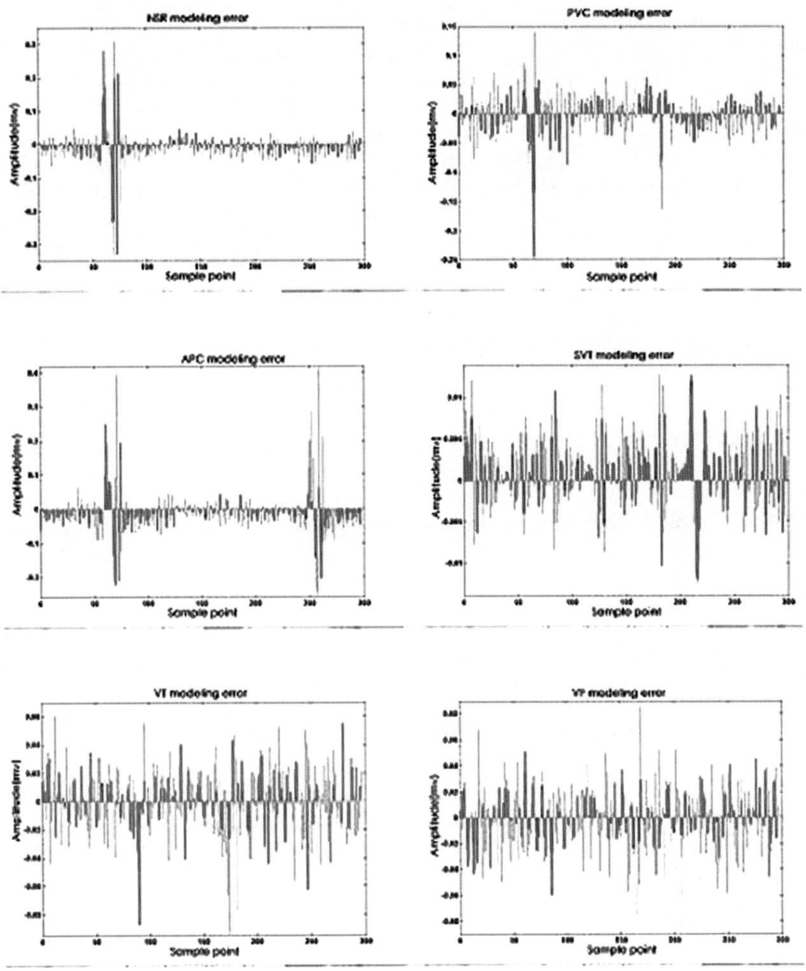

Fig. 8.10. Distribution of modeling error amplitude for the six classes

Table 8.2. Mean feature values

Classes	$a2$	$A3$	$a4$	$a5$	$n1$	$n2$
SVT	−2.7678	3.1868	−1.8599	0.4745	23.8259	24.0149
NSR	−2.1800	1.4756	−0.1027	−0.1684	32.995	50.9000
APC	−2.3384	1.8392	−0.3817	−0.0970	37.3850	52.9750
PVC	−2.2706	2.1637	−1.3440	0.4665	24.1050	33.3100
VT	−1.4278	0.0799	0.3746	−0.0181	63.8650	73.6550
VF	−1.7282	0.4099	0.5037	−0.1703	60.3550	73.8550

Table 8.3. The Euclidean center distance between the classes

Classes	SVT	NSR	APC	PVC	VF	VT
SVT	0	38.095	41.306	19.335	70.212	71.948
NSR	38.095	0	04.8803	19.770	35.738	38.386
APC	41.306	04.8803	0	23.757	31.093	33.665
PVC	19.335	19.770	23.757	0	54.453	56.717
VF	70.212	35.738	31.093	54.453	0	3.5495
VT	71.948	38.386	33.665	56.717	3.5495	0

Table 8.4. QDF-based classification results for a sample training set

Testing data set	Classification results						
	Classes	SVT	NSR	APC	PVC	VF	VT
140	SVT	137	0	1	2	0	0
140	NSR	0	135	3	1	0	1
140	APC	1	8	131	0	0	0
140	PVC	2	4	0	133	0	1
140	VF	0	0	0	1	135	4
140	VT	0	0	0	0	4	136

Table 8.5. Performance of QDF-based classification

Classes	SVT	NSR	APC	PVC	VF	VT
Specificity	97.86%	91.42%	97.14%	97.14%	97.14%	95.71%
Sensitivity	97.86%	96.42%	93.50%	95.0%	96.42%	97.10%

over 20 runs and each run has different training and testing data sets. The classification accuracy for SVT, NSR, APC, PVC, VF and VT were 97.86%, 96.42%, 93.50%, 95.00%, 96.42% and 97.10%, respectively.

8.4 Discussion

The objective of this study is to model the ECG signals for extracting classifiable features using the simpler AR model. Different values of AR modeling orders were tested for the ECG signals. The modeling results show that AR order of 4 was sufficient to model the ECG signals for the purpose of classification of the selected arrhythmias. It has been suggested that increasing the model order would not reduce the AR modeling error implying that a linear predictor order of two is sufficient for fast cardiac arrhythmia detection [25]. However, our experimental results show that an AR model, with the order less than 4 might not be suitable for ECG signals under all conditions. Thus, the AR model of order four was selected for extracting the features in order that more detail information can be incorporated into the model order that might be missing from a lower order model.

The four AR coefficients as well as feature n_1 and n_2 extracted from the amplitude distribution of the modeling error were used to classify the ECG beats into normal and five selected abnormal ECGs. A stage-by-stage QDF based classification algorithm was used to distinguish between the different types of arrhythmias under consideration in this chapter. The classification results show that AR modeling can be used to classify various cardiac arrhythmias. The current study classifies six types of ECG arrhythmias, and some of the proposed techniques use only a smaller numbers of arrhythmias than current study. The normal and abnormal PVC conditions were classified using fuzzy ARTMAP technique [24]. A time sequenced adaptive filter has been proposed for VT and VF alone [3]. A real time discrimination algorithm with a Fourier-transform neural network has been proposed to distinguish between superventricular rhythms and ventricular rhythms in which PVC and VT were lumped together as belonging to a single class of ventricular rhythms [7]. Three classes including NSR, VT and VF have been classified using the complexity measure-based technique [5, 6]. A QRS feature based-algorithm for decimated ECG data using artificial neural networks has been proposed that include various types of beats including APC and PVC, but they do not include the life threatening conditions like VT and VF [31]. The Prony modeling technique has been used to classify SVT, VT and VF, but their study does not include episodes from normal, APC or PVC [4].

The classification results achieved using the proposed method are comparable to the recently published results and to those studies that involve fewer classes on the classification of cardiac arrhythmias. An accuracy of detecting SVT, VT and VF were 95.24%, 96% and 97.78% using the total least squares-based prony modeling technique [4]. Two AR coefficients and the mean-square value of QRS complex segment were utilized as features for the classification of PVC and NSR using a fuzzy ARTMAP classifier with sensitivity of 97% and specificity of 99% were achieved [24]. Accuracy of 93% and 96% has been reported for VT and VF respectively, using a modified sequential probability ratio test algorithm [9].

In this study, the fixed sample size of the various segments used for AR modeling was 1.2 seconds, and it was 3 to 7 seconds and 5 to 9 seconds for the complexity measure-based technique [5] and the prony modeling technique [4], respectively. The algorithms are easy to implement and the AR coefficients can be easily computed. Preprocessing involves the detection of R peaks for which a number of available techniques can be implemented for real-time processing. A detailed comparison of computation times has not been performed, however, it can be noted that computing the AR coefficients is simpler than the proposed measures for arrhythmia recognition.

The generalization capabilities of the AR model and the classification algorithms can be refined by applying the proposed approach to a larger data set. Further work is in progress to extend the proposed approach for classification for other types of cardiac arrhythmias. AR modeling is a linear modeling technique and might not necessarily be suitable for ECG signals under all conditions. Further work can be done to extend the current work to nonlinear parametric models that can capture better the non-linear and non-stationary nature of the ECG. And also, one can try to increase the number of ECG leads to extract better classifiable ECG features using multivariate autoregressive (MAR) modeling.

In addition to their utility in classification and diagnosis, AR coefficients can also be used for the ECG data compression. AR modeling can be used in a low cost, high performance, simple portable telemedicine ECG system with an additional of diagnostic capability of compression.

Importantly, care must be taken while designing the filters in the preprocessing phase in order to extract good classifiable ECG features. Our experimental results show the noises caused by the motion of the electrodes and muscle tremor etc. may affect the performance of the arrhythmias classification.

8.5 Conclusion

The AR modeling technique and QDF based classifier can be used to classify cardiac arrhythmias effectively in critically ill patients, and aid in the diagnosis of heart disease. It is also suitable for real-time implementations and can be used for compression.

References

1. Goldschlager, N., and Goldman, M.J. (1989): 'Principles of Clinical Electrocardiography', *Appleton and Lange.*
2. Tatara, E., and Cinar, A. (2002): 'Interpreting ECG data by integration statistical and artificial intelligence tools', *IEEE Eng in Biomed Biol*, **21** (1) pp. 36–41.
3. Finelli, C.J. (1996): 'The time-sequenced adaptive filter for analysis of cardiac arrhythmias in intraventricular electrograms', *IEEE Trans. Biomed Eng.*, **43**, pp. 811–819.
4. Chen, S.W. (2000): 'Two-stage discrimination of cardiac arrhythmias using a total least squares-based prony modeling algorithm', *IEEE Trans Biomed Eng*, **47**, pp. 1317–1326.
5. Zhang, X.S., Zhu, Y.S., Thakor, N.V., and Wang Z.Z. (1999): 'Detecting ventricular tachycardia and fibrillation by complexity measure', *IEEE Trans Biomed Eng*, **46**, pp. 548–555.
6. Sun, Y., Chan, K.L., and Krishnan, S.M. (2005): 'Life-threatening ventricular arrhythmia recognition by nonlinear descriptor', *Biomed Eng Online*, **4** (1):6.

7. Minami, K.C., Nakajima, H., and Toyoshima, T. (1999): 'Real-time discrimination of ventricular tachyarrythmia with Fourier-transform neural network', *IEEE Trans Biomed Eng.*, **46**, pp. 179–185.
8. Afonoso, V.X., and Tompkins, W.J. (1995): 'Detecting ventricular fibrillation: Selecting the appropriate time-frequency analysis tool for the application', *IEEE Eng Med Biol Mag*, **14**, pp. 152–159.
9. Chen, S.W., Clarkson, P.M., and Fan, Q. (1996): 'A robust sequential detection algorithm for cardiac arrhythmia classification', *IEEE Trans Biomed Eng.*, **43**, pp. 1120–1121.
10. Engin, M. (2004): 'ECG beat classification using neuro-fuzzy network', *Pattern Recognition Letters*, **25**, pp. 1715–1722.
11. Shyu, L.Y., Wu, Y.H., and Hu, W. (2004): 'Using wavelet transform and fuzzy neural network for VPC detection from the holter ECG', *IEEE Transactions on Biomedical Engineering*, **51**, pp. 1269–1273.
12. Tian, L., and Tompinks, W.J. (1997): 'Time domain based algorithm for detection of ventricular fibrillation', *Proceeding EMBS* 19th *Annu Int Conf*, Chico, **3**, pp. 374–377.
13. Owis, M.I., Abou-Zied, A.H., Youssef, A.B., and Kadah, Y.M. (2002): 'Study of features based on nonlinear dynamical modeling in ECG arrhythmia detection and classification', *IEEE Trans. Biomed. Eng.*, **49**, pp. 733–736.
14. Caswell, S.A., Kluge, K.S., and Chiang, C.M.J. (1993): 'Pattern recognition of cardiac arrhythmias using two intracardiac channels', *Proc Comp Cardiol.*, pp. 181–184
15. Zhou, S.H., Rautaharju, P.M., and Calhoun, H.P. (1993): 'Selection of a reduced set of parameters for classification of ventricular conduction defects by cluster analysis', *Proc Comp Cardiol.*, pp. 879–882
16. Thakor, N.V., Zhu, Y.S., and Pan, K.Y. (1990): 'Ventricular tachycardia and fibrillation detection by a sequential hypothesis testing algorithm', *IEEE Trans Biomed Eng.*, **37**, pp. 837–843
17. Minami, K.C., Nakajima, H., and Toyoshima, T. (1999): 'Real-time discrimination of ventricular tachyarrythmia with Fourier-transform neural network', *IEEE Trans Biomed Eng.*, **46**, pp. 179–185.
18. Marple, S.L. (1987): 'Digital spectral analysis with applications', *Prentice Hall*, Englewood Cliffs, New Jersey.
19. Ljung, L. (1999): 'System Identification: Theory for the user', Prentice Hall, Englewood Cliffs, New Jersey.
20. Mukhopadhyay, S., and Sircar, P. (1996): 'Parametric modelling of ECG signal', *Med Biol Eng Comp.*, **34**, pp. 171–173.
21. Pinna, G.D., Maestri, R., and Cesare, A.D. (1996): 'Application of time series spectral analysis theory: analysis of cardiovascular variability signals', *Med Biol Eng Comp.*, **34**, pp. 142–148.
22. Bennett, F.M., Chrisstini, D.J., Ahmed, H., and Lutchen, K. (1993): 'Time series modeling of heart rate dynamics', *Proc Comp Cardiol.*, pp. 273–276
23. Arnold, M., Miltner, W.H.R., and Witte, H. (1998): 'Adaptive AR modeling of nonstationary time series by means of Kalman filtering', *IEEE Trans Biomed Eng.*, **45**, pp. 553–562.
24. Ham, F.M., and Han, S. (1996): 'Classification of cardiac arrhythmias using fuzzy ARTMAP', *IEEE Trans Biomed Eng.*, **43**, pp. 425–430.
25. Lin, K.P., and Chang, W.H. (1989): 'QRS feature extraction using linear prediction', *IEEE Trans Biomed Eng.*, **36**, pp. 1050–1055.

26. Tompkins, W. (1993): 'Biomedical Digital Signal Processing', *Prentice Hall*, Englewood Cliffs, New Jersey.
27. Proakis, J.G., and Manolakis, D.G.V. (1996): 'Digital Signal Processing', *Upper Saddle River*, New Jersey: Prentice-Hall.
28. Giraldo, B.F., Bolea, Y., and Caminal, P. (2002): 'Comparision of different models to analyze the number of patients in waiting-list', *Proceeding of the Second Joint EMBS/MBES Conference*, Houston USA., pp. 23–26.
29. Duda, R.O., and Hart, P.E. (2001): 'Pattern classification', United States of America: A wiley-Interscience Publication, *John Wiley & Sons INC*, pp. 219–223.
30. Barro, S., Ruiz, R., Callello, and Mira, J. (1989): 'Algorithmic sequential decision-making in the frequency domain for life threatening ventricular arrhythmias and imitative artefacts: a diagnostic system', *J Biomed Eng.*, **11**, pp. 320–328.
31. Melo, S.L., Caloba, L.P., and Nadal, J.(2000): 'Arrhythmia analysis using artificial neural network and decimated electrocardiographic data', *Comp Cardiol.*, **27**, pp. 73–76.

9

Classification of Cardiac Abnormalities Using Heart Rate Signals: A Comparative Study

U. Rajendra Acharya, N. Kannathal, P. Subbanna Bhat, Jasjit S. Suri, Lim Choo Min and Jos A.E. Spaan

The electrocardiogram (ECG) is the most important biosignal used by cardiologists for diagnostic purposes. The ECG signal provides key information about the electrical activity of the heart. Continuous ECG monitoring permits observation of cardiac variations over an extended period of time, either at the bedside or when patients are ambulatory, providing more information to physicians. Thus, continuous monitoring increases the understanding of patients' circumstances and allows more reliable diagnosis of cardiac abnormalities. Detection of abnormal ECG signals is a critical step in administering aid to patients. Often, patients are hooked up to cardiac monitors in hospital continuously. This requires continuous monitoring by the physicians. Due to the large number of patients in intensive care units and the need for continuous observation of them, several methods for automated arrhythmia detection have been developed in the past few decades to simplify the monitoring task. These include Bayesian [1] and heuristic approaches [2], expert systems [3], Markov models [4], self-organizing map [5], and Artificial Neural Networks (ANNs) [6].

Ham *et al*, have classified the ECG signal into two classes normal and PVC using fuzzy adaptive resonance theory mapping (ARTMAP) neural network with more than 99% specificity and 97% sensitivity [7]. Recently, Ozbay *et al*, have used multi-layered perceptron (MLP) with backpropagation training algorithm, and a new fuzzy clustering NN architecture (FCNN) for early diagnosis of ten different types of cardiac arrhythmia [8]. Their test results suggest that a new proposed FCNN architecture can generalize better than ordinary MLP architecture and also learn better and faster.

A new fuzzy Kohonen network, which overcomes the shortcomings of the classical algorithm, is presented to classify three cardiac arrhythmia (atrial fibrillation, ventricular fibrillation, and ventricular tachycardia) [9]. The proposed algorithm has achieved high accuracy (more than 97%) and is computationally fast in detection. A three RR-interval sliding window is used to classify the cardiac arrhythmia into four categories of beats (normal, premature ventricular contractions, ventricular flutter/fibrillation and 2 degrees heart block) was proposed [10]. This method indicates high performance: 98% accuracy for arrhythmic beat classification and 94% accuracy for arrhythmic episode detection and classification.

A novel method for detecting ventricular premature contraction (VPC) from the holter system was proposed using wavelet transform (WT) and fuzzy neural network (FNN) [11]. The major advantage of this method is to reuse information that is used during QRS detection, a necessary step for most ECG classification algorithm, for VPC detection. The method classifies the VPC correctly, with an accuracy of 99.79%. ECG arrhythmia classification into five different classes using principal component analysis was proposed [12]. Hebbian neural networks are used for computing the principal components of an ECG signal. An average value for classification sensitivity and positive predictivity were found to be 98.1% and 94.7% respectively. A number of neural network architectures were designed and compared with their ability to classify six different heart conditions [13]. Among different architectures, a proposed multi-stage network named NET_BST showed the highest recognition rate of around 93%. Therefore, this network proved to be suitable in ECG signal diagnosis. Khadra *et al* [14] have classified life threatening cardiac arrhythmias using wavelet transforms. Later, Al-Fahoum *et al* [15], have combined wavelet transformation and radial basis neural networks for classifying cardiac arrhythmias.

Acharya *et al*, have explained all the different types of linear, frequency and non-linear techniques, available for the analysis of heart rate signals [16]. They have also have classified the cardiac abnormality into eight different classes using neural network and fuzzy equivalence relations with an accuracy of more than 80% [17]. A neuro-fuzzy classifier called Fuzzy-Gaussian Neural Network (FGNN) was used to recognize the ECG signals for Ischemic Heart Disease (IHD) diagnosis [18]. The proposed ECG processing cascade had two main stages: (a) feature extraction from the QRST zone of ECG signals using either the Principal Component Analysis (PCA) or the Discrete Cosine Transform (DCT); (b) pattern classification for IHD diagnosis using the FGNN. This method classifies the IHD with an accuracy of almost 100%.

Features are extracted from higher order statistics and the Hermite characterization of QRS complex of the registered electrocardiogram (ECG) waveform. These features are fed to support vector machine (SVM) for the reliable heart beat recognition [19]. The reliable recognition and adequate electrical shock therapy of life-threatening cardiac states depend on the electrocardiogram (ECG) descriptors which were used by the defibrillator-embedded automatic arrhythmia analysis algorithms using twelve parameters [20]. The accuracy for the noise contaminated non-shockable and shockable signals exceeded 93%. Kannathal *et al*, have used an adaptive neuro-fuzzy network to classify heart abnormalities in ten different cardiac states with an accuracy level of more than 94% using heart rate signals [21]. Two different classifiers, decision-tree algorithm based on inductive learning from a training set and a fuzzy rule-based classifier for the identification of premature ventricular contractions (PVCs) in surface ECGs was studied [22]. The decision-tree algorithm showed a gross sensitivity of 85.3% and a positive predictivity of 85.2%. But the gross sensitivity of the fuzzy rule-based system and positive predictivity were 81.3%, and 80.6% respectively. The AR coefficients obtained from Burg's method, were used to classify the cardiac arrhythmia into five different classes using a generalized linear model (GLM) with an accuracy of more than 93% [23]. Mohamed *et al* [24] have used nonlinear dynamical modeling in ECG arrhythmia detection and classification.

The electrocardiogram (ECG) heartbeat recognition using classification enhancible grey relational analysis (GRA) was proposed by Lin [25]. The ECG beat recognition was divided into a sequence of stages, starting with feature extraction and then according to characteristics to identify the cardiac arrhythmias including the supraventricular ectopic beat, bundle branch ectopic beat, and ventricular ectopic beat. Compared with artificial neural network, the test results also showed high accuracy, good adaptability, and faster processing time for the detection of heartbeat signals.

In this chapter, heart rate variability is used as the base signal for classification of cardiac abnormalities into eight classes. The classification is performed using artificial neural network, fuzzy equivalence relation and neuro-fuzzy techniques and their performance is compared. Three parameters extracted from the heart rate signals are used for classification.

9.1 Neural Network Classifier

The ANN used for classification is shown in Fig. 9.1 and Fig. 9.2. The working of the BPA is explained in detail in Chap. 7. In the present case, an near

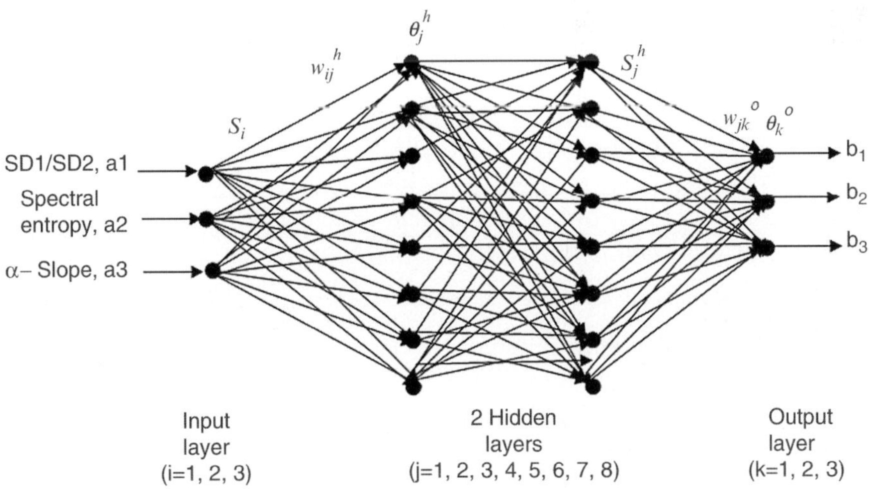

Fig. 9.1. Three-layer feedforward neural network classifier

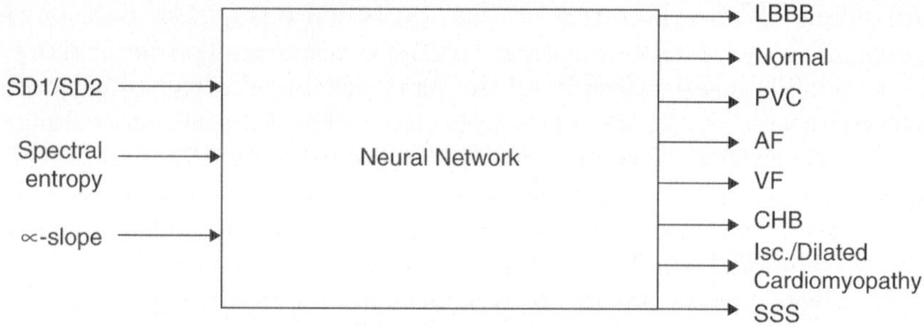

Fig. 9.2. Block schematic for ANN classification model

optimum learning constant $\eta = 0.9$ (which controls the step size), is chosen by trial and error [26, 27]. The input layer consists of nodes to accept data, and the subsequent layers process the data using the activation function. The output layer has three neurons, giving rise to an output domain of sixteen possible classes. However, the network is trained to identify only eight classes given by decoded binary outputs [000, 001, 010, 100, 011, 110, 101, 111].

9.2 Inputs to the Classifier

ECG data for the analysis and classification was obtained from MIT-BIH arrhythmia database, MIT-BIH database. Various ECG segments were selected from the databases for analysis. The data set included around 1000 segments each of normal ECGs, Preventricular Contraction (PVC), Complete Heart Block (CHB), Sick Sinus Syndrome (SSS), Left Bundle Branch Block (LBBB), Ischemic/Dilated Cardiomyopathy, Atrial Fibrillation, and Ventricular Fibrillation. The sampling frequency of the data from the MIT-BIH database is 360 Hz. Prior to analysis, the ECG signals were preprocessed to remove noise due to power line interference, respiration, muscle tremors, spikes etc., to detect the R peaks in the ECG signals. The R peaks of ECG were detected using Tompkins's algorithm [28, 29]. The ANN classifier is fed by three parameters derived from the heart rate signal using Entropy, Poincare plot geometry and detrended fluctuation analysis. These parameters are discussed in Chap. 5 in detail.

The pattern of Poincare plots of HRV data, its position and ranges of SD1/SD2 values are unique for particular type of cardiac abnormality. SD1/SD2 is the ratio of short interval variation to the long interval variation. This ratio is more in the case of PVC, AF, SSS and VF due to more variation in the RR interval. But, this ratio falls (below normal) for the slowly varying signals like CHB, Ischemic/dilated cardiomyopathy.

The fractal scaling ($\propto$) for the normal subjects (healthy young) is closer to 1, and this value falls in different ranges for various types of cardiac abnormalities. This slope is very low for very highly varying signals like PVC, LBBB, AF and VF. But for rhythmically varying signals like SSS, CHB and Ischemic/Dilated cardiomyopathy this value is slightly higher (comparable to 1).

The spectral entropy, is the measure of the disorder in the heart rate signal. It is a measure, quantifying the regularity and complexity of time series. It has higher value in the case of *normal* subjects and this value will have smaller value for abnormal (SSS, CHB, Ischemic/Dilated cardiomyopathy, VF etc) subjects, indicating smaller variability in the beat to beat.

9.3 Surrogate Data

The purpose of surrogate data is to test for any nonlinearity in the original data [30]. To test if the attractor geometry and correlation dimension are truly due to chaotic dynamics one must examine these characteristics for surrogate data sets.

Nonlinear indexes such as α-slope are computed for several surrogate data series. Their values are compared with that assumed by the nonlinear index computed for the original index [30]. The demonstration of statistically significant difference in α-slope between the original and surrogate data are in keeping with the presence of nonlinear dynamics in the original data.

Surrogate data have Fourier decomposition with the same amplitudes as the empirical data decomposition but with random phase components. This is obtained from the Chaos Data Analyzer.

To test for a statistical significance of difference in original α-slope and the surrogate data, 10 surrogate data series were generated to match each original signal. Then it is subjected to Student t test distribution. We found that, the surrogate data α-slope and original data α-slope, are distinct and they differ by more than 50%. This rejects the null hypothesis and hence the original data contain nonlinear features.

9.4 Fuzzy Classifier

A more efficient classifier is developed using Fuzzy Equivalence relation. The process of classification involves obtaining a Fuzzy *equivalence relation* matrix for each class of data, and then comparing a fresh input with each group for classification [31].

The Fuzzy *equivalence relation* requires the properties of reflexivity, symmetry and transitivity, to be satisfied. If it satisfies only the first two – reflexivity and symmetry properties – it is termed as Fuzzy compatible relation. Though it is usually difficult to identify an equivalence relation directly, it is possible to identify a compatible relation in terms of an appropriate 'distance function' of the Minkowski class. The general expression used for the distance function (Minkowski class) is given below:

$$R(x_i, x_j) = 1 - \delta \left(\sum_{l=1}^{n} |x_{il} - x_{jl}|^q \right)^{1/q} \tag{9.1}$$

where $n \rightarrow$ Total dimensionality of the input data point
$l \rightarrow$ Dimensionality index of the input data (1, 2,..n)
$p \rightarrow$ Size of the input data set
$i, j \rightarrow$ input index $i, j \in$ [1..p]
$q \rightarrow$ distance function parameter
$\delta \rightarrow$ Normalizing factor to ensure the resultant $R(x_i, x_j) \in$ [0,1].

n represents the total dimensions of the data, each of which dimension refers to the components of the input data. For example, from Table 9.1, the input data (HRV Signal) is represented by three components, hence $n = 3$ is used.

The Minkowski relation can be evaluated for integer values of q; for q = 1, the 'distance function' happens to be the Hamming distance; for q = 2, it is the Euclidean distance etc. The normalizing factor δ is taken as the inverse of the largest distance among the data pairs. As indicated above, for our purposes, the input data (HRV signal) is represented using the three parameters used for ANN classification earlier (Table 9.1). Thus the data has three components (n = 3). The size of the training data set (defined by $p_k k \in$ [1..4]) is different for each class i. The size of the entire data set is given by p. In the present case, the Euclidean distance *function* of Minkowski class (q = 2) is used as the basis to define mutual relation among the input data belonging to a particular class. Thus Equation (9.1) reduces to:

$$R(x_i, x_j) = 1 - \delta \left(\sum_{l=1}^{3} |x_{il} - x_{jl}|^2 \right)^{1/2} \tag{9.2}$$

Table 9.1. Range of input parameters to ANN classification model

Class	SD1/SD2	Spectral entropy	∝-slope
LBBB	0.7 ± 0.20	1.24±0.047	0.43±0.11
Normal	0.80±0.16	1.63±0.025	0.77±0.076
PVC	1.42±0.54	1.14±0.057	0.27±0.014
AF	2.98±1.56	1.2 ± 0.037	0.13±0.043
VF	1.13±0.47	1.06±0.003	0.34±0.022
Complete Heart Block	0.64±0.024	0.86±0.054	0.54±0.034
Isc./Dil.Cardiomyopathy	0.59±0.37	1.12±0.11	0.97±0.11
SSS	0.96±0.32	1.27±0.135	0.55±0.013

Table 9.2. Results of Neural Network Classifier

Class	No. of data set used for Training	No. of data set used for testing	Percentage (%) of correct classification (10,000 iterations)
LBBB	28	14	85.71
Normal	60	30	90
PVC	45	25	88
AF	30	20	85
VF	28	21	80.95
Complete Heart Block	28	21	80.95
Isc./Dil. Cardiomyopathy	30	18	83.33
SSS	30	18	88.88

The value of δ is taken to be equal to $1/n^{1/2}$, where n is the number dimensionality of the input data points, which is three in our case. The result of the above evaluation can be listed in the form of a symmetrical *pxp* matrix, which satisfies both reflexivity and symmetry conditions. A compatibility relation thus formed, is converted to an equivalence relation by performing the transitive closure operation. The processing algorithm is described in the following discussion. Firstly, a few definitions are in order. For a relation R, we write R(u,v) if the pair <u, v> is a member of the set.

Given a relation R, its transitive closure R^* can be determined as follows. R is transitive if (a, b) $\in R \wedge (\mathrm{b,c}) \in R \wedge (\mathrm{a,c}) \in R$. We may add elements to a relation R and create a new relation that is the transitive closure of R. However, the procedure requires an iterative process. We find the transitive closure by examining every pair of elements of R where the second element of the first pair matches the first element of the second pair. That is, (a, b) $\in R$ and (b, c) $\in R$.

Transitivity requires that (a, c) must also be an element of R. If it is not, then we must add it to the new relation that we are building into the transitive closure. Let's call the new relation R'. (Initially $R' = R$ and when the process of adding edges is over $R' = R^*$). After we have examined all such pairs of members of R and added the required edges to R' where needed we must then begin the same process again. The resultant R^* is the transitive closure of R.

After computing the transitive closure and having satisfied the properties of reflexivity, symmetry and transitivity, the Fuzzy equivalence relation matrix so obtained can now be used for classification of fresh data. The data to be classified is appended to the already classified data and the Fuzzy Equivalence Relation is found. At the end, the class to which the unknown data belongs is the one with whom it has the maximum degree of closeness. The formal algorithm is depicted in Fig. 9.3 and the detailed steps are shown in Table 9.3. And the results of the classification are listed in Table 9.4.

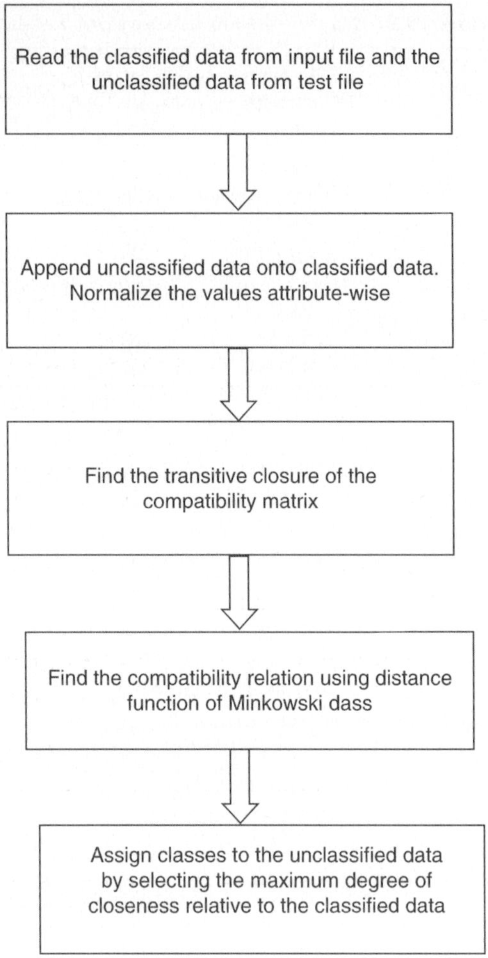

Fig. 9.3. Flowchart for Prediction using Fuzzy Equivalence Relation

9.5 Neuro-Fuzzy Classifier

The neuro-adaptive learning techniques provide a method for the fuzzy modeling procedure to learn information about a data set, in order to compute the membership function parameters that best allow the associated fuzzy inference system to track the given input/output data. This learning method works similarly to that of neural networks. Using a given input/output data set, the toolbox function adaptive neuro-fuzzy inference system (ANFIS) constructs a fuzzy inference system (FIS) whose membership function parameters are tuned (adjusted) using either a backpropagation algorithm alone, or in combination with a least squares type of method. This allows the fuzzy systems to

Table 9.3. Detailed Algorithm for Classifying using Fuzzy Equivalence Relations

Step	Phase	Description
1	Initialization	1.1 Read classified data from an input file $[\mathrm{input\_data}]_{i,j}[\mathrm{classes}]_k \leftarrow$ Buffer i: 1 to m (number of data) j: 1 to n (number of attributes) k: 1 to i; where there are 'o' classes 1.2 Read unclassified data from test file $[\mathrm{unclassified}]_{i,j} \leftarrow$ Buffer i: 1 to p (number of data) j: 1 to n (number of attributes)
2	Pre-processing	2.1 Append unclassified_data onto input_data $[\mathrm{input\_data}]_{l,j} \leftarrow [\mathrm{input\_data}]_{i,j} \Downarrow [\mathrm{unclassified\_data}]_{k,j}$ l: 1 to (m + p) i: m + 1 to m + p k: 1 to p j: 1 to n 2.2 Normalize the data matrix attribute wise selecting the maximum in each attribute $[input\_data]_{l,j} \leftarrow \frac{[input\_data]_{l,j}}{\max\{[input\_data]\}\|_j}$ l: 1 to (m + p) j: 1 to n
3	Compute the Fuzzy Equivalence Relation	3.1 Find the Compatibility relation between the data using distance function of Minkowski class $R_c(x_i, x_j) = 1 - \delta \left(\sum_{l=1}^{n} (x_{il} - x_{jl})^2 \right)^{1/2}$ i, j: 1 to m + p Where $\delta = 1/n^{1/2}$ 3.2 Find the transitive closure of R_c using algorithm given below 1.2 .1. $R' = R_c \cup (R_c \bullet R_c)$ $\cup$ = Max operator $\bullet$ = Max-Min composition where $R \bullet R = r_{ij} = \max_k \min(r_{ik}, r_{kj})$ i, j, k: 1 to m + p 3.2.2 If $R' = R_c$ then Stop. Else make $R_c = R'$ goto step 3.2.1. 3.2.3 $R_t = R'$; R_t is the Transitive closure matrix $(m + p) \times (m + p)$
4	Prediction	Beginning from row m + 1 search for the maximum membership degree until column m and store the corresponding index in matrix Max_membership $[\mathrm{Max\_membership}]_i = \max[R_t]_{i,j}$; i: m + 1 to m + p j: 1 to m corresponding to each unclassified data the class to which it belongs is given by $[\mathrm{classes}]_{[\mathrm{Max\_membershipi}]}$; i: 1 to p

Table 9.4. Results of Fuzzy classifier

Class	No. of data set used for Training	No. of data set used for testing	Percentage (%) of correct classification (10,000 iterations)
LBBB	28	18	88.88
Normal	60	40	92.5
PVC	45	30	90
AF	30	25	88
VF	28	25	84
Complete Heart Block	28	25	88
Isc./Dil. Cardiomyopathy	30	22	86.36
SSS	30	22	90.9

[Table 9.4 is reprinted with permission from U. Rajendra Acharya, Ashwin Kumar, P. Subbanna Bhat, Lim Choo Min, S.S. Iyengar, N. Kannathal, S.M. Krishnan, "Classification of cardiac abnormalities using heart rate signals", IFMBE Journal of Medical & Biological Engineering & Computing Journal, UK, 42(3), 2004, 288–293]

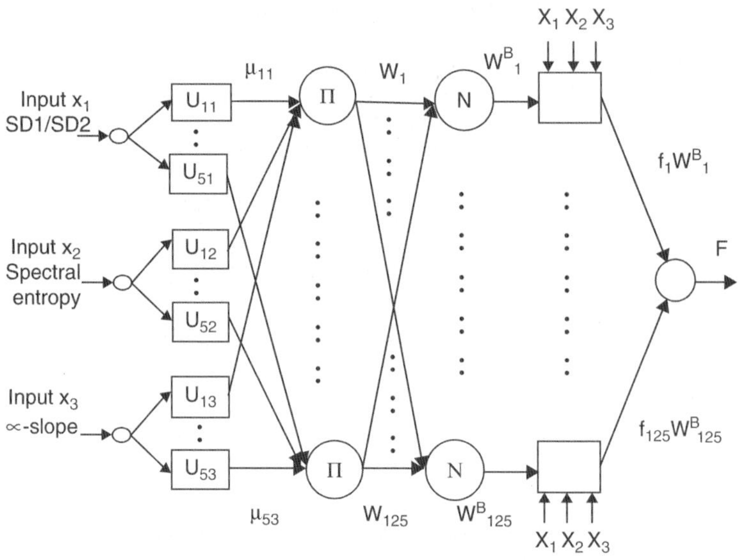

Fig. 9.4. ANFIS Network used for classification

learn from the data they are modeling. The network structure of the ANFIS is similar to neural network. The inputs are fuzzified through input membership functions. The initial and final (after training) input membership function for the input (SD1/SD2) is shown in Fig. 9.5. During the training phase the network converged at 100 epochs with a mean-squared-error of 0.009. After training association rules in the form of if then rules are generated and extracted. The initial and final (after rule extraction) decision surfaces are

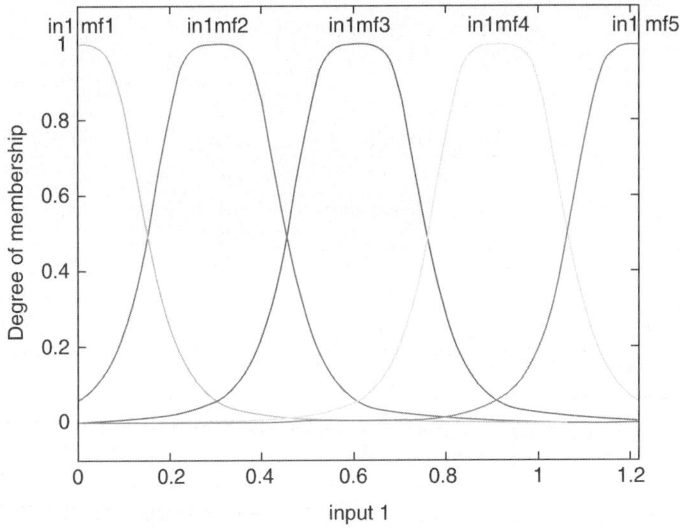

Fig. 9.5(a). Initial Membership Function for input 1 (SD1/SD2)

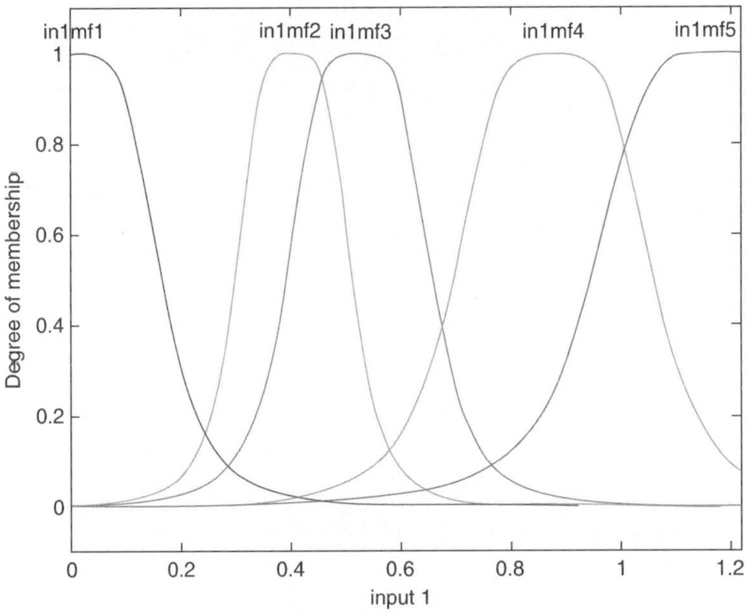

Fig. 9.5(b). Final Membership Function for input 1 (SD1/SD2)

given in Fig. 9.6. The parameters associated with the membership functions will change through the learning process. The computation of these parameters (or their adjustment) is facilitated by a gradient vector, which provides a measure of how well the fuzzy inference system is modeling the input/output

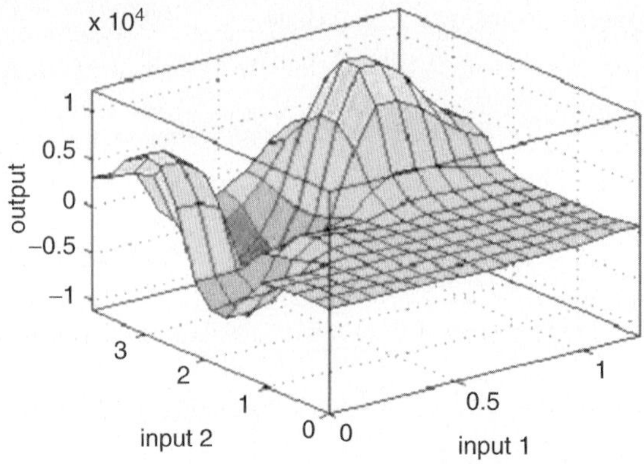

Fig. 9.6(a). Final decision surface for input 1(SD1/SD2) and input 2 (spectral entropy)

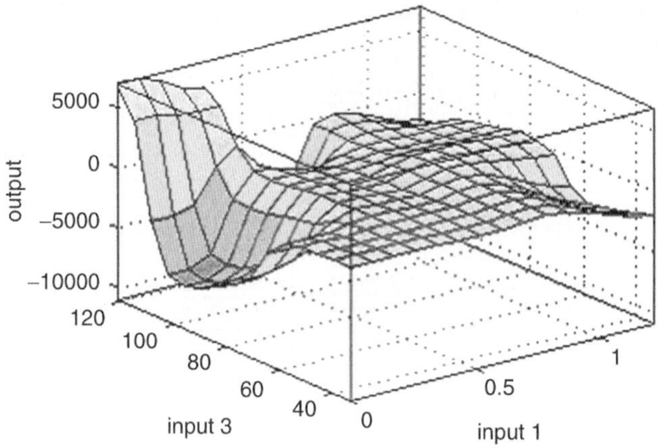

Fig. 9.6(b). Final decision surface for input 1(SD1/SD2) and input 3 (α-slope)

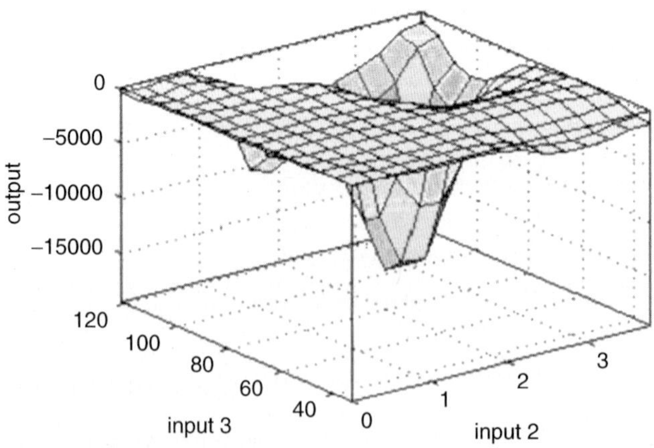

Fig. 9.6(c). Final decision surface for input 3 (α-slope) and input 2 (spectral entropy)

data for a given set of parameters. ANFIS also constructs an input-output mapping based on both human knowledge (in the form of fuzzy if-then rules) and stipulated input-output data pairs.

The ANFIS described by Jang [32] is adopted for the cardiac disease classification purposes. The ANFIS network chosen is as shown in Fig. 9.4 with a first-order Sugeno model. For each input, x_i, five fuzzy sets, U_{ji}, with the corresponding membership functions, $\mu_{ji}(x_i)$, were chosen for $i = 1$ to 3 and $j = 1$ to 5. Thus, the ANFIS network has a total of 125 fuzzy rules and one output, F. The rule structure, for e.g, the nth rule is of the form:

$$\text{If } x_1 \text{ is } U_{i1} \text{ and } x_2 \text{ is } U_{j2} \text{ and } x_3 \text{ is } U_{k3} \text{ then } f_n = d_{i1}x_1 + d_{j2}x_2 + d_{k3}x_3 + d_n \tag{9.3}$$

where $(d_{i1}, d_{j2}, d_{k3}, d_n)$ are adaptable parameters and $n = k + 5(j-1) + 25(i-1)$ for $i, j, k = 1$ to 5.

The membership functions have the form

$$\mu_{ji}(x_i) = \left\{1 + \left[(x_i - c_{ji})/a_{ji}\right]^{2b_{ji}}\right\}^{-1} \tag{9.4}$$

where (a_{ji}, b_{ji}, c_{ji}) are adaptable parameters. The ANFIS output is given by

$$F = \sum_{n=1}^{125} f_n w^{B_n} \tag{9.5}$$

where

$$w^{B_n} = \frac{w_n}{\sum_{n=1}^{125} w_n} \tag{9.6}$$

$w_n = \mu_{i1}(x_1)\mu_{j2}(x_2)\mu_{k3}(x_3)$ and $n = k + 5(j-1) + 25(i-1)$ for $i, j, k = 1$ to 5.

A diagnostic test investigates the statistical relationship between test results and the presence of disease. For all diagnostic tests there are two critical components that determine its accuracy: sensitivity and specificity.

9.6 Results

In this chapter, the heart rate signal of eight different types are classified using neural classifier, fuzzy classifier and neuro-fuzzy classifier. Three features are extracted from the heart rate signals namely, spectral entropy, SD1/SD2 and $\propto$-slope.

In the case of neural network, backpropogation learning algorithm (BPA) is used for learning.

Table 9.1 shows the ranges of the three parameters used to feed as input to the ANN. For the purpose of training and testing the classifier, a data base of

Table 9.5. Results of ANFIS classifier

Class	No. of data set used for Training	No. of data set used for testing	Percentage (%) of correct classification (10,000 iterations)
LBBB	28	24	91.67
Normal	60	29	100.00
PVC	45	27	96.67
AF	30	17	92.00
VF	28	18	96.00
Complete Heart Block	28	22	92.00
Isc./Dil. Cardiomyopathy	30	24	95.83
SSS	30	17	90.91

279 patient samples is divided into two sets – a training set of 167 arbitrarily chosen samples and a test set of 45 samples (Table 9.2). The training consisted of 10,000 iterations.

During the *training* phase, each output of the ANN is an analog value in the range $0 \rightarrow 1.0$, whereas the 'desired' output is either 0 or 1.0. During the *recall* phase, the output signal is approximated to binary levels by comparing it with threshold value of 0.5. The mean square error of the ANN was set to 0.0001.The percentage of correct classification ranges from 80% to 90% (Table 9.2).

Fuzzy classifier works better than the neural network classifier (Table 9.4). And this classification is further improved using neuro-fuzzy classifier (Table 9.5). The classification efficiency will be more than 90% for all disease classes in the neuro-fuzzy classifier. The p-value can be obtained using ANOVA (ANalysis Of VAriance between groups) test. ANOVA uses variances to decide whether the means are different. This test uses the variation (variance) *within* the groups and translates into variation (i.e. differences) *between* the groups, taking into account how many subjects there, in the groups. If the observed differences are high then it is considered to be statistical significant.

9.7 Discussion

The sensitivity of a test is the proportion of people with the disease who have a positive test result. Higher the sensitivity will have greater detection rate and lower false negative (FN) rate. The specificity of the test is the proportion of people without the disease, who have a negative test. The higher specificity indicate, the lower false positive rate and lower the proportion of people with the disease, who will be unnecessarily worried or exposed to unnecessary treatment. The positive predictive value (PPV) of a test is the probability of a patient with a positive test actually having a disease. The

Table 9.6. Results of sensitivity, specificity, positive predictive value, for all the classes using neural network, fuzzy and ANFIS

Classifier	TN	TP	FP	FN	Sensitivity	Specificity	Postive Predictivite Accuracy
ANN	27	119	3	21	85.00%	90.00%	97.54%
FUZZY	37	150	3	20	88.24%	92.50%	98.04%
ANFIS	29	139	0	10	93.29%	100.00%	100.00%

negative predictive value is the probability of a patient with a negative test not having the disease. In our study, we have found the sensitivity of 93%, specificity and positive predictive accuracy were more than 97% (Table 9.6). Hence, it indicates that, these results are clinically significant.

The Neuro-Fuzzy classifier is superior than the neural network and fuzzy classifiers. It yields correct classification of normal subjects and more than 95% of correct classification for the VF, PVC and ischemic/dilated cardiomyoapthy classes. Its sensitivity is about 93%, higher than the neural network and fuzzy classifiers. Sensitivity is the ability to correctly detect the disease and specificity is the ability to avoid calling normal things as disease. The specificity and positive predictive accuracy is 100% for all the classes indicating that, the results are clinically significant. A perfect medical test would have 100% sensitivity and 100% specificity. It would positively identify all the true cases of disease, and it would never mislabel anything normal as disease. The sensitivity is less due to the presence of few false negatives (FN). These false negatives can be decreased by choosing the better features for classification and as a result the sensitivity can be further increased. The accuracy of the system can further be increased by increasing the size and quality of the training set.

9.8 Conclusion

Heart rate variability (HRV) signal can be used as a reliable indicator of heart diseases. In this chapter, the neural network classifier, the fuzzy classifier and neuro-fuzzy classifier are presented as diagnostic tools to aid the physician in the analysis of heart diseases. Our results shown above indicate that the classifiers presented are effective to the tune of more than 90% and the neuro-fuzzy classifier works better than the other two classifiers. However, these tools generally do not yield results with 100% accuracy. The accuracy of the tools depend on several factors, such as the size and quality of the training set, the rigor of the training imparted, and also parameters chosen to represent the input.

References

1. Willems, J.L., and Lesaffre, E., (1987): 'Comparison of multigroup logisitic and linear discriminant ECG and VCG classification', *J. Electrocardiol.*, **20**, 1987, pp. 83–92.
2. Talmon, J.L., (1983): 'Pattern Recognition of the ECG', Berlin, Germany: Akademisch Proefscrift.
3. Hu, Y.H., Palreddy, S., and Tompkins, W.J. (1997): 'A patient-adaptable ECG beat classifer using a mixture of experts approach', *IEEE Trans. Biomed. Eng.*, **44**, pp. 891–900.
4. Coast, A.D., Stern, R.M., Cano, G.G., and Briller, S. A., (1990): 'An approach to cardiac arrhythmia analysis using hidden Markov models', *IEEE Trans. Biomed. Eng.*, **37**, pp. 826–835.
5. Lagerholm, M., Peterson, C., Braccini, G., Edenbrandt, L., and Sornmo, L. (2000): 'Clustering ECG complexes using hermite functions and self-organizing maps', *IEEE Trans. Biomed. Eng.*, **47**, pp. 838–848.
6. Silipo R. and Marchesi, C. (1998): 'Artificial Neural Networks for Automatic ECG Analysis', *IEEE Trans. Signal Processing*, **46**(5), pp. 1417–1425.
7. Ham, F.M., Han, S. (1996): 'Classification of cardiac arrhythmias using fuzzy ARTMAP', *IEEE Trans Biomed Eng.*, **43**(4), pp. 425–430.
8. Ozbay, Y., Ceylan, R., and Karlik, B. (2006): 'A fuzzy clustering neural network architecture for classification of ECG arrhythmias', Comput Biol Med., 2006, **36**(4), pp. 376–388.
9. Wang, Y., Zhu, Y.S., Thakor, N.V., and Xu, Y.H. (2001): 'A short-time multifractal approach for arrhythmia detection based on fuzzy neural network', *IEEE Trans Biomed Eng.*, **48**(9), pp. 989–95.
10. Tsipouras, M.G., Fotiadis, D.I., and Sideris, D. (2005): 'An arrhythmia classification system based on the RR-interval signal', *Artificial Intelligence in Med.*, **33**(3), pp. 237–250.
11. Shyu, L.Y., Wu, Y.H., and Hu, W. (2004): 'Using wavelet transform and fuzzy neural network for VPC detection from the Holter ECG', *IEEE Trans Biomed Eng.*, **51**(7), pp. 1269–1273.
12. Al-Nashash, H. (2000): 'Cardiac arrhythmia classification using neural networks', *Technol Health Care*, **8**(6), pp. 363–372.
13. Hosseini, H.G., Luo, D., and Reynolds, K.J. (2006): 'The comparison of different feed forward neural network architectures for ECG signal diagnosis', *Med Eng Phys.*, **28**(4), pp. 372–378.
14. Khadra, L., Al-Fahoum, A. S., and Al-Nashash, H. (1997): 'Detection of life-threatening cardiac arrhythmias using wavelet transformation', *Medical & Biological Engineering & Computing*, **35**, pp. 626–632.
15. Al-Fahoum, A.S., and Howitt, I. (1999): 'Combined wavelet transformation and radial basis neural networks for classifying life-threatening cardiac arrhythmias', *Medical & Biological Engineering & Computing*, **37**, pp. 566–573.
16. Acharya, U.R., Joseph, P.K., Kannathal, N., Lim, C.M., and Suri, J.S. (2006): 'Heart rate variability: A Review', IFMBE *Journal of Medical & Biological Engineering & Computing Journal.* (in press).
17. Acharya U.R., Kumar, A., Bhat, P.S., Lim, C.M., Iyengar, S.S., Kannathal, N., and Krishnan, S.M. (2004): 'Classification of cardiac abnormalities using heart rate signals', *Med Biol Eng Comput.*, **42**(3), pp. 288–293.

18. Neagoe, V.E., Iatan, I.F., and Grunwald, S. (2003): 'A neuro-fuzzy approach to classification of ECG signals for ischemic heart disease diagnosis', *AMIA Annu Symp Proc.*, pp. 494–498.
19. Osowski S, Hoai Lt, and Markiewicz T. (2004): 'Support vector machine-based expert system for reliable heartbeat recognition', *IEEE Trans Biomed Eng.*, **51** (4), pp. 582–589.
20. Krasteva, V., and Jekova, I. (2005): 'Assessment of ECG frequency and morphology parameters for automatic classification of life-threatening cardiac arrhythmias', *Physiol Meas.* **26**(5), pp. 707–23.
21. Kannathal, N., Lim, C.M., Acharya, U.R., and Sadasivan, P.K. (2006): 'Cardiac state diagnosis using adaptive neuro-fuzzy technique', *Med. Eng Phys.*, **28**(8), pp. 809–815.
22. Wieben, O., Afonso, V.X., and Tompkins, W.J. (1999): 'Classification of premature ventricular complexes using filter bank features, induction of decision trees and a fuzzy rule-based system', *Med Biol Eng Comput.*, **37**(5), pp. 560–565.
23. Dingfei, Srinivasan, N., and Krishnan, S.M. (2002): 'Cardiac arrhythmia classification using autoregressive modeling', *Biomed Eng Online*, **13**; 1:5.
24. Mohamed, I.O., Ahmed, H., Abou-Zied, Abou-Bakr, M., Youssef, and Yasser, M.K. (2002): 'Study of features on nonlinear dynamical modeling in ECG arrhythmia detection and classification', *IEEE transactions on Biomedical Engineering*, **49**(7), pp. 733–736.
25. Lin, C.H. (2006): 'Classification enhancible grey relational analysis for cardiac arrhythmias discrimination', *Med Biol Eng Comput.*, **44**(4), pp. 311–320.
26. Lippman, R.P. (1989): 'Pattern classification using neural networks', *IEEE Communication Magazine*, pp. 47–64.
27. Haykin, S. (1995): 'Neural networks A comprehensive foundation', *New York: MacMillun College Publishing Company.*
28. Wariar, R., and Eswaran, C. (1991): 'Integer coefficient bandpass filter for the simultaneous removal of baseline wander, 50 and 100 Hz interference from the ECG', Medical & Biological Engineering & Computing, pp. 333–336.
29. Pan, J., and Tompkins, W.J. (1985): 'Real Time QRS Detector algorithm', IEEE Transactions on Biomedical Engineering, **32**(3), pp. 230–236.
30. Theiler, J., Eubank, S., Longtin, A., Galdrikian, B., and Farmer, J.D. (1992): "Testing for nonlinearity in time series: the method of surrogate data", *Physica D*, **58**, 1992, pp. 77–94.
31. Klir, G., and Yuan, B. (1995): 'Fuzzy Sets and Fuzzy Logic: Theory and Applications', *Prentice–Hall*, New Delhi, India.
32. Jang, R.J.S. (1993): 'ANFIS – Adaptive-network-based neuro-fuzzy inference systems', *IEEE Transactions on Systems, Man, and Cybernetics*, **20**(3), pp. 665–685.

10

Storage and Transmission of Cardiac Data with Medical Images

U. Rajendra Acharya, P. Subbanna Bhat, U. C. Niranjan, Sathish Kumar, N. Kannathal, Lim Choo Min and Jasjit Suri

The landscape of healthcare delivery and medical data management has significantly changed over the last years, as a result of the significant advancements in information and communication technologies. Complementary and/or alternative solutions are needed to meet the new challenges, especially regarding security of the widely distributed sensitive medical information. Digital watermarking is a technique of hiding specific identification data for copyright authentication.

The DICOM standard is one method to include demographic information, such as patient information and X-ray exposure facilities, in image data. The DICOM standard is a standard that can be used regularly to record demographic information onto the image data header section. Regarding DICOM format images, information on patients and X-ray exposure facilities can be obtained easily from them. On the other hand, general-purpose image formats, such as the JPEG format, offer no standard that can be used regularly to record demographic information onto the header section.

Digital watermark technologies [1–8] can be used to embed demographic information in image data. Digital watermarking have several other uses, such as fingerprinting, authentication, integrity verification purposes, content labeling, usage control and content protection [9, 10]. The efficient utilization of bandwidth of communication channel and storage space can be achieved, when the reduction in data size is done. Recently, Giakoumaki *et al*, have presented a review of research in the area of medical-oriented watermarking and proposed a wavelet-based multiple watermarking scheme. This scheme aimed to address critical health information management issues, including origin and data authentication, protection of sensitive data, and image archiving and retrieval [11]. Their experimental results on different medical imaging modalities demonstrated the efficiency and transparency of the watermarking scheme.

The digital watermarking technique is adapted in this chapter for interleaving patient information with medical images, to reduce storage and transmission overheads. The text data is encrypted before interleaving with images to ensure greater security. The graphical signals are compressed and subsequently interleaved with the image. Differential pulse code modulation and adaptive delta modulation techniques are employed for data compression as well as encryption and results are tabulated for a specific example. Adverse effects of channel induced random errors and burst errors on the text data are countered by employing repetition code, Hamming code and R-S code techniques.

10.1 Concept of Interleaving

With the present trend of using internet as a medium to transmit images and patient data, it is of utmost importance to preserve authenticity of patient information. Exchange of data between hospitals involves large amount of vital patient information such as bio-signals, word documents and medical images. Therefore, it requires efficient transmission and storage techniques to cut down cost of health care. Interleaving one form of data such as 1-D signal, or text file, over digital images can combine the advantages of data security with efficient memory utilization [12]. In this present chapter, the watermarking technique is adapted for interleaving text and graphical signals with medical images (Fig. 10.1).

The watermarking techniques are divided into two basic categories:

- Spatial domain watermarking, in which the least significant bit (LSB) of the image pixels are replaced with that of the *watermark* (authentication text).
- Frequency domain watermarking, in which the image is first transformed to the frequency domain by discrete fourier transform (DFT), discrete cosine transform(DCT) and then the low frequency components are modified to contain the authentication text [13,14]. Since high frequencies will be lost by compression or scaling, the watermark signal is applied to the lower frequencies or applied adaptively to frequencies that contain important information of the original picture. Since watermarks applied to the frequency domain will be dispersed over the entirety of the image upon inverse transformation, this method is not susceptible to defeat by cropping as in the spatial domain.

Many authors have proposed the schemes to protect the ownership rights through the watermarking [12,15–19]. Swanson *et al*, have proposed the robust

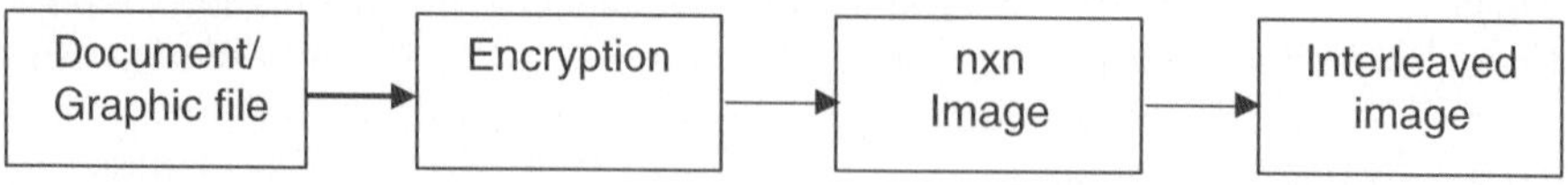

Fig. 10.1. Scheme for data storage

data hiding techniques for images [20–22]. Many authors have implemented adaptive watermarking in the DCT domain [23–26] and wavelet based watermarking techniques [27–31].

10.2 The Interleaving Process

Figure 10.1 indicates the steps involved in interleaving an image with data file. Before interleaving, the data is encrypted to enhance security [12]. Eight bits are usually employed to indicate the grey level of image pixel, and the LSB of each pixel is replaced by the text data. Thus the eight bits of ASCII code replace the LSBs of eight consecutive pixels of the image. (If the data file is a graphic signal having 16-bit word, 16 consecutive pixels are used for interleaving a single word.) Since only the LSB of the pixel is chosen for data interleaving, the resulting quality degradation of image is minimal. Depending upon the type of the signal, (text/graphic), three different encryption algorithms are used [32].

10.3 Encryption of the Text File

A typical text data (patient record) is shown in Fig. 10.2(a). Before interleaving, the document is encrypted by taking the logarithm of the ASCII codes. The encryption algorithm can be mathematically stated as:

$$T_e = (Log\,(T_o * 2) * 100) - 300 \tag{10.1}$$

where, $T_o \rightarrow$ ASCII code of the original text (or graphics file)

$$T_e \rightarrow \text{encrypted text}$$

The encrypted data (T_e) is rounded to the nearest integer for storing/ transmission. The decrypted text is obtained by,

$$T_0 = \exp\left\{(T_e + 300) \Big/ 100 - \log{(2)}\right\} \tag{10.2}$$

The encryption transform pairs given in Eqs. (10.1) and (10.2) yield exact reconstruction, even when T_e is rounded off to the nearest integer. Figure 10.2(b) shows the encrypted document corresponding to the file shown in Fig. 10.2(a). It may be noted that both the files occupy the same amount of memory space.

(a)	(b)
THE MANIPAL HEART FOUNDATION MANIPAL	ÔÅÁ Ì»ÍÆÐ»Ê ÅÁ»ÒÔ ÂÎÖÍ¿»ÔÆÎÍ Ì»ÍÆÐ»Ê
Patient Ref. No: 87906574	Ðãõëçïõ Òçè¯Íð¯¬ª®œ¨¦ª¤
Name of the doctor: Dr.Shyam	Íãîç ðè õêç æðåõðó¯¿ó¯Óêùãî
Name of the patient: Ms.Rani	Íãîç ðè õêç ñãõëçïõ¯Ìô¯Òãïë
Age: 50 years	»éç¯¦œ ùçãóô
Address: 5th cross,Udupi.	»ææóçôô¯ ¦õê åóðôô”ÖæÈñë¯
Case History:	¾ãôç Åëôõðóù¯
Date of admission: 20.09.1997	¿ãõç ðè ãæîëôôëðï ¯¡œ¯œ®¯?□® ®ª
Results: T wave inversion	ÒçôÈúõô¯ Ô ÷ãöç ëïöçóôëðï
Diagnosis: Suspected MI	¿ëãéïðôëô¯ ÓÈôñçåõçæ ÌÆ
Treatment:	Ôóçãõîçïõ¯
Sublingual Nitroglycerin	ÓÈäúëïéÈãú Íëõóðéúùåçóëï

Fig. 10.2. Encryption of patient information: (**a**) Original patient information, (**b**) Encrypted patient information

10.4 Encryption of the Graphic File

The Differential Pulse Code Modulation (DPCM) technique is effectively used to reduce the dynamic range of the graphic signal. It is used here for encrypting the heart rate signal. The differential error output (which is random and uncorrelated) is used as the encrypted version of the original signal. The DPCM is a predictive coding technique where in the present sample $X(n)$ in a signal is expressed as a sum of linearly weighted past sample $X(n-1)$ and error signal e(n) [33].

$$X\,(n) = pX\,(n-1) + e\,(n) \tag{10.3}$$

The predictor coefficient p is determined by the least square technique, as

$$p = r\,(1)\,/r\,(0) \tag{10.4}$$

$$\text{where,}\quad r(m) = \sum_{n=0}^{N-1-m} X\,(n)\,X\,(n+m) \tag{10.5}$$

The differential error $e\,(n)$ is stored along with the first sample $X\,(0)$ and the linear predictor coefficient p. The heart rate signal $X\,(n)$ can be reconstructed from the error signal by auto-regression technique (Eq. 10.3). Thus, the symbol pair p, $X\,(0)$ forms the key for the encrypted heart rate signal $e\,(n)$. This quantified $e\,(n)$ is interleaved with the LSB of image pixels. As the dynamic range of the error signal $e(n)$ is very small, the coding is done with only 4 bits.

Figure 10.3(a) and 10.3(b) display the original and reconstructed heart rate signals, while Fig. 10.3(c) shows the prediction error signal $e\,(n)$.

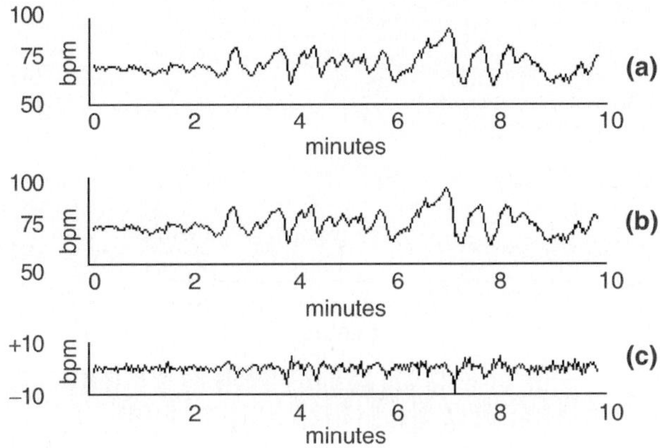

Fig. 10.3. Results of DPCM technique: (**a**) Original signal (**b**) Reconstructed signal (**c**) Error signal

Adaptive Delta Modulation (ADM) is a DPCM scheme in which the error signal $e(n)$ is encoded into a single bit. This single bit, providing just two options, is used to increase or decrease the estimate of $X(n)$ denoted by $\hat{X}(n)$. The amount of change is specified by the step length S(n), which is a multiple of basic step size S_0 and is determined adaptively. At the nth sample instant the step length generated is related to the step length of the previous sampling instant (Eq. 10.6).

$$S(n) = |S(n-1)| .e(n) + S_0.e(n) \tag{10.6}$$

This step length is used to estimate $\hat{X}(n)$ is shown below:

$$\hat{X}(n) = \hat{X}(n-1) + S(n) \tag{10.7}$$

The encrypted signal for storage/transmission is the error signal generated by comparing the actual signal $X(n)$ with its estimated value $\hat{X}(n)$.

$$e(n) = +1 \quad \text{if } X(n) > \hat{X}(n)$$
$$e(n) = -1 \quad \text{if } X(n) < \hat{X}(n)$$

Here $e(n)$ is a single bit and is interleaved with the LSB of the image pixel. The original heart rate signal, reconstructed signal and error signal are shown in Fig. 10.4(a), 10.4(b) and 10.4(c) respectively.

As can be seen from the results in DPCM and ADM, the error signal is random and uncorrelated, and has no resemblance to the original heart rate $X(n)$. This property of $e(n)$ is exploited by using the error signal as encrypted version of $X(n)$. In the case of ADM, $e(n)$ is encoded using a single bit, while in DPCM of $e(n)$ requires 4 bits for encoding. Thus, one sample of encrypted heart rate signal is interleaved into one pixel in case of

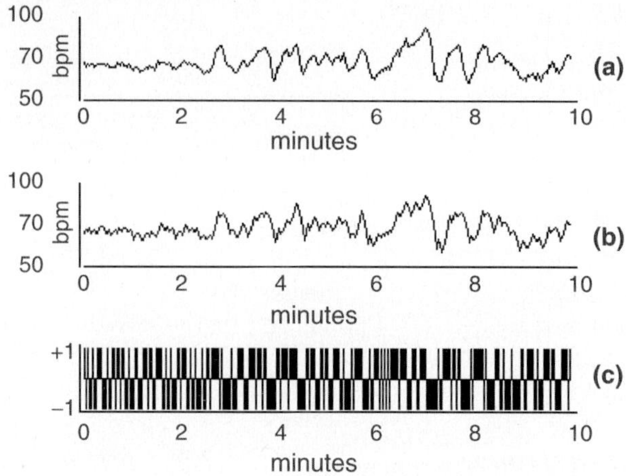

Fig. 10.4. Results of ADM techniques: (**a**) Original signal (**b**) Reconstructed signal (**c**) Error signal
#[Fig. 10.1, Fig. 10.2, Fig. 10.3, Fig. 10.4 are reprinted from Rajendra Acharya U, Deepthi Anand, Subbanna Bhat P, Niranjan UC, "Compact storage of medical images with patient information", IEEE Transactions on Information Technology in Biomedicine, 2001, vol.5(4), pp. 320–323]

ADM, while DPCM requires 4 pixels per sample. The reconstructed heart rate signal, shown in Fig. 10.3(b) and Fig. 10.4(b), (regressive Eqs. (10.4) and (10.5)) are good replica of the original signals.

10.5 Interleaving in DFT Domain

DFT magnitude is robust to translation or shift attacks in the domain since cyclic translation of image in the spatial domain does not affect DFT amplitude. DFT offers the possibility of interleaving either in the magnitude or the phase of the DFT coefficients. The phase is far more important than the magnitude of the DFT coefficients for the intelligibility of an image. Hence the interleaving is done in the magnitude coefficients of the DFT coefficients.

All the DFT coefficients are not modified. The low frequency components of the image are perceptually more significant ones and any modification of them deteriorates the image fidelity. Therefore interleaving should not be carried out in the low frequency region. The high frequency components are less significant in terms of fidelity. Hence the compression techniques use this property and suppress the high frequency coefficients first to reduce the size of images. Hence the interleaving is done in the bandpass magnitude coefficients of the DFT.

10.6 Interleaving in DCT

Blocks of 8 × 8 pixels are selected in a raster fashion from the image to be compressed. This forms the input to the encoder. Applying DCT to these blocks transforms the image from spatial domain to frequency domain. The DCT contains the DC coefficient, which measures the zero frequency, and 63 AC coefficients. The DCT coefficients are quantized according to perceptual criteria. For interleaving, the LSB of each DCT coefficient is replaced by the text data (after the quantization and zigzag encoding). The eight bits of ASCII code in the text file will replace the LSBs of eight consecutive DCT coefficients of the image from the middle frequency range onwards (from 32 to 63 coefficients) [34]. If the data file is a graphic signal having 16-bit word, 16 consecutive DCT coefficients are used for interleaving a single word. The LSB is chosen for data interleaving because, the resulting degradation of image is minimal. In the decoding side, the interleaved text or graphical signal can be obtained by de-interleaving i.e., extracting the LSBs and concatenating the same, before the inverse quantization, zigzag coding and inverse DCT.

10.7 Interleaving in Wavelet Domain

Wavelets are the functions defined over a finite interval and have average value zero. The basic idea of the wavelet transform is to represent any arbitrary function f(t) as a superposition of a set of such wavelets or basis functions. These basis functions are obtained from a single prototype wavelet called the mother wavelet, by dilations or contractions (scaling) and translations (shifts). Besides the usage of DCT in JPEG, a new compression technique called JPEG2000 uses Discrete Wavelet Transform. The blocking artifacts of DCT in JPEG are noticeable and annoying. In wavelet based compression we get higher compression and blocking artifacts can be avoided. The wavelet transform can be implemented using filter banks [35]. The signal is decomposed into various subbands and the octave-band decomposition is most widely used. Figure 10.5 shows three level octave band decomposition. The DWT gives three parts of multiresolution representation and one part of multiresolution approximation [35]. It is similar to hierarchical subband system, where subbands are logarithmically spaced in frequency. The subbands labelled LH1, HL1, HH1 of multiresolution representation represent the finest scale wavelet coefficients. To obtain next coarser scale of the wavelet coefficients, the subband LL1 i.e. multiresolution approximation is further decomposed and critically sub sampled. One can perform three level decomposition of the image and embed the text/graphic file information into high frequency region band (starting from the 32nd coefficient to 64th coefficient). The text and graphic file can be extracted from the DWT coefficients before the inverse quantization, inverse zigzag coding and inverse discrete wavelet transform.

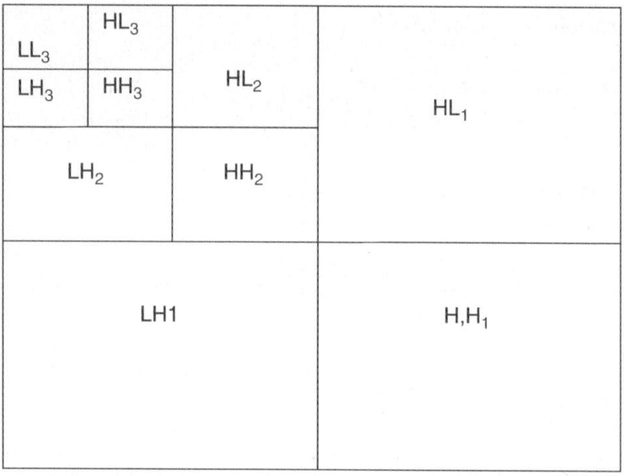

Fig. 10.5. Scheme for DWT decomposition of an image

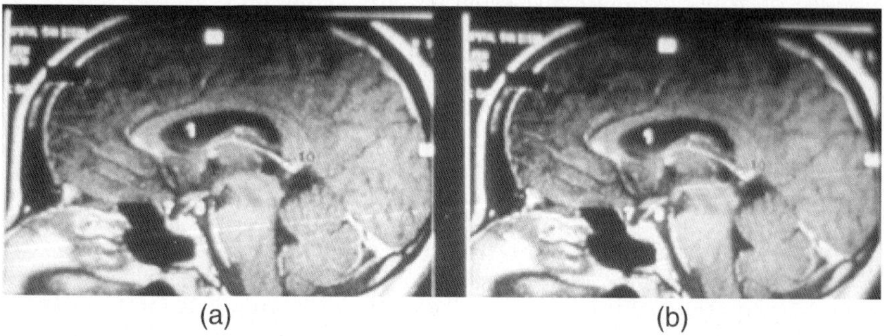

Fig. 10.6. Result of interleaving text in the MRI image: (**a**) Original image (**b**) Interleaved image

10.8 Evaluation of Results

The ASCII codes of the encrypted text shown in Fig. 10.2(b) are broken into bits and interleaved into the pixels of MRI image (Fig. 10.6 (a); size: 196×260 pixels). The resulting image is shown in Fig. 10.6(b). The error signals e_n obtained from DPCM and ADM (Fig. 10.3(c) and 10.4(c)), are interleaved into the Angiogram and the Ultrasound images of Fig. 10.7(a) (size: 200×265 pixels) and Fig. 10.8(a) (size: 270×229 pixels). The resulting interleaved images are shown in Fig. 10.7(b) and 10.8(b). The process affects only the LSB of each pixel, which may cause a change in its brightness by 1 part in 256. As can be seen from these results, it does not affect the picture quality significantly. The text and heart rate information are interleaved into all the pixels of the image.

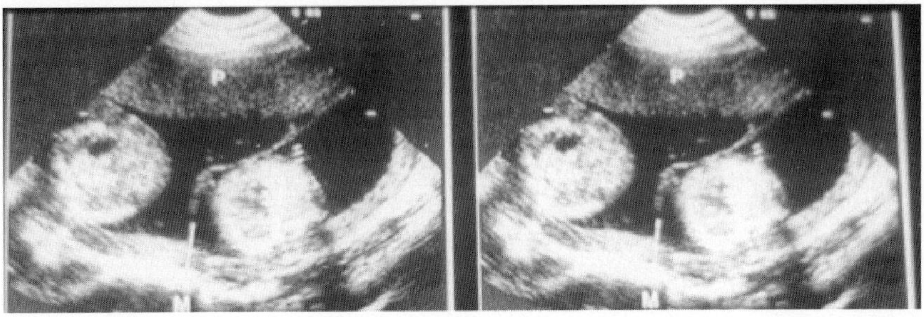

(a) (b)

Fig. 10.7. Result of interleaving ADM error signal in the Ultrasound image: (**a**) Original image (**b**) Interleaved image

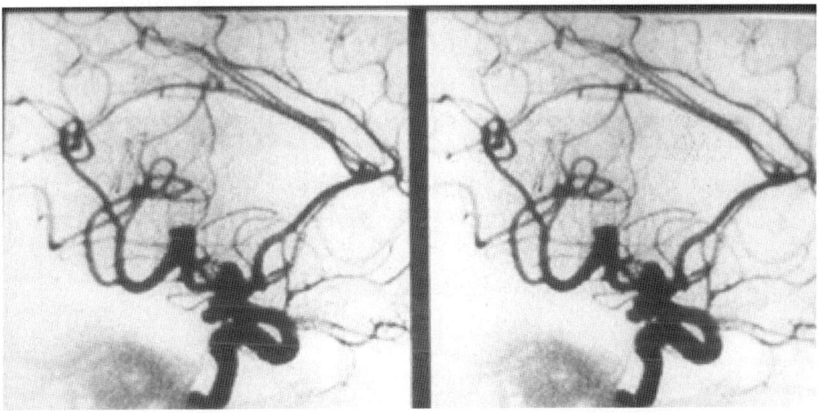

(a) (b)

Fig. 10.8. Result of interleaving DPCM error signal in the Angiogram image: (**a**) Original signal (**b**) Interleaved signal

Figure 10.9(a) and 10.9(b) show the intensity histograms of the original and interleaved ultrasound images (with error signal of Fig. 10.4(c)). It can be seen that the shape of the histogram bears resemblance to that of the original image. The change in the population of pixels of a specific intensity is definite in nature. For example, consider two pixels with grey scale values 0 and 1. When interleaved, the 0 pixel may remain at 0 or change to 1. Similarly the 1 pixel may remain at 1 or change to 0. Thus there is a redistribution of pixels 0 and 1. Such redistribution occurs among all consecutive pairs of even and odd pixels (such as 2 and 3, 4 and 5, and so on.) This pair-wise redistribution of pixel values explains the resemblance between the original and modified histograms.

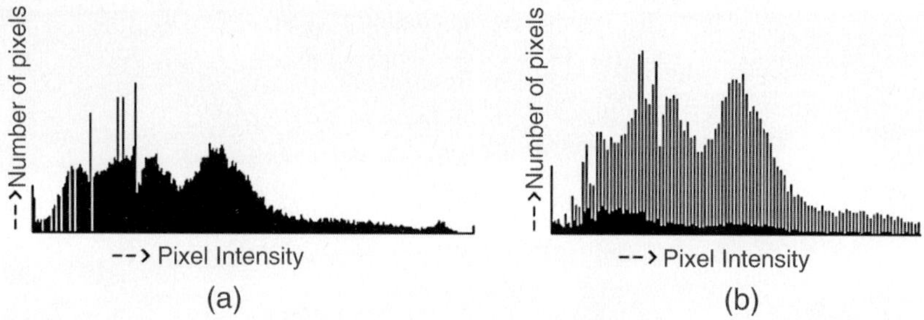

Fig. 10.9. Histogram of Ultrasound images: (**a**) Original image (**b**) Interleaved image

#[Table 10.1, Fig. 10.6, Fig. 10.7, Fig. 10.8, Fig. 10.9 are reprinted with permission from Rajendra Acharya U, Deepthi Anand, Subbanna Bhat P, Niranjan UC, "Compact storage of medical images with patient information", IEEE Transactions on Information Technology in Biomedicine, 2001, vol.5(4), pp. 320–323]

A quantitative assessment of the method is obtained by evaluating the normalized root mean square error (NRMSE) as defined below:

$$NRMSE\,(\%) = \sqrt{\frac{\sum\limits_{Y=0}^{U-1}\sum\limits_{X=0}^{V-1}[f(x,y)-f_w(x,y)]^2}{\sum\limits_{Y=0}^{U-1}\sum\limits_{X=0}^{V-1}[f(x,y)]^2}} \times 100 \qquad (10.8)$$

where

U = Total number of columns
V = Total number of rows in the image
$f(x.y)$ = The original pixel intensity
$f_w(x,y)$ = The modified (interleaved) pixel intensity

From the definition of the NRMSE, it is seen that, the maxima occurs when the LSB of all the pixel values are altered. Then the equation for maximum NRMSE becomes,

$$MNRMSE(\%) = \sqrt{\frac{(U*V)}{\sum\limits_{Y=0}^{U-1}\sum\limits_{X=0}^{V-1}[F(X,Y)]^2}} \times 100 \qquad (10.9)$$

The technique is tested for different images and the NRMSE was found to be less than 0.71 % for 8-bit encoded pixel intensity (Table 10.1). Security of information can be further enhanced by choosing the position of the interleaved bit according to a specific plan known only to the authorised users. In order to assess its effect on reduced intensity images, the images were converted to 128 intensity levels at a coding of 7 bits/pixel and interleaving

Table 10.1. Results of interleaving data with image

Image	Text (NRMSE %)	DPCM (NRMSE %)	ADM (NRMSE %)	MNRMSE (%)
MRI	0.70	0.70	0.69	1.24
Ultrasound	0.53	0.54	0.52	1.43
Angiogram	0.50	0.50	0.50	0.76

operation was carried out. The values of NRMSE for these reduced intensity images doubled since the denominators of Eqs. (10.8) and (10.9) reduce by a factor of two. The proposed technique of interleaving additional data into an image can also be used for transmission purposes, wherein patient information embedded in the medical image can be sent over communication channels, as a single data entity.

10.9 Interleaving Error Correcting Codes

The transmission of interleaved data over noisy channels is examined using standard simulation techniques [36, 37]. The noisy channel is modelled as a simple Binary Symmetric Channel (BSC), where the probability of bit '1' or bit '0' being altered into its complement is the same, say P_e. To simulate this BSC, uniformly distributed random numbers in the range of 0 to 1 are generated using the Linear Congruential Method. For each transmitted bit, the corresponding random number generated is compared against P_e. If it is less than P_e, that bit is inverted. Otherwise the bit is taken as it is. From elementary probability theory it can be easily verified that P_e indeed gives the probability of a bit being inverted in the course of transmission.

A noisy channel is also modelled by assuming zero mean Gaussian noise with variance σ^2. The decision regarding the received bit was assumed to be taken by a 'decision device' whose threshold is set exactly between the *high* and low levels (representing the bits '1' and '0'). If the amplitude of the noise sample added is so large that it takes the received signal amplitude to the wrong side of the threshold, the decision device will make an incorrect decision regarding the received bit [38].

The Gaussian noise can be simulated using the Box-Muller Algorithm [39]. A Gaussian random number generated by the Box-Muller algorithm is added to each bit being transmitted and the resultant signal amplitude is compared with the threshold to simulate the 'decision device'. A decision regarding the bit is then taken depending on whether the bit plus noise signal amplitude is more or less than the threshold.

The major difference between the two approaches discussed above is that in the first case, the statistical nature of the noise affecting the transmissions is not considered, whereas in the second case, noise of a specific distribution is considered. The improvement in the reliability of signal transmission can be demonstrated in both of these cases, by employing repetition codes or (7,4) Hamming codes [36].

Burst error, which is likely to occur while retrieving stored data is modelled as explained below. When a 'burst' strikes, a Poisson distributed random number of data bits with a mean value α, gets inverted. The burst error environment is simulated as follows: Uniformly distributed random numbers in the range of 0 to 1 are generated using the Linear-congruential method. If the generated number is less than P, where P is the probability that a burst strikes the data, a Poisson distributed random number is generated with its mean value α and a corresponding number of bits are inverted.

One of the best techniques to counter burst errors is to resort to interleaving of data words. If the expected length of burst error (α) is large, the depth of interleaving should be correspondingly large for satisfactory performance. In the present work data words are interleaved to the depth of 4. Another very effective way to counter burst errors is to resort to error control coding techniques where the encoding is done on data symbols rather than on data bits. The Reed Solomon (RS) codes are an example of such error control techniques. In the (7,3) RS codes, which is employed here, the encoding is done by adding four parity symbols to a group of three data symbols (each symbol constitutes of three bits) resulting in a total of seven symbols. The (7,3) RS code used in the present work, can correct up to 2 symbol errors in a block of 7 symbols.

Interleaved image data are tested against various types of transmission impurities introduced by the channel. The robustness of the signal is improved by the application of appropriate error correction coding techniques as mentioned in the previous section. As is expected, with an increase in P_e of the BSC or the variance σ^2 of the Gaussian noise, the signal quality deteriorates at the receiving end. This is visible in the form of character loss or distortion in the text data. Application of the repetition code or the (7,4) Hamming code is seen to bring about a remarkable increase in the quality of the received (interleaved) data.

To make a comparative study RS codes have also been employed in a Gaussian noise corrupted channel. The relevant results are as shown in Table 10.2.

From the results reported in Table 10.2, it can be seen that in a Gaussian noise corrupted channel, the (7,4) Hamming code gives the best performance, while the RS code, the poorest. Among the three coding schemes outlined above, the repetition code consumes the maximum bandwidth, whereas the (7,4) Hamming code demands the minimum. Figure 10.10 shows the result of patient information (Fig. 10.2 (a)) with repetition code, corrupted with Gaussian noise ($\sigma^2 = 0.09$). A graph showing the loss of characters due to noise is given in Fig. 10.11.

In the case of burst errors, both the strength and repetition rate of burst can cause distortion in the received data. Interleaving of coded data can be

Table 10.2. Comparison of various codes for Gaussian noise corrupted channel

Variance σ^2	Characters altered with Repetition code	Characters altered with (7,4) Hamming code	Characters altered with (7,3) RS code
0.06	0	0	2
0.07	2	1	2
0.08	3	2	4
0.09	5	2	7

THE MANIPAL HEART FOUNDATION MANIPAL

Patient Ref.No: 87906574
Name of the doctor: Dr.Shyam
Name of the patient: Ms.Rani
Age: 50 years
Address: 5th cross, Udupi.
Case History:
Date of admission: 20.09.1997
Results: T wave inversion
Diagnosis: Suspected MI
Treatment:
Sublingual Nitroglùcerin

Fig. 10.10. Result of patient information (Fig. 10.2(a)) with repetition code, corrupted with Gaussian noise ($\sigma^2 = 0.09$)

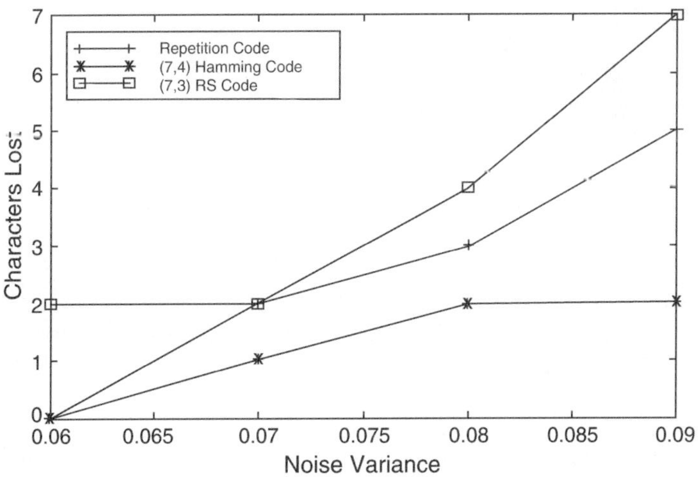

Fig. 10.11. Characters lost due to noise variance (for different types of coding)

effective in combating the ill effects of burst error. As mentioned earlier RS codes can also be seen as an efficient technique to counter burst error. The results are shown in Table 10.3 and Table 10.4. Graphical representations of the results are given in Figs. 10.12 and 10.13 respectively.

Table 10.3. Characters altered due to burst noise (for $\alpha = 2$)

Frequency of burst occurrence (P)	Characters altered with interleaving and depth-4 in combination with (7,4) Hamming code	Characters altered with (7,3) RS code
0.005	4	0
0.007	5	0
0.01	7	0

Table 10.4. Characters altered for different values of P with $\alpha = 2$

Frequency of burst occurrence (P)	Characters altered with interleaving and depth-2 in combination with (7,4) Hamming code	Characters altered with (7,3) RS Code
0.03	0	3
0.04	2	4
0.05	6	5
0.06	9	6
0.07	10	13

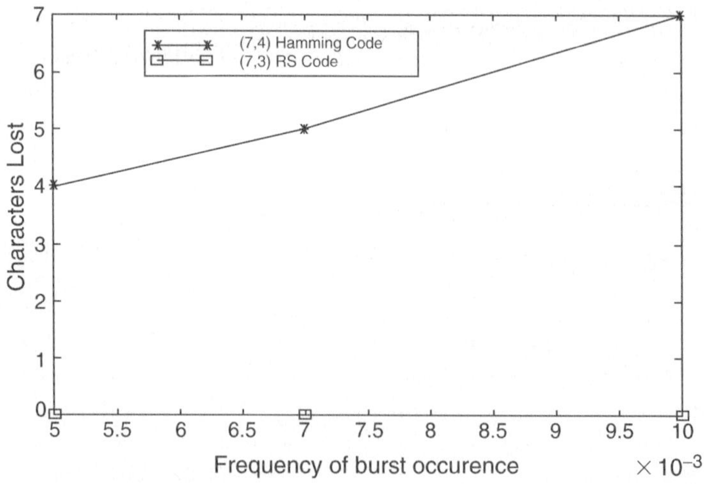

Fig. 10.12. Characters lost as a function of noise variance (interleaving depth = 4)

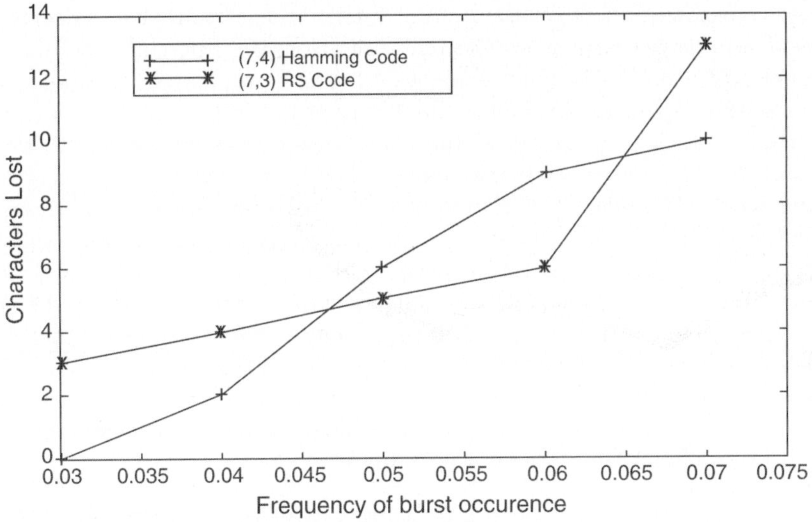

Fig. 10.13. Characters lost as a function of frequency of burst occurrence

These results show that for frequency of burst from .005 to .01, the RS code gives an error free reception of the interleaved data, whereas in a similar environment the (7,4) Hamming code (with interleaving depth = 4) gives poorer performance. For higher values of P, the RS code performance deteriorates.

10.10 Discussion

With the present trend of using the Internet as a medium to transmit images and patient data it is of utmost importance to preserve the authenticity of patient information. The technique of 'watermarking' – interleaving text data with images for copyright authentication – is adapted here to interleave patient data with medical images [39, 40]. Eight bits are usually employed to indicate the grey level of image pixel, and the LSB is replaced by the 'text' information for identification [33]. Thus the LSB is 'sacrificed' for the purpose of authentication, which is usually acceptable. Security of information can be further enhanced by choosing the position of the interleaved bit according to a specific plan known only to the authorized users. Transmission exposes the data to noise, which results in the corruption of data. Therefore, different error correcting codes/techniques are employed to make the data handling more robust with respect to noise [36, 39].

Transmission of cardiac data through a channel, and retrieval of stored data from memory are prone to some kind of errors. The effect of random noise, such as Gaussian noise, and burst errors on interleaved data are studied using standard simulation techniques [39]. In the case of random noise and burst error, the impact on the picture information is tolerable, whereas the loss of interleaved data is quite significant. However, encoding the data using suitable error correction techniques can reduce the impact of these errors significantly. The situation can be further improved if the data is interleaved. As the 'burst' length increases, the depth of interleaving too has to be increased to get satisfactory performance. The results have been demonstrated using repetition code, (7,4) Hamming code (with and without interleaving) and the (7,3) RS code. It is observed that (7,4) Hamming code and the repetition code give good results in a random error channel (corrupted by Gaussian noise); whereas (7,4) Hamming code with interleaving and the (7,3) RS codes give good results in a burst error environment. It is seen that (7,3) RS codes give excellent performance for a range of frequency of burst occurrence ranging from 0.005 to 0.01. In the similar way, we can use these codes to protect the graphical signals like electrocardiogram, heart rate signals etc during transmission through the noisy channels.

10.11 Conclusion

A technique of interleaving patient information such as text documents and physiological signals with medical images is presented for efficient storage. Text files are encrypted using logarithmic technique and heart rate signals are encrypted by DPCM and ADM techniques, prior to interleaving. The technique is tested for different images and the NRMSE was found to be less than 0.71% for 8 – bit encoded pixel intensity. Transmission of cardiac data through a channel and retrieval of stored data from memory are prone to some kind of errors. The effect of Gaussian noise on interleaved data is studied using standard simulation techniques. Encoding the data using suitable error correction techniques can reduce the impact of these errors significantly. The results have demonstrated that (7,4) Hamming code is more superior compared to the repetition code in a random error channel (corrupted by Gaussian noise). Various other type of error correcting codes can be interleaved to overcome the effect of other type of noise.

References

1. Anderson, R. (1997): 'Information Hiding', *Lect. Notes, Comput. Sci.*, **1174**, Springer-Verlag.
2. Cox, I.J., Kilian, J., Leighton, T., and Shamoon, T. (1997): 'Secure spread spectrum watermarking for multimedia', *IEEE Trans. Image. Processing*, **6**, pp. 1673–1687.
3. Marvel, L., Boncelet, C., and Retter, C. (1998): 'Reliable Blind Information Hiding for Images', *Lect. Notes. Comput. Sci.*, **1525**, pp. 48–61.

4. Swanson, M., Kobayashi, M., and Tewfik, A. (1998): 'Multimedia Data-Embedding and Watermarking', Technologies, *Proc. Proceedings of IEEE*, **86(6)**, pp. 1064–1087.
5. Hernandez, J.R., Amado, M., and Perez-Gonzalez, F. (2000): 'DCT-domain watermarking techniques for still images: Detector performance analysis and a new structure', *IEEE.Transactions on Image. Processing*, **9**(1), pp. 55–68.
6. Cho, Y., Ahn, B., Kim, J.S., Kim, I.Y., and Kim, S.I. (2001): 'A Study for Watermark Methods Appropriate to Medical Images', *J. Digit. Imaging*, **14**(2 Suppl 1), pp. 184–186.
7. Miyazaki, A. (2002): 'Digital Watermarking – Protection Technique for Multimedia', *Technical Report of IEICE*, DSP2002-27, pp. 61–66.
8. Chung, Y.Y., and Wong, M.T.(2003): 'Implementation of Digital Watermarking System', *Dig. Tech. Pap. IEEE. Int.Conf. Consum. Electron*, pp. 214–215.
9. Memon, Nasir, and Wong P.W. (1998): 'Protecting Digital Media Content', *Communications of the ACM*, **35**(8), pp. 35–43.
10. Craver, Scott, Boon-Lock, and Yeung M. (1998): 'Technical Trials and Legal Tribulations', *Communications of ACM*, **45**(9), pp. 45–54.
11. Giakoumaki, A., Pavlopoulos, S., and Koutsouris, D. (2006): 'Secure and efficient health data management through multiple watermarking on medical images', *Med Biol Eng Comput.*, **44**(8), pp. 619–631.
12. Berghel, H. (1997): 'Watermarking Cyberspace', *Communications of the ACM*, **40**(11), pp. 19–24.
13. Berghel, H., and Lawrence, O.G. (1996): 'Protecting Ownership Rights Through Watermarking', *IEEE Computer*, pp. 101–103.
14. Johnson, N.F., Duric, Z., and Jajodia, S. (2000): 'Information Hiding: Steganography and Watermarking-Attacks and counterattacks', *Kluwer Acadamic Publishers.*
15. FaBien, A.P., Petitcolas, Anderson, R.A., and Kuhn, M.G. (1999): 'Information hiding: A Survey', *Proceedings of the IEEE*, **85**(6), pp. 1062–1077.
16. Lin, E.T., and Delp, E.J. (1999): 'A review of data hiding in digital images', *Proceedings of the Image Processing, Image Quality, Image Capture Systems Conference*, pp. 274–278.
17. Langelaar, G., Setyawan, I., and Lagendijk, R. (2000): 'Watermarking digital image and video data: a state-of-art overview', *IEEE Signal Processing Magazine*, **17**, pp. 20–46.
18. Cox, J., Miller, M.L., and Bloom, J.A. (2002): 'Digital Watermarking', *San Fransisco, CA: Morgan Kauffman.*
19. Knopp, R., and Robert, A. (2000): 'Detection Theory and Digital Watermarking', *Proceedings of SPIE*, **3971**, pp. 14–23.
20. Swanson, M.D., Zhu, B., Tewfik, A.H. (1996): 'Robust data hiding for images', *Proceedings of the IEEE Digital Signal Processing Workshop*, pp. 37–40.
21. Swanson, M.D., Zhu, B., and Tewfik, A.H. (1997): 'Data Hiding for Video in Video', *Proceedings of International Conference on Image Processing (ICIP'97)*, **2**, pp. 676–679.
22. Swanson, M.D., Zhu, B., and Tewfik, A.H.(1996): 'Transparent robust image watermarking', *Proceedings of IEEE International Conference on Image Processing*, **3**, pp. 211–214.
23. Bros, A.G., and I. Pitas,I. (1996): 'Image Watermarking using DCT domain constraints', *Proceedings of IEEE International Conference on Image Processing*, **3**, pp. 231–234.

24. Barni, M., Bartolini, F., and Piva, A.(1998): 'Copyright protection of digital images by means of frequency domain watermarking', *Mathematics of Data/Image Coding, Compression and Encryption, Proceedings of SPIE*, **3456**, pp. 25–35.
25. Tao, B., and Dickenson, B. (1997): 'Adaptive Watermarking in the DCT Domain', *Proceedings of International Conference on Accoustics, Speech and Signal Processing (ICASSP '97)*, **4**, pp. 676–679.
26. Juan, R.H.M.A., and Perez-Gonzalez, F. (2000): 'DCT-Domain Watermarking Techniques for Still images: Detector Performance Analysis and a New Structure', *IEEE Transactions on Image Processing*, **9**, pp. 55–68.
27. Antonini, M., Barlaud M., Mathieu P., and Daubechies I., "Image Coding using wavelet transform", *IEEE Transactions on Image Processing*, **1**(2), pp. 205–220.
28. Ohnishi, J., and Matsui, K. (1996): 'Embedding a seal into a picture under orthogonal wavelet transform', *Proceeding of International Conference on Multimedia Computing and Systems*, pp. 514–521.
29. Cao, J.G., Fower, J.E., and Younan, N.H. (2001): 'An Image-Adaptive Watermark Based on a Redundant Wavelet Transform',*Proceedings of the International Conference on Image Processing, Greece*, **2**, pp. 277–280.
30. Barni, M., Bartolini, F., and Piva, A. (2001): 'Improved Wavelt-Based Watermarking Through Pixel-Wise Masking', *IEEE Transactions on Image Processing*, **10**(5), pp. 783–791.
31. Tokur, Y., and Ercelebi, E. (2003): 'Wavelet–Based Digital Image Watermarking for Copyright Protection', *IJCI Proceedings of International XII, Turkish Symposium on Artificial Intelligence and Neural Networks*, **1**(1), pp. 47–49.
32. Acharya, U.R., Deepti, A., Bhat, P.S., and Niranjan, U.C. (2001): 'Compact storage of medical images with patient information', *IEEE Transactions on Information Technology in Biomedicine*, **5**(4), pp. 320–323.
33. Gonzalez, R.C., and Wintz, P. (1987): 'Digital Image Processing', *Addison-Wesley Publishing Co.*
34. Acharya, U.R., Niranjan, U.C., Iyengar, S.S., Kannathal, N., and Min, L.C. (2004): 'Simultaneous Storage of Medical Images with Patient Information',*Computer methods and programs in Biomedicine*, **76**(1), pp. 13–19.
35. Mallat, S. (1989): 'A theory of multiresolution signal decomposition', *IEEE Transactions on Pattern Analysis and Machine Intelligence*, **11**, pp. 674–694.
36. Sweeney, P. (1991): 'Error Control Coding – An Introduction', *Prentice Hall.*
37. Proakis, J.G. (1996): 'Digital Communications', *McGraw Hill*, New York.
38. Haykin, S. (1996): 'Communication Systems', Wiley Eastern.
39. Banks, J., and Carson, J.S. (1991): 'Discrete Event System Simulation', *Prentice Hall*, 1984.Peter Sweeney, Error Control Coding-An Introduction', Prentice Hall.
40. Acharya, U.R., Bhat, P.S., Sathish, K., and Min, L.C. (2003): 'Transmission and Storage of Medical Images with Patient data', *Computers in Biology and Medicine*, **33**(4), pp. 303–310.

11

Assessment of Cardiac Function in Filling & Systolic Ejection Phases: A Mathematical and Clinical Evaluation

Liang Zhong, Dhanjoo N. Ghista, Eddie Y.K. Ng, Ru San Tan, Soo Teik Lim and Terrance S.J. Chua

11.1 Introduction and Scope

The chapter is made up of three sections. The first section discusses various works delineating indices of left ventricular (LV) function, namely the LV myocardial contractile element's shortening velocity (V_{max}), end-systolic pressure-volume relationship (ESPVR), ventricular preload-adjusted maximal power and strain rate.

The second part deals with passive and active elastances (E_p and E_a), and demonstrates how they contribute to left ventricular (LV) pressure dynamics, as a response to volume changes during filling and ejection phases.

The third part deals with a new concept of cardiac contractility in the form of the maximal circumferential wall stress normalized with respect to LV pressure. This index can be employed noninvasively without requiring monitoring of LV pressure. We have further shown that this (easy to compute index) correlates well with the traditional index of dP/dt_{max}. Hence, it may be employed as a substitute for dP/dt_{max}.

11.2 Known Characterization of Left Ventricular Function

Heart failure is a leading cause of death in developed and developing countries. This clinical syndrome may be a manifestation of filling abnormalities and/or inadequate left ventricular (LV) myocardial contraction (due to myocardial ischemia or infarct and/or electrical contraction abnormality). Cardiologists have been using simple clinical measurements such as heart rate, ECG patterns, blood pressure and ejection fraction to diagnose cardiac pathologies. However, these measurements do not provide adequate insight into the heart's complex electro-mechanical machinery involved in its functional performance and clinical syndromes. In this context, this chapterconstitutes a study of biomechanical mechanisms and phenomena of the heart function, leading to the development of new indices of assessment of heart function and impairment.

In a bioengineering sense, the heart can be described as a blood pump and its biomechanical behavior can be understood in terms of the time-varying relationship between ventricular blood pressure and volume. For many years, clinicians have used this relationship as a measure of cardiac function [1–14]. Concomitant with this approach (and resulting from these measurements) has been the concept of LV myocardial stress and modulus [15–17], to characterize both diastolic and systolic performance.

It is apparent that an understanding of LV functional mechanisms in terms of its myocardial properties is important to characterize the underlying ventricular pump function. Pressure and volume are the fundamental measurable physical variables, from which to derive and express these biomechanical properties of the LV. This theme has been extensively incorporated clinically [4, 18–21].

Characterizing the LV as a pump is not simple, as the LV action is a complex phenomenon. It depends both on its myocardial properties and chamber geometry, as well as on the preload (i.e., end-diastolic pressure) and afterload (i.e., aortic pressure). Therefore, it is necessary to know the characteristics of LV shape-dynamics and LV muscle contraction parameters during physiological and pathological conditions, in order to characterize LV functional performance. In an intact beating heart, it is difficult to measure these parameters and to relate them with the different diseases. Hence, we need to develop appropriate LV models, to characterize and simulate its in-vivo anatomy and physiology as well as its sarcomere properties, and link them to its clinical performance.

Analytical modeling will be necessary for such a quantitative understanding of the cardiac function and dysfunction during both diastolic-filling and ejection phases. In this regard, the primary objectives of this chapter are: (i) develop time-dependent active-elastance and volume-dependent passive-elastance model; (ii) develop a novel noninvasive determinable cardiac LV contractility index, based on LV normalized wall stress with respect to LV intracavitary pressure, and (iii) provide a preliminary validation for the contractility indices developed in patients. Our work is intended to enable comprehensive, convenient and reliable diagnosis of cardiac performance and disorders, in terms of these functional model parameters.

11.2.1 Left Ventricular Diastolic Dysfunction Characterization

From a physiological point of view, the LV is an integrated electro-mechanical muscle-pump system. The term diastole is interpreted as division, notch, or separation between two contraction-relaxation cycles [22]. In this interpretation, its meanings is restricted to the passive properties of the LV [23]. LV Diastole starts when active relaxation has been completed, and includes the diastasis and the left atrial (LA) contraction phase. In the English medical literature, diastole is interpreted as "the dilatation or period of dilatation of the LV, coinciding with the interval between the second and first heart sounds" [18]. In this interpretation, it is that part of the cardiac cycle, which starts with the isovolumic relaxation phase and ends with cessation of mitral flow [24].

The functional properties of the LV during diastole have mostly been described in terms of either the rate at which it relaxes in early diastole and its stiffness, when it is fully relaxed in the later part of diastole. Hence, diastolic dysfunction can be characterized by relaxation abnormalities (influenced by ischemia, cardiomyopathy), decreased compliance (i.e., hypertrophy, myocardial fibrosis) and abnormal high heart rate [25, 26].

However, the filling pattern of the LV depends on a complex and continuous interaction of multiple factors, of which only some relate directly to the diastolic properties of the LV itself. Other factors relate purely to hemodynamic conditions imposed on the LV, including venous return, suction, resistance-to-filling and LA contraction and so on. Therefore, diastolic dysfunction can be regarded as the result of diversity of physiologic abnormalities of myocardial relaxation and/or ventricular compliance. Signs and symptoms of patients with diastolic dysfunction are related to decreased cardiac output. To date, many diagnostic tools are available for the diagnosis of diastolic function, based on echocardiography (i.e., e and a wave), cineventriculography (for end-diastolic pressure-volume relationship) and MRI (ratio of myocardial volume and intracavitary LV volume). None of the tests however can diagnose diastolic dysfunction unambiguously. Hence, predictive values based on each of these techniques and indices can only provide partial information for the diagnosis of diastolic dysfunction.

11.2.2 Left Ventricular Systolic Dysfunction Assessment

In LV systolic function, the LV myocardium must generate adequate coordinated contractile force to provide sufficient LV pressure to open the aortic valve, and then contract and shorten the wall to pump an appropriate stroke volume. This means systole is a process of converting chemical energy into mechanical energy, by means of actin-myosin crossbridges, for contraction. The functional properties of the LV during systole have generally been described in terms of contractility terms on different levels. Hence, systolic dysfunction can be defined as the inability of the LV to adequately contract and eject the requisite stroke volume, given a normal diastolic function.

'How to intrinsically assess the systolic function of the LV' has long been the major problem in physiology and cardiology, and this problem remains yet unsolved. The systolic function of the LV has been viewed at different levels [27]. Way back in 1895, Frank characterized ventricular function in terms of the pressure-volume diagram. In 1918, Starling first viewed the LV as a pump to generate cardiac output (proportional to its filling) against an afterload. The loss of cardiacinotropy (i.e., decreased contractility) causes a downward

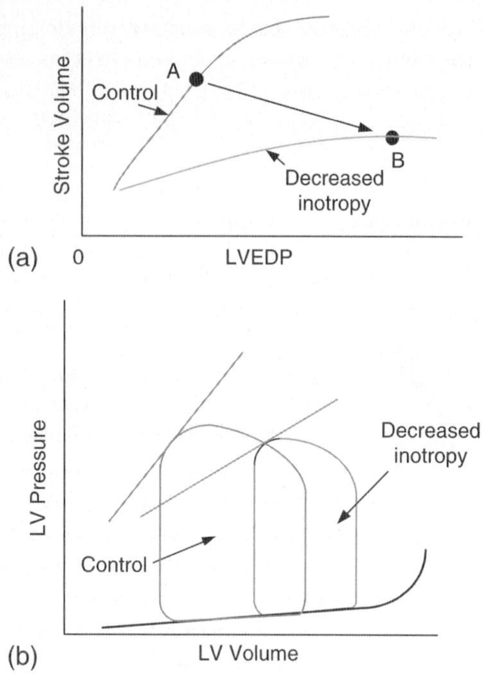

Fig. 11.1. (**a**) The Frank-Starling relationship showing the effects of systolic dysfunction on stroke volume and left ventricular end-diastolic pressure (LVEDP). (**b**) Effects of left ventricular systolic dysfunction on left ventricular pressure-volume loop

shift in the Frank-Starling curve (Fig. 11.1). This results in a decrease in stroke volume and a compensatory rise in preload (i.e., end-diastolic volume, end-diastolic pressure). The rise in preload is considered compensatory, because it activates the Frank-Starling mechanism to help maintain stroke volume despite decreased inotropy. Also the effects of a decreased inotropy on stroke volume and end-diastolic volume are depicted in Fig. 11.1, using LV pressure-volume loop. Hill [28] and Huxley [29, 30] investigated muscle contraction by means of muscle force-velocity relationship at the micro-structural level (using the cross-bridge theory). Thus, as also indicated by Suga [31], the systolic function of the LV is a process of converting the actin-myosin cross-bridge formation chemical energy into the mechanical energy during contraction.

11.3 Justification for Clinical Indices of Cardiac Function

The use of indices, mathematically derived from cardiac (pressure-volume) data measurements, facilitates the monitoring and the follow-up of cardiac function. In this section, we will discuss some of these indices.

11.3.1 Indices Characterizing the "Passive" Ventricle

The LV end-diastolic pressure remains the most commonly used clinical parameter to describe the passive elastance of the ventricle. A higher-than-normal filling pressure is considered as an index of LV dilatation, stiffness and systolic failure. This parameter is easily measured by catheterization. However, elevated filling pressure does not distinguish between overfilling, venoconstriction, absence of left atrial contraction or a changed compliance of the left ventricle. Therefore, there is need to develop some other indices to describe the passive property of the LV, especially during filling.

In this regard, chamber stiffness and myocardial stiffness have been investigated [32–36]. Chamber stiffness is derived from the relationship between pressure and volume during diastole. The operational stiffness at any point along a given pressure-volume curve is equal to the slope of a tangent drawn on the curve at that point. Thus, the shape of the entire pressure-volume relationship can be used to calculate the chamber stiffness during diastole and systole. As shown in Fig. 11.2, the operating chamber stiffness changes throughout the filling phase; stiffness (dP/dV) is less with a smaller volume (point 'a') and greater with larger volume (point 'b').

Since diastolic pressure-volume data can be assumed by an exponential relationship, the modulus of chamber stiffness (k_c) can thus be derived as the slope of the relationship between dP/dV and pressure [12, 36]. When the overall chamber stiffness increases, the pressure-volume curve shifts to the left, the slope of the dP/dV versus pressure relation becomes steeper, and k_c

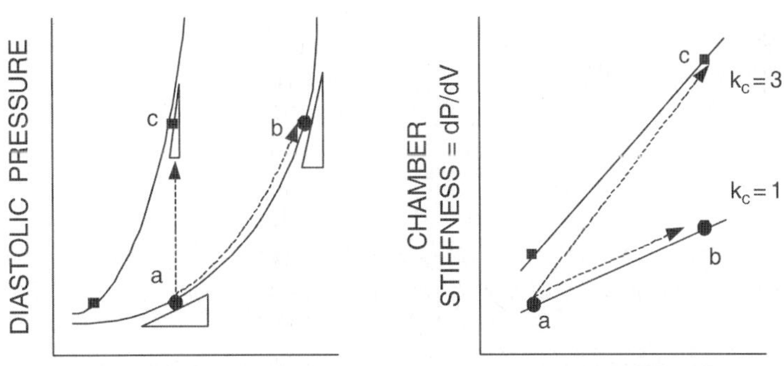

Fig. 11.2. Schematic diagram of LV diastolic pressure-volume relation. On the left, a progressive increase in volume will increase chamber stiffness (dP/dV). On the right, a leftward shift of the pressure-volume relation will also increase chamber stiffness. As the pressure-volume relation is assumed exponential, the relation between dP/dV and pressure is linear; the slope of this relation represents the chamber stiffness constant (k_c)

increases. In order to decrease complexity, Lisauskas *et al* [37] developed the averaged passive ventricular diastolic stiffness from transmitral flow using a simple equation $(\Delta P/\Delta V)_{avg}$.

Corresponding with chamber stiffness, the myocardial muscle of the LV behaves as an elastic material, developing a resisting force as it is stretched by ventricular filling. The greater the change in muscle length (strain), the greater the increase in the force (wall stress) that resists this deformation, and the stiffer the myocardium becomes. Myocardial stiffness can be estimated by examining the relation between LV wall stress and strain during diastole. At any given strain throughout filling, myocardial stiffness is equal to the slope $(d\sigma/d\varepsilon)$ of a tangent drawn to the stress-strain curve at that strain value.

Most investigators have considered the diastolic pressure-volume relation and stress-strain relation as curvilinear and exponential; hence the relationships between dP/dV versus pressure and between $d\sigma/d\varepsilon$ versus stress can be comsidered to be linear (34). It is common practice to curve-fit diastolic P-V or stress-strain data from minimum pressure (stress) to end diastole in exponential or power form, and then use the exponent as an index of chamber or myocardial stiffness. Such approaches have their limitations, because (1) the exponent used is often LV size dependent and therefore not always suitable for patient-to-patient comparisons, and (2) events that occur during the early rapid filling suction phase are ignored in the analysis (35).

11.3.2 Indices Characterizing Contractility

So many cardiac models have dealt with the measurement of contractility (12, 13, 38, 39, 40, 41, 42). Then what is the meaning of contractility? Conceptually, cardiac contractility is the potential for contraction that cardiac muscle possesses by virtue of the physic-cochemical environment of the muscle cells, e.g., calcium handling and contractile proteins. It is what the muscle (heart) is capable of doing (43). But it is impossible to measure the total physic-cochemical environment of the LV muscle cells in order to measure contractility. Thus, to define contractility for evaluating LV performance, one needs an operational definition of contractility, as will be shown later on in this chapter.

V_{max} as a Measure of Cardiac Muscle Contractility

Based on the Hill's theory on activated muscle, Sonnebblick (44, 45) has extended Hill's theory on activated muscle to cardiac muscle. He suggested that a hyperbolic relation exists between the velocity-of-shortening of the contractile element and the developed force based on the afterload. In particular, the results indicate that V_{max} (the contractile element shortening velocity at zero load) is independent of the initial muscle length or preload, and seems be represent a good contractility index.

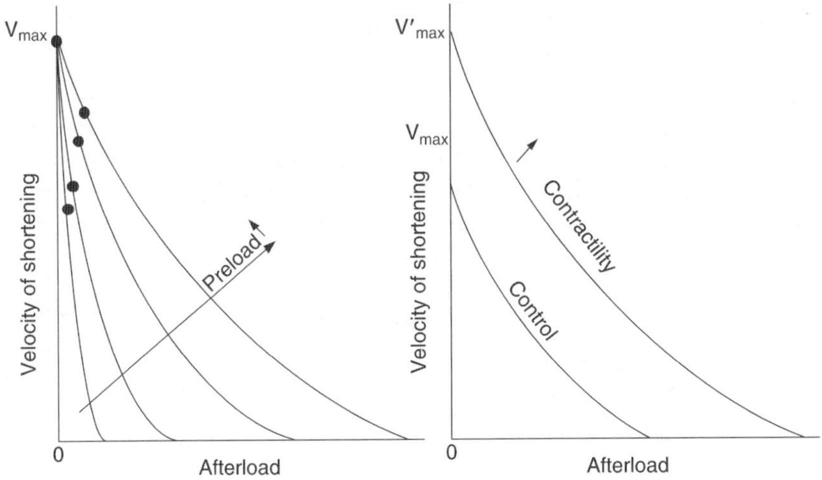

Fig. 11.3. Force (afterload)-velocity-of-shortening relationship

As shown in Fig. 11.3, as the afterload is reduced, the velocity of shortening increases. When the afterload is zero, the velocity of shortening is maximum (V_{max}). However, V_{max} does not depend on the length of the muscle from which contraction is initiated (left part); V_{max} (velocity at zero afterload) increases with an increase in inotropic state (i.e., increased contractility), even when the initial length (preload) is unchanged. The parameter V_{max} is therefore proposed as a measure of contractility (46, 47).

Mirsky and Ghista (48) have given a summary review of the existing formulae available for force-velocity analyses derived from animal experimental studies, and have employed them for development of indices for the assessment of heart muscle function. On the basis of this study, it may be seen that V_{max} is sensitive to relatively high preloads, and may not be a reliable index of contractility. A more reliable parameter for assessing cardiac muscle function can then be taken to be the traditional dP/dt_{max} index.

dP/dt_{max} as a Measure of Cardiac Contractility

The index dP/dt means the rate of development of the left ventricular intracavitary pressure with respect to time. Figure 11.4 illustrates two separate recordings of the ventricular intracavitary pressure wave as well as simultaneous value of dP/dt.

For the past three decades, dP/dt_{max} has been considered to be the most sensitive ventricular index of inotropicity and the current "gold standard" (49, 50). Unfortunately, the quantitative value of dP/dt_{max} is affected by other factors such as preload and afterload. Therefore, it is often difficult to use dP/dt_{max} as a measure of contractility in comparing ventricle from one

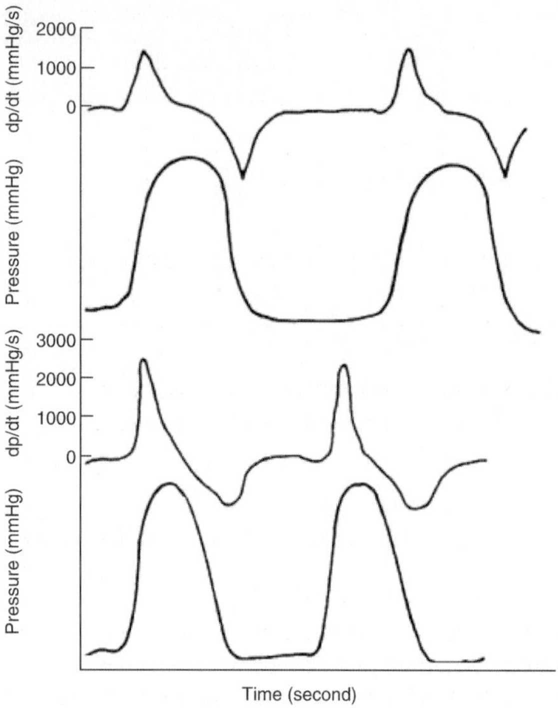

Fig. 11.4. Simultaneous recordings of ventricular pressure and dP/dt. The upper part shows results from a normal heart, and the below part shows results from a heart stimulate by isoproterenol

person to another. For this reason, other quantitative measures have also been used in attempts to assess cardiac contractility.

E_{max} as a Measure of Cardiac Contractility

Conceivably, LV pressure-volume relationship and elastance reflect LV contractile function more accurately (3, 19, 20, 21, 51, 52, 53, 54). Suga and Sagawa (51) formalized this concept as the time-varying elastance of the ventricle, by defining elastance, E, as: $E(t) = P(t)/(V(t) - V_d)$, where $P(t)$ and $V(t)$ are ventricular pressure and volume that vary with time t, respectively. V_d is the LV volume corresponding to zero LV pressure, obtained by drawing a tangent to the pressure-volume curves at end-ejection, as shown in Fig. 11.5.

It has been shown by them that the end-systolic pressure volume (ESPV) relationship, which is the loci of pressure and volume points at end-systole, is insensitive to variations of both the end-diastolic volume (preload) and the mean arterial pressure (afterload). ESPV relationship is usually a straight line with a slope of E_{max} as the maximal elastance. Apparently, E_{max} remains

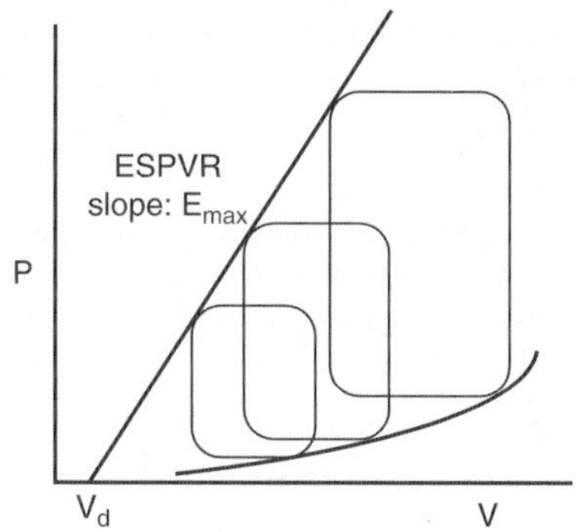

Fig. 11.5. Defining the concept of ESPVR as a measure of LV contractility (corresponding to zero pressure) is obtained by drawing a tangent to P-V curves at end ejection, according to Suga [51]

essentially constant if the preload and afterload are allowed to vary within the physiologic range, but is sensitive to inotropic agents and ischemia. Hence, E_{max} has been proposed as a "load independent" index of contractility of the ventricle (4, 55). However, the loading conditions do affect E_{max}, if ventricular pressure and volume are made to vary over a wider range (56, 57). Under these conditions, the ESPV relationship becomes curvilinear and concave towards the volume axis. A simple calculation of the elastance of a ventricle for different arterial loads reveals that its elastance changes and is load dependent. Hence, derived indices such as E_{max} and ESPV relationship loose credibility for their clinical applicability (57). Further, elastance measures also require cardiac catheterization for measurement of pressure which further reduces their utility.

An additional limitation of E_{max} is that it is not easy to change afterload and obtain multiple pressure-volume data points in a given subject, while maintaining a constant contractility. Hence, it is impractical to use E_{max} clinically for patient-specific LV catheterization-ventriculography data.

Preload-Adjusted Maximal Power as a Measure of Cardiac Contractility

Preload-adjusted maximal power divided by end-diastolic volume squared, has been shown to be relatively independent of preload, afterload and resistance (58, 59). The maximal value of power is calculated by determining the instantaneous values of pressure and flow, and obtaining their product during ventricular ejection. In order to get an accurate value of power, it is necessary to obtain the ventricular intracavitary pressure by catheterization. In fact, by combining arterial pressure tracing and flow (measured as velocity) obtained with continuous wave Doppler echocardiography through the aortic valve, it is possible to acquire the data necessary for power estimation [60–62].

Although preload-adjusted maximal power has these appealing characteristics, there is still an important limitation to its use in clinical practice. This limitation is mainly practical in nature, namely that a series of off-line and time-consuming analyses (of pressure, volume and flow-rate) are necessary to obtain it. Another limitation is that instead of ventricular intracavitary pressure, the brachial arterial pressure is employed. Due to damping and resonance in the arterial waveform, peripheral pressure can be a poor estimate of true ventricular intracavitary pressure. However, recently Lester *et al.* (2004) found that preload-adjusted maximal power was not preload independent in anesthetized humans [63], and this may be termed as a small limitation of this power index.

Strain Rate as a Measure of Cardiac Contractility

Recently, with the advent of new technology, a noninvasive index characterizing contractility in terms of myocardial fiber strain rate, obtained by Tissue Doppler echocardiography (TDE) and strain rate imaging (SRI), have been proposed by Greenberg *et al* (41) and Costa *et al* (64), respectively. Myocardial strain reflects the deformation of tissue in response to an applied force. The first temporal derivative of strain, namely strain rate, is the velocity-change in myocardial length. One limitation of TDE, however, is that regional TDE velocities are affected by heart translation and tethering of adjacent myocardial segments. The method of SRI has eliminated the tethering effect as a tool for assessing segmental LV dysfunction, such as coronary disease.

Intra-Cardiac Blood-Flow Velocity and Pressure Distributions Using Finite-Element Analysis

In general, LV dysfunction involves a number of interrelated events both in systole and diastole. All of these events ultimately contribute to intra-LV flow velocity and pressure gradients, which are the primary determinants of LV performance. Hence, a detailed investigation of intra-ventricular flow patterns (obtained by finite element analysis of blood flow within the LV) can provide practical insight into how the LV is functioning in terms of the flow velocity and pressure gradients within it; this can help to facilitate the diagnosis and treatment of LV dysfunction. The early computational models developed in the 70–80's were confined to 2D flow patterns, pressure waveforms and transmitral flow in simplified geometries [65]. On the other hand, Subbaraj and Ghista [66] computed instantaneous distributions of intra-LV flow and differential pressure during the filling and ejection phases, and derived the pressure gradient fields during diastolic and systolic phases.

A uniform pressure gradient towards the aortic outflow tract will contribute to efficient pumping. On the other hand, a non-uniform pressure gradient, caused by asynchronous myocardium contraction due to coronary lesions or infarcts leads to poor pumping performance. Later on, these models have been improved to incorporate more realistic LV geometries and fluid/ventricular wall interactions, in order to obtain velocity and pressure distribution patterns in the LV, as well as the stress distributions within its wall [67–70].

11.4 Passive and Active Elastances of the Left Ventricle

As an alternative to the P-V loop as a measure of LV function, a new approach for quantifying ventricular performance is proposed. Herein, we have come up with a new concept of dual passive and active elastances operating throughout the cardiac cycle. The passive elastance (E_p) represents the LV pressure response to LV volume change (i.e., to LV volume increase during LV filling phase and to LV volume decrease during LV ejection phase). Additionally, we have also defined active elastance (E_a) to represent the contraction of the left ventricle due to its sarcomeric activation (and the development of force between the sarcomere actin-myosin units) and relaxation (due to disengagement of the actin-myosin units). The index E_p governs LV stiffness variation due to volume increase during filling phase and due to volume decrease during ejection phase, while E_a governs LV pressure increase due to LV myocardial contraction.

Definition of Passive and Active Elastances of the LV

At the start of diastolic-filling phase, the LV decreasing pressure is the response to LV E_a continuing to decrease due to the sarcomere continuing to relax well into the filling phase. Thereafter, LV increamental pressure of the supposedly passive LV is the response to the rapid inflow of blood (due to left atrial contraction) and the corresponding increase in LV volume, along with the associated increase in LV E_p. The corresponding governing differential equation, relating LV pressure and volume, can be put down by referring to our paper [12] for its derivation as:

$$M(d\dot{V}) + d(EV) = M(d\dot{V}) + VdE + EdV = dP_{LV} \tag{11.1}$$

where t represents the time variable (s) from the start of filling phase;

V represents the volume of LV (ml) during the filling phase;
P_{LV} represents pressure of the LV, in mmHg (hereafter symbolized by P) (mmHg);
M represents the inertia term =[LV wall-density (ρ)/(LV surface-area/wall-thickness)] = $\rho h/4\pi R^2$ for a spherical LV model, in mmHg/(ml/s^2);
E represents LV elastance (mmHg/ml).

Likewise during ejection, the LV pressure variation (dP_{LV}) is caused by both E_a variation as well as E_p decrease (due to LV volume decrease). The instantaneous time-varying ventricular elastance (E) is the sum of (i) the volume-dependent passive elastance (E_p) and (ii) the active elastance (E_a) due to the activation of the LV sarcomere. Hence,

$$E = E_a + E_p \tag{11.2}$$

Together, they influence the LV pressure dynamics during the cardiac cycle. We will now provide the expressions for E_p and E_a.

Expression for Passive Elastance (E_p) of the LV

The passive (unactivated) myocardium exhibits properties of an elastic material, developing an increasing stress as strain increases, as occurs during ventricular filling. The passive stress-strain relation of a myocardial muscle strip is nonlinear, and follows an exponential relationship [34, 71, 72]. Likewise, the relation between LV passive pressure and volume is adopted to be exponential, as:

$$P = P_0 e^{z_p V}, \tag{11.3}$$

so that,

$$E_p = (dP/dV) = E_{p0} e^{z_p V} \tag{11.4}$$

where E_{p0} is the passive elastance coefficient ($= P_0 z_p$), z_p is the passive elastance exponent parameter, and V is the LV volume; its evaluation for a clinical case is provided in a subsequent section. During the latter part of the diastolic phase, we use equation (11.3) to fit the LV pressure-volume relation to determine the corresponding parameters, P_0 and z_p (or E_{p0} and z_p), and to hence obtain the passive elastance E_p.

Figure 11.7 displays the characteristics of the passive elastance E_p for a patient, using left ventriculography data shown in Fig. 11.6 [12].

Expression for Active Elastance (E_a) of the LV

During isovolumic contraction, $dV = 0$. Hence $d\dot{V} = 0$, and E_p is constant and equal to E_{ped} (the value of E_p at end-diastole). As a result, the governing equation (11.1) becomes $VdE = dP_{LV}$, which can be discretized as:

$$\begin{aligned} V_i(E_i - E_{i-1}) &= V_i[(E_{a,i} + E_{p,i}) - (E_{a,i-1} + E_{p,i-1})] \\ &= V_i(E_{a,i} + E_{ped} - E_{a,i-1} - E_{ped}) \\ &= dP_{LV,i} = P_i - P_{i-1} \end{aligned}$$

Hence,

$$E_{a,i} = \frac{(P_i - P_{i-1})}{V_i} + E_{a,i-1} \tag{11.5}$$

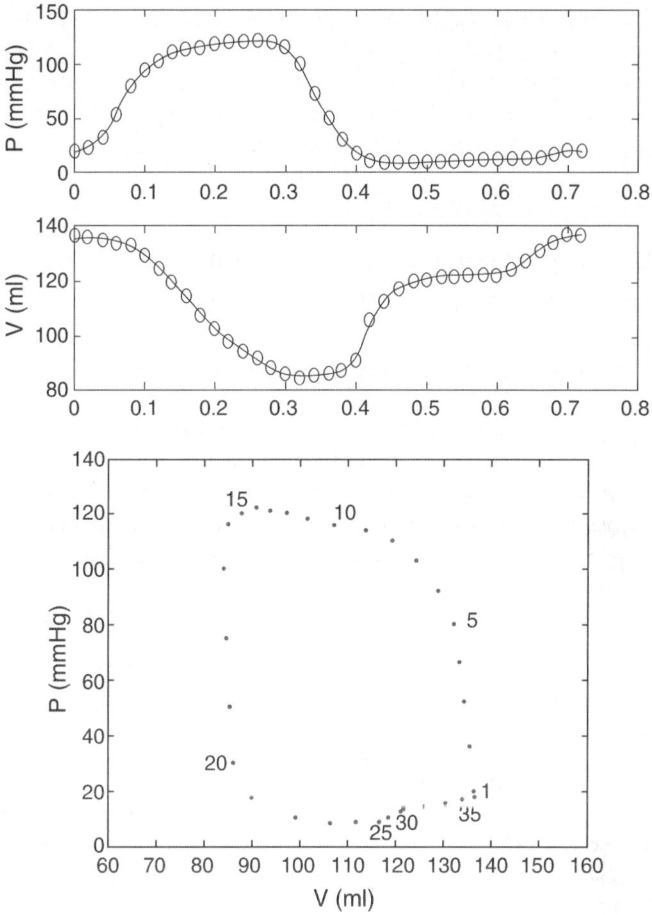

Fig. 11.6. (**a**) An example of a patient's measured LV pressure and volume during a cardiac cycle; t = 0–0.08s is the isovolumic contraction phase, t = 0.08s–0.32s is the ejection phase, t = 0.32s–0.40s is the isovolumic relaxation phase, and t = 0.40s–0.72s is the filling phase. (**b**) Points (21–36) constitute the filling phase, points (1–5) constitute the isovolumic contraction phase, points (5–17) constitute the ejection phase, and points (17–21) constitute the isovolumic relaxation phase. Note that after point 21 (the start of LV filling), the LV pressure decreases; this characterizes the 'so-called' LV suction effect

where i is a time instant during the isovolumic contraction and relaxation, V_i and $P_{LV,i}$ are the monitored LV volume and pressure at this instant, and E_{ped} is the passive elastance at the end-diastolic phase.

During the ejection phase, the governing equation (11.1) can be discretized as:

$$E_{a,i} = \frac{(P_i - P_{i-1}) - M(\dot{V}_i - \dot{V}_{i-1}) - V_i(E_{p,i} - E_{p,i-1}) - E_{p,i}(V_i - V_{i-1}) + V_i E_{a,i-1}}{2V_i - V_{i-1}} \tag{11.6}$$

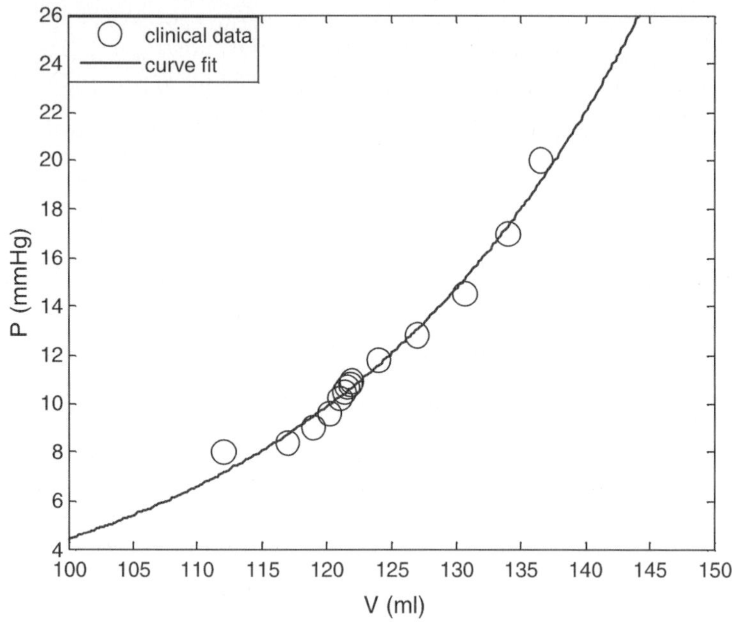

Fig. 11.7. Representation of LV E_p, using equation (11.3) to fit the pressure volume data during filling phase. We obtain: $E_p = 3.20 \times 10^{-3} e^{0.040V}$. The volume 100ml corresponding to the start of filling phase, and the volume 150 ml corresponds to the end of filling phase

Also, during isovolumic relaxation, because $dV = 0, d\dot{V} = 0$ and E_p is constant and equal to its end-systolic value of E_{pes}. Hence, the governing equation (11.1) again becomes $VdE = dP_{LV}$, which can be represented as:

$$V_i[(E_{a,i} + E_{p,i}) - (E_{a,i-1} + E_{p,i-1})]$$
$$= V_i(E_{a,i} + E_{pes} - E_{a,i-1} - E_{pes}) = dP_{LV,i} = P_i - P_{i-1}$$

Therefore,

$$E_{a,i} = \frac{P_i - P_{i-1}}{V_i} + E_{a,i-1} \tag{11.7}$$

where E_{pes} is the passive elastance at the end of systole.

During the diastolic phase, the formula for computing active elastance is the same as equation (11.6). Hence, from equation (11.5–11.7), we can calculate the values of active elastance from LV pressure-volume data during the cardiac cycle. After thereby calculating the values of active elastance (E_a), we then adopt the following expression for E_a [12] (to represent the variation of E_a during the cardiac cycle):

$$E_a = E_{a0}[1 - e^{-\left(\frac{t}{\tau_C}\right)^{Z_c}}][e^{-\left(\frac{(t-d)u(t-d)}{\tau_R}\right)^{Z_R}}] \tag{11.8}$$

where (i) t is measured from the start of isovolumic contraction, (ii) the parameter E_{a0} is the active elastance coefficient; (iii) the time-coefficient (τ_C) describes the rate of elastance rise during the contraction phase, while (τ_R) describes the rate of elastance fall during the relaxation phase; (iv) the exponents "Z_C" and "Z_R" are parameters of the E_a curve during isovolumic contraction and relaxation phases; (v) the parameter d is a time constant, and (vi) $u(t-d)$ is the unit step function, so that $u(t-d) = 0$ for $t < d$.

The computed values of the active elastance E_a during a cardiac cycle for the same patient (whose data is shown in Fig. 11.6) are shown in Fig. 11.8.

How E_p and E_a Can Explain LV Pressure Volume Dynamics

(a) Pressure Dynamics During Filling Phase

The pressure variation during filling is a combination of pressure changes due to the action of both active elastance (E_a) and passive elastance (E_p) response to blood filling caused by LA contraction. In equation (11.1), by neglecting the term ($Md\dot{V}$) because it is small [12], the pressure dynamics is expressed as:

$$P_i - P_{i-1} = (E_{p,i} + E_{a,i})(V_i - V_{i-1}) + V_i(E_{a,i} - E_{a,i-1} + E_{p,i} - E_{p,i-1}) \quad (11.9)$$

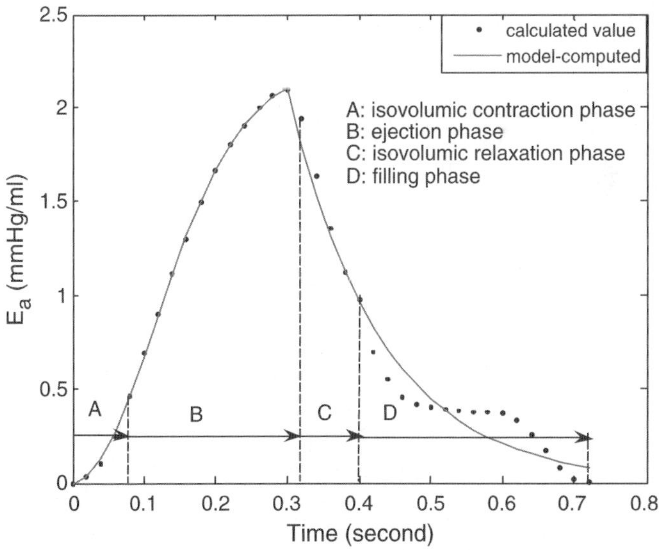

Fig. 11.8. The calculated values of active elastance E_a during cardiac cycle are shown dotted. Then using equation (11.8) to fit the calculated values, thereby obtaining: $E_{a0} = 2.20$mmHg/ml, $\tau_C = 0.17$s; $Z_C = 1.96$; d $= 0.3$s, $\tau_R = 0.12$s; $Z_R = 0.96$. It is noted (from the figure) that active elastance Ea continues into the filling phase, corresponding to our tacit assumption that LV myocardial contraction decreases during filling

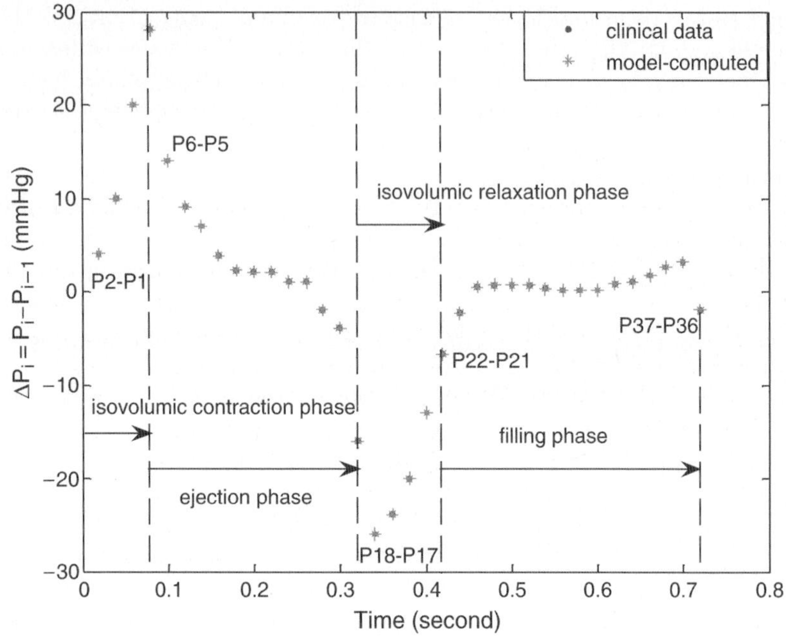

Fig. 11.9. Pressure dynamics during ejection and filling phases. Note the pressure decrease (i.e., negative ΔP_i) during early filling (from frames 21 to 23), representing LV suction phenomenon. The LV pressure-increase keeps decreasing during the first third of ejection phase, remains constant in the middle third phase of ejection, and becomes negative in late ejection phase

By employing the monitored LV volume values and the computed values of E_p and E_a, we can compute the values of LV pressure. In other words, if we monitor the LV volume values, and if somehow the E_p and E_a functions were known as intrinsic properties of the LV, then we could compute the LV pressure variation from equation (11.9).

Figure 11.9 shows the pressure drops ΔP_i are plotted from the start of filling phase, i.e., from t = 0.42s.

(b) Pressure Dynamics During Ejection Phase

We can likewise determine the pressure variation during the ejection phase, by neglecting the term $(Md\dot{V})$ because it is small [12], as:

$$P_i - P_{i-1} = (E_{p,i} + E_{a,i})(V_i - V_{i-1}) + V_i(E_{a,i} - E_{a,i-1} + E_{p,i} - E_{p,i-1}) \quad (11.10)$$

Let us take the calculated values of (E_p & E_a, V_i & V_{i-1}) during early and late ejection, and compute the pressure drop $(P_7 - P_6)$ and $(P_{16} - P_{15})$, as follows

$$\begin{aligned} P_7 - P_6 &= (E_{p,7} + E_{a,7})(V_7 - V_6) + V_7(E_{a,7} - E_{a,6} + E_{p,7} - E_{p,6}) \\ &= 9mmHg \end{aligned} \tag{11.11}$$

$$\begin{aligned} P_{16} - P_{15} &= (E_{p,16} + E_{a,16})(V_{16} - V_{15}) + V_{16}(E_{a,16} - E_{a,15} + E_{p,16} - E_{p,15}) \\ &= -4mmHg \end{aligned} \tag{11.12}$$

We note that the pressure difference $(P_7 - P_6)$ is positive, while the pressure difference $(P_{16} - P_{15})$ is negative.

In Fig. 11.9, these computed pressure differences $(P_i - P_{i-1})$ are plotted for the entire cycle. This graph illustrates how (i) E_a increase (due to force development in the myocardial sarcomere) and constant E_p during isovolumic contraction contribute to LV pressure increase, (ii) E_a increase during ejection (due to increase in sarcomeric force development) and E_p decrease (due to blood volume decrease) contribute to LV pressure dynamics during ejection phase, and (iii) E_a decrease and E_p increase (due to blood volume increase) contribute to the pressure dynamics during the filling phase.

It is seen that, LV E_a develops at the start of isovolumic contraction, becomes maximum some time during late ejection, and thereafter decreases and becomes zero during diastolic filling. On the other hand, LV E_p starts increasing after the initiation of LV filling as the LV volume increases. It reaches its maximum value at the end-of-filling phase, remains constant during isovolumic contraction, and thereafter decreases during ejection (as the LV volume decreases). The generation (and increase) of E_a helps us to explain the development of the LV pressure increase during isovolumic contraction, while the decrease of E_a during diastole helps us to explain the decrease in LV pressure during early filling to create LV suction of the blood (even before the onset of left-atrial contraction). The incorporation of both E_p and E_a helps us to explain the LV pressure changes during the filling and ejection phases [73].

11.5 LV $d\sigma^*/dt_{max}$ as Noninvasive Contractility Index Based on Wall-Normalized Stress

11.5.1 Background

The traditional index of LV contractility is dP/dt_{max}. However, this index requires the measurement of LV pressure. Alternatively, dP/dt_{max} involves LV pressure, which in turn is generated by LV wall stress. Hence, it is logical that LV wall stress be employed as a contractility index. However, in order to make it a noninvasive index, we normalize it with respect to LV pressure, in the form of $(d(\sigma/P)/dt)_{max}$.

11.5.2 Derivation of $d\sigma^*/dt_{max}$

During LV systole, LV wall stress is generated intrinsically by sarcomere contraction, and results in the development of extrinsic LV pressure. LV wall stress is dependent on wall thickness, LV geometry, chamber pressure and sarcomere contraction. Hence, it is rational to characteirze this normalized LV wall stress as an intrinsic measure of myocardial contractility. We have validated a new LV contractility index, $d\sigma^*/dt_{max}$, based on the maximum rate of development of LV wall stress with respect to LV pressure [74].

The expression for normalized wall stress in a thick-walled LV spherical model is given by:

$$\sigma^* = \sigma_\theta/P = \left[\frac{r_i^3/r_e^3 + 1/2}{1 - r_i^3/r_e^3}\right] \tag{11.13}$$

Based on cineventriculography measurements of LV volume (V) and myocardial volume, we can express σ^* as follows:

$$\sigma^* = \sigma_\theta/P = \left(\frac{3V}{2V_m} + \frac{1}{2}\right) \tag{11.14}$$

$$d\sigma^*/dt_{max} = \left|\frac{d(\sigma_\theta/P)}{dt}\right|_{max} = \frac{3}{2V_m}\left|\left(\frac{dV}{dt}\right)\right|_{max} \tag{11.15}$$

where P and V are the LV intracavitary pressure and volume, respectively, V_m is myocardial volume, and σ_θ is the wall stress. It is noted that this index requires the determination of LV volume and the myocardial volume.

11.5.3 Clinical Evaluation

Thirty subjects in this study were studied in a resting recumbent (baseline) state, after premedication with 100–500 mg of sodium pentobarbital by retrograde aortic catheterization. Angiography was performed by injecting 30–36 ml of 75% sodium diatrizoate into the LV at 10 to 12 ml/s. During ventriculography, left ventricular chamber pressure was measured by a pigtail catheter and Statham P23Eb pressure transducer. From biplane angiocardiograms, it is seen that the orthogonal chamber diameters are nearly identical [75]. These findings are used to justify the use of single-plane cine techniques, which allow for beat-to-beat analysis of the chamber dimensions.

For our study, monoplane cineangiocardiograms were recorded in a RAO 30° projection from a 9 in image intensifier using 35 mm film at 50 frames/s, using INTEGRIS Allura 9 system at the Nation Heart Centre (NHC), Singapore. Therefrom, automated analysis was carried out to calculate LV volume and myocardial wall thickness. The LV data consists of measured volume as well as the corresponding pressure (Fig. 11.6); this monitored pressure is employed to compute the traditional contractility index dP/dt_{max}. All measurements are corrected for geometric distortion due to the respective recordings systems. All patients provided informed consent, and the relevant institutional review board approved the protocol.

In the 30 clinical subjects studied, $d\sigma^*/dt_{max}$ correlates well with the traditional and invasively determined dP/dt_{max} as well as with the computed maximal active elastance $E_{a,max}$ (Fig. 11.10). These data provide preliminary

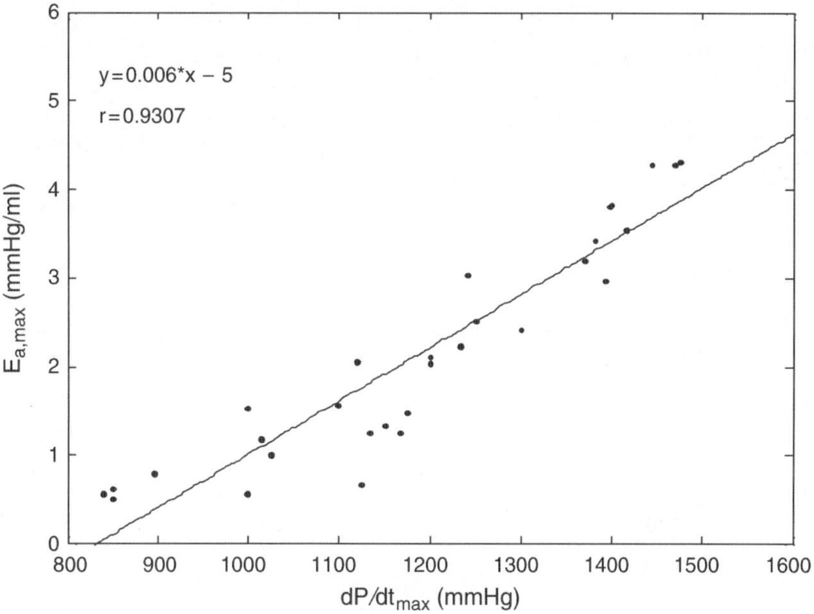

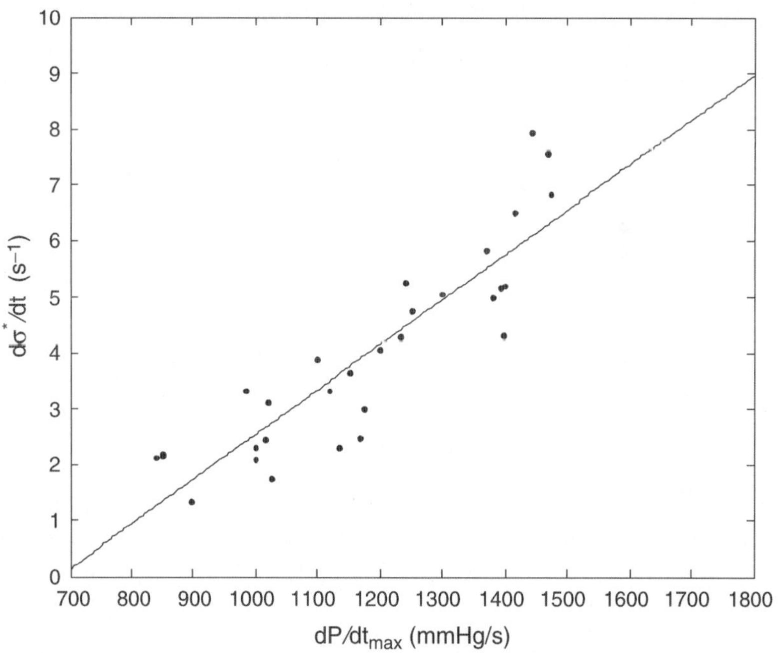

Fig. 11.10. Linear regression analysis demonstrates good correlation between (a) $d\sigma^*/dt_{max}$ and $dP/dt_{max}(d\sigma^*/dt_{max} = 0.0083dP/dt_{max} - 5.50, \mathrm{r} = 0.893)$, (b) $d\sigma^*/dt_{max}$ and $E_{a,max}(d\sigma^*/dt_{max} = 1.30E_{a,max} + 1.20, \mathrm{r} = 0.924)$

validation of the proposed contractility index $d\sigma^*/dt_{max}$, and may as an effective clinical substitutive for dP/dt_{max}. Additionally, this index $d\sigma^*/dt_{max}$ has the dual advantage of requiring relatively simple computations, with no requirement for invasive LV pressure measurement. It is to be noted that clinical evaluation of $d\sigma^*/dt_{max}$ only requires the monitoring of LV volume and myocardial volume.

11.6 Concluding Remarks

An attempt has been made in this chapter to give a summary of the existing indices available for assessing cardiac function in filling and systolic ejection phases. This chapter deals with two major concepts. The first concept is that of passive and active elastances. While Ep is related to LV stiffness, Ea is related to LV myocardial contractility. We have shown how Ep and Ea can together explain LV pressure variation during a cardiac cycle. The second concept is that of a new contractility index in the form of LV maximal change rate of normalized wall stress with respect to intracavitary pressure $d\sigma^*/dt_{max}$, which can be viewed as a novel LV contractility indices.

In determining $d\sigma^*/dt_{max}$ as a contractility index, we have assumed the LV as a sphere which is not an accurate representation of the LV geometry. A better representation of the LV geometry would be an ellipsoidal model. We have in fact determined the normalized wall stress for an ellipsoidal shaped LV model [76]; however this expression is relatively complex and not easy for clinical employment. The spherical assumption underlying our $d\sigma^*/dt_{max}$ index can be justified on the basis of (i) this index correlating well with the traditional contractility index of dP/dt_{max}, (ii) its being obtainable noninvasively, and (iii) its simplicity for clinical application.

We are of course admitting that the diagnostic indices proposed here have not been applied extensively in the clinical situation. Hence, it is expected that their extensive application along with their correlation with the histological and pathological findings will help to incorporate them into clinical cardiology practice.

References

1. Grodins, F.S. (1959): 'Integrative cardiovascular physiology: A mathematical synthesis of cardiac and blood vessel hemodynamics', *Quart. Rev. Biol.*, **34**, pp. 93–116.
2. Elzinga, G., Westerhof, N. (1972): 'Pressure and flow generated by the left ventricle against different impedances', *Circ. Res.*, **32**, pp. 178–186.
3. Suga, H. (1969): 'Time course of left ventricular pressure-volume relationships under various enddiastolic volume', *Jpn Heart J.*, **10**, pp. 509–515.
4. Sagawa, K.(1978): 'The ventricular pressure-volume diagram revisited', *Circ Res.*, **43**, pp. 677–687.

5. Vaartjes, S.R., and Boom, H.B.K. (1987): 'Left ventricular internal resistance and unloaded ejection flow assessed from pressure-flow relations: A flow-clamp study on isolated rabbit hearts', *Circ. Res.*, **60**, pp. 727–737.
6. Shoucri, R.M. (1990): 'The pressure-volume relation and mechanics of left ventricular contraction', Jpn Heart J, 31, pp. 713–729.
7. Shoucri, R.M. (1998): 'Studying the mechanics of left ventricle contraction', *IEEE Eng Biomed Eng.*, **17**, pp. 95–101.
8. Shoucri, R.M. (2000): 'Active and passive stresses in the myocardium', *Am J Physiol Heart Circ Physiol.*, **279**, pp. 2519–H2528.
9. Palladino, J.L., Mulier, J.P., and Noordergraaf, A. (1998): 'Defining ventricular elastance', *Proc. 20th Int. Conf. IEEE Eng. Med. Biol. Soc., Hong Kong*, pp. 383–386.
10. Kass, P., Ding, Y.A., and Chen, C.H. (2002): 'Assessment of left ventricular end-systolic elastance from aortic pressure-left ventricular volume relations', *Heart Vessels*, **16**, pp. 99–104.
11. Ishida, M., Tomita, S., and Nakatani, T. (2004): 'Acute effects of direct cell implantation into the heart: A pressure-volume study to analyze cardiac function', *J. Heart Lung Transplant*, **23** (7), pp. 881–888.
12. Zhong, L., Ghista, D.N., Ng, E.Y.K., and Lim, S.K. (2005): 'Passive and active ventricular elastance of the left ventricle', *Biomedical Engineering Online*, **4**, 10.
13. Zhong, L., Ghista, D.N., Ng, E.Y.K., Lim, S.T., and Chua, S.J.T. (2005): 'Measures and indices for intrinsic characterization of cardiac dysfunction during filling and systolic ejection phase', *Journal of Mechanics in Medicine and Biology*, **5** (2), pp. 307–332.
14. Ghista, D.N., Zhong, L., Chua, L.P., Ng, E.Y.K., Lim, S.T., Tan, R.S., and Chua, T.S.J. (2005): 'Systolic modeling of the left ventricle as a mechatronic system: determination of myocardial fiber's sarcomere contractile characteristics and new performance indices', *Molecular & Cellular Biomechanics*, **2** (4), pp. 217–233.
15. Ghista, D.N., Advani, S.H., Gaonkar, G.H., Balachandran, K., and Brady, A.J. (1971): 'Analysis and physiological monitoring of human left ventricle', *J. Basic Eng.*, **93**, pp. 147–161.
16. Ghista, D.N., Brady, A.J., and Radhakrishnan, S. (1973): 'A three-dimensional analytical (rheological) model of the human left ventricle in passive-active states', *Biophy. J.*, **13**, pp. 832–854.
17. Ghista, D.N., and Sandler, H. (1975): 'Elastic modulus of the human intact left ventricle-determination and physiological interpretation', *Med. Biol. Eng.*, **13** (2), pp. 151–160.
18. Burkhoff, D., Yue, D.T., Franz, M.R., Hunter, W.C., and Sagawa, K. (1984): 'Mechanical restitution of isolated perfused canine left ventricles', *Am J Physiol Heart Circ Physiol*, **246**, pp. H8–H16.
19. Sagawa, K., Maughan, W.L., Suga, H., and Sunagawa, K. (1988): 'Cardiac contraction and the pressure-volume relationship', *Oxford Univ. Press.*
20. Suga, H., and Sagawa, K. (1974): 'Instantaneous pressure-volume relationships and their ratio in the excised, supported canine left ventricle', *Circ Res*, **35**, pp. 117–126.
21. Van Der Velde, E.T., Burkhoff, D., Steendijk, P., Karsdon, J., Sagawa, K., and Baan, J. (1991): 'Non-linearity and load sensitivity of the end-systolic pressure-volume relation of canine left ventricle in vivo', *Circulation*, **83**, pp. 315–327.

22. Brutsaert, D.L., Rademakers, F.E., and Sys, S.U. (1984): 'Triple control of relaxation: implication in cardiac disease', *Circulation*, **69**, pp. 190–196.
23. Gillebert, T.C., and Sys, S.U. (1994): 'Physiologic control of relaxation in isolated cardiac muscle and intact left ventricle', in Gaasch, W.H., and LeWinter, M.M. (1994): 'Left ventricular diastolic dysfunction and heart failure', *Philadelphia: Lea & Febiger*, pp. 25–42.
24. Arrighi, J.A., and Soufer, R. (1995): 'Left ventricular diastolic function: physiology, methods of assessment, and clinical significance', *J. Nucl. Cardiol.*, **2**, pp. 525–543.
25. Brutsaert, D.L., Sys, S.U., and Gillebert, T.C. (1993): 'Diastolic failure: pathophysiology and therapeutic implications', *J. Am. Coll. Cardiol.*, **22**, pp. 318–325.
26. Rusconi, C.U., Ghizzoni, G.L., Sabatini, T., Oneglia, C.A., and Faggiano, P.M. (1998): 'Pathophysiology of diastole and left ventricular filling in humans: noninvasive evaluation', in Drzewiecki, G. M., and Li, J. K. J. (1998): 'Analysis and assessment of cardiovascular function', *Springer-Verlag, New York Inc.*, 172–192.
27. Frank, O. (1959): Zur Dynamik des Herzmuskels, *Z. Biol.* 32: 370–447, Translated into English by Chapman, C.B., and Wasserman, E. (1959): 'On the dynamics of cardiac muscle', *Am. Heart J.*, **58**, 282–317 and 467–478.
28. Hill, A.V. (1938): 'The heat of shortening and dynamic constants of muscle', *Pro. R. Soc.* London, Ser. B, pp. 126–136.
29. Huxley, A.F., and Niedergerke, R. (1954): 'Structural changes in muscle during contraction', *Nature*, **173**, pp. 971–973.
30. Huxley, A.F. (1974): 'Muscular contraction, A review lecture', *J. Physiol.*, **243**, 1974, 1–43.
31. Suga, H. (1979): 'Total mechanical energy of a ventricle model and cardiac oxygen consumption', *Am J Physiol Heart Circ Physiol*, **236**, pp. H498–H505.
32. Gaasch, W.H., Levine, H.J., and Quinones, M.A. (1976): 'Left ventricular compliance: Mechanisms and clinical implications', *Am. J. Cardiol.*, **38**, pp. 645–653.
33. Mirsky, I., and Rankin, J.S. (1979): 'The effects of geometry, elasticity and external pressure on the diastolic pressure-volume and stiffness-stress relations. How important is the pericardium?', *Circ. Res.*, **44**, pp. 601–611.
34. Mirsky, I. (1984): 'Assessment of diastolic function: Suggested methods and future considerations', *Circulation*, **69**, pp. 836–841.
35. Mirsky, I., and Pasipoularides, A. (1990): 'Clinical assessment of diastolic function', *Prog. Cardiovasc. Dis.*, **32**, pp. 291–318.
36. Smith, V.E., and Zile, M.R. (1992): 'Relaxation and diastolic properties of the heart', in Fozzard, H.A. (1962): 'The heart and cardiovascular system', *Raven Press*, NY, pp. 1353–1367.
37. Lisauskas, J.B., Singh, J., Bowman, A.W., and Kovács, S.J. (2001): 'Chamber properties from transmitral flow: prediction of average and passive left ventricular diastolic stiffness', *J. Appl. Physiol.*, **91**, pp. 154–162.
38. Mason, D.T. (1969): 'Usefulness and limitations of the rate of rise of intraventricular pressure (dP/dt) in the evaluation of myocardial contractility in man', *Am. J. Cardiol.*, **23**, pp. 516–527.
39. Little, W.C. (1989): 'Comparison of measures of left ventricular contractile performance derived from pressure-volume loops in conscious dogs', *Circulation*, **71**, pp. 994–1009.

40. Kara, S., Okandan, M., Usta, G., and Tezcaner, T. (2004): 'Investigation of a new heart contractility power parameter', *Comput. Methods Programs Biomed.*, **76**, pp. 177–180.
41. Greenberg, N.L., Firstenberg, M.S., Castro, P.L., Main, M., and Travaglini, A.T. (2002): 'Doppler-derived myocardial systolic strain rate is a strong index of left ventricular contractility', *Circulation*, **105**, pp. 99–105.
42. Xiao, S.Z., Guo, X.M., Wang, F.L., Xiao, Z.F., Liu, G.C., Zhan, Z.F., and Sun, X.B. (2003): 'Evaluating two new indicators of cardiac reserve', *IEEE Eng. Med. Biol.*, **22**, pp. 147–152.
43. Slinker, B.K. (1995): 'Searching for indices of contractility is counterproductive', in Ingels, N.B., Daughters, Jr., G.T., Baan, J., Covell, J.W., Reneman, R.S., and Yin, F.C.P. (1995): 'Systolic and diastolic function of the heart', *IOS press and Ohmsha.*
44. Sonnenblick, E.H. (1962): 'Force-velocity relations in mammalian heart muscle', *Am. J. Physiol.* **202** (5), pp. 931–939.
45. Sonnenblick, E.H. (1969): 'The contractile state of the heart as expressed by force-velocity relations', *Am. J. Cardiol.*, **23**, pp. 488–503.
46. Mason, D.T., Spann, J.F., and Zelis, R. (1970): 'Quantification of the contractile state of the human heart. Maximal velocity of contractile element shortening determined by the instantanenous relation between the rate of pressure rise and pressure in the left ventricle during isovolumic systole', *Am. J. Cardiol.*, **26**, pp. 248–257.
47. Wolk, M.J., Keefe, J.F., Bing, P.H.L., and et al. (1971): 'Estimation of V_{max} in auxotonic systoles from the rate of relative increase of isovolumic pressure: dP/dt/KP', *J Clin Invest*, **50**, pp. 1276–1285.
48. Mirsky, I., Ghista, D.N. (1972): 'A critical evaluation of force-velocity analyses', *J Assoc Adv Med Instrum*, **6** (6), pp. 374–379.
49. Kass, D.A., and Beyer, R. (1991): 'Evaluation of contractile state by maximal ventricular power divided by the square of end-diastolic volume', *Circulation*, **84**, pp. 1698–1708.
50. Kass, D.A., Maughan, W.L., Guo, Z.M., Kono, A., Sunagawa, K., and Sagawa, K. (1987): 'Comparative influence of load versus inotropic states on indices of ventricular contractility: experimental and theoretical analysis based on pressure-volume relationships', *Circulation*, **76**, pp. 1422–1436.
51. Suga, H., Sagawa, K., and Shoukas, A.A. (1973): 'Load independence of the instantaneous pressure-volume ratio of the canine left ventricle and effects of epinephrine and heart rate on the ratio', *Circ Res.*, **32**, pp. 314–322.
52. Sagawa, K. (1981): 'The end-systolic pressure-volume relation of the ventricles: definition, modification and clinical use', *Circulation*, **63**, pp. 1223–1227.
53. Mirsky, I., Tajimi, T., and Peterson, K.L. (1987): 'The development of the entire end-systolic pressure-volume and ejection fraction-afterload relations: a new concept of systolic myocardial stiffness', *Circulation*, **76**, pp. 343–356.
54. Burkhoff, D., De Tombe, P.P., and Hunter, W.C. (1993): 'Impact of ejection on magnitude and time course of ventricular pressure-generation capacity', *Am. J. Physiol Heart Circ Physiol.*, **265**, pp. H899–H903.
55. Sagawa, K., Suga, H., Shoukas, A.A., and Bakalar, K.M. (1977): 'End-systolic pressure-volume ratio: a new index of ventricular contractility', *Am. J. Cardiol.*, **40**, pp. 748–753.

56. Noda, T., Cheng, C.P., De Tombe, P.P., and Little, W.C. (1993): 'Curvilinearity of LV end-systolic pressure-volume and dP/dt_{max}-end-diastolic volume relations', *Am J Physiol.*, **34**, pp. H910–H917.
57. Mulier, J.P. (1994): 'Ventricular pressure as a function of volume and flow', *PhD thesis*, University of Leuven.
58. Glower, D.D., Spratt, J.A., Snow, N.D., Kabas, J.S., Davis, J.W., Olsen, C.O., Tyson, G.S., Sabison, D.C., and Rankin, J.S. (1985): 'Linearity of the Frank-Starling relationship in the intact heart: the concept of preload recruitable stroke work', *Circulation*, **71**, pp. 994–1009.
59. Kass, D.A., and Beyar, R. (1991): 'Evaluation of contractile state by maximal ventricular power divided by the square of end-diastolic volume', *Circulation*, **84**, pp. 1698–1708.
60. Sharir, T., Feldman, M.D., Haber, H., and et al. (1994): 'Ventricular systolic assessment in patients with dilated cardiomyopathy by preload-adjusted maximal power: validation and noninvasive application', *Circulation*, **89**, pp. 2045–2053.
61. William, A.M., Michael, R.P., and John, G. (1998): 'Assessment of left ventricular contractile state by preload-adjusted maximal power using echocardiographic automated border detection', *J. Am. Coll. Cardiol.*, **31** (4), pp. 861–868.
62. Schmidt, C., Roosens, C., Struys, M., and et al. (1999): 'Contractility in humans after coronary artery surgery. Echocardiographic assessment with preload-adjusted maximal power', *Anesthesiology*, **91**, pp. 58–70.
63. Lester, S.J., Shin, H., Lambert, A.S., Miller, J.P., Cahalan, M.K., Seeberger, M.D., Foster, E., and Schiller, N.B. (2004): 'Is PA-PWRmax truly a preload-independent index of myocardial contractility in anesthetized humans', *Cardiology*, **102**, pp. 77–81.
64. Costa, S.P., Sam, F., Falk, R.H., Colucci, W.S., and Davidoff, R. (2004): 'Strain rate imaging in idiopathic cardiomyopathy: more sensitive than tissue Doppler and potential application as a contractility index', (abstract), *J. Am. Coll. Cardiol.*, **6** (2), pp. 416.
65. Gordon, D.G. (1976): 'The physics of left ventricular ejection and its implications for muscle mechanics', *European J Cardiology*, **4**, pp. 87–95.
66. Subbaraj, K., Ghista, D.N., and Fallen, E.L. (1987): 'Intrinsic indices of the left ventricle as a blood pump in normal and infarcted left ventricles', *J. Biomed. Eng.*, **9**, pp. 206–215.
67. Chahboune, B., and Crolet, J.M. (1998): 'Numerical simulation of the blood-wall interaction in the human left ventricle', *European Physical J Applied Physics*, **2**, pp. 291–297.
68. Peskin, C.S., and McQueen, D.M. (1992): 'Cardiac fluid dynamics', *Crit Rev Biomed Eng*, **20**, pp. 451–459.
69. Redaelli, A., and Montevecchi, F.M. (1998): 'Intraventricular pressure drop and aortic blood acceleration as indices of cardiac inotropy: a comparison with the first derivative of aortic pressure based on computer fluid dynamics', *Medical Engineering & Physics*, **20**, pp. 231–241.
70. Saber, N.R., Gosman, A.D., Wood, N.B., Kilner, P.J., Charier, C., and Firmin, D.N. (2001): 'Computational flow modeling of the left ventricle based on in vivo MRI data-initial experience', *Annals Biomed Eng*, **29**, pp. 275–283.
71. Mirsky, I. (1976): 'Assessment of passive elastic stiffness of cardiac muscle: mathematical concepts, physiologic and clinical consideration, direction of future research', *Prog. Cardiovasc. Dis.*, **XVIII**, pp. 277–308.

72. Gibert, J.C., and Glantz, S.A. (1989): 'Determinants of left ventricular filling and of the diastolic pressure-volume relation', *Circ. Res.*, **64** (5), pp. 827–852.
73. Zhong, L., Ghista, D.N., Ng, E.Y.K., Lim, S.T., Tan, R.S., and Chua, L.P. (2006): 'Explaining left ventricular pressure dynamics in terms of LV passive and active elastance', *Journal of Engineering in Medicine*, **220**, pp. 647–655
74. Zhong, L., Tan, R.S., Ghista, D.N., and Kassab, G.S. (2006): 'Validation of a novel cardiac left ventricular contractility index in patients', *American Journal of Physiology Heart Circulatory Physiology* (Revised).
75. Sandler, H., and Dodge, H.T. (1968): 'The use of single plane angiocardiograms for the calculation of left ventricle volume in man', *Am. Heart. J.*, **75**, pp. 325–334.
76. Zhong, L., Ghista, D.N., Ng, E.Y.K., Lim, S.T., Chua, T.S.J., and Lee, C.N. (2006): 'Left ventricular shape-based contractility index', *Journal of Biomechanics*, **39**, pp. 2397–2409.

12

Arterial Wave Propagation and Reflection at a Bifurcation Site

Dhanjoo N Ghista, Liang Zhong, Eddie Y.K Ng and Ru San Tan

In the circulatory system, the pulse wave travels along, with exchange of energy between the flowing blood and the elastic vessel walls. In the aorta, the pressure increases with left ventricular ejection, and blood flows through it, as it dilates and relaxes (or contracts). The pulse wave travels, with this interchange of energy through the system. In situations associated with arterial stiffening (such as arteriosclerosis, which causes the arterial wall to become stiffer), the speed of the pulse wave increases [1]. Hence, the state of the vessel can be described by the pulse wave velocity (PWV) [2]. Pulse wave velocity (PWV) describes how quickly a blood pressure pulse wave travels from one point to another in the human body. The time difference between these two locations is known as the pulse transit time (PTT).

In the circulatory system, wave travels along, with exchange of energy between the flowing blood and the elastic vessel walls. In the aorta, the pressure increases with left ventricular ejection, and blood flows through it, as it dilates and relaxes (or contracts). The pulse wave travels, with this interchange of energy through the system. In situations associated with arterial stiffening (such as arteriosclerosis, which causes the arterial wall to become stiffer), the speed of wave increases [1]. Hence, the state of the vessel can be described by the pulse wave velocity (PWV) [2]. Pulse wave velocity (PWV) describes how quickly a blood pressure pulse travels from one point to another in the human body. The time difference between these two locations is known as the pulse transit time (PTT). PWV is typically measured between the carotid and the femoral artery [3]. The increased stiffness of the arterial wall serves to increase PWV, because the energy of the blood pressure pulse cannot be stored in an inflexible wall. Hence, PWV can be used as an index of arterial distensibility [4]. In past years, a number of papers have been published on the diagnosis of cardiovascular diseases and mortality risk prediction [5–9], using PWV.

A pressure or flow waveform consists of forward and backward (or reflected) waves. In the aorta, the forward pressure and flow waveform comes from the heart, and the reflected wave comes from various locations in the arterial system (such as from bifurcations). The pulsatile pressure and flow can thus be viewed as the sum of forward and reflected pulse wavefronts. The relationship between blood pressure and wave reflections has long been investigated, both theoretically and experimentally [10–15]. Traditionally, reflection is believed to increase pressure (depending on its phase relative to the forward propagating wave) and may cause hypertension. Hence, investigators have suggested that reducing reflection should be a clinical goal for the reducing systolic hypertension [16, 17]. However, based on Berger and his colleagues' work, depending on the phase difference between the forward-propagating and reflected waves, reflection can even decrease stroke work and mean arterial pressure [18–20].

Arterial stiffness is recognized as a major driver of cardiovascular disease. An increase in arterial stiffness elevates systolic pulse pressure, as well as left ventricular afterload and decreases coronary artery perfusion pressure. These effects increase the risk of stroke, heart failure and myocardial infarction [21,22]. Pulse wave velocity (PWV) is a well established technique for measuring the arterial stiffness parameters (because PWV is a function of the arterial wall elastic modulus), mostly at the carotid and femoral artery sites as well as in the aorta, where it is convenient to monitor the arterial cross-sectional changes as the pulse wave propagates past an arterial cross-section [23–27]. There are two types of arterial stiffness properties: (i) the arterial wall stress-strain relationship or the elastic modulus (E) as a function of arterial wall stress (σ), and (ii) arterial impedance z, the ratio of the amplitudes of the pressure and flow-rate pulses, representing the incremental pressure pulse response to the incremental flow-rate pulse.

This chapter provides the mathematically derivation of PWV, and the analytical basis for determining these arterial elasticity parameters. We have also provided herein, the concept of impedance and wave reflection. Based on that, we have analyzed what occurs at an arterial bifurcation. We have shown that the reflected wave can be either in phase or out of phase with the incident wave, and hence either enhance or decrease the afterload on the left ventricle.

12.1 Analysis for Pulse Wave Propagation Velocity

In this section, we present the analysis for expressing the pulse wave velocity (PWV) in terms of the arterial elastic properties and its geometrical dimensions. We then present the manner in which the derived expression for PWV can be employed in conjunction with in vivo measurements (of pulse wave velocity and arterial diameters) to obtain the values of the arterial elasticity parameters.

Figure 12.1 illustrates the arterial pressure-profile, the arterial wall displacement and the intra-arterial flow-velocity profile due to the propagating pressure and flow-rate pulse waves.

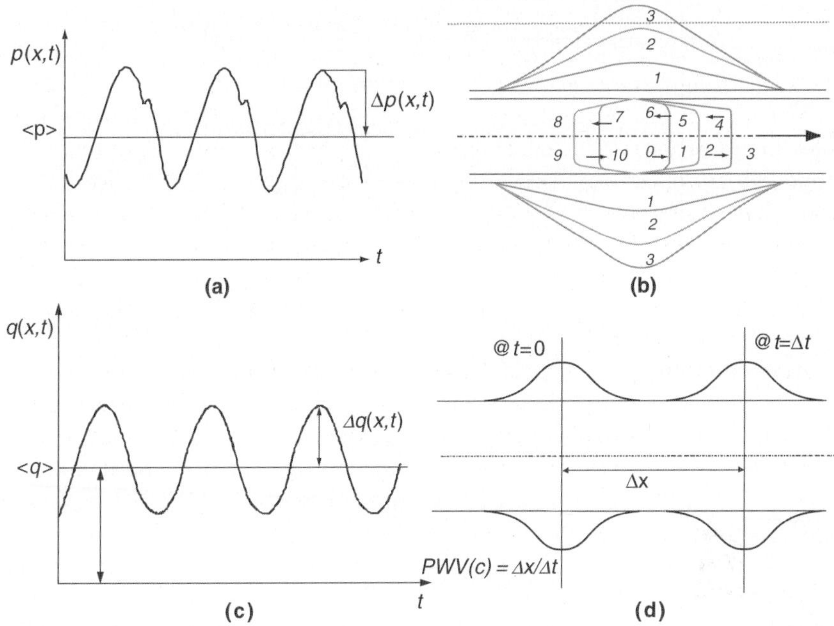

Fig. 12.1. (**a**) Schematics of arterial pressure profile; (**b**) Schematics of arterial wall displacement and flow-velocity profiles at a site during the passage of the pulse wave; (**c**) Schematics of arterial flow profile; (**d**) Traveling pressure profile at two sites Δx apart, at times t = 0 & Δt

Let us consider a segment of a circular, cylindrical, elastic vessel containing blood, as an incompressible fluid. When the fluid in this tube is disturbed by the pulse wave entering it at one place, the disturbance will be propagated as a wave along the tube at a finite speed, denoted as the pulse wave velocity (or PWV). The derivation of PWV will involve (i) governing equations of fluid-flow, (ii) governing equations of the arterial-wall motion, (iii) macthing of fluid velocity at the wall with the wall motion, and (iv) the constitute property of the arterial wall.

In general, in response to a pressure pulse wave, $u = f(x, r, t)$, we have the averaged velocity $\overline{u}$, given by:

$$\overline{u}(x,t) = (1/A) \int_0^{a+\eta} u(x,r,t) 2\pi r dr \tag{12.1}$$

The Conservation-of-mass Equation (refer Fig. 12.2):

$$-\overline{u} \cdot \pi a^2 + \pi a^2(\overline{u} + \partial \overline{u}/\partial x dx) + (2\pi a \cdot dx) u_w = 0$$

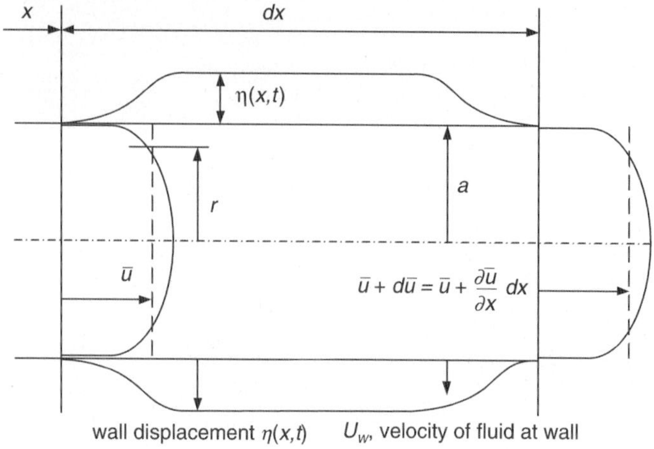

Fig. 12.2. Flow-velocity within an arterial element (of length dx)

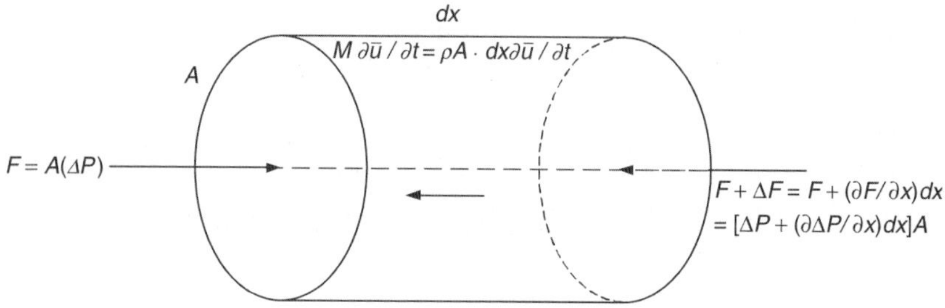

Fig. 12.3. Forces acting on a fluid element of length dx

Hence,

$$u_w = -a/2(\partial \overline{u}/\partial x) \tag{12.2}$$

Note: $@t = 0, \Delta p = 0, \overline{u} = 0$, because the wave has not arrived at site x. Also, $\overline{u}$ (and the associated flow rate Δq) will propagate as a wave, in conjunction with the pressure pulse (Δp) wave.

Momentum equations (refer fig. 12.3):

a) $\Delta F + M\partial\overline{u}/\partial t = (\partial F/\partial x)dx + \rho A dx \partial\overline{u}/\partial t = 0$

Hence, $-\partial F/\partial x + m\partial\overline{u}/\partial t = 0$;

where $m = \mathit{fluid\text{-}element\ mass\ per\ unit\ length} = M/dx = \rho A$, and A is the cross-sectional area of the artery.

Since $\partial F/\partial x = A(\partial\Delta p/\partial x)$, we have

$$A\partial\Delta p/\partial x + \rho A\partial\overline{u}/\partial t = 0$$

and,

$$\partial\Delta p/\partial x = -\rho\partial\overline{u}/\partial t \tag{12.3}$$

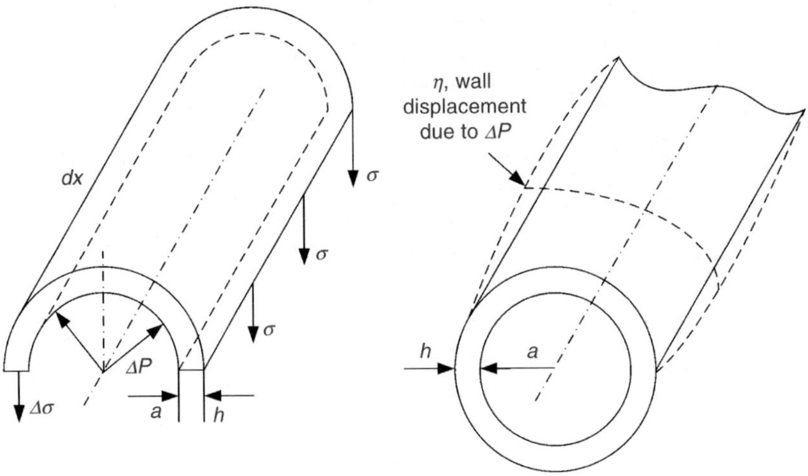

Fig. 12.4. Schematics of wall displacement and forces acting on a wall element

i.e. the pressure gradient = inertia force per unit volume.

b) $\partial \Delta p/\partial r = 0$

Arterial wall motion equation:

From Fig. 12.4, the Equilibrium Equation is:

$$\Delta p(2a \cdot dx) - 2 \cdot \Delta\sigma(h \cdot dx) - \rho_w 2\pi a \cdot hdx\ddot{\eta} = 0$$

where η is the wall displacement.

If we take representative values for the aorta as: $a = 1cm, h = 3mm, \eta = 2mm,\ and\ \ddot{\eta} = 2nm/s^2$, it can be shown that the inertia force term $(\rho_w 2\pi ahdx\ddot{\eta})$ can be neglected, and hence we get[1]:

$$\Delta\sigma = \Delta p \cdot a/h \tag{12.4}$$

Arterial $\sigma - \varepsilon$ relation (Fig. 12.5):

$$\Delta\sigma = E[(2\pi(a+\eta) - 2\pi a)/2\pi a] = E\eta/a \tag{12.5}$$

where E is the incremental modulus of elasticity.

η response to ΔP:

Combining Eqs. (12.4) & (12.5), we get

$$\eta = (a^2/hE)\Delta p$$

[1] Note: Strictly speaking we should have $\therefore\ \Delta\sigma = \Delta Pa'/h'$ where the deformed internal radius $a' = a + \eta$ and the deformed wall thickness $h' = h - \Delta h$
Then, by the virtue of the incompressibility condition $2\pi ah = 2\pi(a+\eta)h' = 2\pi(a+\eta)(h - \Delta h)$ we get: $\eta h = a(\Delta h), or\ \Delta h = \eta h/a$ so, we have
$\Delta\sigma = \Delta P(a'/h') = \Delta P(a+\eta)/(h-\Delta h) = \Delta P(a+\eta)/(h-(\eta h/a))$
$= \Delta P(a+\eta)a/h(a-\eta) \cong \Delta Pa/h(\because \eta \ll a)$

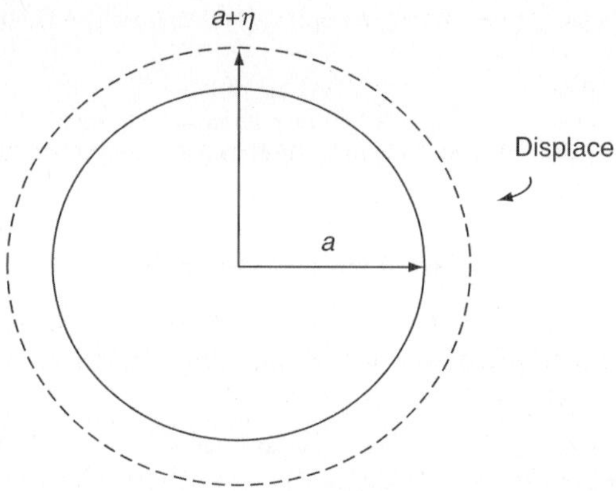

Fig. 12.5. Arterial wall displacement

i.e.

$$\eta(x,t) = (a^2/hE)\Delta p(x,t)$$

or,

$$\eta/a = (a/hE)\Delta p \tag{12.6}$$

Kinematics-matching relations

At the wall, the wall-velocity $\dot{\eta}$ must equal the fluid-velocity at the wall (u_w):

$$u_w(x,t) = \dot{\eta}(x,t) \tag{12.7}$$

Integration of fluid and arterial-wall motion equations:

From Eqs. (12.2, 12.6 and 12.7), we get:

$$u_w = -a/2\partial\overline{u}/\partial x = (a^2/hE)\partial\Delta p/\partial t \tag{12.8}$$

$$\therefore \partial\overline{u}/\partial x = (2a/hE)\cdot\partial\Delta p/\partial t \tag{12.9}$$

Also, from Eq. (12.3)

$$-1/\rho\partial\Delta p/\partial x = \partial\overline{u}/\partial t \tag{12.10}$$

Differentiating Eqs. (12.9 and 12.10) (with respect to t and x), as ∂ (12.9)/∂t – ∂ (12.10)/∂x, *we obtain:*

$$\partial^2\Delta p/\partial x^2 = [1/(Eh/2a\rho)]\partial^2\Delta p/\partial t^2 \tag{12.11}$$

and

$$\partial^2\overline{u}/\partial x^2 = [1/(Eh/2a\rho)]\partial^2\overline{u}/\partial t^2 \tag{12.12}$$

Hence the pressure-pulse wave velocity (PWV) is given by:

$$c = (Eh/2a\rho)^{1/2} \tag{12.13}$$

In other words, Δp and $\overline{u}$ are propagating as waves. From Eqs. (12.6 & 12.13), we can also put down:

$$c = (Eh/2a\rho)^{1/2} = \Delta Pa/2\eta\rho \tag{12.14}$$

In other words, the PWV(= c) is (i) proportional to arterial elasticity (E) and (h/a) or (ii) directly proportional to the magnitude of the pressure pulse (ΔP) and inversely proportional to (η/a).

For example, in a normal adult, whose E = 4.0 × 105 N/m^2, h=1.5 mm, a = 1.15 cm, we get c = 4.96 m/s. In an atherosclerotic patient whose E = 4.0 × 105 N/m^2, h = 1.5 mm, a = 0.81 cm (due to 50% artery occlusion), we get c = 5.91 m/s. For an arteriosclerotic patient whose E = 8.0 × 105 N/m^2 (due to stiffer aortic artery), h = 1.5 mm, a = 1.15 cm, we get c = 7.02 m/s. The schematic of pulse wave velocity in a normal subject, atherosclerotic subject and arteriosclerotic subject are shown in Fig. 12.6. Herein, representative

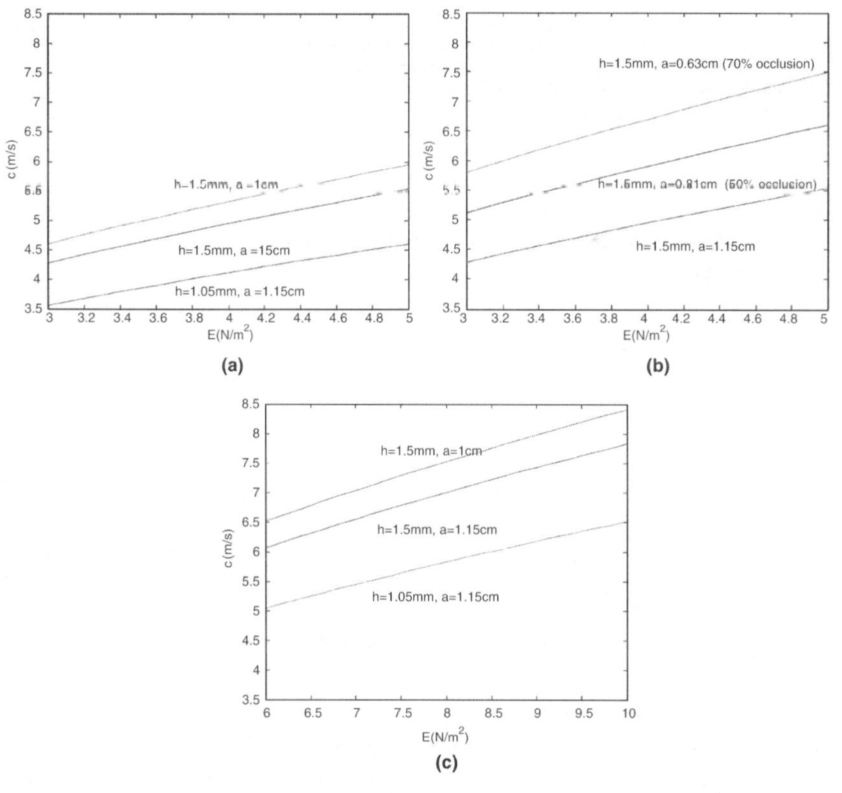

Fig. 12.6. Schematics of pulse wave velocity in (**a**) normal subject, (**b**) atherosclerotic subject and (**c**) arteriosclerotic subject

values of aortic diameter and wall thickness are assumed based on the echocardiography in clinical practice [28].

12.2 Depiction of Pulse Pressure Wave Propagation

We can represent the propagated pressure pulse solution to Eq. (12.11), as:

$$\Delta p(x,t) = A\sin[2\pi(x-ct)/\lambda] + B\sin[2\pi(x+ct)/\lambda] \tag{12.15}$$

where the first term represents a right-propagating wave and the second term represents a left-propagation wave. By substituting Eq. (12.15) in Eq. (12.11), we can check that Eq. (12.11) is satisfied.

Refering to Fig. 12.7, if for a right propagation pulse wave, we have

$$f(x-ct) = A\sin 2\pi(x-ct)/\lambda,$$

Then, f(x−ct) = A Sin2πx/λ at t = 0, when the UVW segment of the wave has emerged into the arterial segment, and f(x − ct) = A Sin2π(x − ct_1)/λ,

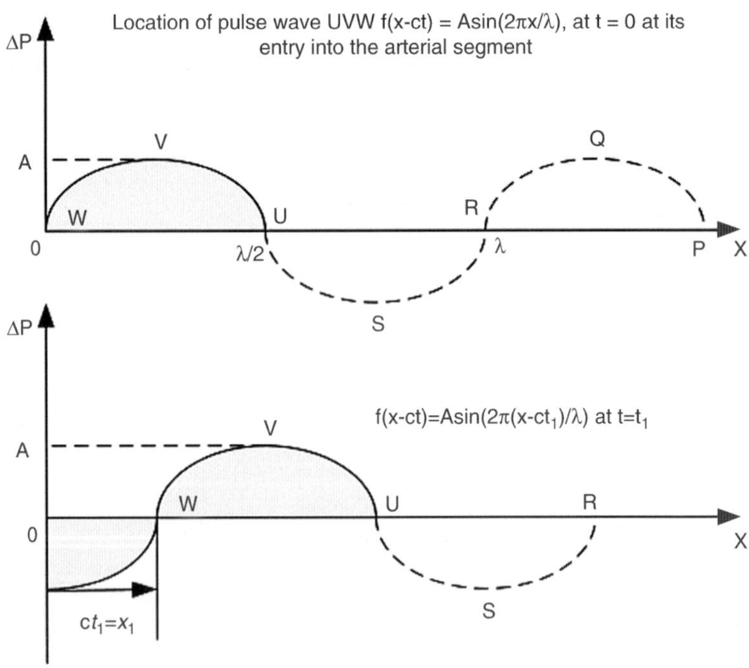

Fig. 12.7. Top: the ΔP wave-segment UVW (from x = 0 to λ/2) has entered the arterial segment, while the ΔP wave segment RSU is the tail end of the diastolic portion of the previous wave PQRSU (shown dotted). Bottom: the wave segment UVW has traveled by x_1(= ct_1) distance to the right in time t_1

after an additional time t_1, obtained by shifting the UVW segment by ct_1 units to the right. (i.e. $f(x - ct)$ represents a wave traveling at speed c from $t = 0$ to t_1). In other words, the wave segment UVW has moved to the right by $x_1 = ct_1$. (i.e., in time $t = x_1/c = t_1$ (because $c = x_1/t_1$).

In Fig. 12.8 (top), the UVW wave has traveled further along a distance λ in time T, i.e $t = T = \lambda/c$ (where T is the time required by wave to travel λ). In Fig. 12.8 (bottom), at $t = 5T/4$, the wave has traveled further to the right by an additional amount $= \lambda/4$; note that at $x = \lambda$, $\Delta p = -A$

In Fig. 12.9, (i) At $x = 0$, the entered wave $UVWXY = A\ Sin(2\pi ct/\lambda)$ *is depicted in the top part of the figure;* (ii) at $x = \lambda$, the wave UVWXY is displayed in the bottom part of the figure.

12.3 Determination of Arterial Elasticity Parameters

In Fig. 12.10, let $\langle P \rangle$ be the mean pressure before the arrival of the pulse wave. Then, the pulse wave ΔP arrives at the site.

Based on fig. 12.11, we have the relationship

$$\sigma = \frac{\langle P \rangle a}{h}, \quad \text{or} \frac{h}{a} = \frac{\langle P \rangle}{\sigma} \tag{12.16}$$

where $\langle P \rangle$ is the mean arterial pressure and σ is the arterial wall stress due to $\langle P \rangle$,

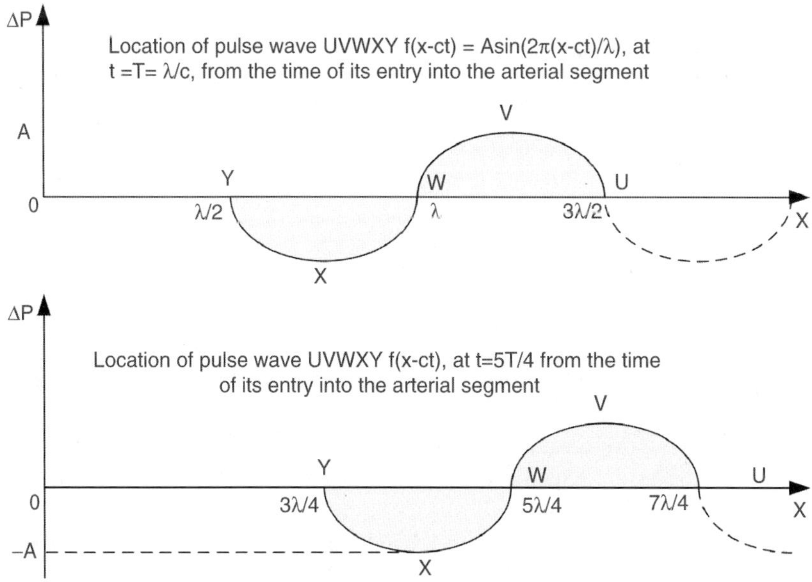

Fig. 12.8. Locations of UVWXY pulse wave at t = T and 5T/4

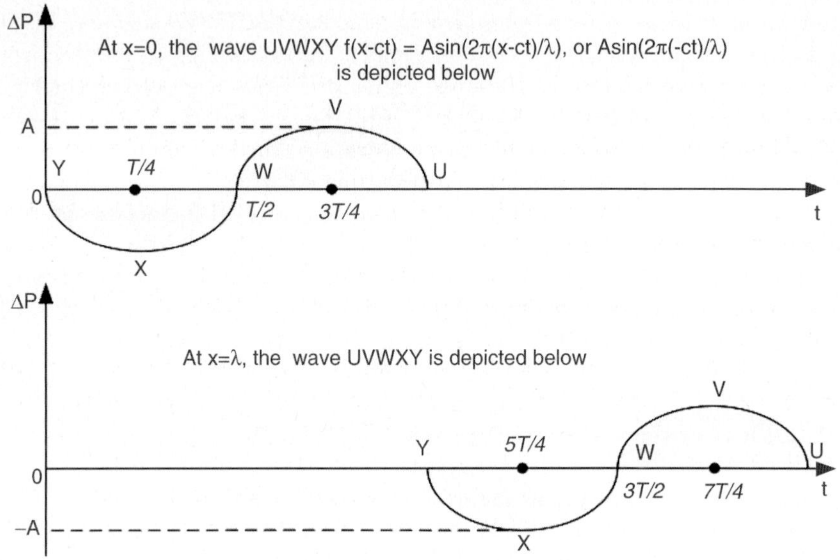

Fig. 12.9. The wave $A \sin 2\pi(x - ct)$ displayed on the time scale at x = 0 and x = λ

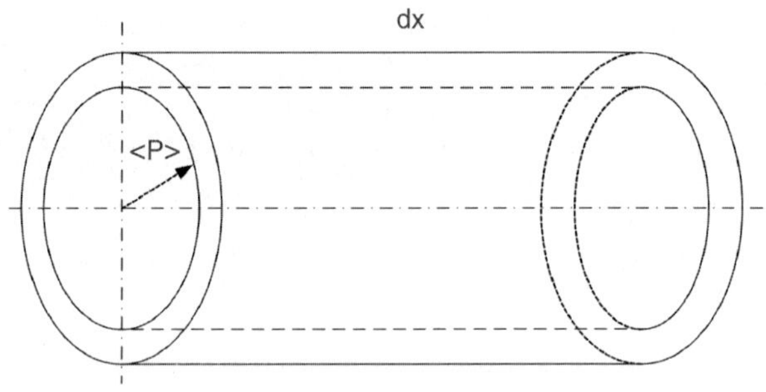

Fig. 12.10. The arterial tube prior to arrival of the pressure pulse

The arterial wall elasticity E_t (as depicted by fig. 12.12) can be expressed as

$$E_t = E_0 + E_1 e^{b\sigma} \tag{12.17}$$

where E_0 is the residual elastic-modulus, and E and b are the constitutive parameters of the artery.

By comparing the E and σ forms of our derived arterial modulus and stress expressions (based on Eqs. 12.13 and 12.16), we note that if the values of the deformed arterial dimensions and the corresponding pressures are known, both the arterial modulus E and wall stress σ can be calculated. The experimental in-vitro arterial data of Simon et al [29] provides the values of pressure, (a & h), and the pulse velocity c, as shown in Table 12.1, corresponding to three different instants. Using these values, we compute, by means of Eq. (12.13)

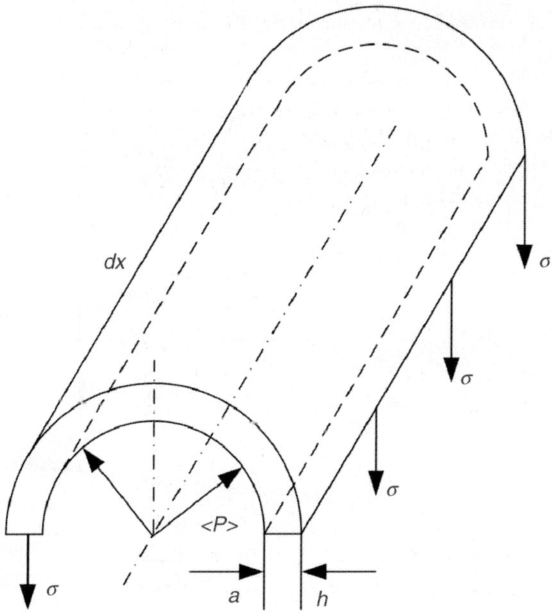

Fig. 12.11. Relation between wall stress and internal pressure

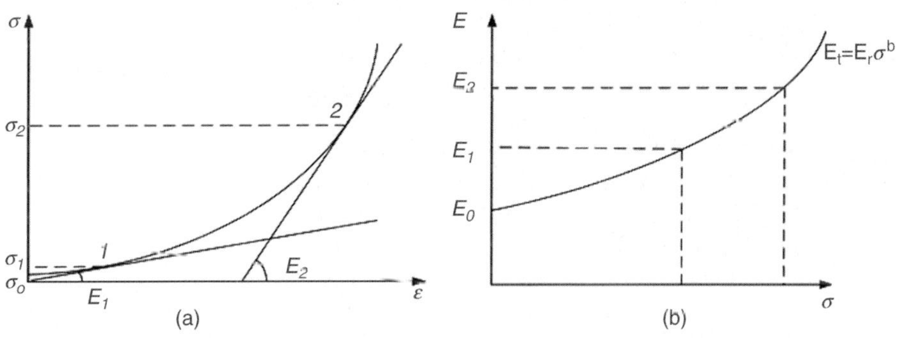

Fig. 12.12. (**a**) Schematics of the arterial stress-strain curve, depicting the incremental elastic modulus E = E_1 and E_2 corresponding to wall stresses σ_1 and σ_2; (**b**) Schematics of E-σ property of an artery corresponding to Figure (**a**)

and (12.16), the arterial modulus E and stress σ, and hence the E against σ curve in Fig. 12.13. By fitting Eq. (12.17) to this curve, we obtain the values of the constitutive parameters (E_0, E and b), show in the Table 12.1. It may be noted that this constitute property can be obtained noninvasively if the diastolic pressure is obtained by cuff sphygmomanometry and the arterial dimension are obtained by ultrasound.

Table 12.1. Calculating elastic modulus (E) parameters from the experimental data of Simon et al [29]

	c (m/s)	Diastolic pressure (mmHg)	a (cm)	h (cm)	σ (Pa) $\times 10^5$	E (Pa) $\times 10^5$	b	E_0	E_1
1	1.899	50	4.29	0.43	0.66	0.73	1.003×10^{-4}	5.63×10^4	22.96
2	2.364	56	4.37	0.41	0.79	1.20			
3	3.745	63	4.45	0.40	0.93	3.15			

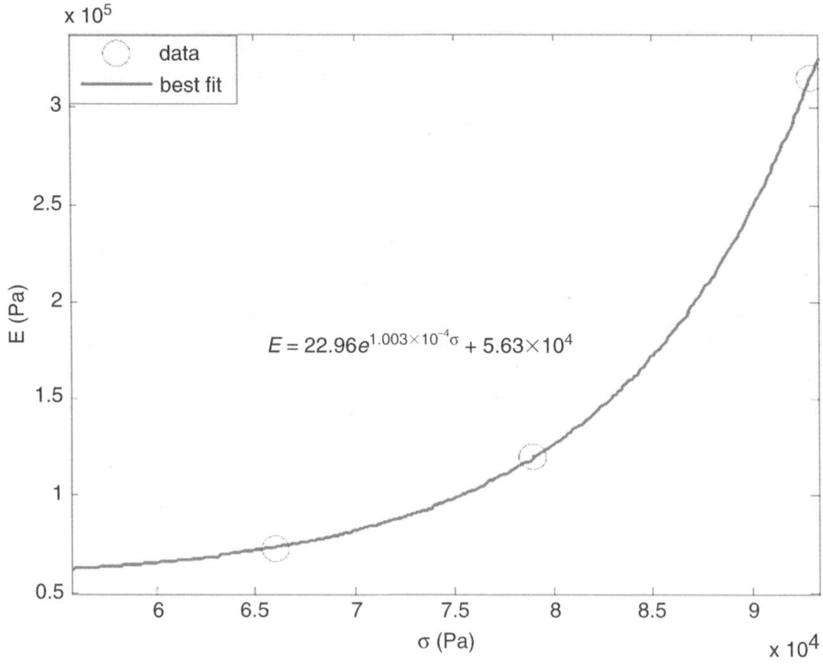

Fig. 12.13. The curve of arterial modulus (E) against arterial wall-stress (σ), as obtained by Eqs. (12.13 & 12.16) from the experimental data of Simon et al [29]

12.4 Determination of Arterial Impedance from PWV and Arterial Cross-Section Area

The arterial pulse waveform is derived from the complex interaction of the left ventricular stroke volume, the physical properties of the arterial tree and the characteristics of the fluid in the system [30]. The principal components of blood pressure, flow rate (q) and velocity (u) comprise both a steady component (mean arterial pressure & flow-rate) and a pulsatile component (pulse pressure & flow-rate), schematically in Fig. 12.14 [31].

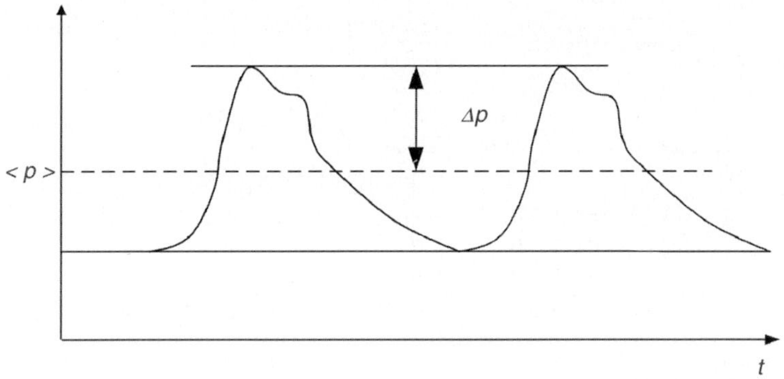

Fig. 12.14. Schematics of an arterial pressure profile, showing $\langle p \rangle$ and ΔP

$$\begin{aligned} p &= \langle p \rangle + \Delta P \\ q &= \langle q \rangle + \Delta q \end{aligned} \tag{12.18}$$

The pulsatile component of pressure is determined by the pattern of left ventricular ejection, the stroke volume and the compliance characteristics of the arterial circulation [32]. Arterial compliance is defined as the change in area or volume of an artery (or arterial bed) for a given change in pressure [33]. The pulse pressure for a given ventricular ejection and heart rate will depend on arterial compliance, as well as the timing and magnitude of peripheral pulse wave reflection.

Impedance, a term borrowed from electrical engineering theory, describes the opposition to flow presented by a system. The impedance load of the arterial tree can be quantified by analyzing pulse pressure-flow relationships, produced through the effects of disease on the structural and functional components of the arterial system [34, 35]. Input impedance relates simultaneously recorded pressure and flow waveforms under specific mathematical conditions.

12.4.1 Peripheral Resistance (*R*)

We define

$$\langle q \rangle = \frac{\langle p \rangle}{R}, \text{ where } R = \frac{8\mu L}{\pi a^4} = TPR \tag{12.19}$$

With atherosclerosis, the arterial radius (a) decreases, and resistance (R) increases. Hence for a given $\langle q \rangle$, $\langle p \rangle$ is increased.

12.4.2 Arterial Impedance (Z_0)

a) We define $\Delta q = \dfrac{\Delta p}{z_0}$, where

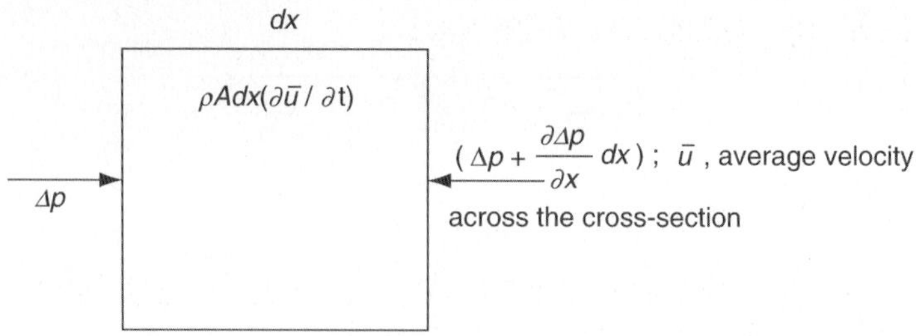

Fig. 12.15. Force equilibrium of a fluid element

$$z_0(impedance) = \frac{\rho c}{A} = \frac{\rho}{A}\sqrt{\frac{Eh}{2a\rho}} \tag{12.20}$$

For a given Δq, Δp is high if z_0 (impedance) is high. In other words, the impedance (z_0) is a direct measure of arterial hardening or stiffness. We will now derive Eq. (12.20).

For force equilibrium (as shown in Fig. 12.15):

$$-A\frac{\partial \Delta p}{\partial x}dx - \rho A dx\frac{\partial \overline{u}}{\partial t} = 0 \tag{12.21}$$

$$A\rho\frac{\partial \overline{u}}{\partial t} = -A\frac{\partial \Delta p}{\partial x} \tag{12.22}$$

Now, $\Delta q(x,t) = A\overline{u}(x,t)$

$$\therefore \rho\frac{\partial \Delta \mathrm{q}}{\partial \mathrm{t}} = -A\frac{\partial \Delta p}{\partial x} \tag{12.23}$$

As a consequence of the pulse wave velocity equation:

$$\frac{\partial^2 \Delta p}{\partial x^2} = \frac{1}{(Eh/2a\rho)}\frac{\partial^2 \Delta p}{\partial t^2} = \frac{1}{c^2}\frac{\partial^2 \Delta p}{\partial t^2}$$

we can put down:

$$\Delta p = f_1(x - ct) \tag{12.24a}$$

Also, since

$$\frac{\partial^2 \Delta q}{\partial x^2} = \frac{1}{c^2}\frac{\partial^2 \Delta q}{\partial t^2}$$

we can put down:

$$\Delta q = \phi_1(x - ct) \tag{12.24b}$$

For instance, we can have $\Delta p = \Delta p_1 \sin\frac{2\pi}{\lambda}(x - ct)$, and $\Delta \mathrm{q} = \Delta \mathrm{q}_1 \sin\frac{2\pi}{\lambda}(x - ct)$

Upon substituting (12.24) in (12.23), we get:

$$-\rho c\phi_1'(x-ct) = -Af_1'(x-ct) \tag{12.25}$$

Integrating, we get

$$\rho c\phi_1 = Af_1 \tag{12.26}$$

Hence, for right-propagating waves $f(x-ct)$ and $\phi(x-ct)$, the impedance is given by:

$$\therefore \mathrm{z}_{01} = \frac{\Delta\vec{p}}{\Delta\vec{q}} = \frac{\rho c}{A} = \frac{\rho}{A}\left(\frac{Eh}{2a\rho}\right)^{1/2} = \frac{4\rho}{\pi}\left(\frac{Eh}{2a^5\rho}\right)^{1/2} \tag{12.27}$$

Now, for a left-propagating wave $\Delta p_2 = f_2(x+ct)$ and $\Delta q_2 = \phi_2(x+ct)$, and we have $-\rho c\phi_2 = +Af_2$

Hence, for a left-traveling wave,

$$z_{02}\left(=\frac{\rho c}{A}\right) = -\frac{f_2}{\phi_2} = -\frac{\Delta\overleftarrow{p}_2}{\Delta\overleftarrow{q}_2} \tag{12.28}$$

b) **Implication:** If the arterial stiffness E is high (as in arteriosclerosis), then (as per Eqs. 12.20 & 12.27), both z_0 and Δp will be high. Based on Eqs. (12.19) & (12.27), (i) if a person smokes or has atherosclerosis, $\langle p\rangle$ will be elevated (ii) if a person has hardened artery (arteriosclerosis), Δp will be elevated. Note that R (in Eq. 12.19) is primarily affected by high μ and low a. On the other hand, z_0 is affected by high E as well as low a and high h.

Reiterating, for a right-propagating wave, in eqn. (24),

$$\text{let } \Delta p = f_1 = \Delta p_1 \sin\frac{2\pi}{\lambda}(x-ct), \text{and } \Delta q = \phi_1 = \Delta q_1 \sin\frac{2\pi}{\lambda}(x-ct)$$

Then from Eq. (12.26)

$$\rho c\Delta q_1 \sin\frac{2\pi}{\lambda}(x-ct) = A\Delta p_1 \sin\frac{2\pi}{\lambda}(x-ct) \tag{12.29}$$

and hence

$$\mathrm{z}_{01} = \frac{\rho c}{A} = \frac{\Delta p_1}{\Delta q_1} \tag{12.30}$$

For a left-propagating wave,

$$\Delta p = f_2 = \Delta p_2 \sin\frac{2\pi}{\lambda}(x+ct), \text{and } \Delta q = \phi_2 = \Delta q_2 \sin\frac{2\pi}{\lambda}(x+ct)$$

Then from Eq. (12.23),

$$\rho\frac{\partial\Delta q}{\partial t} = -A\frac{\partial\Delta p}{\partial x},$$

we get:

$$\frac{2\pi\rho c}{\lambda}\Delta q_2 \sin\frac{2\pi}{\lambda}(x+ct) = -\frac{A2\pi}{\lambda}\Delta q_2 \sin\frac{2\pi}{\lambda}(x+ct) \tag{12.31}$$

and hence,

$$\rho c\Delta \mathrm{q}_2 = -A\Delta p_2 \tag{12.32}$$

so that,

$$\mathrm{z}_{02} = \frac{\rho c}{A} = -\frac{\Delta p_2}{\Delta q_2} \tag{12.33}$$

c) What about the constant of Integration in Eq. (12.26)
In general, we can put down:

$$\rho c[\phi(x-ct)+\phi(0)] = A[f(x-ct)+f(0)] \tag{12.34}$$

(i) If $f(x-ct) = \Delta p_1 \sin\frac{2\pi}{\lambda}(x-ct)$, and $\phi(x-ct) = \Delta q_1 \sin\frac{2\pi}{\lambda}(x-ct)$ then,

$$f(0) = 0,\ \text{and also } \phi(0) = 0 \tag{12.35}$$

(ii) If $f(x-ct) = \Delta p_1 \cos\frac{2\pi}{\lambda}(x-ct)$ and $\phi(x-ct) = \Delta q_1 \cos\frac{2\pi}{\lambda}(x-ct)$ then,

$$f(0) = \Delta p_1 \text{ and } \phi(0) = \Delta q_1 \tag{12.36}$$

Hence, we have:

$$\rho c[\phi(x-ct)+\Delta q_1] = A[f(x-ct)+\Delta p_1]$$

or,

$$\rho c\left[\Delta q_1 \cos\frac{2\pi}{\lambda}(x-ct)+\Delta q_1\right] = A\left[\Delta p_1 \cos\frac{2\pi}{\lambda}(x-ct)+\Delta p_1\right] \tag{12.37}$$

i.e., $\rho c\Delta q + \rho c\Delta q_1 = A\Delta p + A\Delta p_1$

Hence, we can put down: $\rho c\Delta \mathrm{q} = \mathrm{A}\Delta \mathrm{p}$, for $\mathrm{z}_0 = \frac{\Delta p}{\Delta q} = \frac{\rho c}{A}$

$$\text{and, } \rho c\Delta \mathrm{q}_1 = \mathrm{A}\Delta \mathrm{p}_1,\ \text{for } z_0 = \frac{\Delta p_1}{\Delta q_1} = \frac{\rho c}{A} \tag{12.38}$$

i.e., $\rho c\phi(0) = Af(0)$

$$\text{So, either, } \phi(0) = f(0) = 0,\ \text{based on equation (12.35)} \tag{12.39a}$$

$$\text{Or, } \rho c\phi(0) = Af(0),\ \text{based on equation (12.38), for which } z_0 = \frac{f(0)}{\phi(0)} = \frac{\rho c}{A} \tag{12.39b}$$

d) The arterial pressure characteristics in normal, atherosclerosis and arterioslcerosis states are depicted in Fig. 12.16.

Summarizing, we have $\langle p \rangle = R\langle q \rangle$; where R is the arteriolar bed resistance and, $\Delta p = Z_0 \Delta q$; where Z_0 is the arterial impedance. So in a hypertensive person, we can reduce $\langle p \rangle$ by reducing R, and reduce Δp by reducing Z_0 (or increasing compliance), by appropriate medication.

Since, $\dfrac{\langle p_1 \rangle}{\langle p_2 \rangle} = \dfrac{R_1}{R_2}$,

if we want $\langle p_1 \rangle$ to be 0.8 $\langle p_2 \rangle$, then R_1 is also to be $0.8R_2$.

Since, $\dfrac{\Delta p_1}{\Delta p_2} = \dfrac{z_{01}}{z_{02}}$,

if we want Δp_1 to be 0.8 Δp_2, then z_{01} to be 0.8 z_{02}.

12.5 Reflection at Arterial Bifurcation

a) **Analysis:**

Based on Fig. 12.17, we can put down:

$$\Delta p = \Delta p' = \Delta p_1{}' = \Delta p_2{}' \tag{12.40}$$

$$\begin{aligned}\Delta q' &= \Delta q_1{}' + \Delta q_2{}' = \frac{\Delta p'}{z_{01}{}'} + \frac{\Delta p'}{z_{02}{}'} = \Delta p'\left(\frac{1}{z_{01}{}'} + \frac{1}{z_{02}{}'}\right) \\ &= \frac{2\Delta p'}{z_0{}'} (\text{if } z_{01}{}' = z_{02}{}')\end{aligned} \tag{12.41}$$

Also,

$$\Delta p = \Delta p_1 + \Delta p_2 = \Delta p' = \Delta p_1{}' = \Delta p_2{}' \tag{12.42}$$

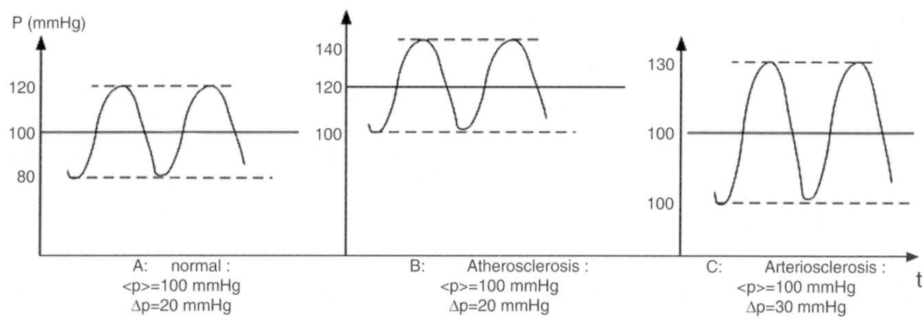

Fig. 12.16. Schematic of arterial pressure waveform under different situations: A: normal; B: atherosclerosis, the artery is stiff (i.e., E is high in Eq. 12.20); C: arteriosclerosis, where there is vessel occlusion and high peripheral resistance (i.e. R is high in Eq. 12.19)

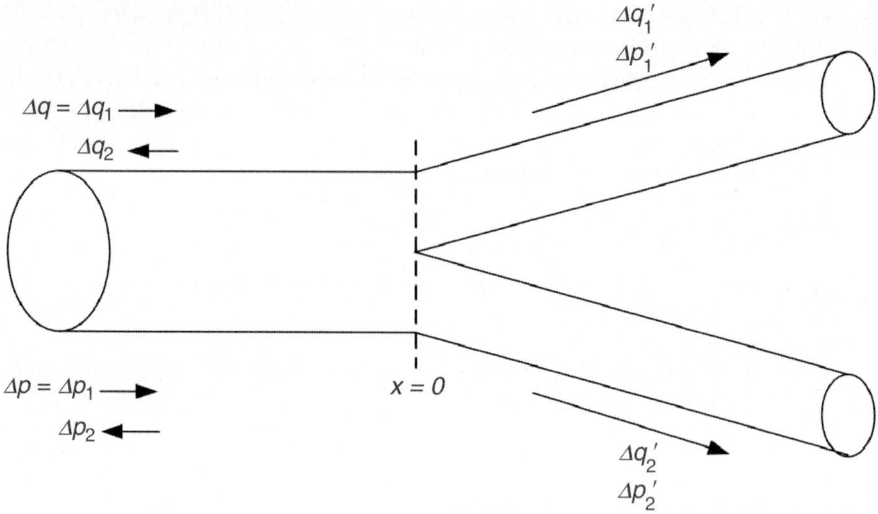

Fig. 12.17. Schematics of pressure and flow at an arterial bifurcation

and,

$$\Delta q = \Delta q_1 + \Delta q_2 = \Delta q' = \Delta {q_1}' + \Delta {q_2}' \tag{12.43}$$

Hence, from Eq. (12.43),

$$\frac{\Delta p_1}{z_0} - \frac{\Delta p_2}{z_0} = \frac{\Delta p'}{{z_0}'} + \frac{\Delta p'}{{z_0}'} = \frac{2\Delta p'}{{z_0}'} \tag{12.44}$$

Then, from Eqs. (12.44) & (12.42),

$$\frac{{z_0}'}{z_0}\Delta p_1 - \frac{{z_0}'}{z_0}\Delta p_2 = 2\Delta p' = 2\Delta p_1 + 2\Delta p_2$$

or,

$$\Delta p_1 \left(\frac{{z_0}'}{z_0} - 2\right) = \Delta p_2 \left(\frac{{z_0}'}{z_0} + 2\right) \tag{12.45}$$

Let λ be defined as

$$\frac{z_0}{{z_0}'} = \left(\frac{E}{E'} \times \frac{h}{h'}\right)^{0.5} \times \left(\frac{a'}{a}\right)^{2.5}. \tag{12.46}$$

Then,

$$\frac{\Delta p_2}{\Delta p_1} = \frac{{z_0}' - 2z_0}{{z_0}' + 2z_0} = \frac{1 - 2\lambda}{1 + 2\lambda} \tag{12.47}$$

Let R_f (reflection coefficient) be defined as

$$R_f = \frac{\Delta p_2}{\Delta p_1} = \frac{1 - 2\lambda}{1 + 2\lambda} \tag{12.48}$$

b) **Applications:**

(i) So, for no reflection (as expected in nature), $\Delta p_2 = 0$, and $\mathrm{R_f} = 0$

$$\text{i.e, } \lambda = \frac{z_0}{z_0{}'} = 0.5 \tag{12.49}$$

$$\text{i.e, } \lambda = \left(\frac{E}{E'} \times \frac{h}{h'}\right)^{0.5} \times \left(\frac{a'}{a}\right)^{2.5} = 0.5 \tag{12.50}$$

Suppose

$$a' = 0.75a. \tag{12.51}$$

Then, if we can assume conservation of mass between the parent vessel and its bifurcations, we will have

$$2\pi a h = 2(2\pi a' h') \tag{12.52}$$

Hence, from Eqs. (12.51 & 12.52),

$$\frac{h}{h'} = 2\frac{a'}{a} = 2 \times (0.75) = 1.5 \tag{12.53}$$

Then, from Eqs. (12.50, 12.51 and 12.53),

$$\frac{\mathrm{E}}{\mathrm{E'}} = \frac{0.25}{(a'/a)^5 \times (h/h')} \tag{12.54}$$

Now since $(a'/a)^5 = 0.237$ and $h/h' = 1.5$, based on Eqs. (12.51 and 12.53),
Then, from Eq. (12.54),

$$E/E' = 0.703 \tag{12.55}$$

Hence, we have:
For $\mathrm{R_f} = 0, \Delta p_2 = 0$, or no reflection and $\lambda = 0.5$
i.e.,

$$\lambda = 0.5 \text{ means } \Delta p_2 = 0, \text{ and } \mathrm{R_f} = 0 \text{ (i.e., no reflection)} \tag{12.56}$$

(ii) For

$$\lambda < 0.5, \Delta \mathrm{p}_2 = R_f \Delta p_1, \text{ and } 0 < \mathrm{R_f} < 1 \tag{12.57}$$

In this case, there is reflection with no phase change, i.e., an incident expansion wave at a site will be superimposed by a reflected expansion wave (of less magnitude) at that site.

Let us represent Eq. (12.57) by

$$A_2 \sin(x + ct) = R_f A_1 \sin(x - ct); \;\; 0 < \mathrm{R_f} < 1$$

and adopt $R_f = 0.3$, so that

$$A_2 \sin(x + ct) = 0.3 A_1 \sin(x - ct)$$

This means that the reflected pulse wave amplitude will add to the amplitude of the incident wave, and increase the afterload on the LV.

(iii) For $\lambda > 0.5$, we have from Eq. (12.47), $\Delta p_2 = -R_f \Delta q_2$; i.e., $R_f < 0$
Let $R_f = -0.2$, then,

$$A_2 \sin(x + ct) = -\frac{1}{5} A_1 \sin(x - ct) = \frac{1}{5} A_1 \sin[(x - ct) - \pi], \text{i.e.}, \mathrm{A}_2 = -\frac{A_1}{5}$$

This means that the reflected pulse wave will be 180° out of phase with the incident way, and contribute to decreasing the amplitude of the combined incident and reflected wave. This aspect has some important clinical inference. It would be interesting to verify that for normal persons, (if $a' = 0.75a$ and $h = 1.5h'$, as per Eqs. 12.51 and 12.53), then from Eq. **??**, by putting $\lambda = 0.75$, we get $\mathrm{E/E'} = 0.703$. These persons will be intrinsically normaltensive persons. In other words, persons for whom $\lambda < 0.5$ (and say $R_f = 0.3$), will be intrinsically hypertensive. On the other hand, persons, for whom $\lambda > 0.5$ (and say $R_f = -0.2$), will have intrinsically low blood pressure.

12.6 Conclusing Remarks

In this chapter, we have derived the governing equation for pulse wave velocity. We have then depicted the locations of the pulse wave as it travels along the artery. After that, we have defined the concepts of (i) total peripheral resistance in Eq. (12.19) and of (ii) arterial impedance in Eq. (12.20). Lastly, we have depicted what happens at an arterial bifurcation, and provided the concept of wave reflection at a bifurcation, resulting in the reflected wave being either in phase or 180° out of phase with the incident wave depending on the ratio of modulus of the parent and the bifurcated vessels.

We have arrived at an interesting observation. It is shown that if the reflected wave is in phase with the incident wave, then it will increase the afterload on the heart. On the other hand, if the reflected wave is out of phase with the incident wave, then it will decrease the afterload on the heart. If we assume conservation of mass between the parent and daughter vessels, we have shown that for $E/E' = 0.703$, the afterload is minimized.

References

1. Ross, R. (1999): 'Atherosclerosis an inflammatory disease', *New England Journal of Medicine*, **340**, pp. 115–126.
2. Nichols, W.W., O'Rourke, M.F. (1998): 'McDonald's blood flow in arteries: theoretical, experimental and clinical principles', 4th *edn, London, Amold.*
3. Kelly, R.P., Hayward, C.S., Ganis. J., Daley, J.M., Avolio, A.P., and O'Rourke, M.F. (1989): 'Noninvasive registration of the arterial pulse waveform using high fidelity applanation tonometry', *J Vasc Med Biol*, **1**, pp. 142–149.

4. Asmar, R., Benetos, A., Topouchian, J., Laurent, P., Pannier, B., Brisac, A.M., Target, R., and Levy, B.I. (1995): 'Assessment of arterial distensibility by automatic pulse wave velocity measurement', *Hypertension*, **26**, pp. 485–490.
5. Blacher, J., Asmar, R., Djane, S., London, G.M., and Safar, M.E. (1999): 'Aortic pulse wave as a marker of cardiovascular risk in hypertensive patients', *Hypertension*, **33**, pp. 1111–1117.
6. Laurent, S., Boutouyrie, P., Asmar, R., Gautier, L., Guize, L., Ducimetiere, P., and Benetos, A. (2001): 'Aortic stiffness is an independent predictor of all-cause and cardiovascular mortality in hypertensive patients', *Hypertension*, **37**, pp. 1236–1245.
7. Popele, Van, N.M., Grobbee, D.E., Bots, M.L., Asmar, R., Topouchian, Reneman, R.S., Hoeks, A.P.G., Van Der, K.D.A.M., Hofman, A., and Witteman, J.C.M. (1998): 'Association between arterial stiffness and atherosclerosis', *Stroke*, **32**, pp. 454–457.
8. Safar, M.E., London, G.M., Asmar, R., and Frohlich, E.D. (1998): 'Recent advances on large arteries in hypertension', *Hypertension*, **32**, pp. 156–161.
9. Madhavan, S., Ooi, W.L., and Cohen, H. (1994): 'relation of pulse pressure and blood pressure reduction to the incidence of myocardial infarction', *Hypertension*, **23**, pp. 395–401.
10. Noordergraaf, A. (1978): 'Circulatory system dynamics', New York, NY: *Academic Press*, pp. 105–151.
11. Wemple, P.R., and Mockros, L.F. (1972): 'Pressure and flow in the systemic arterial system', *J of Biomechanics*, **5**, pp. 629–641.
12. Westerhof, N., Sipkema, P., Van Den Bos, G.C., and Elzinga, G. (1972): 'Forward and backward waves in the arterial system', *Cardiovasc Res*, **6**, pp. 648–656.
13. O'Rourke, M.F. (1982): 'Vascular impedance in studies of arterial and cardiac function', *Physiol Rev*, **62**, pp. 570–623.
14. Papageorgiou, G.L., and Jones, N.B. (1988): 'Wave reflection and hydraulic impedance in the healthy arterial system: a controversial subject', *Med Bio Eng Comput*, **26**, pp. 237–242.
15. Quick, C.M., Berger, D.S., and Noordergraaf, A. (2001): 'Construction and destructive addition of forward and reflected arterial pulse waves', *Am J of Physiol Heart and Circulatory Physiol*, **280**(4), pp. H1519–H1527.
16. O'Rourke, M. (1990): 'Arterial stiffness, systolic blood pressure, and logical treatment of arterial hypertension', Hypertension, **15**, pp. 339–347.
17. Westerhof, N., and O'Rourke, M.F. (1995): 'Haemodynamic basis for the development of left ventricular failure in systolic hypertension and for its logical therapy', *J Hypertens*, **13**, pp. 943–952.
18. Berger, D.S., Li, J.K., and Noordergraaf, A. (1995): 'Arterial wave propagation phenomena, ventricular work, and power dissipation', *Ann Biomed Eng*, **23**, pp. 804–811.
19. Berger, D.S., Li, J.K., and Noordergraaf, A. (1994): 'Differential effects of wave reflections and peripheral resistance on aortic blood pressure: a model-based study', Am J Physiol Heart *Circ Physiol*, **266**, pp. H1626–H1642.
20. Berger, D.S., Robinson, K.A., and Shroff, S.G. (1996): 'Wave propagation in coupled left ventricle-arteiral system: Implications for aortic pressure', *Hypertension*, **27**, pp. 1079–1089.

21. Fang, J., Madhavan, S., and Cohen, H. (1995): 'Measures of blood pressure and myocardial infarction in treated hypertensive patients', J Hypertens, **13**, pp. 413–419
22. Weber, T., Auer, J., O'Rourke, M.F., Kvas, E., Lassnig, E., Berent, R., and Eber, B. (2004): 'Arterial stiffness, wave reflections and the risk of coronary artery disease', *Circulation*, **109**, pp. 184–189.
23. Westerhof, N., Sipkema, P., Van Den Bos, G.C., and Elzinga, G. (1972): 'Forward and backward waves in the arterial system', *Cardiovascular Research*, **6**(6), pp. 648–656.
24. Liu, Z., Brin, K.P., and Yin, F.C. (1986): 'Estimation of total arterial compliance: an improved method and evaluation of current methods', *Am J Physiol*, **251**(3), pp. H588–H600.
25. Laskey, W.K., Parker, H.G., Ferrari, V.A., Gussmaul, W.G., and Noordergraaf, A. (1990): 'Estimation of total systemic arterial compliance in humans', *J Appl Physiol*, **69**, pp. 112–119.
26. Stergiopulos, N., Meister, J.J., and Westerhof, N. (1994): 'Simple and accurate way for estimating total and segmental arterial compliance: the pulse pressure method', *Annals of Biomedical Engineering*, **22**, pp. 392–397.
27. Stergiopulos, N., Segers, P., and Westerhof, N. (1999): 'Use of pulse pressure method for estimating total arterial compliance in vivo', *American Journal of Physiology Heart Circulation Physiology*, **276**, pp. H424–H428.
28. Chambers, J. (2002): 'Echocardiography in clinical practice', London UK, *The Parthenon publishing group.*
29. Simon, B.R., Kobayashi, A.S., Strandness, D.E., and Wiederhielm, C.A. (1971): 'Large deformation analysis of the arterial cross-section', *Journal of Basic Engineering*, **93**, pp. 138–145.
30. McVeigh, G.E., Bank, A.J., and Cohn, J.N. (2000): 'Arterial compliance', in *Cardiovascular Medicine*, (Willerson, J.T., Cohn, J.N. (2000):) *Churchill Livingstone*, Philadelphia, pp. 1479–1496.
31. Smulyan, H., and Safar, M.E. (1997): 'Systolic blood pressure revisited', J. Am. Coll. Cardiol, **29**, pp. 1407–1413.
32. McVeigh, G.E., Finklestein, S.M., and Cohn, J.N. (1996): 'Pulse contour and impedance parameters derived from arterial waveform analysis', in Functional abnormal of the aorta (Boudoulas H, Toutouzas P, Wooley CF, eds) *Futura, Armonk*, pp. 183–193.
33. Simon, A., and Levenson, J. (1991): 'Use of arterial compliance for evaluation of hypertension', *Am. J. Hypertensions*, **4**, pp. 97–105.
34. Finkelstein, S.M., and Collins, V.R. (1982): 'Vascular hemodynamic impedance measurement', *Prog Cardiovascular Dis*, **24**, pp. 401–418.
35. Nichols, W.W., Pepine, C.J., Geiser, E.A., and Conti, C.R. (1980): 'Vascular load defined by the aortic input impedance spectrum', *Fed. Proc*, **39**, pp. 196–201.

13

ECG Signal Conditioning by Morphological Filters

S.M. Krishnan, Kwoh Chee Keong, Sun Yan and Chan Kap Luk

Electrocardiography (ECG) signals are often contaminated by noise of diverse forms: 50/60 Hz power line interference, motion artifact from the electrode-skin interface, muscle activities [1, 2]. In addition, baseline drift caused by the respiration, radio frequency surgical noise and motion of the subject [3], degrades ECG signals significantly. Therefore, signal conditioning for baseline correction and noise suppression is typically the first step in the analysis of ECG signals. The objective of ECG signal conditioning is to produce an output that can facilitate the subsequent processing, such as ECG episode characterization for life-threatening arrhythmia recognition, or the characteristic wave detection for non-life-threatening ECG signals. It is important to minimize the distortion of the ECG signal caused by signal conditioning algorithms before analysis tasks can be performed.

Conventionally used technique for noise suppression is band-pass filtering [4–6]. However, band pass type of linear filtering techniques has a fixed cut-off frequency which distorts the ST segment as well as the QRS complex significantly. It is also not adaptive and hence cannot track the changing characteristics of the time-varying ECG signals, which tend to vary quasi-periodically, with each period corresponding to one heart beat. Recently, adaptive filtering techniques have been developed for the purpose of noise suppression in ECG signals. Most adaptive noise-removal methods [7–9] are based on the least mean squares (LMS) principle or on the recursive least squares (RLS) principle. They gradually reduce the mean squared error between the input signal and some reference signal. However, in some cases, these techniques experience the problem of not being able to obtain a suitable reference signal, which limits the wide application of this kind of approach. Wavelet transform, being a very promising technique for joint time-frequency analysis, provides an interesting solution to ECG signal conditioning [10, 11]. By decomposing signals into the transform domains, a number of coefficients at different scales can be obtained. By selecting suitable scales and disregarding the coefficients below predefined thresholds, additive noise and baseline drift can be separated from the ECG signal components. However, in this kind of techniques, the scales and the thresholds for the nonstationary baseline correction and noise suppression cannot be selected adaptively.

Morphological operators have been widely used in the signal and image processing fields because of their robust and adaptive performance in extracting the shape information in addition to their simple and quick set computation [12–15]. Chu and his collaborators used the combined opening and closing operators for baseline correction and noise suppression of ECG signals and good filtering performance was obtained [16]. However, the proposed morphological filtering (MF) algorithm smoothes the sharp variations of ECG signals, such as the R peak, the Q wave, the S wave, the onsets and offsets of the P and T waves. This makes it difficult for the subsequent processing to reliably detect the above mentioned characteristic waves. In this chapter, a modified morphological filtering (MMF) algorithm is proposed for baseline correction and noise suppression of ECG signals. For baseline correction, same operators are used in MF algorithm and the MMF algorithm. The results for baseline correction using the MMF algorithm are compared with those obtained by the wavelet filtering (WF) algorithm [17]. For noise suppression, modified morphological operators are used in the MMF algorithm. Compared with MF algorithm, a structuring element pair is used instead of a single structuring element. Performance for noise suppression using these two morphological techniques is compared. For noise suppression, the WF algorithm is not used for comparison in this chapter because there are some parameters needed to be selected empirically, making fair comparison difficult. Noise reduction in ECG is accomplished applying filtering techniques. However, such filtering may mutate the original wave making difficult the interpretation of pathologies [18]. To overcome this problem an adaptive neural method able to filter ECGs without causing the loss of important information is proposed. In 96% of the cases the signal processed by the network is coherent with the original one within a coherence value of 0.92, whereas this values for the morphological filter is 0.70. Moreover, the adaptability of the neural method does not require estimating appropriate filter parameters for each ECG segments. Clinically obtained electrocardiographic (ECG) signals are often contaminated with different types of noise and baseline drifting commonly occurs. A modified morphological filtering (MMF) technique is used for signal conditioning in order to accomplish baseline correction and noise suppression with minimum signal distortion [19]. Compared with existing methods for ECG signal conditioning, MMF performs well in terms of the filtering characteristics, low signal distortion ratio, low computational burden as well as good noise suppression ratio and baseline correction ratio.

This chapter is organized as follows. Mathematical background on morphological operators is introduced in Sect. 13.1. The proposed algorithm scheme for the conditioning of the ECG signal is described in Sect. 13.2. Section 13.3 covers the experimental results and discussions. Lastly, concluding remarks of this chapter are given in Sect. 13.4.

13.1 Mathematical Morphology Operators

Mathematical morphology (MM), which is based on sets operations, provides an approach to the development of non-linear signal processing methods, in which the shape information of a signal is incorporated [20]. In mathematical morphology operations, the result of a set transformed by another set depends on the shapes of the two sets involved. The shape information of a signal can be extracted by using a structuring element to operate on the signal. Depending on the shape characteristics of the signal that is to be extracted, a specific structuring element has to be designed.

There are two basic morphological operators: erosion (Θ) and dilation ($\oplus$). Opening ($\circ$) and closing ($\bullet$) are derived operators defined in terms of erosion and dilation. These operators are described in detail below with corresponding mathematical expressions. Throughout this chapter, $f(n), \{n = 0, 1, \ldots, N-1\}$ denotes a discrete signal consisting N points, and $B(m), \{m = 0, 1, \ldots, M-1\}$ is a symmetric structuring element of M points.

$$erosion : (f\Theta B)(n) = \min_{m=0,..M-1} \left\{ f\left(n - \frac{M-1}{2} + m\right) - B(m) \right\} \quad (13.1)$$

$$for\ n = \left\{ \frac{M-1}{2}, \ldots, N - \frac{M+1}{2} \right\}$$

Erosion is a 'shrinking' operator in which the values of $f\Theta B$ are always less than those of f

$$dilation : (f \oplus B)(n) = \min_{m=0,..M-1} \left\{ f\left(n - \frac{M-1}{2} + m\right) - B(m) \right\} \quad (13.2)$$

$$for\ n = \left\{ \frac{M-1}{2}, \ldots, N - \frac{M+1}{2} \right\}$$

The dilation operation is an 'expansion' operation in which the values of $f \oplus B$ are always greater than those of f.

$$\text{opening} : \ f \circ B = f \,\Theta\, B \oplus B \quad (13.3)$$

$$\text{closing} : \ f \bullet B = f \oplus B \,\Theta\, B \quad (13.4)$$

The opening of a data sequence can be interpreted as sliding structuring element along the data sequence from beneath and the result is the highest points reached by any part of the structuring element. Similarly, the closing of a data sequence can be interpreted as sliding a 'flipped-over' version of the structuring element along the data sequence from above and the result is the set of lowest points reached by any part of the structuring element. In most applications, opening is used to suppress peaks, while closing is used to suppress pits.

In Chu's MF algorithm [16], the baseline correction and noise suppression are performed as follows:

$$f_b = f_o \circ B_o \bullet B_c$$

$$f = \frac{1}{2}[(f - f_b) \circ B \bullet B + (f - f_b) \bullet B \circ B] \qquad (13.5)$$

where, f_o is the original ECG signal, f_b is the detected baseline drift, f is the resultant signal after signal conditioning. B_o and B_c are structuring elements for opening and closing, respectively. B is another structuring element for opening and closing in noise suppression.

13.2 Proposed MMF Algorithm for ECG Signal Conditioning

The proposed MMF algorithm, *i.e.*, the morphological filtering algorithm with modified opening and closing operators, for baseline correction and noise suppression in the conditioning of the ECG signal, is shown in the Fig. 13.1.

ECG signal is conditioned through a sequence of opening and closing operations. Based on the different characteristics of the baseline drift and the noise contamination in the ECG signals, different structuring elements and different morphological operators are used. For baseline correction, an opening operator followed by a closing operator is defined; for noise suppression, modified opening and closing operators are used. They are described in detail in the following subsections.

13.2.1 Baseline Correction

The correction of baseline is performed by removing the drift in background from the original ECG signal. The signal is first opened by a structuring

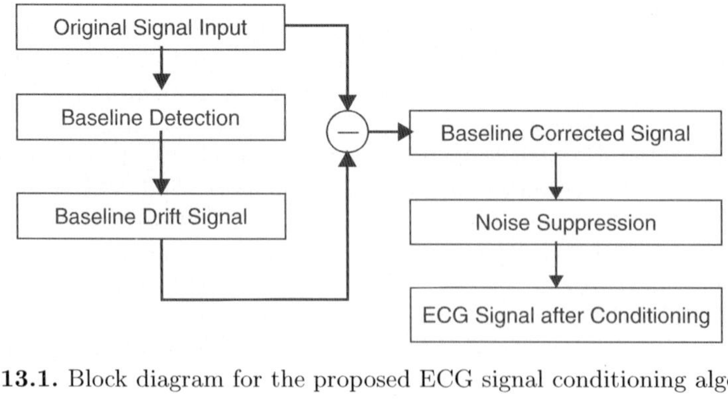

Fig. 13.1. Block diagram for the proposed ECG signal conditioning algorithm

element B_o, for removing peaks in the signal. Then the resultant waveform with pits is removed by a closing operation using the other structuring element B_c. B_o and B_c are selected as two horizontal line segments of zero amplitude, but with different length. The final result is then an estimate of the baseline drift f_b, which is described by Eq. (13.6).

$$f_b = f_o \circ B_o \bullet B_c \tag{13.6}$$

where, f_b is the baseline drift signal, f_o is the original signal. The correction of the baseline is then done by subtracting f_b from the original signal f_o, which is stated in Eq. (13.7).

$$f_{bc} = f_o - f_b \tag{13.7}$$

where f_{bc} is the signal after baseline correction.

The reasons for using different lengths in B_c and B_o are as follows: The baseline drift signal is estimated by removing the ECG signal from the test signal. Hence, the construction of the structuring element for baseline correction depends on the duration (or width) of the characteristic wave and the sample frequency (F_s Hz) of the ECG signal. If the width of a characteristic wave is T_w sec, the number of samples of the wave is $T_w F_s$. In order to extract the characteristic wave, the structuring element B_o should have a length larger than $T_w F_s$. Since the subsequent closing operation is used to remove the pit left by the opening operation, the length of the structuring element B_c must be longer than the length of B_o. The width of the characteristic wave in the ECG signal, such as the P wave, the T wave, and the QRS complex, is generally less than 0.2s. Hence, L_o, the length of B_o, is selected as $0.2F_s$, and L_c, the length of B_c, is typically selected to be longer than B_o, about $1.5L_o$.

13.2.2 Noise Suppression

After baseline correction, noise suppression is performed by processing the data through an opening and a closing operation concurrently, and then the results are averaged. In this study, the opening and closing operations for noise suppression use a structuring element pair, B_{pair}, not a single structuring element as in Chu's MF algorithm shown in Eq. (13.5). B_{pair} is defined as $B_{pair} = \{B_1, B_2\}$, where $B_1 \neq B_2$ i.e., different shape but the same length. The sequence of B_1 and B_2 corresponds to the order of dilation and erosion in the opening and closing operations. The process of signal conditioning is illustrated in the following equation:

$$\begin{aligned} f &= \frac{1}{2}\left(f_{bc} \bullet B_{pair} + f_{bc} \circ B_{pair}\right) \\ &= \frac{1}{2}(f_{bc} \oplus B_1 \ominus B_2 + f_{bc} \ominus B_1 \oplus B_2) \end{aligned} \tag{13.8}$$

where f is the resultant signal after noise suppression and f_{bc} is the signal after baseline correction. The B_{pair} is selected based on the usage and the morphological properties of the ECG signal. B_1 is selected to be a triangular shape, and B_2 is a line segment. A triangular structuring element is used to retain the peaks and valleys of the characteristic waves, such as the QRS complex, the P wave and T wave, while a short line segment structuring element is used for removing noise in the ECG signal. In order to minimize the distortion to the ECG signal, the length of B_1 is selected to be same as that of B_2. Since the processing is discrete, the selection of the structuring element length is related to the bandwidth of ECG signal, and the sampling rate. Given that the sampling frequency is fixed, a shorter structuring element can be used to reduce the distortion of the waveform. Based on the sample frequency F_s, the lengths of B_1 and B_2 are selected as 5 sample units each in this chapter, with values $B_1 = (0, 1, 5, 1, 0)$, $B_2 = (0, 0, 0, 0, 0)$.

Using the proposed structuring element pair, noise can be suppressed while minimizing the smoothing of the significant peaks and valleys in the ECG signal, which are essential for the subsequent reliable detection of the characteristic waves.

13.2.3 Filtering Performance Evaluation

The performance of any filter is judged by the level of noise reduction it achieves and on the signal distortion it causes. In this study, three parameters are used for algorithm evaluation, namely, the *baseline distortion ratio* (BDR), the *noise distortion ratio* (NDR), and the *signal distortion ratio* (SDR). They are defined as follows.

$$\begin{aligned} BDR &= \frac{\|b_o - b\|}{\|b\|} \\ NDR &= \frac{\|n_0 - n\|}{\|n\|} \\ SDR &= \frac{\|d_0 - d\|}{\|d\|} \end{aligned} \tag{13.9}$$

where b is the baseline component in the original input signal, b_o is the detected baseline drift using one of the filtering algorithms, n is the noise component in the original input signal, n_o is the detected noise, d is the clean signal component in the input signal and d_o is the filtered signal. BDR is defined for measuring the degree of baseline not being corrected. NDR is defined for measuring the degree of noise not being suppressed, and SDR is defined for measuring the degree of signal being distorted after the conditioning.

13.3 Experimental Results and Discussion

In order to evaluate the proposed algorithm, a number of experiments were performed using simulated data and known MIT-BIH arrhythmia database. Experimental results for noise and baseline drift suppression using the proposed mathematical morphology operators are presented and discussed in this section. In addition, they are compared with those obtained using the WF method.

13.3.1 Algorithm Testing Using Simulated Data

Experiments using simulated data were performed so that the proposed algorithms could be tested in an absolutely controlled environment. The performance of the algorithms could be evaluated by starting with a known signal, corrupting it by adding noise and baseline drift, performing signal conditioning operations, obtaining the recovered signals and comparing the recovered signal with the known signal. Then, *BDR* is used for evaluating the performance of baseline correction by the MMF algorithm and the WF algorithm. *NDR* and *SDR* are computed for evaluating the performance of noise suppression by the MMF algorithm and the MF algorithm.

A noisy ECG signal can be modelled as:

$$S(n) = I(n) + N(n) + B(n) \tag{13.10}$$

where $S(n)$ is the corrupted ECG signal (discrete time series), $I(n)$ is the clean ECG signal generated from the simulator, $N(n)$ is the noise component and $B(n)$ is the baseline drift.

In this study, clean ECG signal is generated from a *PROPAQ encore* ECG simulator manufactured by Welch Allyn Protocol System Inc. Company.

Noise is modelled by a mixture of Gaussian noise according to [16], which has a probability distribution function of:

$$N(n) = (1 - \varepsilon)G_1\left(\frac{n}{\sigma_1}\right) + \varepsilon G_2\left(\frac{n}{\sigma_2}\right) \tag{13.11}$$

where G_1 and G_2 are the probability distribution functions of Gaussian random variable. σ_1 and σ_2 are standard deviations of G_1 and G_2. σ_2 is typically much larger than σ_1. As σ_1 and σ_2 increase, the noise amplitude increases. ε is a weight, which controls the distribution of G_1 and G_2, the background noise and the impulsive noise.

Baseline drift is simulated by adding a slanted line to a sinusoidal signal [16]:

$$B(n) = B + m \times n + A \times \cos\left(2\pi\frac{n}{N} + \phi\right) \tag{13.12}$$

The period of the sinusoid N controls the severity of the baseline roll. m controls the slope of the baseline drift, which can be represented as

$m = \tan(\theta), \theta$ is the angle of slope. A controls the amplitude of upward or downward drift. Using different values for ϕ allows different baseline drift sequences to be generated with similar characteristics. The bias term B is set so that the sequence values do not get out of range.

Data Set 1

As shown in Fig. 13.2, the top four plots are the generated test data, which include the following:

- clean signal: normal ECG time series generated from the simulator
- noise: $\varepsilon = 0.2, \sigma_1 = 0.1$ and $\sigma_2 = 1$
- baseline drift: $m = 0.01$, $B = -0.6\,mV$, and $A = 0.2\,mV$
- corrupted signal: clean signal corrupted with baseline drift and noise

Data Set 2

As shown in Fig. 13.3, the test data set includes the following:

- clean signal: ECG time series with abnormal beats generated from the simulator
- ε mixture noise: $\varepsilon = 0.1, \sigma_1 = 0.15$ and $\sigma_2 = 1.8$
- baseline drift: $m = 0.02, B = 0.0\,mV$, and $A = 0.8\,mV$
- corrupted signal: clean signal corrupted with baseline drift and noise

Based on the test results of data sets 1 and 2 shown in Fig. 13.2 and Fig. 13.3, the values of *BDR* for the WF algorithm and the MMF algorithm, the *NDR* and *SDR* for the MF and MMF algorithms, are computed and listed in the Tables 13.1–13.3 for comparison.

BDR is used to evaluate the performance of baseline correction by the WF algorithm and the MMF algorithm. The smaller the value of *BDR*, the better is the performance of baseline correction. The computational burden for the WF algorithm is $O(N \log_2 N)$ while for the MMF algorithm is $O(MN)$, where N is the length of the test signal and M is the length of structuring element. Usually, M is far shorter than N. For baseline correction, M is larger than $\log_2 N$, hence the computational burden of the WF algorithm is less than that of the MMF algorithm. For noise suppression, M is much smaller than $\log_2 N$, hence the computational burden of the WF algorithm is larger than that of the MMF algorithm. Because of this the WF algorithm is not compared with the MMF algorithm in the following test of noise suppression and the problem of setting empirical thresholds.

The value of *NDR* is computed for evaluating the performance of noise suppression by the two morphological algorithms. The smaller the value of *NDR*, the better is the performance of noise suppression. *SDR* is computed to measure the degree of the clean signal being contaminated after signal

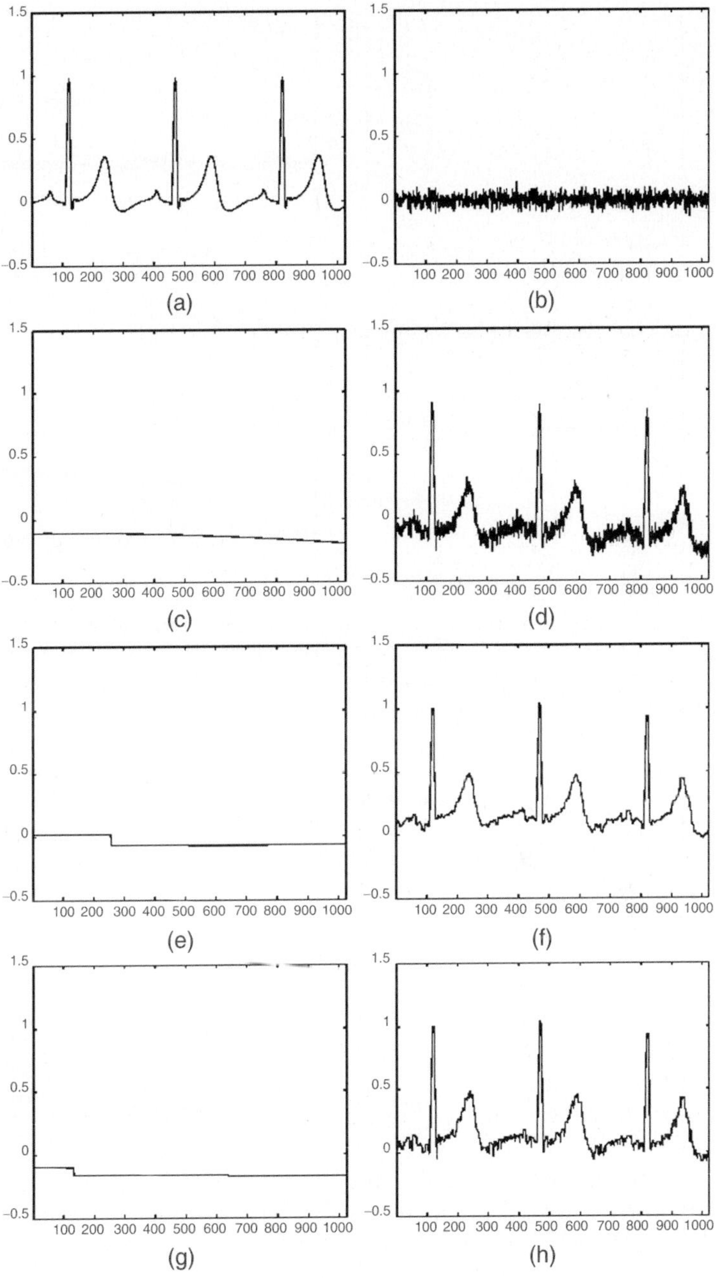

Fig. 13.2. Results of data set 1: (**a**) simulated clean ECG signal (**b**) simulated noise signal (**c**) simulated baseline drift signal (**d**) simulated contaminated ECG signal (**e**) detected baseline drift by the WF algorithm (**f**) denoised signal by the MF algorithm (**g**) detected baseline drift by the MMF algorithm (**h**) denoised signal by the MMF algorithm

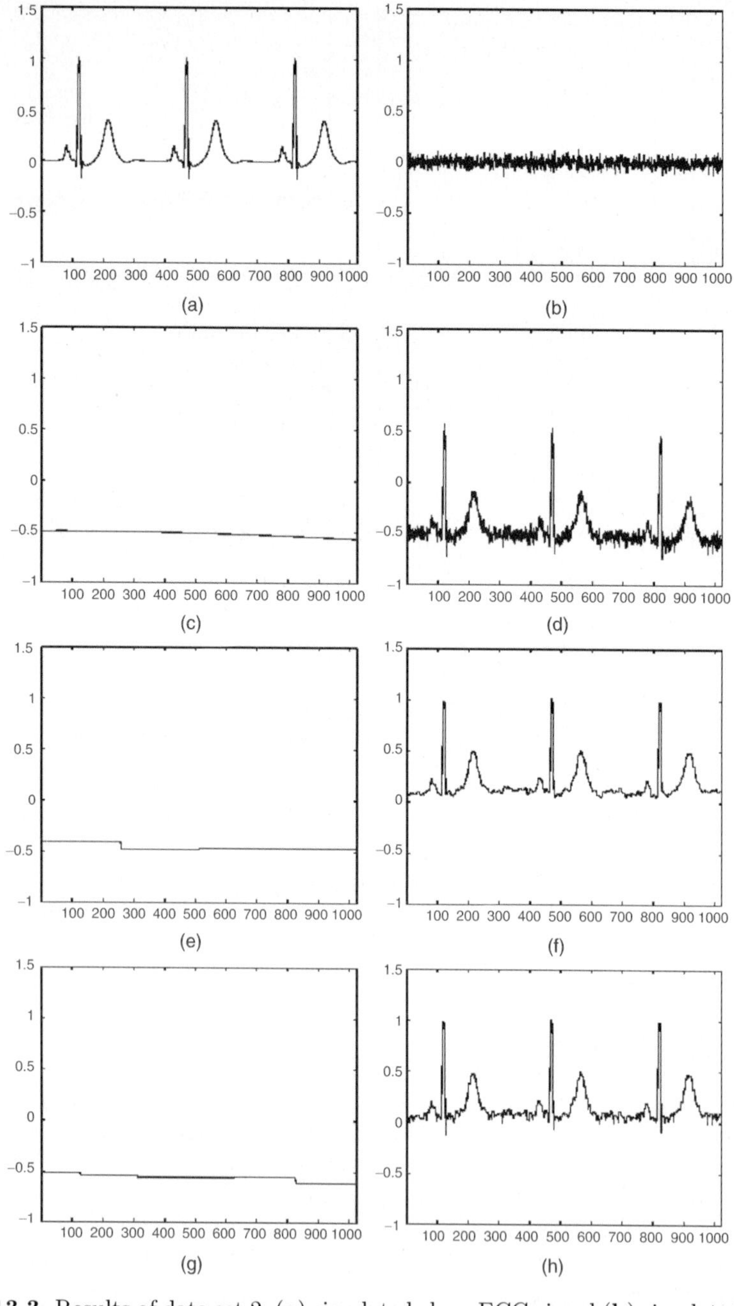

Fig. 13.3. Results of data set 2: (**a**) simulated clean ECG signal (**b**) simulated noise signal (**c**) simulated baseline drift signal (**d**) simulated contaminated ECG signal (**e**) detected baseline drift by the WF algorithm (**f**) denoised signal by the MF algorithm (**g**) detected baseline drift by the MMF algorithm (**h**) denoised signal by the MMF algorithm

Table 13.1. *BDR* for performance evaluation of baseline correction

Data set	*BDR* values for baseline correction	
	WF	MMF/MF
Data set 1	0.1630	0.0164
Data set 2	0.1140	0.0101

Table 13.2. *NDR* for performance evaluation of noise suppression

Data set	*NDR* values for noise suppression	
	WF	MMF/MF
Data set 1	0.2030	0.2144
Data set 2	0.2036	0.2097

Table 13.3. *SDR* for performance evaluation of signal conditioning

Data set	*SDR* value before conditioning	*SDR* values after conditioning	
		MF	MMF
Data set 1	1.9562	0.1132	0.0612
Data set 2	1.8441	0.1065	0.0584

conditioning. The smaller the value of *SDR*, the better is the performance of noise suppression.

In the simulated signal described by Eq. (13.10), two parts are included, viz, the baseline drift described by Eq. (13.12) and the noise described by Eq. (13.11). Simulations were conducted to evaluate the performance of all these algorithms. For baseline correction, fixing $\theta = 0$, the *BDR* values for the WF algorithm and the MMF algorithm, as A varies from 0.2 to 1, are plotted in Fig. 13.4(a). Fixing $A = 0$, the *BDR* values varied with the θ, from 15 to 75 are plotted in Fig. 13.4(b).

The results shown in Fig. 13.4 are consistent with the results obtained in Data sets 1 and 2. The *BDR* values for the MMF algorithm are lower than those for the WF algorithm. Hence, better performance of baseline correction can be obtained by the proposed MMF algorithm than by the WF algorithm.

For noise suppression, defining $K = \sigma_2/\sigma_1$, fixing $K = 10$, $\sigma_1 = 0.1$, and allowing ε to vary from 0.1 to 0.5, the *NDR* values obtained by the MF and the MMF algorithms are shown in Fig. 13.5(a). The *SDR* values obtained by the MF and the MMF algorithms are shown in Fig. 13.6(a). Fixing $\varepsilon = 0.2, \sigma_1 = 0.1$ as K varies from 5 to 25, the *NDR* values obtained by the MF and the MMF algorithms are shown in Fig. 13.5(b), the *NDR* values obtained by the MF and the MMF algorithms are shown in Fig. 13.6(b).

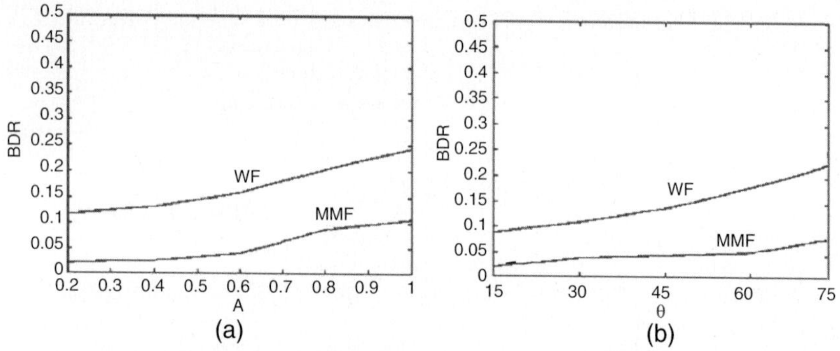

Fig. 13.4. Comparison of *BDR* values for the WF algorithm and the MMF algorithm with the variation of A and θ: (**a**) Fixing $\theta = 0$, A varies from 0.2 to 1 (**b**) Fixing $A = 0$, θ varies from 15 to 75

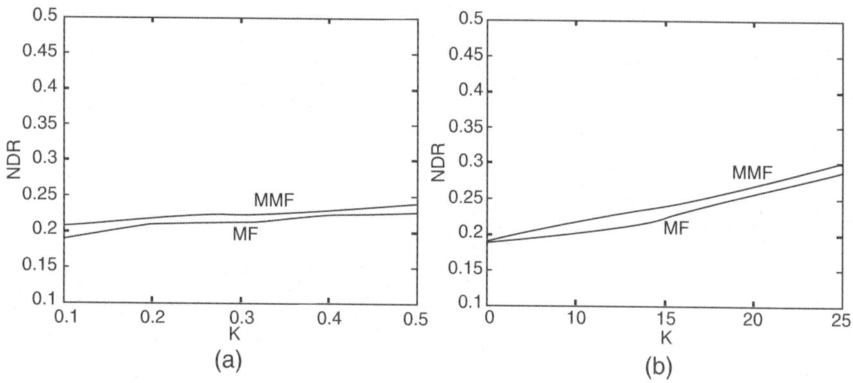

Fig. 13.5. Comparison of *NDR* values for the MF algorithm and the MMF algorithm with the variation of ε and K: (**a**) Fixing $K = 10, \sigma_1 = 0.1, \varepsilon$ varies from 0.1 to 0.5 (**b**) Fixing $\varepsilon = 0.2$, $\sigma_1 = 0.1, K$ varies from 5 to 25

The results shown in Fig. 13.5 and Fig. 13.6 are also consistent with those obtained in the above two cases (data sets 1 and 2). The *NDR* values for the MF algorithm are slightly lower than those for the MMF algorithm. Also, the *SDR* values obtained by the MMF algorithm are lower than those obtained by the MF algorithm. The MMF algorithm achieves lower signal distortion ratio by sacrificing the noise reduction ratio.

The following conclusions can be drawn from the results so far:

- For baseline correction, the MMF algorithm works better than the WF algorithm.
- For noise suppression, the MF is a little better than MMF algorithm. However, the MF algorithm distorts the clean ECG signal more than the

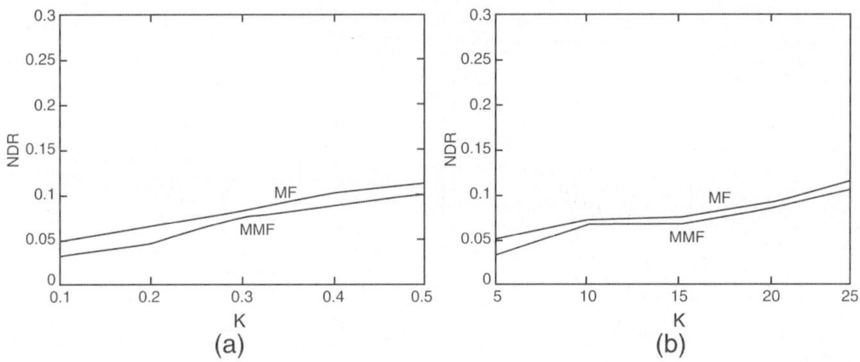

Fig. 13.6. Comparison of *SDR* values for the MF algorithm and the MMF algorithm with the variation of ε and K: (**a**) Fixing $K = 10, \sigma_1 = 0, 1, \varepsilon$ varies from 0.1 to 0.5 (**b**) Fixing $\varepsilon = 0.2, \sigma_1 = 0.1, K$ varies from 5 to 25

MMF algorithm does. The MMF algorithm retains the significant singular points by sacrificing the noise reduction ratio. Since the significant singular points are very important in the subsequent ECG analysis steps, the MMF algorithm is preferred.

13.3.2 Algorithm Testing Using MIT-BIH Arrhythmia Database

Selected ECG data from MIT-BIH arrhythmia database were used to evaluate the performance of the proposed algorithm. Comparative results using MF and MMF algorithms are given. Each set of data was digitized at 360 *Hz* from a single patient using "lead II". Figure 13.7 shows a signal exhibiting bursts of baseline wander (record 31).

The signal conditioning results of record 31 are shown in Fig. 13.7. The plot (a) shows the original ECG signal; (b) shows the signal after baseline correction; (c) is the denoised signal obtained by MF algorithm; and (d) is the denoised signal obtained by the proposed MMF algorithm. Figure 13.8 shows another example of ECG signal conditioning. As shown in Fig. 13.7 and Fig. 13.8, good performance of baseline correction can be observed. The characteristic waves in the ECG signal are shown to be greatly smoothed in (c), the plot of the noise suppression results obtained by the MF algorithm. In (d), the plot of the result obtained by the MMF algorithm, the significant variations in the characteristic waves can be retained well.

Selected ECG time series with over 5000 QRS complexes were tested for detecting the QRS complex using a peak-valley-extractor defined using MM operators proposed in [21]. The correct detection rate (*CDR*) of the QRS complexes is used to evaluate the performance of different conditioning techniques. The values of *CDR* for the original signal without conditioning, the signals conditioned using Chu's MF algorithm and the proposed MMF

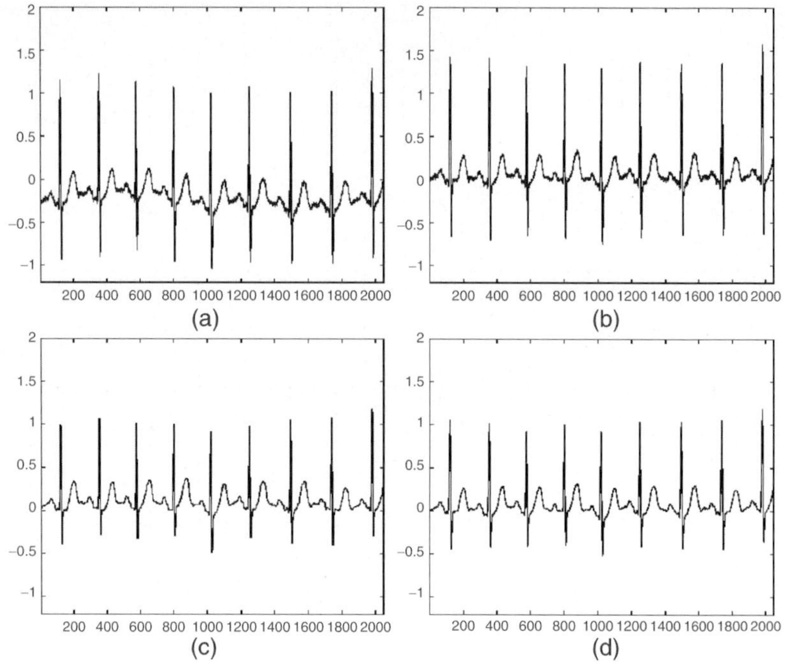

Fig. 13.7. Conditioning results of record 31: (**a**) original ECG signal (**b**) signal after baseline correction (**c**) denoised signal obtained by MF algorithm (**d**) denoised signal obtained by the proposed MMF algorithm

algorithm, were computed respectively and compared. The comparative results are reported in Table 13.4.

Compared with the MF algorithm, better detection performance using the proposed MMF algorithm has been observed as shown in Table 13.4. Therefore, as an ECG conditioning algorithm, MMF is preferable.

13.4 Conclusion

In conclusion, a morphological filtering algorithm using modified morphological operators, called the MMF, is proposed for baseline correction and noise suppression in ECG signals. By using a structuring element pair in closing and opening operations, significant variations in ECG signal can be retained and the corresponding computational burden is lessened. The MMF algorithm can retain the significant variations in the ECG signal, which is more important for subsequent processing, such as the ECG episode characterization, and the characteristic wave detection. The performance of the proposed algorithm was evaluated by using simulated signals and clinically acquired ECG data from a standard set. A comparison of the correct detection rate of

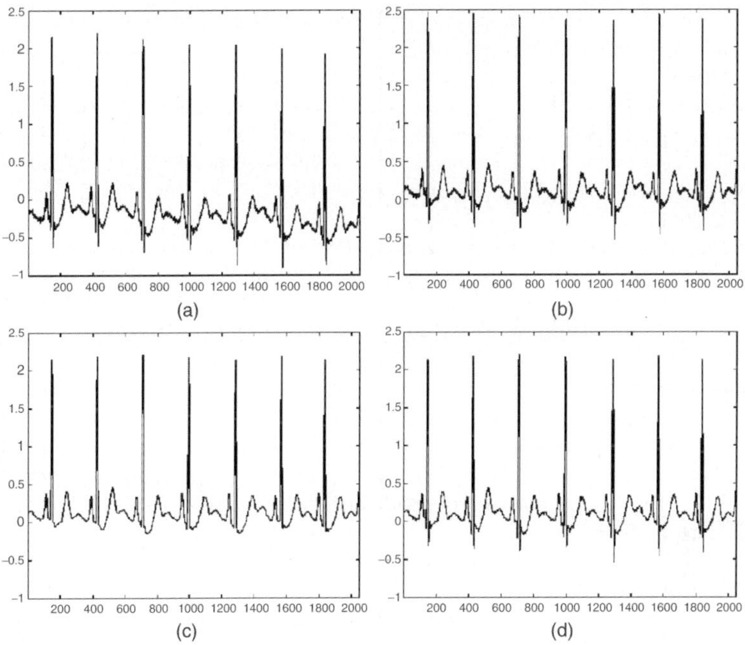

Fig. 13.8. Conditioning results of record 37: (**a**) original ECG signal (**b**) signal after baseline correction (**c**) denoised signal obtained by MMF algorithm (**d**) denoised signal obtained by the proposed MF algorithm

Table 13.4. The comparison of *CDR* for three algorithms

	without conditioning	with conditioning	
		MF	MMF
CDR	96.7%	98.9%	99.4%

the QRS complexes for the original signal, for the signal after conditioning using MF algorithm as well as using the proposed MMF algorithm, shows that the proposed MMF algorithm is more suitable for the conditioning of the ECG signal, after considering the requirement of subsequent processing. Subsequent to filtering and preprocessing, novel features are extracted from cardiovascular signals for arrhythmia recognition.

References

1. Philips, W. (1996): 'Adaptive noise removal from biomedical signals using warped polynomials', *IEEE Trans. Biomed. Engg.*, **43**, pp. 480–492.
2. Moody, G.B., Muldrow, W.K., and Mark, R.G. (1984): 'A noise stress test for arrhythmia detectors', *Comp. Cardiol.*, pp. 381–384.

3. Ahlstrom, M.L., and Tompkins, W.J. (1985): 'Digital filters for real-time ECG signal processing using microprocessors', *IEEE Trans. Biomed. Engg.*, **32**, pp. 708–713.
4. Pahlm-Sornmo, (1984): 'Software QRS detection in ambulatory monitoring: A review', *Med. Biol. Eng. Comput.*, **22**, pp. 289–297.
5. Christov-Dotsinsky, I.I., Dotsinsky, I.A., and Daskalov, I.K. (1992): 'High pass filtering of ECG signals using QRS elimination', *Med. Biol. Engg. Comput.*, **30**, pp. 253–256.
6. Pei, S., and Tseng, C. (1995): 'Elimination of AC interference in electrocardiogram using {IIR} notch filter with transient suppression', *IEEE Trans. Biomed. Engg.*, **42**, pp. 1128–1132.
7. Ferrara, E., and Widrow, B. (1981): 'The time-sequenced adaptive filter', *IEEE Trans. Circuits & System.*, **28**, pp. 519–523.
8. Thakor, N.V., and Zhu, Y.S. (1991): 'Applications of cardiac filtering to ECG analysis: Noise cancellation and arrhythmia detection', *IEEE Trans. Biomed. Engg.*, **39**(8), pp. 785–794.
9. Strobach, P., Fuchs, K.A., Harer, W. (1994): 'Event-synchronous cancellation of the heart interference in biomedical signals', *IEEE Trans. Biomed. Engg.*, **41** (4), pp. 343–350.
10. Dai, W.W., Yang, Z., Lim, S.L., Mikhailova, O., and Chee, J. (1998): 'Processing and analysis of ECG signal using nonorthogonal wavelet transform', *Proc. 20th Annual International Conference of the IEEE Engg. Med. Biol. Soc.*, **1**, pp. 139–142.
11. Nikolaev, N., Zikolov, Z., Gotechev, A., and Egiazarian, K. (2000): 'Wavelet domain wiener filtering for ECG denoising using improved signal estimate', *Proc. IEEE Internatl Conf. Acoustics, Speech, and Signal Proc.*, **6**, pp. 3578–3581.
12. Chen, C.S., Wu, J.L., and Hung, Y.P. (1999): 'Theoretical aspects of vertically invariant gray-level morphological operators and their application on adaptive signal and image filtering', *IEEE Trans. Sig. Proc.*, **47**(4), pp. 1049–1060.
13. Park, K.R., and Lee, C.N. (1996): 'Scale-space using mathematical morphology', *IEEE Trans. Patt. Anal. Mach. Intell.*, **18**(11), pp. 1121–1126.
14. Maragos, P., Schafer, R.W. (1987): 'Morphological filters: Part i and ii', IEEE Trans. Acoust., Speech, Signal Processing, **35**, pp. 1170–1184.
15. Skolnick, M.M., and Butt, D. (1985): 'Cellular array algorithms for the analysis of EKG signals', *Proc. IEEE Workshop Comp. Architecture Pattern Anal. Image Database Manag.*, pp. 438–443.
16. Chu, C.H., and Delp, E.J. (1989): 'Impulsive noise suppression and background normalization of electromagnetism signals using morphological operators', *IEEE Trans. Biomed. Engg.*, **36**, pp. 262–272.
17. Li, C., Zheng, C., and Tai, C.F. (1995): 'Detection of ECG characteristic points using wavelet transforms', *IEEE Trans. Biomed. Engg.*, **42**, pp. 21–29.
18. Esposito, A., and Drsquoandria, P. (2003): 'An Adaptive Learning Algorithm for ECG Noise and Baseline Drift Removal', *Lecture Notes in Computer Science*, Springer Berlin/Heidelberg, **2859**.
19. Sun, Y., Chan, K., and Krishnan, S.M. (2002): 'ECG signal conditioning by morphological filtering', *Comput Biol Med.*, **32**(6) pp. 465–479.
20. Serra, J. (1982): 'Image analysis and mathematical morphology', 1, *Academic Press.*
21. Trahanias, P.E. (1993): 'An approach to QRS complex detection using mathematical morphology', IEEE Trans. Biomed. Engg., **40**, pp. 262–272.

14

Multivariate Analysis for Cardiovascular and Respiratory Signals

Narayanan Srinivasan and S.M. Krishnan

The previous chapters on analysis focused on individual signals like ECG, ABP and HRV signals. It is important to analyse the relationship between cardiovascular and respiratory signals to understand various abnormalities. The current chapter looks at the interaction between cardiovascular and respiratory signals using multivariate analysis. Since the time of the earliest measurements of arterial blood pressure (ABP) and of the electrical activity of the heart (ECG) it was noticed that signals of cardiovascular origin, though almost periodical, were characterized by slight cycle-by-cycle variations (oscillations) in both amplitude and time duration. With the development of ABP recording techniques, it was noticed that wave amplitude variations had different cyclical patterns not only synchronous with breathing activity but also with longer periods of about 10–20 beat duration, which are frequently referred to as Mayer waves [2]. Respiratory system influence is either considered a source of ABP variability inducing a reflex respiratory sinus arrhythmia [3] or a source of a direct modulation of the sinus node [4]. The contributions to cardiovascular variability may be various and change in different experimental conditions [5].

The analysis of biological signals often requires the comparison of multiple recordings which are differently affected by the same oscillation sources [6, 7]. Parametric spectral analysis [8, 9] permits only the recognition and quantification of the oscillatory components in the single signals [1]; on the contrary, multivariate (MV) parametric identification [10–12], provides further information about the casual interactions among the signals [13, 14] and about the cross-spectral patterns [15]. However, studies have been made mostly to the analysis of single-channel signals. Therefore, in order to investigate the interactions between the cardiopulmonary variable signals, a model of multi-channel series that is able to consider multiple inputs simultaneously is needed.

There is a general class of multivariate dynamic adjustment (MDA) models, which includes monovariate autoregressive (AR), multivariate autoregressive (MAR) models. Kalli *et al.* used a black box method to resolve the casual interactions between the cardiovascular variability signals; multivariate autoregressive

(MAR) modelling [16]. The MAR model describes the composition of the signals from each other via linear relationship. As a black-box model, the MAR model requires no presumptions of the system structure. In the model identification, standard techniques, like the multivariate Levinson algorithm [17] can be used.

Spectral analysis which characterizes the frequency content of the measured signals had been performed on the heart rate (or equivalently of heart period) variability [15], and on other cardiovascular signals (mainly arterial pressure). It has revealed the presence of spectral peaks which carry information about the sympathovagal interaction that governs autonomic cardiovascular control [18, 19].

The PSD analysis of the cardiovascular signals seems capable of contributing to the functional investigation into various pathophysiological states (e.g., hypertension, diabetes, etc) [20] or during patient treatment with drugs [15]. Furthermore, studies carried out on both animals and humans clearly indicate the potential importance of this type of analysis for a quantitative evaluation of the role of the autonomic nervous system in the genesis of these rhythms as clearly visible in the spectra (power and frequency of variability components) [1, 9, 13]. Such an approach will improve knowledge of the complex neural mechanisms which will provide an insight into the function of the autonomic nervous system.

In the comprehension of the mechanisms involved in the regulation of the cardiovascular function, the simultaneous observation of ABP and HR variability spectra reveals (under normal conditions) the presence of the 10-sec rhythm(as well as respiratory rhythm) in both the signals [21–23]. Furthermore, animal experiments have indicated that they vary in power under particular experimental conditions, such as pharmacological neural blockade or cardiac pacing, etc [24].

Parametric multivariate analysis of several cardiovascular variability signals is able not only to extract phase, coherence and gain relationships [25,26], but also to assess the casual relationship between two or more signals [27]. The cross spectral analysis of ABP and HR variability signals and the study of the relevant coherence and phase spectra could importantly contribute in the description of the amount of power interchanged between the signals, of the delays by which the rhythms propagate [3, 28, 29] and also of the role of respiration.

The model developed in this chapter is based on the MAR model which is driven by monovariate and uncorrelated autoregressive random inputs. The MAR model is applied to clinical data. The parameters extracted from the developed model are intended for the power spectral analysis and the spectral parameters found may be sensitive enough to differentiate between normal and pathological conditions, particularly for cardiac patients. Indices have been proposed for the detection of cardiac abnormality.

14.1 Method

The MAR model is a black-box method and thus requires no presumptions of the system structure [27]. The model is able to describe the composition of the signals from one another via linear relationships. A multichannel linear system of m variables is represented as

$$X(k) = -\sum_{i=0}^{p} A(i)X(k-i) + E(k) \tag{14.1}$$

The above equation represents the general MAR model of a stationary discrete time m-variate process where $X(k) \in \mathrm{R^m}$ is the observed (multivariate) measurement signal; and $E(\mathrm{k}) \in \mathrm{R^m}$ is the (multivariate) disturbance signal consisting of m white noise processes with Gaussian probability; k is the sample number; and p is the order of the MAR model. Equation 14.1 may be expressed in the matrix form:

$$\mathrm{A(q)\ X(k) = E(k)} \tag{14.2}$$

where $A(q) \in \mathrm{R^{mxm}}$ and q is the delay operator. A(q) can be represented as

$$A(q) = \mathrm{I_m + A_1q^{-1} + \ldots\ldots\ldots\ldots + A_pq^{-p}} \tag{14.3}$$

as well as in the matrix form:

$$A(q) = \begin{bmatrix} a_{11}(q) & a_{12}(q) & \cdots & \cdots & a_{1m}(q) \\ a_{21}(q) & a_{22}(q) & \cdots & \cdots & a_{2m}(q) \\ \vdots & \vdots & \ddots & & \vdots \\ \vdots & \vdots & & \ddots & \vdots \\ a_{m1}(q) & a_{m2}(q) & \cdots & \cdots & a_{mm}(q) \end{bmatrix} \tag{14.4}$$

where the entries a_{kj} are polynomials in the delay operator $\mathrm{q^{-1}}$. This polynomial describes how old values of the output number j affect the output number k.

Equation 14.1 shows that the MAR model can be considered as a one-step-ahead prediction model, where the present value of the system output X(k) is a linear combination of the p past values of X(k) and the prediction error E(k). Thus, a MAR model describes a system where all the signals involved explain themselves and each other via certain linear transfer functions defined by A(q). The MAR model is very flexible as it enables interactions between all the involved signals in any direction. Many criteria have been proposed as objective functions for selection of the model order, p. An aid in the determination of the order of the multichannel AR model is the multichannel version of the Akaike AIC criterion [27].

14.2 Multichannel Spectral Analysis

The motivation for parametric models is the ability to achieve better PSD estimators based upon the model than calculated by classical spectral estimators. The goal of multichannel spectral analysis of m channels of data is the estimation of the Hermitian PSD matrix. For a bivariate spectral estimation, the Hermitian matrix P is given as

$$\mathrm{P}\ (f) = \begin{bmatrix} P_{xx} & P_{xy} \\ P_{yx} & P_{yy} \end{bmatrix} \tag{14.5}$$

The diagonal elements are the single channel autospectral densities, P_{xx} and the non-diagonal elements are the cross spectral densities, P_{xy} between the two-channels. The complex dimensionless expression

$$\varphi_{\mathrm{xy}}(f) = \frac{P_{xy}(f)}{\sqrt{P_{xx}(f)P_{yy}(f)}} \tag{14.6}$$

is termed the coherence function which can be described in terms of the magnitude squared coherence known as the k^2 function

$$\mathrm{k}^2 = |\phi_{xy}(f)|^2 = \frac{|P_{xy}(f)|^2}{P_{xx}(f)P_{yy}(f)} \tag{14.7}$$

The magnitude of k^2 function lies between 0 and 1 and the function may be used to measure, as a function of frequency, the similarity between a pair of signals.

The classical FFT-based methods for estimating the coherence function suffer from an inherent bias towards an over-estimation of the k^2 function. The problem is more pronounced if the averaging process involved in these methods is ignored. For signals of short duration, it is only possible to have limited number of segments for averaging. Therefore, a bias in the k^2 function is inevitable, which results in an over estimation of the degree of coherence between the two channels. As a result, the classical methods for multi-channel spectral estimation are not capable of providing an efficient tool for this purpose. In contrast, parametric methods are known to be capable of estimating the coherence function without introducing bias into the resultant k^2 function.

Indices for abnormality detection

New indices are proposed which are intended for the diagnosis of abnormal state in cardiac patients. Early detection of signals of abnormality is crucial in the assessment of a cardiac patient's condition, as it may indicate the onset of catastrophic physiologic events such as respiratory failure or even leading to sudden cardiac death. Therefore, in such critical cases, a fast and accurate analysis of the simultaneously recorded signals is required in order for prompt action to be taken to save life. The indices proposed here are effective in discriminating between normal and abnormal signals.

These indices are obtained after spectral and cross-spectral analysis by obtaining the fraction of power of the ECG, which is coherent (or not coherent) with respiration (or ABP signal). The total power of the ECG which is coherent with respiration is obtained by multiplying the autospectrum of the ECG by the k^2 function and integrating on the frequency axis. This first index is termed as the ρ.

$$\rho = \Sigma \left(\mathrm{P}_{11}{}^{*} \frac{|P_{12}(f)|^2}{P_{11}(f)P_{22}(f)} \right) \tag{14.8}$$

where P_{11} represents the autospectrum of the ECG, P_{22} represents the autospectrum of the respiratory signal and P_{12} is the cross spectrum between the two signals. $\bar{\rho}$ is the power not coherent with respiratory signal obtained by subtracting ρ from the autospectrum of the ECG.

$$\bar{\rho} = P_{11} - \rho \tag{14.9}$$

Finally, the percentage of the ρ with respect to the total power of the ECG is tabulated and defined as the Respiration Coherent Index (RCI).

$$\mathrm{RCI}\,(\%) = (\rho/P_{11})^{*}100 \tag{14.10}$$

Similarly, the bivariate spectral analysis is also applied to the ECG-ABP pair. The procedures stated above are repeated to obtain the total power of the ECG which is coherent with ABP. This index is termed as the α

$$\alpha = \Sigma \left(P_{11}{}^{*} \frac{|P_{13}(f)|^2}{P_{11}(f)P_{33}(f)} \right) \tag{14.11}$$

where P_{11} represents the autospectrum of the ECG, P_{33} represents the autospectrum of the ABP and P_{12} is the cross spectrum between the two signals. Then, $\bar{\alpha}$ will be the power not coherent with ABP and is obtained as

$$\bar{\alpha} = P_{11} \quad \alpha \tag{14.12}$$

Lastly, the Pressure Coherent Index (PCI) is obtained as

$$\mathrm{PCI}\,(\%) = (\alpha/P_{11})^{*}100 \tag{14.13}$$

14.3 Results and Discussion

The three signals ECG, ABP and respiration, which are obtained from the MIT-BIH database are used as inputs into the proposed MAR model. Cross spectrum analysis is performed on the two selected signals via bivariate spectral estimation. The various indices are obtained after spectral and cross spectral analysis of the signals.

14.3.1 ECG and ABP Signal Analysis

Normal Signals: Three consecutive peaks (equivalent to two R-R intervals) of the ECG signal and the corresponding interval in the ABP signal is considered for analysis. Figure 14.1 shows the graph of the coherence (k^2) function. It can be seen that the relatively high values of the coherence function at higher frequencies indicate a strong correlation between the two signals at higher frequencies. However, at low frequencies, the coherence between the two signals is less. In other words, there is greater similarity between the two signals at higher frequencies than at lower frequencies. The total ECG power coherent with ABP is shown in Fig. 14.2. The total power of the ECG is highly coherent to the ABP signal as indicated in Fig. 14.2, ρ and $\overline{\rho}$ values of 1800 and 48 are obtained with a PCI index value of 97.3%.

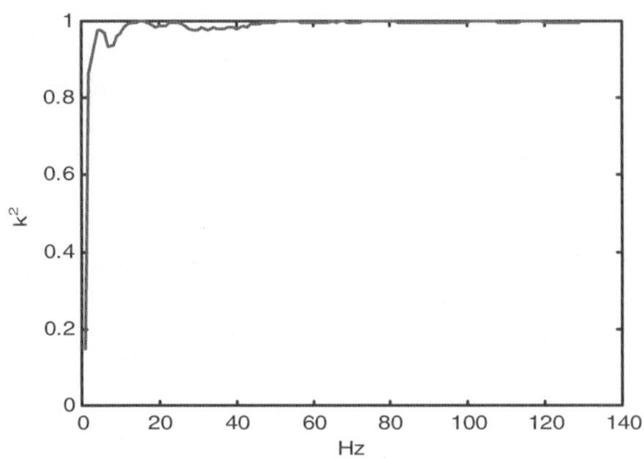

Fig. 14.1. Coherence between normal ECG and ABP signals

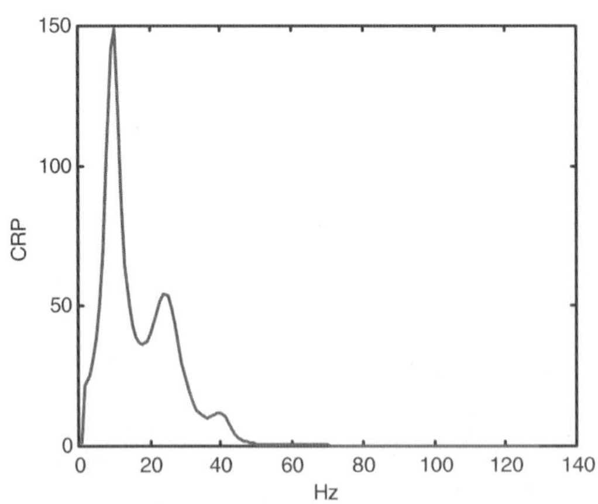

Fig. 14.2. Total power of normal ECG coherent with ABP

Abnormal Signals: Similarly a three peaks interval of the abnormal ECG signal and the corresponding ABP signal is extracted for the power spectrum analysis. Figure 14.3 shows the coherence function after performing cross spectrum analysis between the two signals. Total ECG power coherent with ABP under abnormal condition is shown in Fig. 14.4. The total power of the ECG which is coherent with ABP is much lesser as indicated by the drop in amplitude in the ρ graph.

In the normal signals, the total power coherent with ABP is very much higher as indicated by the PCI index (97.3%). However, in the abnormal signals the total power coherent with ABP is much lower, at only 46.9%. Therefore, it can be concluded that the percentage of the autospectrum of the ECG which is coherent with ABP decreases as the signals go into abnormality

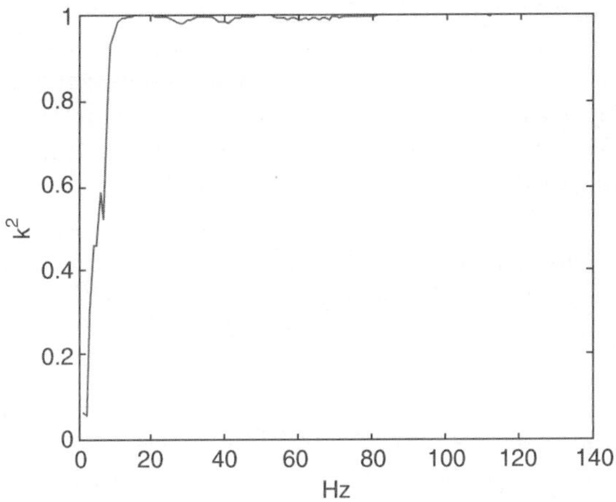

Fig. 14.3. Coherence between abnormal ECG and ABP signals

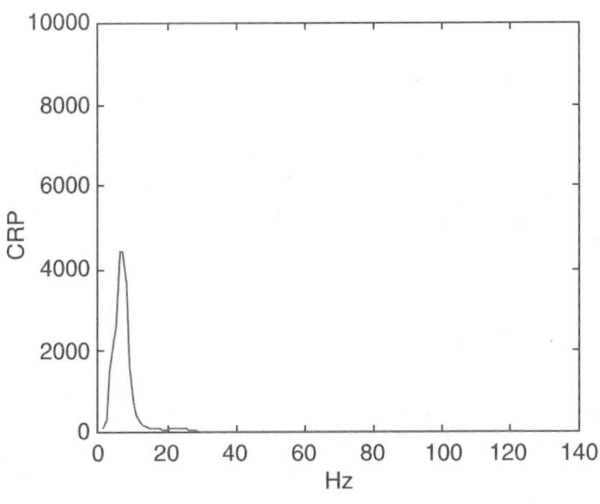

Fig. 14.4. Coherence between abnormal ECG and ABP signals

and increases during normal state. Hence the PCI index can also constitute a quantitative mean for discriminating between normal and abnormal signals in stable and unstable condition of the patient.

The above analysis is then performed on a range of data which is obtained from a cardiac patient whose initial condition was stable but became unstable at the end due to cardiac problem resulting in respiratory failure. The range of PCI index for the signals under stable condition is between 90% and 97%.

The line graph is also obtained for the PCI index for transition period from stable to unstable condition. There is a drop in the PCI index from above 90% for normal stable signals to below 80% when the signals go into abnormality. When the patient is in total unstable condition, the range of PCI index is smaller than that of the RCI index for abnormal signals.

14.3.2 ECG and Respiratory Signal Analysis

Three consecutive peaks (equivalent to two R-R intervals) of the ECG signal and the corresponding interval in the respiratory signal is considered for analysis each time. In this case also, the graph of the k^2 function shows there is greater similarity between the two signals at higher frequencies than at lower frequencies.

Analysis similar to the one done previously for ABP signals has been made and similar results have been obtained. In the normal signals, the total power coherent with respiration is very much higher as indicated by the RCI index (96%). However, in the abnormal signals the total power coherent with respiration is very low, at only 35%. Therefore, it can be concluded that the percentage of the autospectrum of the ECG which is coherent with respiration decreases as the signals go into abnormality and increases during normal state. Hence the RCI index can constitute a quantitative measure for discriminating between normal and abnormal signals.

The above analysis is then performed on a range of data from a cardiac patient whose initial condition was stable but became unstable towards the end. The range of RCI index for the signals under stable condition is between 90% and 97%. In the transition period, there is a drop in the RCI index from above 90% for normal stable signals to below 80% when the signals go into abnormality. When the condition of the patient further deteriorates, the RCI index falls from a high of 90% to a low of 2%.

14.4 Conclusion

ECG, blood pressure and respiratory signals can provide important information on the pathophysiology of the cardiovascular regulatory mechanisms. Spectral and cross-spectral analysis of these signals gives quantitative information which can be of potential interest in clinical studies. This has been done in this chapter using a methodology based upon multivariate autoregressive identification and parametric power spectral density estimation. Autospectra and coherence, which completely characterize the physiological relations between the signals in terms of exchanging powers and statistically consistent phase relations, has been used to obtain useful indices. These indices, called respiration coherence index and ABP coherence index, have been found very useful in differentiating normal signals from abnormal ones and they act as a fast and convenient tool for the purpose. The indices proposed have the advantage of being automatically calculated on the PC in an ICU setup. Results on a few test cases show that the multivariate autoregressive model developed is able to differentiate accurately the condition of a patient in an ICU changing from stable to unstable condition.

It is known that various systems involved in the generation of ECG, ABP and respiration are interrelated and hence one can expect a good coherence between these signals under normal conditions. The results obtained here indicate that this coherence is somehow reduced when the condition of body becomes abnormal. It will be interest to study the mechanisms leading to such a decrease in coherence.

References

1. Akselrod, S., Gordon, D., Ubel, F.A., Shannon, D.C., Baeger, A.C., and Cohen, R.J. (1981): 'Power spectrum analysis of heart rate fluctuation: a quantitative probe of beat-to-beat cardiovascular control', *Science*, **213**, pp. 220–222.
2. Penaz, J. (1978): 'Mayer waves: History and methodology', *Automedica*, **2**, pp. 135–141.
3. De Boer, R.W., Karemaker, J.M., and Strackee, J. (1985): 'Relationships between short-term blood pressure fluctuations and heart-rate variability in resting subjects', *Med. Biol. Eng. Comput.*, **23**, pp. 352–364.
4. Akselrod, S., Gordon, D., Madwed, J.B., Snidam, N.C., Shannon, D.C., and Cohen, R.J. (1985): 'Hemodynamic regulation: investigation by spectral analysis', *Am. J. Physiol.*, **249**, pp. 867–875.
5. Koepchen, H.P. (1984): 'History of studies and concepts of blood pressure waves', In Miyakawa, K., Polosa, C., and Koepchen, H.P., (Eds.): *Mechanisms of blood pressure waves*, Japan Sc. Soc. Press, Tokyo, Springer-Verlag, Berlin, pp. 3–23.
6. Madwed, J.B., Sands, K.E.F., Saul, J.P., and Cohen, R.J. (1986): 'Spectral analysis of beat-to-beat variability in heart rate and arterial blood pressure during hemorrhage and aortic constriction', *In Neural Mechanisms and Cardiovascular Disease*, Berlin: Springer, pp. 291–302.
7. Sayers, B.M. (1973): 'Analysis of heart rate variability', *Ergonomics*, **16**, pp. 17–32.
8. Hyndman, B.W., Kitney, R.I., and Sayers, B. (1974): 'Spontaneous rhythms in physiological control systems', *Nature*, **233**, pp. 339–341.
9. Kitney, R.I., and Rompelman, O. (1980): 'The study of heart rate variability', *Oxford: Clarendon.*

10. Bartoli, F., Baselli, G., and Cerutti, S. (1982): 'Application of identification and linear filtering algorithms to R-R interval measurement', *Proc. IEEE, Comp, Card. Conf.*, Seattle, WA.
11. Bartoli, F., Baselli, G., and Cerutti, S. (1985): 'AR identification and spectral estimate applied to R-R interval measurement', Integrative J. Bio. Med. Comput., 16, pp. 201–215.
12. Baselli, G., and Cerutti, S. (1985): 'Identification techniques applied to processing of signals from cardiovascular systems', *Med. Inf.*, **10**, pp. 223–235.
13. Pagani, M., Lombardi, F., Guzzetti, S., Rimoldi, O., Furlan, R., Pizzinelli, P., Sandrone, G., Dell'Orto, S., Picalunga, E., Turiel, M., Baselli, G., Cerutti, S., and Malliani, A. (1986): 'Power spectral analysis of heart rate and arterial pressure variability as a marker of sympatho-vagal interactions in man and conscious dog', *Cir. Res.*, **59**, pp. 178–193.
14. Baselli, G., Cerutti, S., Civardi, S., Lombardi, F., Malliani, A., Merri, M., Pagani, M., and Rizzo, G. (1987): 'Heart rate variability signal processing: a quantitative approach as an aid to diagnosis in cardiovascular pathologies', *Int. J. Biomed. Comp.*, **20**, pp. 51–70.
15. Pomeranz, B., Macaulay, R.J.B., Caudill, M.A., Kutz, I., Adam, D., Gordon, D., Kilborn, K.M., Barger, A.C., Shannon, D.C., Cohen, R.J., and Benson, H. (1985): 'Assessment of autonomic function in humans by heart rate analysis', *Amer. J. Physiol.*, **248**, pp. H151–153.
16. Kalli, S., Suoranta, R., and Jokipii, M. (1986): 'Applying a multivariate autoregressive model to describe interactions between blood pressure and heart rate', In Measurement in clinical medicine, pp. 77–82.
17. Marple, S.L. (1987): 'Digital spectral analysis with applications', Prentice-Hall Inc., Englewood Cliffs, New Jersey, (USA).
18. Pagani, M., Lombardi, F., Guzzetti, S., Sandrone, G., Rimoldi, O., Malfatto, G., Cerutti, S., and Malliani, A. (1984): 'Power spectral density of heart rate variability as an index of sympatho-vagal interaction in normal and hypertensive subjects', *J. Hypertens.*, **2**, pp. 383–385.
19. Lombardi, F., Sandrone, G., Perpuner, S., Sala, R., Garimoli, M., Cerutti, S., Baselli, G., Pagani, M., and Malliani, A. (1987): 'Heart rate variability as an index of sympatho-vagal interaction after acute myocardial infarction', *Am. J. Cardiol.*, **60**, pp. 1239–1245.
20. Lombardi F., et al. (1985): 'Heart rate variability in the first year after myocardial infarction', Proc. Diagnosis of myocardial ischemia in man, Pisa.
21. Pagani, M., Furlan, R., Dell'Orto, S., Pizzinelli, P., Lanzi, G., Baselli, G., Santoli, C., Cerutti, S., Lombardi, F., and Malliani, A. (1986): 'Continuous recording of direct high fidelity arterial pressure and ECG in ambulatory patients', *Cardiovasc. Res.*, **XX**, pp. 384–388.
22. Kitney, R.I., and Rompelman, O. (1982): 'The analysis of heart rate variability and blood pressure fluctuation', Int. Workshop, Delft.
23. Kitney, R.I., Fultron, T., McDonald, A.H., and Linkens, D.A. (1985): 'Transient interactions between blood pressure, respiration and heart rate in man', J. Biomed. Eng., **7**, pp. 217–224.
24. Akselrod, S., Gordon, Madwed, D.J.B., Snidam, N.C., Shannon, D.C., and Cohen, R.J. (1986): 'Beat-to-beat variability in hemodynamic parameters', Am. J. Physiol., **253**.

25. De Boer, R.W., Karemaker, J.M., and Strackee, J. (1987): 'Hemodynamic fluctuations and baroflex sensitivity in humans: a beat-to-beat model', Am.J. Physiol., **253**, pp. H680-H689.
26. Robbe, H., Mulder, L., Ruddel, H., Langewitz, W., Veldman, J., and Mulder, G., 'Assessment of baroceptor reflex sensivity by means of spectral analysis', *Hypertension*, **10**, pp. 538–543.
27. Baselli, G., Cerutti, S., Livarghi, M., Meneghin, C.I., Pagani, M., and Rimoldi, O. (1988): 'Casual relationship between heart rate and arterial blood pressure variability signal', *Med. Biol. Eng. Comp.*, **26**, pp. 374–378.
28. Zwiener, U. (1978): 'Physiological interpretation of autospectra. Coherence and phase spectra of blood pressure, heart rate and respiration waves in man', *Automedica*, **2**, pp. 161–169.
29. Bianchi, A., Bontempi, B., Cerrutti, S., Gianoglio, P., Comi, G., and Natali Sora, M.G. (1990): 'Spectral Analysis of heart rate variability signal and respiration in diabetic subjects', *Med. Biol. Eng. Comput.*, **28**, pp. 205–211.

15

Phase Space Analysis for Cardiovascular Signals

S.M. Krishnan, D. Narayana Dutt, Y.W. Chan and V. Anantharaman

The previous chapters focused on the detection of a prevalent cardiac arrhythmia, namely, PVC. The current chapter focuses on a nonlinear analysis based technique that can be applied to detect abnormalities in various biomedical signals including cardiovascular and respiratory signals. The main focus of the chapter is on the analysis of cardiovascular signals. However, the proposed analysis can be applied to other biomedical signals as well.

The cardiovascular system is a complex system that includes the heart and blood vessels and signals like electrocardiogram (ECG), arterial blood pressure (ABP) and heart rate variability (HRV) are known to reflect the changes in the cardiovascular system. Hence, many attempts have been made to analyze these signals and extract information about the cardiovascular system. Most of the methods used are linear and it has been recognized that nonlinear methods may be more suitable for analyzing signals that originate from very complex nonlinear living systems [1]. Recent developments in nonlinear dynamics have provided various methods for the study of the complex cardiovascular system [2]. It is now generally recognized that many processes generated by the biological system can be described in an effective way by using the methods of nonlinear dynamics. The nonlinear dynamical techniques are based on the concept of chaos, which was first introduced with applications to complicated dynamical systems in meteorology [43]. Since then they have been applied to many areas including the area of medicine and biology [44,45]. A particularly active area for the application of chaos theory has been cardiology [42,46,47], where many aspects have been addressed including whether chaos can be used to represent healthy or diseased state [48].

It is well known that the electrical activity of the heart, as manifested in ECG, is a good diagnostic tool in detecting abnormal condition of the heart. The concept of cardiac rhythm as a periodic oscillator is now being changed to that of the output of a chaotic system [49]. It has been demonstrated that externally stimulated cardiac tissue develops bifurcation patterns like period doubling that are characteristic of nonlinear dynamic system [50,51]. It has also been hypothesized that strictly periodic cardiac dynamics are accompanied by pathological states rather than healthy condition [52]. The theory of chaos has been used to detect some cardiac arrhythmia such as ventricular fibrillation [53]. These developments have led to the idea that these methods might be more effective in reflecting the normal as well as pathological functioning of the heart and this in turn has stimulated increasing interest in the application of nonlinear dynamics to the analysis of ECG signals [8].

It has been observed that the complex neural mechanisms controlling the cardiovascular system generate different rhythms in heart rate. It has also been recognized that some nonlinear processes are involved in the dynamics of these signals. Effort has been made in determining nonlinear dynamical parameters, like correlation dimension, of ECG and HRV signals and it has been shown that they are very useful indicators of pathologies [4,5]. Methods based on chaos theory have been applied in tracking HRV signals and predicting the onset of events such as Ventricular Tachycardia and detecting congestive heart failure situations [3]. Nonlinear dynamical analysis of heart rate and respiratory signals has been made so as to improve the understanding of the underlying physiological processes of the autonomic nervous system [55].

With the recognition of significant relationship between the autonomic nervous system and cardiovascular mortality, efforts for development of autonomic activity have led to the use of HRV as one of the most promising markers. The association of high risk of post-infarction mortality with reduced HRV was first shown by Wolf et. al., [56] and later Akselrod introduced power spectral analysis of heart rate fluctuations to evaluate beat-to-beat cardiovascular control [57]. HRV gained more clinical importance when it was confirmed that HRV is a strong and independent predictor of mortality after an acute myocardial infarction. Studies have been made regarding the chaoticity of HRV signals [58] and nonlinear dynamical analysis of HRV has shown promise in studying the status of human cardiac transplant recipients [59].

Another signal of importance is the blood pressure signal. It is known that blood flow is brought about by pressure differences between the various vascular regions and the resistance to flow depends on various factors like vascular architecture, blood viscosity and neuronal control. Hence, one can expect to get a reliable evaluation of the cardiovascular control by studying the ABP signal. It is a sensitive marker of a variety of autonomic disorders [60]. While some work on the application of the nonlinear dynamical techniques to ECG and HRV signals are reported in literature, not much attention seem to have been given to the application of the technique to ABP signals. Only recently some applications to ABP signals have been reported in literature [61,62] and it has been shown that it provides additional insight [63].

The description of a non-linear dynamic system begins with the reconstruction of its phase space trajectory [64]. The phase space of a dynamical system is a mathematical space with orthogonal coordinate directions representing each of the vectors needed to specify the instantaneous state of the system. Usually takens method of delays is used to construct an attractor of dynamical system in a multidimensional state space from only the knowledge of a one-dimensional time sequence that describes the system behavior [65].

The objective of the present work is to develop a novel method of multi-dimensional phase space technique using a weighted spatial filling index for the analysis of cardiovascular signals. The new approach is based on the fact that the distribution of points in phase space depends on the type (normal or abnormal) of signal and it is quantified by calculating a weighted index in multi-dimensional space for the assessment of cardiac dysfunction.

15.1 Method

Let the signal be represented by the coordinates of a point $\mathbf{X}(\mathrm{k})$ in phase space. Then the dynamical behavior of the signal is reconstructed by succession of these points $\mathbf{X}(\mathrm{k})$ in the phase space. Phase space reconstructions are based on the analysis of dynamic systems by delay maps. The vectors $\mathbf{X}(\mathrm{k})$ in the multidimensional phase space are constructed by time delayed values of the time series, which determine the coordinates of the phase space plot.

$$\mathbf{X}(\mathrm{k}) = \{\mathrm{x}(\mathrm{k}),\ \mathrm{x}(\mathrm{k}+\tau), \mathrm{x}(\mathrm{k}+\mathrm{i}\tau), \ldots\ \mathrm{x}\ (\mathrm{k}+(\mathrm{E}-1)\tau)\} \quad \mathrm{k} = 1, 2, \ldots\ldots (\mathrm{N}-\mathrm{E}) \tag{15.1}$$

where $\mathbf{X}(\mathrm{k})$ is one point of the trajectory in the phase space at time k, $\mathrm{x}(\mathrm{k}+\mathrm{i}\tau)$ are the coordinates in the phase space corresponding to the time delayed values of the time series, τ is the time delay between the points of the time series considered and E is the embedding dimension, which is the number of coordinates of the phase space plot. The attributes of the reconstructed phase space plot depend on the choice of value of τ. One way to choose τ is to take it as the time it takes the autocorrelation function of the data to decay to 1/e [55]. Another method is to take the first minimum in the graph of average mutual information [66], which appears to be better since it considers the nonlinear structure in the signal. It has been established using this method that the value of 7 for τ is the best choice for ECG signals and 5 for HRV signals [4].

From the given signal x(1), x(2) ... x(n), a matrix A is obtained as

$$\begin{bmatrix} x(1) & x(1+\tau) & \ldots & x(1+(E-1)\tau) \\ x(2) & x(2+\tau) & \ldots & x(2+(E-1)\tau) \\ . & . & . & \\ & & & . \\ . & . & . & \\ & & & . \\ . & . & . & \\ x(m) & x(m+\tau) & \ldots & x(n) \end{bmatrix} \tag{15.2}$$

where E is the number of dimensions and m is related to n by the equation:

$$m = n - (E - 1)\tau \tag{15.3}$$

By plotting column 2 of matrix A against column 1 (for the case E = 2), the phase space plot for two dimensions is obtained.

$$\begin{bmatrix} x(1) & x(1+\tau) \\ x(2) & x(2+\tau) \\ \cdot & \cdot \\ x(m) & x(n) \end{bmatrix} \tag{15.4}$$

Similarly, the first three columns of matrix A represent a phase space plot in three dimensions. Now, a normalized matrix B is obtained by dividing each element of A by q where

$$q = \max|x(k)| \quad 1 \le k \le n \tag{15.5}$$

The matrix B(in two dimensions) is hence represented as

$$\begin{bmatrix} x(1)/q & x(1+\tau)/q \\ x(2)/q & x(2+\tau)/q \\ \cdot & \cdot \\ x(m)/q & x(n)/q \end{bmatrix} \tag{15.6}$$

In two dimensions, the phase space plot corresponding to the normalized matrix spans from -1 to $+1$ on either axis. The phase space area is now divided into small square areas of size R $\times$ R, where R is chosen such that $2/R$ is an integer. Then the number of grids in the normalized phase space is $N = 2/R$. A matrix C is now obtained with its elements $c(i, j)$ equal to the number of phase space points falling in a grid $g(i, j)$. The matrix C is called the phase space matrix and its elements are divided by M, where

$$M = \sum_{i,j=1}^{N} c(i, j) \tag{15.7}$$

This division yields $P(i, j)$, the probability of a phase space point falling in a grid $g(i, j)$. A matrix R is now formed by squaring each element of P to get $r(i, j)$ as the elements of R. The sum of elements of matrix R is calculated as

$$S = \sum_{i,j=1}^{N} r(i, j) \tag{15.8}$$

The spatial filling index η is defined as:

$$\eta = S/N^2 \tag{15.9}$$

Now the value of η is used to quantify the degree of variability in the test signals.

15.2 Extension to Higher Dimension

For higher dimensions, the equations have to be modified to accommodate spatial coordinates of higher dimensions. For example, in the third dimension the Eqs. (15.7), (15.8) and (15.9) are modified as:

$$M = \sum_{i,j,k=1}^{N} c(i,j,k) \tag{15.10}$$

$$S = \sum_{i,j,k=1}^{N} r(i,j,k) \tag{15.11}$$

$$\eta = S/\mathrm{N}^3 \tag{15.12}$$

The modification in the equations is attributed to the change in the area to cubes in the third dimension (hypercubes in higher dimensions) and the grid $g(i,j)$ becomes $g(i,j,k)$. Subsequently, the spatial coordinates get extended with increasing dimensions. The filling index η is a measure of non-zero elements in the matrix derived from the phase portraits. The parameter η shows considerable differences in value for normal and abnormal signals and is therefore of interest.

15.3 Weighted Spatial Filling Index

The manner in which the phase space plot of a particular data occupies the phase space depends on the condition under which the data is obtained. If the data is obtained from a normal subject, then the phase space plot of this normal data is expected to be distributed in phase space in a particular fashion. The distribution is different for an abnormal data obtained from a cardiac patient. Since the spatial filling index η quantifies the distribution of points in phase space, it can be effectively used to distinguish abnormal signals from normal ones. A suitable range is established by studying the phase plots of various normal and abnormal signals. This range isolates a domain of the phase space that encompasses phase points corresponding to the abnormalities of the time signal. By giving suitable weights to these phase space points, the abnormality of the signal can be magnified and reflected in the spatial filling index. This weighted spatial filling index has been used to get better detection of abnormality in patients with cardiac abnormalities. Computation of this weighted index in higher dimensions is found to improve identification of cardiac dysfunction.

The phase space region is divided into small hypercubes and the number $\mathrm{c(i,j,k)}$ of phase space points falling in each hypercube are weighted by a factor which has different distribution functions in the phase space region for different signals. The normalized sum of weighted $[\mathrm{c(i,j,k)}]^2$ is used as a weighted spatial filling index in higher dimension to discriminate between normal and abnormal signals. Once the range of the normal signal is established, uniform weightage of the entire phase space does not serve to enhance the difference in index values between the normal and abnormal signals. By giving weights, this disparity becomes explicit. For example, by giving a high weight to signal values outside the normal range and a lower weight to values in the normal range, the index value for an abnormal signal would be higher. Along the same lines, by giving a lower weight to signal values outside the normal range and a higher weight to values in the normal range, the index value is expected to be higher for the normal signals. Either approach is acceptable for analysis.

15.4 Results and Discussion

15.4.1 ECG Signals

The plot of electrical activity of the heart is obtained in multidimensional phase space using Takens method of delays. Phase space plots of normal and abnormal ECG signals are shown in Fig. 15.1 and Fig. 15.2 respectively. It may be observed that the distribution of phase space points of the abnormal signal is clearly different from that of the normal signal. The range for application of weights is established by a comparative study of phase plots of normal and abnormal signals. The concentration of signal values occurs in different regions of the phase space for different signals.

For the range highlighted in Fig. 15.1 and Fig. 15.2, the normal signal has very few signal points outside the range as opposed to the abnormal signal which has more signal values outside the range. By giving a lower weight outside the range compared to the weight inside the range, the two signals show differences in weighted spatial index which can be used for abnormality detection.

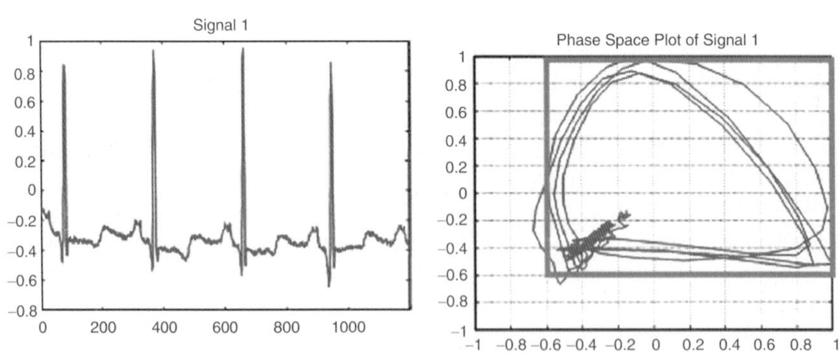

Fig. 15.1. Phase space plot of normal ECG

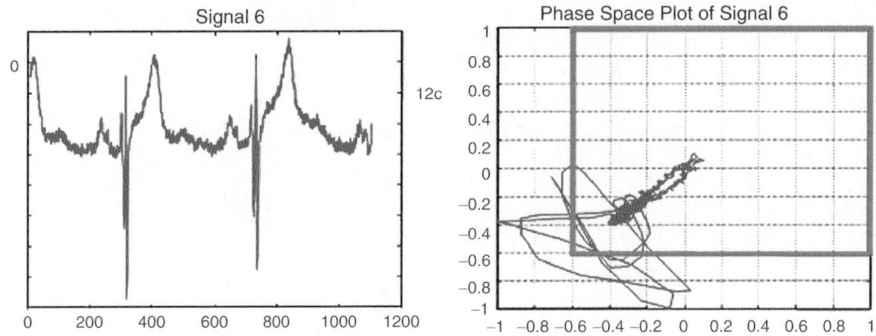

Fig. 15.2. Phase space plot of PVC

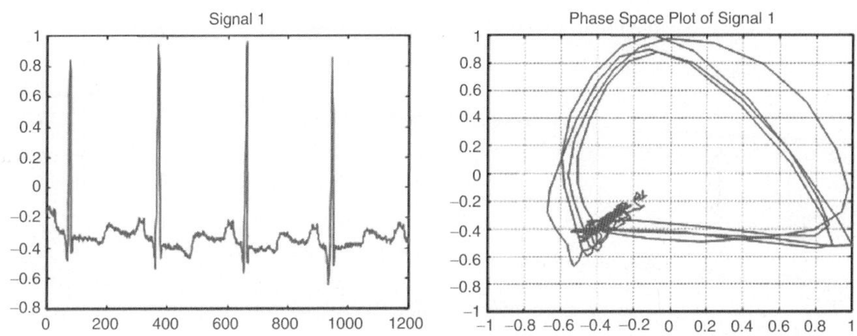

Fig. 15.3(a). Normal ECG signal and the corresponding phase space plot

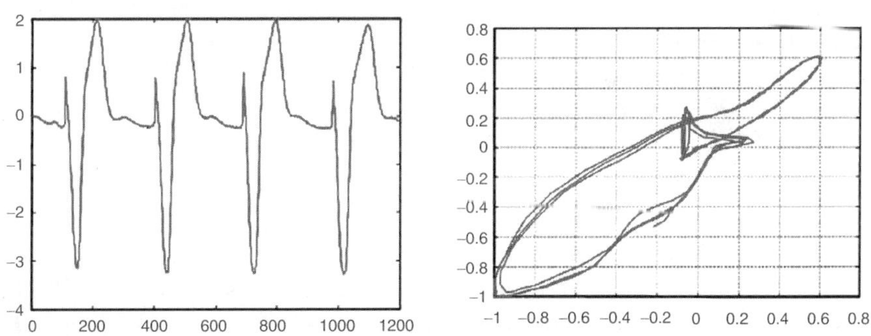

Fig. 15.3(b). Abnormal ECG signal and the corresponding phase space plot

The proposed technique was applied to a number of different signals, both normal and abnormal. Some of the normal and abnormal signals used in the analysis, along with their two dimensional phase space plots are shown in the Figs. 15.3–15.5. Signal 1 shown in Fig. 15.3 is a normal signal while Signals B and C shown in Fig. 15.4 and Fig. 15.5 are abnormal signals. The unweighted spatial filling index for these signals is shown in Table 15.1.

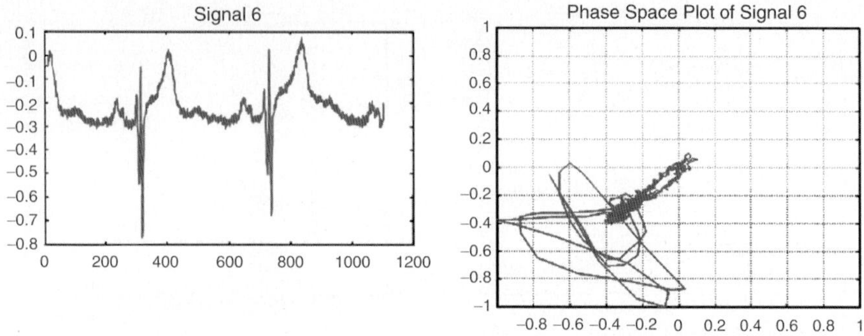

Fig. 15.4. Abnormal ECG Signal and its Phase Space Plot

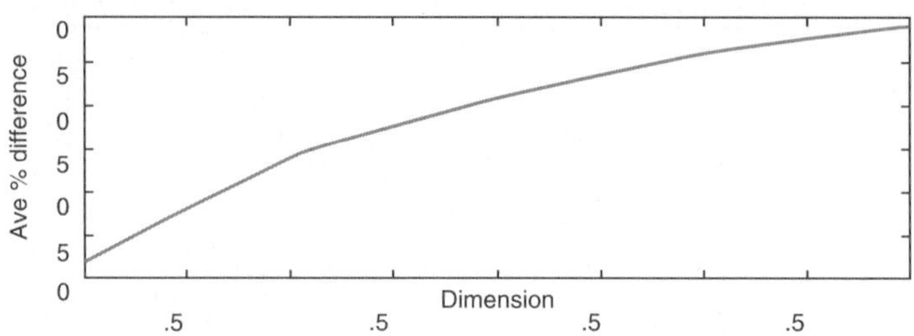

Fig. 15.5. Variation in Percentage Difference of weighted η with Dimension

Table 15.1. Weighted and unweighted spatial filling index for ECG signals

η	Signal 1	Signal B	Signal C
Unweighted	0.4702	0.2083	0.1601
Weighted	46.750	10.1329	7.9228

The method of applying different weights to the signal in different ranges gave promising results that can be used to assess the abnormalities. Giving a higher weight for the signal in the normal range and lower weight outside the range helps in differentiating abnormal signals from normal ones. This is evident from Table 15.1 that shows that the weighted spatial filling index for an abnormal signal is much lower than that for a normal signal. The percentage change is more for weighted η than unweighted η showing that the weighted index is a better measure of abnormality than unweighted one.

The extension of the spatial filling index into higher dimensions was explored with consistent results. The spatial filling index was calculated for various higher dimensions and the index was observed to have different values for abnormal and normal signals. As the dimension increases, the trend in the change in spatial filling index was different for normal and abnormal signals. The percentage change in spatial filling index for abnormal signals was found to be higher than that for normal signals on approaching higher dimensions.

Table 15.2. Trend of weighted η for normal ECG signals

Dimension	Signal 1	Signal 2	Signal 3
2	49.0590	43.2541	41.9745
3	43.1829	38.3366	37.8043
4	38.4410	33.7945	34.3735
5	33.9529	29.6211	31.3529
6	30.0019	25.8763	28.3439

Table 15.3. Trend of weighted η for abnormal ECG signals

Dimension	Signal 4	Signal 5	Signal 6
2	17.7258	11.4759	35.4476
3	12.1316	4.4173	25.5638
4	9.1967	1.7769	20.0569
5	7.1645	0.7525	15.3428
6	5.3257	0.3792	12.2710

Tables 15.2 and 15.3 show the trend of weighted η for normal and abnormal signals for various dimensions and it may be observed that a better distinction between normal and abnormal signals can be obtained with increase in dimension.

The average percentage change in weighted spatial index from the second to the sixth dimension is calculated with reference to the weighted η in the second dimension. This average change was found to be 38% for normal signals and 72% for abnormal signals. The difference in the weighted spatial filling index between normal and abnormal signals increases for higher dimensions of the phase space as shown in Fig. 15.5.

Data obtained from a patient in the ICU was used to test the technique developed. Figure 15.6 shows the recording of the ECG signal. It can be seen that the patient's condition has deteriorated with time, which is expected to be reflected in the phase space plots and the value of the spatial filling index. The normal and the abnormal recordings have different phase space distributions and hence give different values of the spatial filling index. It can be seen from Fig. 15.7 that there is a drop in the value of the weighted spatial filling index that reflects the deterioration in the patient's condition.

15.4.2 HRV and ABP Signals

Besides the ECG signal, other signals like the heart rate variability and blood pressure signals were also analyzed using the same technique. The HRV signals required for the analysis were derived from the ECG signal. The peaks in an ECG are determined by setting a threshold for the signal strength above which signal values are compared consecutively to determine the highest signal

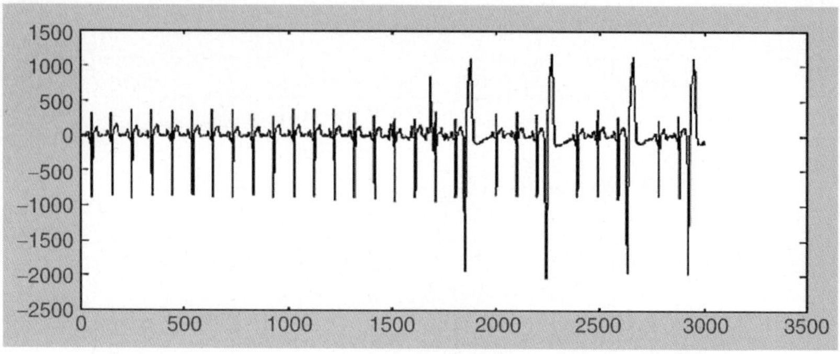

Fig. 15.6. Recording of ECG Signal of a Patient in ICU

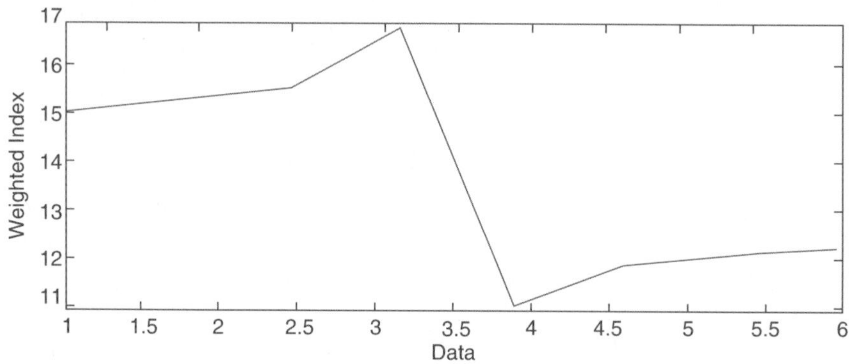

Fig. 15.7. Variation of weighted Spatial Index for the above signal

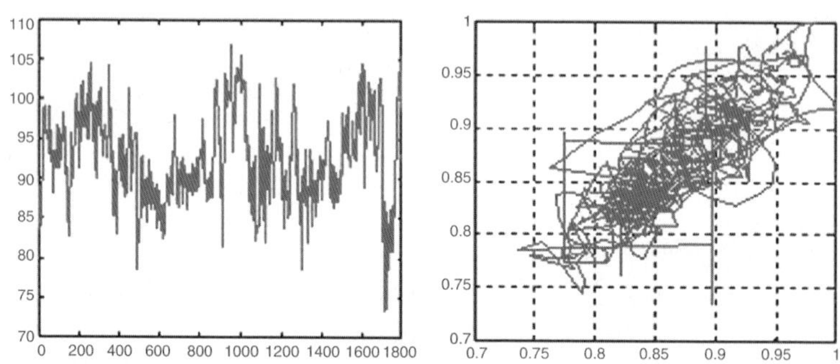

Fig. 15.8. Abnormal HRV Signal and its Phase Space Plot

value. Once the peaks are established, their indices are used for calculation of variability. Using a value of $\tau = 5$, the algorithm was applied on derived HRV signals with similar results. The derived HRV signals and the corresponding two dimensional phase space plots are shown in Fig. 15.8 and Fig. 15.9.

The effect of increase in dimensions is favourable on the value of the weighted η parameter. With increase in dimensions, the difference in η between

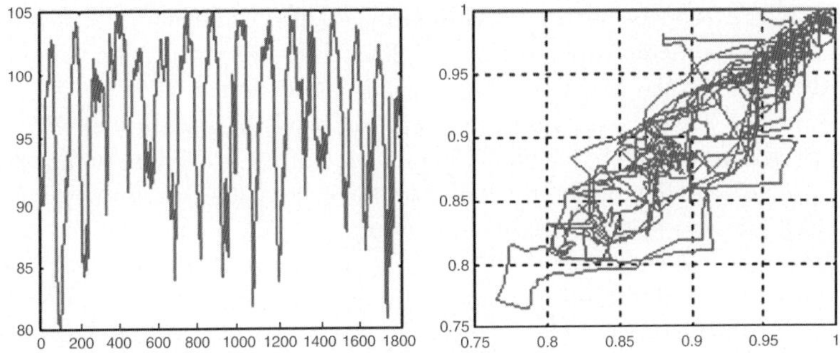

Fig. 15.9. Abnormal HRV Signal and its Phase Space Plot

Table 15.4. Difference in weighted η between normal and abnormal HRV signals

Dimension	Normal	Abnormal	% change
2	20.18	23.58	17%
3	19.00	23.28	23%
4	17.92	22.98	28%
5	16.88	22.70	34%
6	15.86	22.41	40%

Table 15.5. Difference in weighted η between normal and abnormal ABP signals

Dimension	Normal	Abnormal	Change(%)
2	172.4844	224.4463	30%
3	94.9698	135.9043	43%
4	60.5892	88.5284	46%
5	39.1325	58.9844	51%
6	25.5556	41.5321	63%

normal and abnormal signals increases, thus making it easy to detect abnormality in the signal. Table 15.4 shows the variation of the difference in weighted spatial filling index between normal and abnormal HRV signals, with increase in dimensions. It can be noticed that the percentage change is more for higher dimensions and hence better results may be expected with increase in dimension.

These techniques were also explored on arterial blood pressure signals (ABP). The effects of increase in dimension are shown in Table 15.5. It can be seen from the table that the difference in spatial index between normal and abnormal signals increases with increase in dimension.

15.4.3 Graphical Representation in Higher Dimensions

A method to represent the information of the third dimension in a 2D plot is achieved through the use of colour as a dimension. Each colour pixel signifies a value, which reflects the signal value in the third dimension. This representation has the same features as the 3D plot, in the sense that the values of the three dimensions are featured. The use of colour as a dimension introduces novelty in the sense that abnormality is easily visible to the naked eye. For example, the number of red pixels in the plot of an abnormal signal may be much more than that in the normal one and hence the distribution of colour pixels can be used in abnormality detection.

It has been shown earlier that the difference between normal and abnormal signals increases with the dimension of analysis. Graphical analysis limits the number of dimensions to three. However, the fourth dimension can be represented by colour as well. For the case of a Premature Ventricular Contraction (PVC), the proportion of high wavelength pixels to the total pixels is found to be greater than in the normal ECG. Other abnormality specific scans can be implemented to check for major cardiac arrhythmia.

15.5 Conclusion

Considering heart as a nonlinear complex system and processing various cardiovascular signals like ECG, HRV and ABP seems to provide very useful information for detection of abnormalities in the condition of the heart than is possible by conventional means. In this chapter, a new multidimensional phase space analysis of these cardiovascular signals using weighted spatial filling index has been proposed for detecting cardiac dysfunction. The successful analysis of planar vector-cardiographic loops in various planes suggests that phase space technique could be a promising approach. The proposed technique introduces a new quantitative measure, called weighted spatial filling index, and our finding show that this measure is very useful in differentiating abnormal cardiovascular signals from normal ones. It was also found that the percentage change in the weighted index with dimension is larger for abnormal signals than for normal ones, thus showing its use in the detection of abnormality. It should be noted that evaluation of the proposed technique on a larger data set is required for ensuring the efficacy of the technique. It is hoped that the graphical representation along with its corresponding analytical index proposed here will find potential applications in computer analysis of cardiac patients' status in intensive care units.

References

1. Abarbanel, H.I. (1996): 'Analysis of Observed Chaotic Data', *Springer Verlag*, Berlin.
2. Karrakchou, M., Rheymer, K.V., Vesin, J.M., Pruvot, E., and Kunt, M. (1996): 'Improving cardiovascular monitoring through modern techniques', *IEEE Engg. Med. Biol. Mag.*, **15**, pp. 68–78.

3. Cohen, M.E., Hudson, D.L., and Deedwania, P.C. (1996): 'Applying continuous chaotic modeling to cardiac signal analysis', *IEEE Engg. Med. Biol. Mag.*, **15** (5), pp. 97–102.
4. Fojt, O., and Holcik, J. (1998): 'Applying non-linear dynamics to ECG signal processing', *IEEE Engg. Med. Biol. Mag.*, **17**, pp. 96–101.
5. Khadra, L., Fahoum, A.S., and Nashash, H.A. (1997): 'Detection of life-threatening cardiac arrhythmia using the wavelet transformation', *Med. Biol. Eng. Comput.*, **35**, pp. 626–632.
6. Bezerianos, A., Bountis, T., Papaioannou, G., and Polydoropoulos, P. (1995): 'Non-linear time series analysis of electrocardiograms', *Chaos*, **5**, pp. 95–101.
7. Zhang, X.S., Zhu, Y.S., and Zhang, X.J. (1997): 'New approach to studies on ECG dynamics: extraction and analysis of QRS complex irregularity time series', *Med. Biol. Engg. Comput.*, **35**, pp. 467–473.
8. Ravelli, F., and Antolini, R. (1992): 'Complex dynamics underlying the human electrocardiogram', *Biol. Cybern.*, **67**, pp. 57–65.
9. Jenkins, J.M., and Caswell, S.A. (1996): 'Detection algorithms in implantable cardioverter defibrillators', *Proc. IEEE*, **84**, pp. 428–445.
10. Zheng, X.S., Zhu, Y.S., Thakor, N.V., and Wang, Z.Z. (1999): 'Detecting ventricular tachycardia and fibrillation by complexity measure', *IEEE Trans. Biomed. Engg.*, **46** (5), pp. 548–555.
11. Mit-Bih Arrhythmia Database. (1980): *Available from Beth Israel Hospital*, Biomedical Engineering Division Room.
12. Flandrin, P. (1992): 'Wavelet analysis and synthesis of fractal Brownian motion', *IEEE Trans. Inf. Theory*, **38** (2), pp. 910–917.
13. Mandelbrot, B.B., and Vanness, J.W. (1968): 'Fractal Brownian motions, fractional noises and applications', *SIAM review*, **10** (4), pp. 422–437.
14. Falconer, K.(1990): 'Fractal Geometry', *J. Wiley and Sons*, New York.
15. Mallat, S. (1992): 'Characterization of signals from multiscale edges', *IEEE Trans. Patt. Anal. Mach. Intelli.*, **14**, pp. 1019–1033.
16. Mallat, S. (1991): 'Zero-crossing of a wavelet transform', *IEEE Trans. Inf. Theory*, **37**, pp. 1019–1033.
17. Hamilton, P.S., and Tompkins, W.J. (1996): 'Quantitative investigation of QRS detection rules using the MIT-BIH arrhythmia database', *IEEE Trans. Biomed. Engg.*, **33**, pp. 1157–1187.
18. Coast, A.D., Stern, R.M., Casio, G.G., and Britter, S.A. (1990): 'An approach to Cardiac arrhythmia analysis using hidden Markov models', *IEEE Trans. Biomed. Engg.*, **37**, pp. 826–835.
19. Gritzali, F., Frangakis, G., and Papakonstantinous, G. (1989): 'Detection of the P and T waves in ECG', *Comp. Biomed. Res.*, **22**, pp. 83–91.
20. Xie, Q.Z., Hu, Y.H., and Tompkins, W.J. (1992): 'Neural network based adaptive matched filtering of QRS detection', *IEEE Trans. Biomed. Engg.*, 39, pp. 317–329.
21. Emilio, E.S., and Sober, M.M. (1998): 'Application of adaptive signal processing for determining the limits of P and T waves in an ECG', *IEEE Trans. Biomed. Engg.*, **45** (8), pp. 1077–1080.
22. Sahambi, J.S., Tandon, S.N., and Bhatt, R.K. (1997): 'Using wavelet transform for ECG characterization', *IEEE Engg. Med. Biol. Mag.*, **16**, pp. 77–83.
23. Bangham, J.A., and Marshall, S. (1998): 'Image and signal processing with mathematical morphology', *Electronics & Communication Engineering Journal*, **10** (3), pp. 117–128.

24. Schmid, P. (1999): 'Segmentation of digitized dermatoscopic images by two-dimensional color clustering', *IEEE Trans. Medical Imaging,* **18** (2), pp. 164–171.
25. Bloch, I. (1999): 'Fuzzy relative position between objects in image processing: A morphological approach', *IEEE Trans. Patt. Mach. Intell.*, **21** (7), pp. 657–664.
26. Pesaresi, M., and Benediktsson, J.A. (2001): 'A new approach for the morphological segmentation of high-resolution satellite imagery', *IEEE Trans. Geosci. Remote Sensing*, **39** (2), pp. 309–320.
27. Mallat, S., and Zhong, S. (1992): 'Characterization of signals from multiscale edges', *IEEE Trans. Patt. Anal. Mach. Intell.*, **14** (7), pp. 710–732.
28. Hibbard, L.S., and McKeel, D.W. (1995): 'Multiscale detection and analysis of the senile plaques of alzheimer's disease', *IEEE Trans. Biomed. Engg.*, **42** (12), pp. 1218–1225.
29. Setarehdan, S.K., and Soraghan, J.J. (1999): 'Automatic cardiac {LV} boundary detection and tracking using hybrid fuzzy temporal and fuzzy multiscale edge detection', *IEEE Trans. Biomed. Engg.*, **46** (11), pp. 1364–1378.
30. Bangham, J.A., Chardaire, P., Pye, C.J., and Ling, P.D. (1996): 'Multiscale non-linear decomposition: the sieve decomposition theorem', *IEEE Trans. Patt. Anal. Mach. Intelli.*, **18** (5), pp. 529–539.
31. Brockett, R.W., and Maragos, P. (1994): 'Evolution equations for continuous-scale morphological filtering', *IEEE Trans. Sig. Proc.*, **42** (11), pp. 3377–3386.
32. Chen, M.H., and Yan, P.F. (1989): 'A multiscaling approach based on morphological filtering', *IEEE Trans. Patt. Anal. Mach. Intelli.*, **11** (7), pp. 694–700.
33. Haralick, R.M., Sternberg, S.R., and Zhuang, X. (1987): 'Image analysis using mathematical morphology: Part I', *IEEE Trans. Patt. Anal. Mach. Intelli.*, **9**, pp. 532–550.
34. Babaud, J., Baudin, M., Witkin, A., and Duda, R. (1986): 'Uniqueness of the gaussian kernel for scale-space filtering', *IEEE Trans. Patt. Anal. Mach. Intell.*, **8**, pp. 26–33.
35. Marr, D., and Hildreth, E. (1980): 'Theory of edge detection', *Proc. Royal Soc. London B*, **207**, pp. 187–217.
36. Froment, J., and Mallat, S. (1992): 'Wavelets – A tutorial in theory and applications', *Second generation compact image coding with wavelets*, Academic, NY, pp. 655–678.
37. Witkin, A.P. (1984): 'Scale-space filtering: A new approach to multiscale description', *Image understanding*, Ablex, Norwood, N.J., pp. 79–95.
38. Witkin, A.P. (1983): 'Scale-space filtering', *Proc. Int'l Joint Conf. Artificial Intelligence*, Kaufmann, Palo Alto, Calif, Augest, pp. 1019–1022.
39. Wilson, R., and Granlund, G.H. (1984): 'The uncertainty principle in image processing', *IEEE Trans. Patt. Anal. Mach. Intelli.*, **6**, pp. 758–767.
40. Jackway, P.T., and Deriche, M. (1996): 'Scale space properties of the multiscale morphological dilation-erosion', *IEEE Trans. Patt. Anal. Mach. Intelli.*, **18** (1), pp. 38–51.
41. Boomgaard, R.V., and Smeulders, A. (1993): 'Morphological multiscale image analysis', *Proc. First Int'l Workshop on Math. Morphology and Its Applications to Signal Processing*, Univ. Politecnica de Catalunya, Barcelona, Spain.
42. Depree, J.D., and Swartz, C.W. (1988): 'Introduction to Real Analysis', *John Wiley*, New York.
43. May, R.M. (1976): 'Simple mathematical models with very complicated dynamics', *Nature*, **261** (4), pp. 459–467.

44. Tsonis, P.A., and Tsonis, A.A. (1989): ‘Chaos: Principles and implications in biology’, *Computer Applications in Biosciences*, **5**, pp. 27–32.
45. Eberhart, R.C. (1989): ‘Chaos theory for biomedical engineer’, *IEEE Engg. Med. Biol. Mag.*, **8**, pp. 41–45.
46. Glass, L., and Zeng, W.Z. (1990): ‘Complex bifurcations and chaos in simple theoretical models of cardiac oscillations’, *Annals NY Academy of Sc.*, **591**, pp. 316–327.
47. Denton, T.A., and Diamond, G.A. (1990): ‘Fascincating rhythms: a primer on chaos theory and its applications to cardiology’, *Am. Heart Jour.*, **120**, pp. 1419–1440.
48. Lipsitz, L.A., and Goldberger, A.L. (1992): ‘Loss of complexity and aging’, *Jour. Am. Med. Assoc.*, 267, pp. 1806–1809.
49. Babloyantz, P., and Destexhe, A. (1988): ‘Is the normal heart a periodic oscillator?’, *Biol. Cybern.*, **58**, pp. 203–211.
50. Chialvo, D.R., and Jalife, J. (1987): ‘Nonlinear dynamics of cardiac excitation and impulse propagation’, *Nature*, **330**, pp. 749–752.
51. Chialvo, D.R., Gilmour, R.F., and Jalife, J. (1990): ‘Low dimensional chaos in cardiac tissue’, *Nature*, **343**, pp. 653–657.
52. Pool, R. (1989): ‘Is it healthy to be chaotic?’, *Science*, **243**, pp. 604–607.
53. Kaplan, D.K., and Cohen, J.R. (1991): ‘Searching for chaos in fibrillation’, *Annals NY Acad. Sci.*, pp. 367–374.
54. Skinner, J.E., Pratt, C.M., and Vybiral, T. (1993): ‘A reduction of correlation dimension of heart beat intervals precedes imminent ventricular fibrillation in human subjects’, *Am. Heart Jour.*, 125, pp. 731–743.
55. Hoyer, D., Schmidt, K., Bauer, R., Zweiner, U., Kohler, M., Luthke, B., and Eiselt, M. (1997): ‘Nonlinear analysis of heart rate and respiratory dynamics’, *IEEE Engg. Med. Biol. Mag.*, **16**, pp. 31–39.
56. Wolf, M.M., Varigos, G.A., Hunt, D., and Soloman, J.G. (1978): ‘Sinus arrhythmia in acute myocardial infarction’, *Medical Jour. Aust.*, **2**, pp. 52–53.
57. Akselrod, S., Gordon, D., Ubel, F.A., Shannon, D.C., Baeger, A.C., and Cohen, R.J. (1981): ‘Power spectrum analysis of heart rate fluctuation: a quantitative probe of beat-to-beat cardiovascular control’, *Science*, **213**, pp. 220–222.
58. Goldberger, A.L. (1991): ‘Is normal heart beat chaotic or homostatic?’, *News Physiol. Sci.*, **6**, pp. 87–91.
59. Khadra, L.M., Maayah, T.J., and Dickhaus, H. (1997): ‘Detecting chaos in HRV signals in human cardiac transplant recipients’, *Comp. Biomed. Res.*, **30**, pp. 188–199.
60. Adams, M.A., Bobik, A., and Korner, P.I. (1989): ‘Differential development of vascular and cardiac hypertension in genetic hypertension. Relation to sympathetic function’, *Hypertension*, **14**, pp. 191–202.
61. Wagner, C.D., and Persson, P.B. (1996): ‘Nonlinear chaotic dynamics of arterial blood pressure and renal flow’, *Amer. Jour. Physiol.*, **37**, pp. H621–H627.
62. Zweiner, U., Hoyer, D., Bauer, R., Luthke, B., Walter, B., Schmidt, K., Hallmeyer, S., Kratzsch, B., and Eiselt, M. (1996): ‘Deterministic-chaotic and periodic properties of heart rate and arterial pressure fluctuations and their mediation in piglets’, *Cardiovasc. Res.*, **31**, pp. 455–465.
63. Almog, Y., Oz, O., and Akselrod, S. (1999): ‘Correlation dimension estimation: can this nonlinear description contribute to the characterization of blood pressure control in rats’, *IEEE Trans. Biomed. Engg.*, **46**, pp. 535–546.

64. Hilborn, R.C. (1994): 'Chaos and nonlinear dynamics', *New York: Oxford University Press.*
65. Takens, F. (1981): 'Detecting Strange Attractors in Turbulence', In Rand D, Young L.S. (Eds.), *Dynamical Systems and Turbulence*, Berlin: Springer.
66. Fraser, A.M., and Swinney, H.L. (1986): 'Independent coordinates for strange attractors from mutual information', *Physical Review A*, **33**, pp. 1134–1140.

16

Linear, Non-Linear and Wavelet Analysis of Cardiac Health Using Heart Rate Signals

U. Rajendra Acharya, N. Kannathal, Lim Choo Min and Jasjit S. Suri

Heart rate variability (HRV) study refers to the characterization and measurement of changes in heart rate. Disease and affliction influence the heart rate, and therefore, the pattern and range of heart rate variability (HRV) contain important information about the robustness of health and type of disease. Classification based on features that capture the spread and pattern of this parameter can provide useful insight about the type and intensity of the affliction.

There are two main approaches for analysis: time domain analysis of HRV for standard deviation of normal to normal intervals (SDNN); and frequency domain analysis for power spectrum density (PSD). By 1991, frequency domain measures, based on the fast Fourier transform or similar mathematical analyses, had been applied to 24-hour Holter recordings, with seemingly even better results [1]. Large datasets of Holter recordings began to be collected, especially by the group at St. George's in London who developed their own geometric measures of HRV that could be applied to Holter recordings without the requirement of painstaking and exacting characterization of all interbeat intervals [2]. By the mid-90's, there were a number of studies verifying the predictive value of HRV in various cardiac patient populations, and the Task Force of the European Society of Cardiology and the North American Society of Pacing and Electrophysiology had issued a set of formal recommendations for the measurement and clinical uses of HRV [3]. The analysis tools derived from time and frequency domains were the main focus of that report, but in recent years new developments flourished in the field of nonlinear dynamics and deterministic chaos [4–12].

In 1982 Kobayashi and Musha have reported on the frequency dependence of the HRV power spectrum a in normal young man [13]. In 1995, CK Peng and colleagues proposed a novel non-linear measure of HRV called the short-term fractal scaling exponent [14]. Unlike many of the other proposed non-linear measures, this new measure was relatively straightforward to compute and not hampered by any mathematical assumptions.

A nonlinear deterministic approach appears to be more appropriate to describe more complex phenomena, showing that apparently erratic behavior can be generated even by a simple deterministic system with nonlinear structure [15, 16].

It is therefore not surprising that a specific subtype of nonlinear dynamics, chaos theory and fractals, has recently been applied to the study of HRV signal. Initially it was assumed that chaotic fluctuations could explain cardiac electrical activity during atrial or ventricular fibrillation. Later, it was proposed that the fluctuations of heartbeats during normal sinus rhythm could be partially attributed to deterministic chaos and that a decrease in this type of nonlinear variability could be observed in different cardiovascular diseases and before ventricular fibrillation [3, 15, 16].

The importance of the biological time series analysis, which exhibits typically complex dynamics, has long been recognized in the area of non-linear analysis. Several features of these approaches have been proposed to detect the hidden important dynamical properties of the physiological phenomenon. As the statistical characteristics of biological signals often change with time and are typically both highly irregular and non-stationary in many cases, such analysis is complicated. The nonlinear dynamical techniques are based on the concept of chaos and it has been applied to many areas including the areas of medicine and biology [17–20]. The theory of chaos has been used to detect the cardiac arrhythmia such as ventricular fibrillation [21]. Efforts have been made in determining nonlinear parameters like correlation dimension for pathological signals and it has been shown that they are useful indicators of pathologies. Methods based on chaos theory have been applied in tracking HRV signals and predicting the onset events such as Ventricular Tachycardia and congestive heart failure situations [22]. A novel method based on phase space technique to distinguish normal and abnormal cases has been proposed for cardiovascular signals [23]. The technique has been extended here to identify the abnormalities of different types.

A complex system like cardiovascular system is not linear in nature and by considering it as a nonlinear system can lead to better understanding of the system dynamics. Recent studies have also stressed the importance of nonlinear techniques to study HRV in both health and disease. The progress made in this field using measures of chaos has attracted scientific community applying these tools in studying physiological systems, and HRV is no exception. There have been several methods of estimating invariants from nonlinear dynamical systems reported in the literature. Fell *et al* and Radhakrishna *et al* have tried the nonlinear analysis of ECG and HRV signals respectively [24,25]. Also, Paul *et al* showed that coordinated mechanical activity in the heart during ventricular fibrillation may be made visible in the surface ECG using wavelet transform [26]. Sun *et al* [27] have proposed an arrhythmia detection technique using nonlinear techniques by taking ECG as the base signal. Khadra *et al* [28] have proposed a classification of life-threatening cardiac arrhythmias using wavelet transforms. Later, Al-Fahoum *et al* [29], have combined wavelet transformation and radial basis neural networks for classifying life-threatening cardiac arrhythmias. Then, Mohamed *et al* [30] have studied features based on nonlinear dynamical modeling in ECG arrhythmia detection and classification. Acharya *et al*, have explained all the different types of linear and non-linear techniques, available for the analysis of heart rate signals [31]. They have also classified the HRV signals using ANN and Fuzzy equivalence relation into four groups [19]. Dingfie *et al* have classified cardiac arrhythmia into six classes using autoregressive modeling [32].

Power spectral analysis of beat-to-beat heart rate variability (HRV) has provided a useful means of understanding the interplay between autonomic and cardiovascular functionality. Mager *et al*, have developed an algorithm that utilizes continuous wavelet transform (CWT) parameters as inputs to Kohonen's self-organizing map (SOM), for providing a method of clustering subjects with similar wavelet transform signatures [33]. Maja *et al*, have analyzed human blood flow in the time-frequency domain, and used the wavelet transform (Morlet) which gives good time resolution for high frequency components and good frequency resolution for low-frequency components [34]. Recently, continuous wavelet transform (using Morlet wavelet as mother wavelet) for different cardiac arrhythmias was proposed [35]. Hisa *et al*, have used Morlet mother wavelet to evaluate the performance of frequency power spectrum during QRS in intraventricular conduction abnormalities (IVCA) [36]. They have observed that there is reduction of the low frequency power in IVCA and the increased power and number of peaks in high frequency range in IVCA with MI.

This chapter uses the heart rate variability as the base signal for the linear, frequency, nonlinear and wavelet analysis. Eight different classes have been analyzed in this chapter and the range of linear, frequency and nonlinear parameters are tabulated below.

16.1 Data Used for Analysis

ECG data for the analysis was obtained from MIT-BIH arrhythmia database. Prior to recording, the ECG signals were processed to remove noise due to power line interference, respiration, muscle tremors, spikes etc. The R peaks of ECG were detected using Tompkins's algorithm [37]. The number of dataset chosen for each of the eight classes is given in Table 16.1. Each dataset consists

Table 16.1. Number of subjects in various groups

Type	Normal	PVC	CHB	SSS	LBBB	Ischemic/ Dilated	AF	VF
Number of datasets	60	60	20	20	40	20	35	45

of around 10,000 samples and the sampling frequency of the data is 360 Hz. The interval between two successive QRS complexes is defined as the RR interval (t_{r-r} seconds) and the heart rate (beats per minute) is given as:

$$HR = 60/t_{r-r} \tag{16.1}$$

A brief description of the eight different kinds of cardiac states are discussed in Chapter 1.

16.2 Methods Used for Analysis

The heart rate is analyzed in the time domain, in frequency domain, using nonlinear parameters and wavelet transformation. Its detailed analysis is given below.

16.2.1 Time Domain Analysis

Six standard measures are SDNN, SENN, SDSD, RMSSD, pNN50% and HRV triangular index [3] are used for this work. The results of the time domain analysis are tabulated in Table 16.2.

16.2.2 Frequency Domain Analysis

Frequency domain analysis is carried out using AR (Autoregressive) method for this work. The Burg method is used to get the AR model parameter. In this work the order of the AR model is taken as p = 16 [38–40].

Table 16.2. Range of statistical parameters for various classes

TYPE	NSR	PVC	LBBB	AF	VF	CHB	SSS	ISCH-EMIC	p value
SDNN msec	28.7 ± 25.32	432. ± 8.91	30.6 ± 11.48	254.85 ± 57.48	22.4 ± 14.066	81.4 ± 49.47	274.85 ± 74.69	103.2 ± 151.79	0.051
SENN msec	4.11 ± 4.01	55.6 ± 18.66	3.92 ± 0.70	20.8 ± 004.29	1.71 ± 1.10	2.7 ± 1.62	10.51 ± 3.26	3.73 ± 5.46	0.082
SDSD msec	32.2 ± 24.5	714 ± 156.1	32.8 ± 6.17	467.9 ± 155.65	31.0 ± 22.23	99.9 ± 67.95	369.08 ± 87.7	3.73 ± 5.46	0.071
RMSSD msec	31.9 ± 24.2	708 ± 152.0	32.5 ± 6.17	466.4 ± 155.21	30.9 ± 22.17	99.8 ± 67.92	368.81 ± 87.62	140.7 ± 24.62	0.052
pNN50 %	7.01 ± 11.6	40.8 ± 18.14	4.41 ± 2.18	48.4 ± 0.73	7.52 ± 13.03	12.5 ± 14.8	35.1 ± 10.3	2.48 ± 3.83	0.071
TINN	5.34 ± 3.13	9.33 ± 3.25	7.53 ± 2.55	5.21 ± 0.47	7.02 ± 2.13	2.49 ± 0.53	14.57 ± 4.13	4.57 ± 1.69	0.034

16.2.3 Nonlinear Methods of Analysis

The nonlinear parameters like correlation dimension (CD), largest Lyapunov exponent (LLE), Hurst exponent (H), SD1/SD2 of Poincare plot, approximate entropy (ApEn), fractal dimension, and α slope of detrended fluctuation analysis have been used in this study.

16.3 Results

The result of statistical analysis, frequency domain analysis, and nonlinear analysis of the heart rate variability for various types of cardiac abnormalities are listed in Tables 16.2–16.4 ($p < 0.09$) respectively.

The statistical parameters SDNN, SENN, SDSD, RMSSD, pNN50% and TINN have bigger value for the classes like PVC, SSS, and AF due to higher RR variation. And for the slowly varying signal like CHB, LBBB and Ischemic/dilated cardiomyopathy, these parameters will be lesser because of the smaller RR variation.

In the frequency domain, ratio of low frequency to the high frequency is high for the complete heart block (CHB) and Ischemic/Dilated cardiomyopathy abnormalities are high, because the RR variation is very small. This RR variation will be more (as compared to the normal) in the case of left bundle branch block (LBBB), atrial fibrillation (AF), sick sinus syndrome (SSS), Pre-ventricular contraction (PVC) and ventricular fibrillation (VF). Hence, this ratio will be more. So, for the abnormal cases, this ratio either decreases or increases from the normal range.

In this chapter, we have evaluated non-linear parameters SD1/SD2, ApEn, LLE, α slope, CD, H, FD, for the analysis of various cardiac abnormalities. And the results of these were subjected to 'ANOVA' test with more than 89% confidence interval giving excellent 'p' values in all cases.

For the normal subjects shape of the plot is an ellipse and lies at the center of the quadrant and this position and shape of the plot shifts depending on the abnormality. The pattern of Poincare plots of HRV data, its position and ranges of SD1/SD2 values are unique for particular type of cardiac abnormality. SD1/SD2 shows the ratio of short interval variation to the long interval

Table 16.3. Range of frequency domain parameters for various classes

TYPE	NSR	PVC	LBBB	AF	VF	CHB	SSS	ISCHEMIC	p value
LF/HF	0.86 ± 0.33	1.34 ± 0.53	0.24 ± 0.17	0.54 ± 0.56	0.28 ± 0.19	1.15 ± 0.34	0.41 ± 0.21	2.99 ± 1.21	0.072

#[Tables 16.1–16.3, are reprinted with permission from Rajendra Acharya U, Kannathal N, S.M. Krishnan "Comprehensive analysis of cardiac health using heart rate signals" Physiological Measurement Journal, UK, vol. 25, 2004, pp. 1130–1151]

Table 16.4. Range of nonlinear parameters for various classes

TYPE	NSR	PVC	LBBB	AF	VF	CHB	SSS	ISCH-EMIC	p value
$\propto$-slope	0.77 ± 0.076	0.27 ± 0.014	0.43 ± 0.11	0.13 ± 0.043	0.34 ± 0.022	0.54 ± 0.034	0.55 ± 0.013	0.97 ± 0.11	0.076
SD1/SD2	0.80 ± 0.16	1.42 ± 0.54	0.7 ± 0.20	2.98 ± 1.56	1.13 ± 0.47	0.64 ± 0.024	0.96 ± 0.32	0.59 ± 0.37	0.011
ApEn	1.75 ± 0.077	1.51 ± 0.091	1.47 ± 0.137	1.57 ± 0.23	1.09 ± 0.173	0.97 ± 0.15	1.57 ± 0.097	0.76 ± 0.065	0.065
CD	3.58 ± 0.23	2.29 ± 0.099	3.20 ± 0.415	2.58 ± 0.033	2.90 ± 0.039	2.72 ± 0.139	2.35 ± 0.448	3.30 ± 0.142	0.032
LLE	0.50 ± 0.058	0.62 ± 0.003	0.47 ± 0.044	0.56 ± 0.112	0.42 ± 0.036	0.17 ± 0.011	0.82 ± 0.102	0.193 ± 0.066	0.056
HE	0.611 ± 0.019	0.873 ± 0.032	0.643 ± 0.011	0.796 ± 0.043	0.706 ± 0.021	0.748 ± 0.011	0.821 ± 0.023	0.654 ± 0.021	0.081
$D^{Higuchi}$ (FD1)	1.36 ± 0.043	1.19 ± 0.043	1.31 ± 0.032	1.21 ± 0.036	1.27 ± 0.039	1.24 ± 0.042	1.21 ± 0.021	1.32 ± 0.024	0.072
D^{Katz} (FD2)	1.58 ± 0.016	1.31 ± 0.019	1.53 ± 0.021	1.39 ± 0.023	1.46 ± 0.021	1.41 ± 0.033	1.36 ± 0.011	1.52 ± 0.017	0.046

variation. This ratio is more in the case of PVC, AF, SSS and VF due to more variation in the RR interval. But, this ratio falls (below normal) for the slowly varying signals like CHB, Ischemic/dilated cardiomyopathy.

The importance of ApEn lies in the fact that it is measure of the disorder in the heart rate signal. It is a measure, quantifying the regularity and complexity of time series. It has higher value in the case of normal and SSS subjects and it falls as the RR variation decreases. Hence, the ApEn will have smaller value for cardiac abnormal cases, indicating smaller variability in the beat to beat. But, for SSS this RR variation will be higher than the normal subjects.

Largest Lyapunov exponent's (LLE) quantify sensitivity of the system to initial conditions and gives a measure of predictability. This value decreases for slowly varying signals like CHB and Ischemic/dilated cardiomyopathy and will be higher for the other cases as the variation of RR is more.

The fractal scaling ($\propto$) for the normal subjects (healthy young) is closer to 1, and this value falls in different ranges for various types of cardiac abnormalities. This slope is low for highly varying signals like PVC, LBBB, AF andVF. But for rhythmically varying signals like SSS, CHB and Ischemic/Dilated cardiomyopathy this value is slightly higher (comparable to 1).

The CD gives the measure of the spread of the phase space. This value is high for the normal subjects ($CD = 3.58 \pm 0.23$) due to the high variation in the HR. This value gradually falls as the heart rate variability decreases. In the case of PVC, SSS, AF and VF the variability is more. But these variations are sometimes periodic or rhythmic. Hence, the CD value is lower than the *normal* subjects. These values are lower for CHB, Ischemic/dilated cardiomyopathy, LBBB subjects due to reduced heart rate variability between consecutive heart rate values.

The Hurst-exponent that characterizes the nonstationary behavior of the signals is calculated. From Table 16.4 it can be seen there is clearly a negative correlation between the values of correlation dimension and Hurst exponent. This is the expected behavior of a stochastic system with power-law spectra, D2 = max (1-H2, M), where M is the embedding dimension. Increase in the value of the Hurst exponent indicates less complexity and more synchronization. The cardiac abnormalities like AF, SSS, VF, PVC, CHB, Ischemic/dilated cardiomyopathy have slightly higher values as compared to the normal heart rate. This is because of the presence of the inherent rhythm in these heart rate data.

FD2 $= 1.41 \pm 0.033$ slightly low value, indicating low variation in the heart rate data. In *Ischemic/Dilated cardiomyopathy,* the variation between the consecutive heart rates is low (FD1 $= 1.32 \pm 0.024$ and FD2 $= 1.52 \pm 0.017$). For *Sick Sinus Syndrome – III (SSS – III),* the FD is low (FD1 $= 1.21 \pm 0.021$ and FD2 $= 1.36 \pm 0.017$) indicating the inherent periodicity, for *Atrial Fibrillation (AF)* has too much variation in the heart rate data (FD1 $= 1.21 \pm 0.036$ and FD2 $= 1.39 \pm 0.011$). During *Ectopic beat* variation in the heart rate is high (FD1 $= 1.19 \pm 0.043$ and FD2 $= 1.31 + -0.019$). And, for the *Normal* subjects have variation in their heart rates (FD1 $= 1.36 + -0.043$ and FD2 $= 1.58 + -0.016$). This value is high due to the high variation in the heart rate. The FD for the subjects with *LBBB* (FD1 $= 1.31 \pm 0.032$; FD2 $= 1.53 \pm 0.021$) is slightly less then *normal* subjects. The FD (FD1 $= 1.27 \pm 0.039$; FD2 $= 1.46 \pm 0.021$) for subjects with VF is high due to high variability. But, however it is less than the *normal* subjects. For normal subjects, the FD is high due to the variation being chaotic. And for CHB and Ischemic/dilated cardiomyopathy, this FD decreases because the R-R variation is low. And for AF and SSS, LBBB, VF this FD value falls further, because the R-R variation becomes erratic or periodic respectively.

The resulting phase space plot and corresponding wavelet scalograms, for various types of diseases are shown in Figs. 16.1–16.16. In the CWT plots shown, white color indicates high value of (wavelet) coefficient and black corresponds to low value. As can be seen in the figures, the patterns show continuity in the patterns indicating a continuous variation of heart rate. For Normal cases, the heart rate is continuously varying. Hence the phase space plot is spread randomly at the center of the plot (Fig. 16.2). The CWT pattern appears to be flowery and regular (Fig. 16.1). In the *Ectopic beat* abnormality; there would be a sudden impulsive jump in the heart rate. This may be due to a Pre-Ventricular beat in the ECG signal. This is indicated as a sudden surge of radial white lines in the CWT plot (Fig. 16.3), and spikes (ectopic beats) in the phase space plot (Fig. 16.4). The black patches indicate the *Bradycardia* and the rest is normal. In the Atrial *Fibrillation* (AF), heart rate signal records highly erratic variability; this is depicted as sudden changes in colors

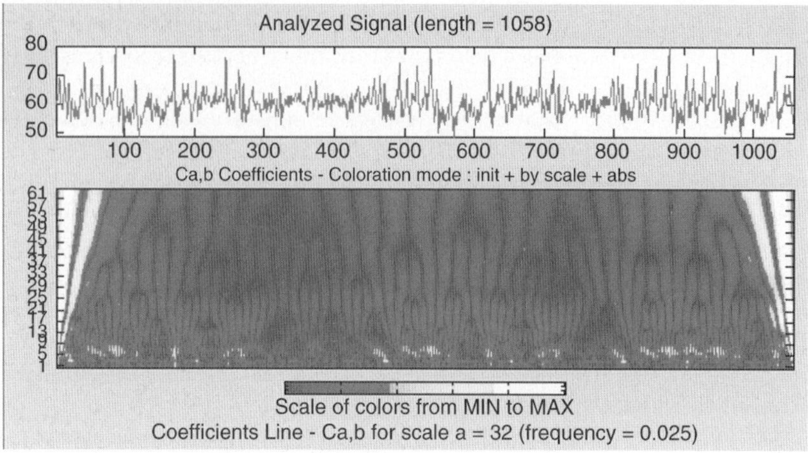

Fig. 16.1. CWT plot of normal heart rate

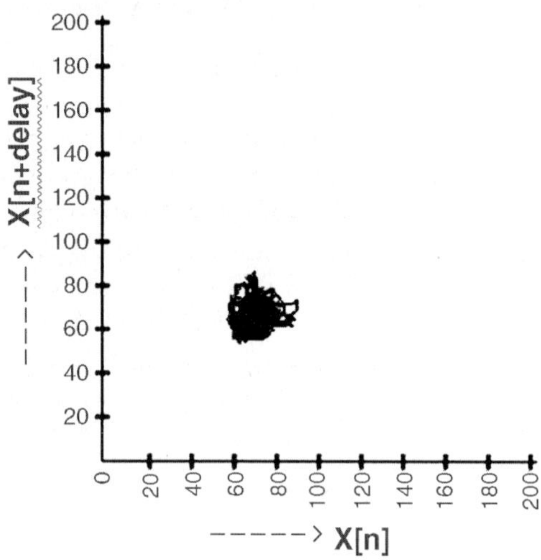

Fig. 16.2. Phase space plot of Normal subject CD = 3.58 ± 0.23

(Fig. 16.5). These abrupt changes in the heart rate are shown clearly in the phase space plot by a big spread in the phase space plot (Fig. 16.6). In *Complete Heart Block* (CHB) cases as the A_V node fails to send electrical signals rhythmically to the ventricles, the heart rate remains low. The pattern is predominantly red (low coefficient value) with very little change in color intensity (Fig. 16.7). The phase space plot reduces almost to a point, indicating very little change with time (Fig. 16.8). In *SSS – III* (*Sick Sinus Syndrome – III, Bradycardia-Tachycardia*) there is a continuous variation of heart rate between *Bradycardia* and *Tachycardia*, which shows up by way of alternating patches

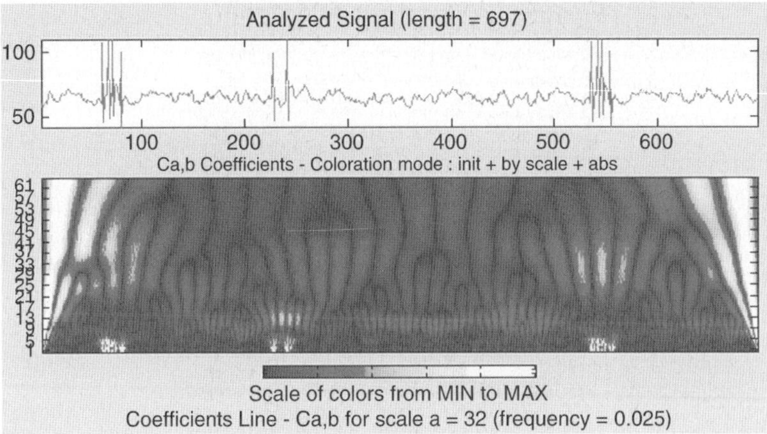

Fig. 16.3. CWT plot of ectopic beat

of black (*Brady*) and colored (*Tachy*) patterns (Fig. 16.9). The phase space plot spreads over a larger area (Fig. 16.10). In the case of *Ischemic/Dilated cardiomyopathy*, the ventricles are unable to pump out blood to the normal degree. Here the heart rate variation is very small. Correspondingly, the color variation too is gradual and periodic (Fig. 16.11). The resulting phase space plot looks like a small spread in the phase space plot (Fig. 16.12). Heart rate variation in *LBBB* is not as high as in *normal* subjects. Hence, the CWT plot (Fig. 16.13), resembles that of the *normal* CWT. The phase space plot is a spread smaller than the *normal* subjects (Fig. 16.14). The variation of the heart rate in *VF* is high and rhythmic. The CWT plot of *VF* (Fig. 16.15) resembles that of the *Ischemic/Dilated* cardiomyopathy indicating rhythm in the heart rate. The resulting phase space plot is shown in Fig. 16.16 and it is spread at the top corner of the plot due to the high heart rate values.

16.4 Discussions

Results of statistical, geometric, frequency domain, time-frequency and non-linear parameters have been discussed in detail in the previous sections. Conventionally, heart rate fluctuation has been assessed by calculating indices based on statistical operations on RR intervals (means and variance). The most widely used time domain indices are the average heart rate, SDNN and RMSSD. All the time domain measure indices could be affected by artifacts and outliers, and these measures therefore require data from which artifacts and ectopic beats have been carefully eliminated.

Geometrical methods present RR intervals in geometric patterns and triangular index. Poincare plots have been used to derive measures of heart rate

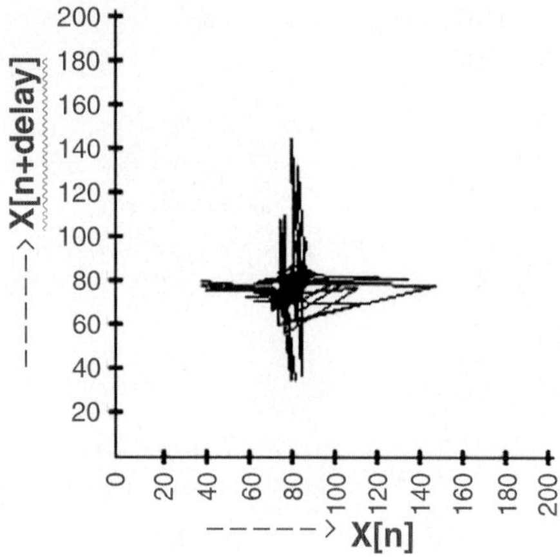

Fig. 16.4. Phase space plot of subject with ectopic beat CD = 2.29 ± 0.099

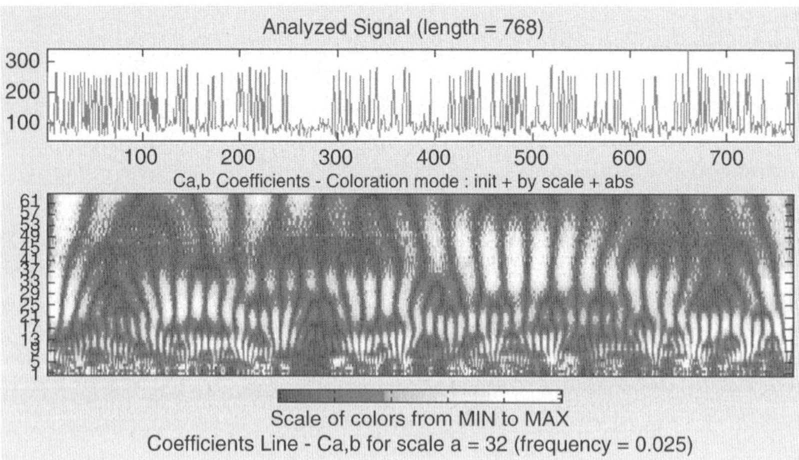

Fig. 16.5. CWT plot of AF heart rate

signals. This triangular index had a high correlation with the standard deviation of all RR intervals, but it is highly insensitive to artifacts and ectopic beats, because they are left outside the triangle.

Experience with frequency domain analysis over the past two decades strongly suggests that it represents a unique, noninvasive tool for achieving a more precise assessment of autonomic function in both the experimental and clinical settings. Available studies indicate that the significance of the HF component is far better understood than that of the lower frequency components. Respiratory pattern also can significantly influence HF power. The use of controlled breathing minimizes these problems, improves reproducibility of test findings, and also facilitates quantitative comparisons. The situation with

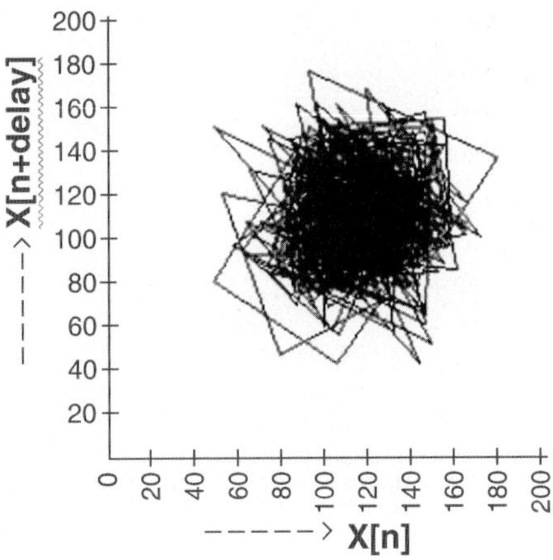

Fig. 16.6. Phase space plot of subject with AF CD = 2.58 ± 0.033

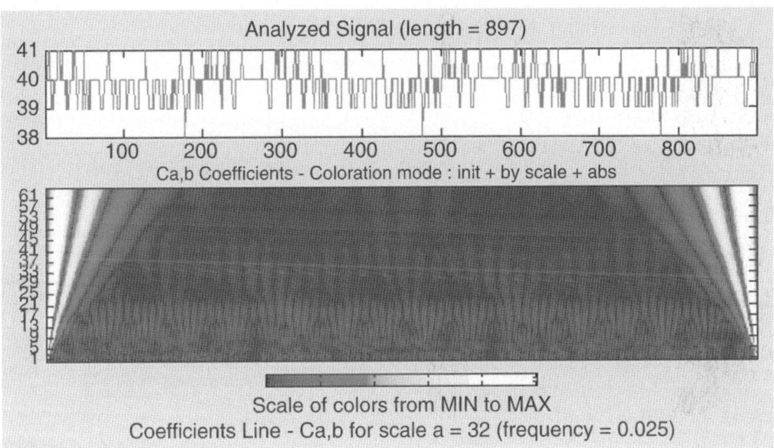

Fig. 16.7. CWT plot of CHB heart rate

respect to LF power is more complicated because it is modulated by both sympathetic and parasympathetic outflows as well as by other factors, including baroreceptor activity. Therefore, LF analysis per se cannot afford a precise delineation of the state of sympathetic activation. The meaningful determinations of VLF and ULF power may be difficult because decrease in frequency to such low levels are associated with an increasing propensity to violate the rules governing power spectral determinations, violations that diminish reliability despite the most sophisticated preprocessing. It is also noteworthy that the reliability of spectral power determinations diminishes with decrease in the power of the signal and of the signal-to-noise ratio.

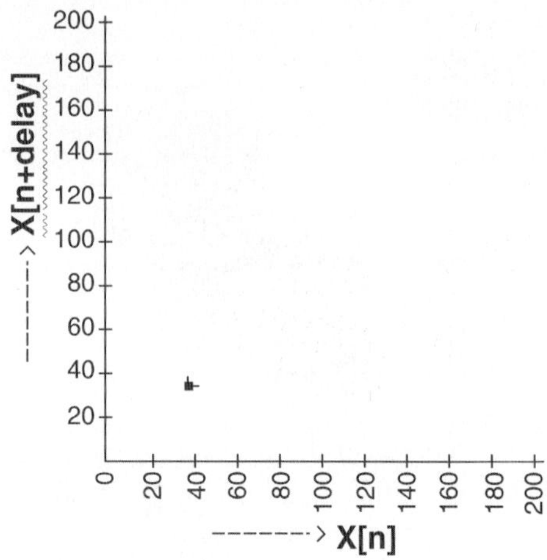

Fig. 16.8. Phase space plot of subject with CHB CD = 2.72 ± 0.139

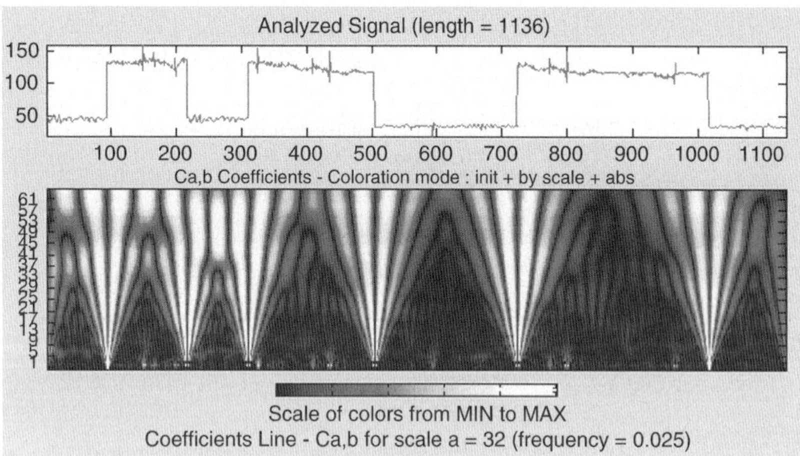

Fig. 16.9. CWT plot of SSS III heart rate

The wavelet transform has become a valuable analysis tool due to its ability to elucidate simultaneously both spectral and temporal information within the signal. This overcomes the basic shortcoming of Fourier analysis, which is that the Fourier spectrum contains only globally averaged information, so leading to location specific features in the signal being lost.

There is increasing evidence to suggest that the heart is not a periodic oscillator under normal physiologic conditions [41], and the commonly employed moment statistics of heart rate variability may not be able to detect subtle, but important changes in heart rate time series. Therefore several new analysis method of heart rate behavior, motivated by nonlinear dynamics

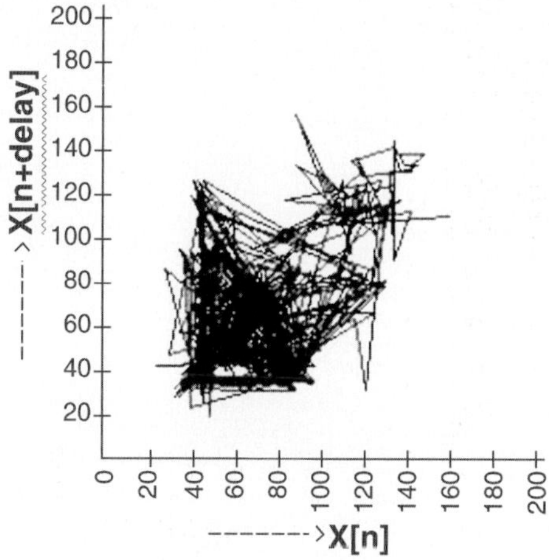

Fig. 16.10. Phase space plot of subject with SSS III CD $= 2.35 \pm 0.448$

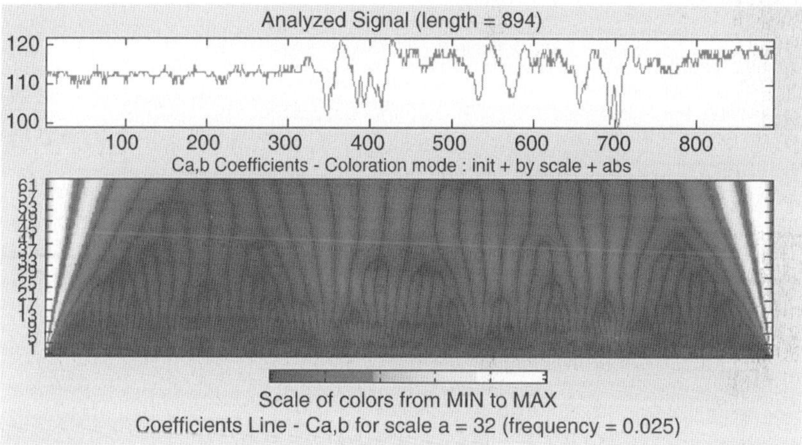

Fig. 16.11. CWT Poincare plot of Ischemic/Dilated Cardiomyopathy heart rate

and chaos theory, have been developed to quantify the dynamics of heart rate fluctuations [41, 42].

Heart rate ApEn has demonstrated the capacity to predict atrial arrhythmias, including spontaneous [43] and postoperative atrial fibrillation after cardiac surgery [44], and to differentiate ventricular arrhythmias [45]. Heart rate ApEn is decreased in infants with aborted sudden infant death syndrome [46]; among adults, postoperative patients with ventricular dysfunction [47] and healthy individuals infused with endotoxin [48] exhibit reduced heart rate ApEn.

The detrended fluctuation analysis technique is a measurement which quantifies the presence or absence of fractal correlation properties and has

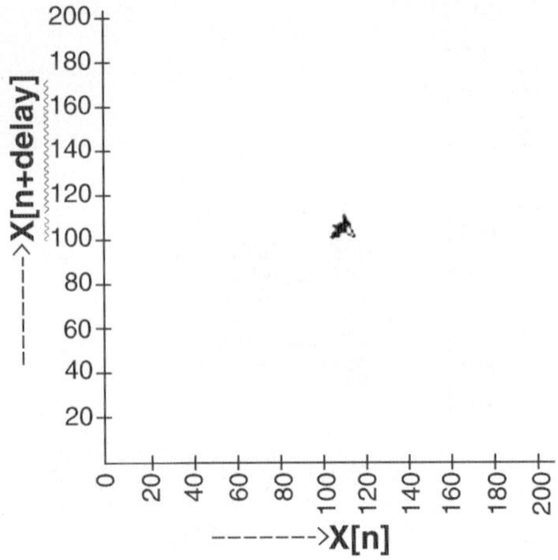

Fig. 16.12. Phase space plot of subject with Ischemic/Dilated Cardiomyopathy CD = 3.30 ± 0.142

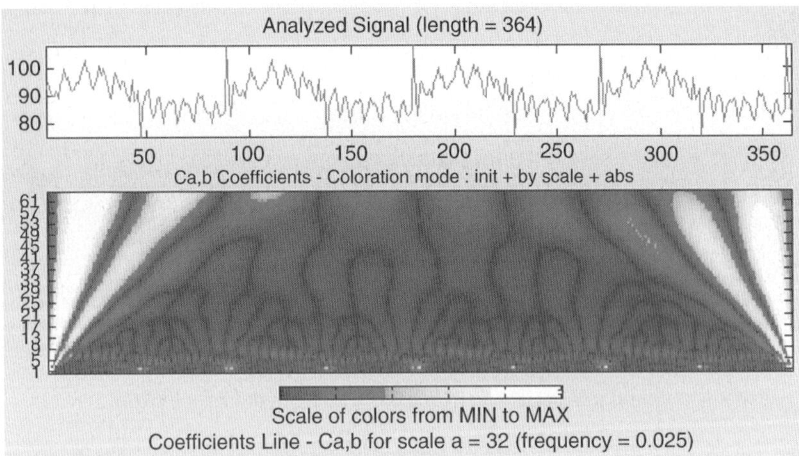

Fig. 16.13. CWT plot of LBBB heart rate

been validated for time series data [49]. It was developed to characterize fluctuations on scales of all lengths. The self-similarity occurring over a large range of time scales can be defined for a selected time scale with this method. The fractal scaling (∝) for the normal subjects (healthy young) is closer to 1, and this value falls in different ranges for various types of cardiac abnormalities. This slope is very low for very highly varying signals like PVC, LBBB, AF and VF. But for rhythmically varying signals like SSS, CHB and Ischemic/Dilated cardiomyopathy this value is slightly higher (comparable to 1) [20]. The Lyapunov exponent is a quantitative measure of separation of

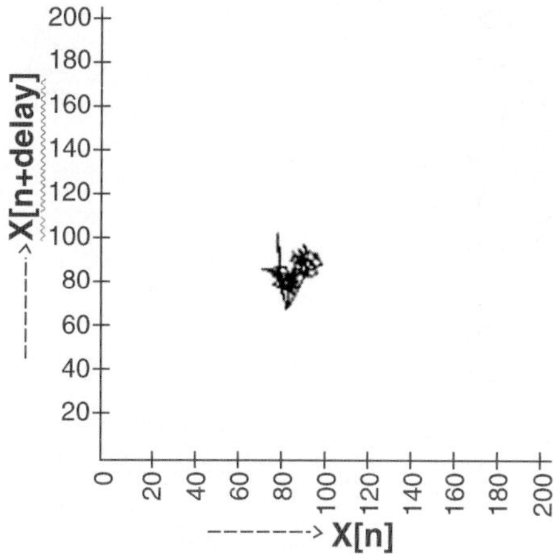

Fig. 16.14. Phase space plot of subject with LBBB CD = 3.20 ± 0.415

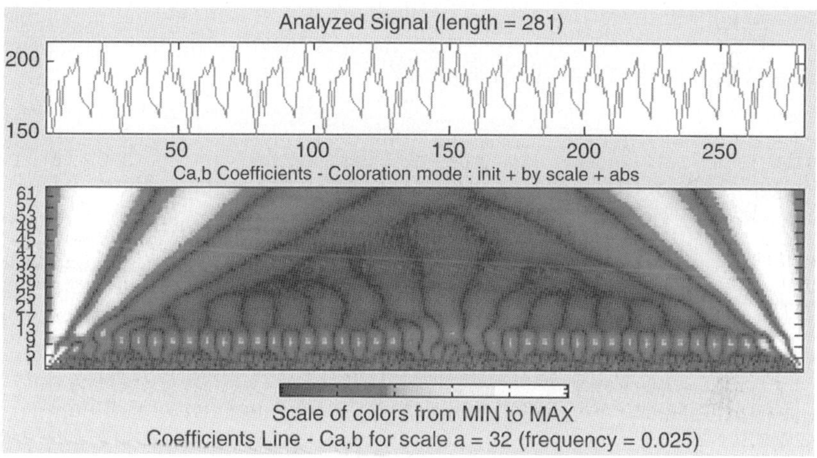

Fig. 16.15. CWT plot of VF heart rate

the divergence of the trajectories from their initial close positions. The magnitude of this exponent indicates the intensity of the chaotic system. This value decreases for slowly varying signals like CHB and Ischemic/dilated cardiomyopathy and will be higher for the other cases as the variation of RR is more [20].

When the heart rate is steady and unchanging, the phase-space plot reduces to a point, but otherwise, the trajectory spreads out to give some patterns on the screen. The pattern that emerges can be interpreted for finer details – such as whether the heart rate is periodic, chaotic, or random etc. A Correlation Dimension factor is defined to obtain a quantitative measure of the nature of trajectory of the phase space plot. It has higher values for

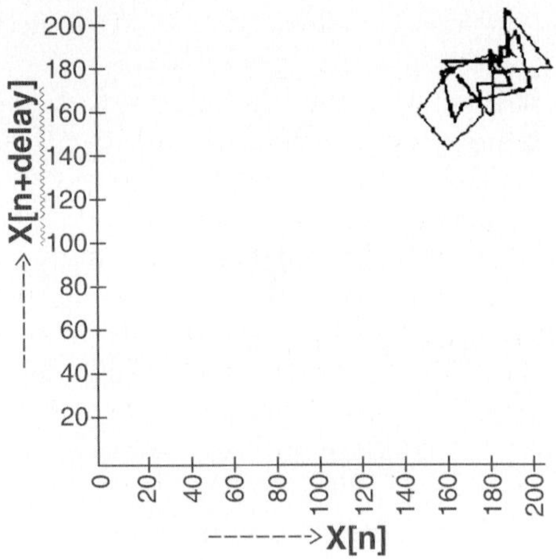

Fig. 16.16. Phase space plot of subject with VF CD $= 2.90 \pm 0.039$
#[Fig. 16.2, Fig. 16.4, Fig. 16.6, Fig. 16.8, Fig. 16.10, Fig. 16.12, Fig. 16.14, Fig. 16.16, are reprinted from Rajendra Acharya U, Kannathal N, S.M.Krishnan "Comprehensive analysis of cardiac health using heart rate signals" Physiological Measurement Journal, UK, vol. 25, 2004, pp. 1130–1151]

normal heart rate signals and this value falls as the beat to beat variation falls [20]. Hurst Exponent is the measure of the smoothness of a fractal time series based on the asymptotic behavior of the rescaled range of the process. The cardiac abnormalities like AF, SSS, VF, PVC, CHB, Ischemic/dilated cardiomyopathy have slightly higher values as compared to the normal heart rate. This is because of the presence of the inherent rhythm in these heart rate data.

Nikhil Iyengar *et al*, have applied DFA to the *normal* heart rate variability signal [50]. Recently, Echeverria *et al* [51], have applied detrended fluctuation analysis to the *normal* and *Congestive Heart Failure (CHF)* subjects. Short range correlation α_s value is low in the abnormal signals. Similarly, the long range correlation α_l values are more for the normal signal indicating the fractalness in the data and this range decreases as the subject grows old. The α slope and FD quantifies the variability in the heart rate data. This will have high value for normal subjects and decreases in different ranges for the various abnormalities.

16.5 Surrogate Data

The purpose of surrogate data is to test for any nonlinearity in the original data. Nonlinear indices such as ApEn are computed for several surrogate data series. Their values are compared with the nonlinear index computed for the original index [52]. The demonstration of significant difference in nonlinear indices between the original and surrogate data are in keeping with the presence of nonlinear dynamics in the original data.

Surrogate data have Fourier decomposition with the same amplitudes as the empirical data decomposition but with random phase components. This is obtained using the Chaos Data Analyzer. 20 sets of surrogate data are generated for each of the eight classes. ApEn is obtained for both the original and surrogate data sets. We found that, the surrogate data ApEn and original data ApEn, are different from each other by more than 50%. The similar procedure is repeated for Correlation Dimension. The surrogate data Correlation Dimension and original data Correlation Dimension are different from each other by more than 58%. This rejects the null hypothesis and hence the original data contain nonlinear features.

16.6 Conclusion

Heart rate variability (HRV) signal can be used as a reliable indicator of heart diseases. Ever since the birth of 'non linear science' chaoticians of physiology, biomedical engineering and theoretical biology are searching for meaningful chaotic parameters in physiological processes. Here we have evaluated linear parameters and nonlinear parameters: SD1/SD2, ApEn, scaling exponent, correlation dimension and LLE. And the results of these were subjected to 't' test with more than 90% confidence interval giving significant 'p' values in all cases. The importance of ApEn lies in the fact that it is measure of disorder in the heart rate signal. For the more regular and predictable heart rate values like complete heart block and ischemic/dilated cardiomyopathy type of abnormalities, the lower is the ApEn. On the other hand for the normal subjects where the heart rate is more random, ApEn has higher value. Hence the ApEn is used to find the unpredictability of fluctuations in the heart rate signals. This value decreases for the abnormal cardiac beats. ApEn for various types of cardiac disorders are listed in this chapter. Similarly, Correlation Dimension, LLE, FD and the scaling exponent $\propto$ decreases as the R R interval decreases. In this work, we have proposed a set of ranges for these linear and nonlinear parameters for various cardiac abnormalities.

A CWT scalogram of HRV signal can provide a visual pattern, which may provide considerable insight into the nature and pattern of the disease. The scalogram pattern is dependent upon the type of the wavelet used for analysis; which is not examined in this chapter.

References

1. Bigger, J.T., Fleiss, J.L., Rolnitzky, L.M., Steinman, R.C., and Schneider, W.J. (1991): 'Time course of recovery of heart period variability after myocardial infarction', *J Am Coll Cardiol.*, **18**, pp. 1643–1649.

2. Cripps, T.R., Malik, M., Farrell, T.G., and Camm, A.J. (1991): 'Prognostic value of reduced heart rate variability after myocardial infarction: clinical evaluation of a new analysis method', *Br Heart J.*, **65**, pp. 14–19.
3. Task Force of the European Society of Cardiology and the North American Society of Pacing and Electrophysiology (1996): 'Heart rate variability standards of measurement, physiological interpretation, and clinical use', *Circulation*, **93**, pp. 1043–1065.
4. Goldberger, A.L. (1990): 'Nonlinear dynamics, fractals and chaos: applications to cardiac electrophysiology', *Annals of Biomedical Engineering*, **18**, pp. 195–198.
5. Mansier, P., Clairambault, J., Charlotte, N., Mdigue,C., Vermeiren, C., Lepape, G., Carr, F., Agounaropoulou, A., and Swyndghedauw, B. (1996): 'Linear and nonlinear analyses of heart rate variability: a mini review', *Cardiovascular Research*, **31**, pp. 371–379.
6. Guzzetti, S., Signorini, M.G., Cogliati, C., Mezzetti, S., Porta, A., Cerutti, S., and Malliani, A. (1996): 'Non-linear dynamics and chaotic indices in heart rate variability of normal subjects and heart transplanted patients', *Cardiovascular Research*, **31**, pp. 441–446.
7. Hoyer, D., Scmidt, K., Bauer, R., Zwiener, U., Khler, M., Lthke, B., and Eiselt, M. (1997): 'Nonlinear analysis of heart rate and respiratory dynamics', *IEEE Engineering in Medicine and Biology*, **16**, pp. 31–39.
8. Poon, C.S., and Merrill, C.K. (1997): 'Decrease of cardiac chaos in congestive heart failure', *Nature*, **389**, pp. 492–495.
9. Wagner, C.D., and Persson, P.B. (1998): 'Chaos in the cardiovascular system: an update', *Cardiovascular Research*, **40**, pp. 257–264.
10. Makikallio, T. (1998): 'Analysis of heart rate dynamics by methods derived from nonlinear mathematics-clinical applicability and prognostic significance', *Ph.D Thesis*, University of Oulu.
11. Pikkujamsa, S.M., Makikkallio, T.H., Sourander, L.B., Raiha, I.J., Puukka, P., Skytta, J., Peng, C.K., Goldberger, A.L., and Huikuri, H.V. (1999): 'Cardiac interbeat interval dynamics from childhood to sensescence-Comparison of conventional and new measures based on fractals and chaos theory', *Circulation*, **100**, pp. 393–399.
12. Porta, A., Guzzetti, S., Montano, N., Pagani, M., Somers, V., Malliani, A., Baselli, G., and Cerutti, S. (2000): 'Information domain analysis of cardiovascular variability signals: evaluation of regularity, synchronization and coordination', *Medical & Biological Engineering & Computing*, **38**, pp. 180–188.
13. Kobayashi, M., and Musha, T. (1982): '1/f fluctuation of heartbeat period', *IEEE Trans Biomed Eng.* **29**, pp. 456–457.
14. Peng, C.K., Havlin, S., Stanley, H.E., and Goldberger, A.L. (1995): 'Quantification of scaling exponents and crossover phenomena in nonstationary heartbeat time series', *Chaos*, **5**, pp. 82–87.
15. Denton, T.A., Diamond, G.A., Helfant, R.H., Khan, S., and Karagueuzian, H. (1990): 'Fascinating rhythm: a primer on chaos theory and its application to cardiology', *Am Heart J.*, **120**, pp. 1419–1440.
16. Goldberger, A.L. (1996): 'Non-linear dynamics for clinicians: chaos theory, fractals, and complexity at the bedside', *Lancet*, **347**, pp. 1312–1314.
17. Pomeranz, B., Macaulay, R.J.B., Caudill, M.A., Kutz, I., Adam, D., Kilborn, K.M., Barger, A.C., Shannon, D.C., Cohen, R.J., and Benson, H. (1985):

'Assessment of autonomic function in humans by heart rate spectral analysis', *American journal of Physiology*, **248**, pp. H151–H153.

18. Acharya, U.R., Min, L.C., and Joseph, P. (2002): 'HRV analysis using Correlation Dimension and DFA', *Innovations and Technology in Biology and Medicine (ITBM-RBM)*, France, 23, pp. 333–339.
19. Acharya, U.R., Bhat, P.S., Iyengar, S.S., Rao, A., and Dua, S. (2003): 'Classification of heart rate using artificial neural network and fuzzy equivalence relation', *Pattern Recognition*, **36**, pp. 61–68.
20. Acharya, U.R., Kannathal, N., and Krishnan, S.M. (2004): 'Comprehensive analysis of cardiac health using heart rate signals', *Physiological Measurement Journal*, **25**, pp. 1130–1151.
21. Kaplan, D.K., and Cohen, J.R. (1991): 'Searching for Chaos in fibrillation', *Annals NY Academic Science*, pp. 367–374.
22. Cohen, M.E., Hudson, D.L., and Deedwania, P.C. (1996): 'Applying continuous chaotic modeling to cardiac signal analysis', *IEEE Engg. Med. Biol.*, **15**, pp. 97–102.
23. Dutt, N.D., and Krishnan, S.M. (1999): 'Application of Phase space techniques to the analysis of cardiac signals', *Proceedings of IEEE EMBS Conference*, Atlanta, U.S.A.
24. Fell, J., Mann, K., Roschke, J., and Gopinathan, M.S. (2000): 'Nonlinear analysis of continuous ECG during sleep I. Reconstruction', *Biological Cybernatics*, **82**, pp. 477–483.
25. Radhakrishna, R.K.A., Yergani, V.K., Dutt, N.D., and Vedavathy, T.S. (2001): 'Characterizing Chaos in heart rate variability time series of panic disorder patients', *Proceedings of ICBME, Biovision 2001*, Bangalore, India, pp. 163–167.
26. Paul, S.A., James, N.W., Gareth, R.C., Petter, A.S., and Colin, E.R. (2002): 'Finding Coordinated Atrial Activity During Ventricular Fibrillation Using Wavelet Decomposition', *IEEE Engineering In Medicine and Biology Magazine*, **21(1)**, pp. 58–61.
27. Sun, Y., Chan, K.L., and Krishnan, S.M. (2000): 'Arrhythmia detection and recognition in ECG signals using nonlinear techniques', *Annals of Biomedical Engineering*, **28**(1), S-37.
28. Khadra, L., Al-Fahoum, A.S., and Al-Nashash, H. (1997): 'Detection of life-threatening cardiac arrhythmias using wavelet transformation', *Medical & Biological Engineering & Computing*, **35**, pp. 626–632.
29. Al-Fahoum, A.S., and Howitt, I. (1999): 'Combined wavelet transformation and radial basis neural networks for classifying life-threatening cardiac arrhythmias', *Medical & Biological Engineering & Computing*, **37**, pp. 566–573.
30. Mohamed, I.O., Ahmed, H., Abou-Zied, Abou-Bakr, M., Youssef, and Yasser, M.K. (2002): 'Study of features on nonlinear dynamical modeling in ECG arrhythmia detection and classification', *IEEE transactions on Biomedical Engineering*, **49(7)**, pp. 733–736.
31. Acharya, U.R., Joseph, P.K., Kannathal, N., Lim, C.M., and Suri, J.S. (2006): 'Heart rate variability: A Review', IFMBE *Journal of Medical & Biological Engineering & Computing Journal.* (in press).
32. Dingfei, G.E., Narayanan, S., and Krishnan, S.M. (2002): 'Cardiac arrhythmia classification using autoregressive modeling', *BioMedical Engineering OnLine*, **1**(1): 5.

33. Mager, D.E., Merritt, M.M., Kasturi, J., Witkin, L.R., Urdiqui-Macdonald, M., Sollers, J.J. (2004): 3rd, Evans, M.K., Zonderman, A.B., Abernethy, D.R., and Thayer, J.F. (2004): 'Kullback-Leibler clustering of continuous wavelet transform measures of heart rate variability', *Biomed Sci Instrum.*, **40**, pp. 337–342.
34. Bracic, M., and Stefanovska, A. (1998): 'Wavelet-based Analysis of Human Blood-flow Dynamics', *Bulletin of Mathematical Biology*, **60**, pp. 919–935.
35. Faust, O., Acharya, U.R., Krishnan, S.M., and Min, L.C. (2004): 'Analysis of Cardiovascular signals using spatial filling index and time-frequency domain', *Biomedical online journal* USA, **3**:30.
36. Shimojima, H., Tsutsumi, T., Yanagisawa, F., Komukai, M., Zenda, N., Higashi, Y., Takeyama, Y., and Okamoto, Y. (2003): 'Application of Wavelet Transform for Analysis of QRS Complex in Intraventricular Conduction Abnormalities', *International Journal of Bioelectromagnetism*, **5**(1), pp. 279–281.
37. Pan, J., and Tompkins, W.J. (1985): 'Real Time QRS detector algorithm', *IEEE transaction on Biomedical Engineering*, **32**(3), pp. 230–233.
38. Akaike, H. (1969): 'Fitting autoregressive models for prediction', *Annals of the Institute of Statistical Mathematics*, **21**, pp. 243–247.
39. Akaike, H. (1974): 'A new look at statistical model identification', IEEE transaction on Automatics Control, 19, pp. 716–723.
40. Anita, B., Fernando, S.S., Rocha, A.P., and Argentina, L. (2002): 'A study on the optimum order of autoregressive models for heart rate variability', *Physiological Measurements*, **23**, pp. 324–336.
41. Goldberger, A.L., and West, B.J. (1987): 'Application of nonlinear dynamics to clinical cardiology', *AnnNYAcadSci.*, **504**, pp. 195–213.
42. Pincus, S.M. (1991): 'Approximate entropy as a measure of system complexity', Proceed*ings of National Acadamic Science*, **88**, pp. 2297–2301.
43. Vikman, S., Makikallio, T.H., Yli-Mayry, S., Pikkujamsa, S., Koivisto, A.M., Reinikainen, P., Airaksinen, K.E., and Huikuri, H.V. (1999): 'Altered complexity and correlation properties of R-R interval dynamics before the spontaneous onset of paroxysmal atrial fibrillation', *Circulation*, **100**, pp. 2079–2084.
44. Hogue, C.W., Domitrovich, P.P., Stein, P.K., Despotis, G.D., Re., L., Schuessler, R.B., Kleiger, R.E., and Rottman, J.N. (1998): 'RR interval dynamics before atrial fibrillation in patients after coronary artery bypass graft surgery', *Circulation*, **98**, pp. 429–434.
45. Schuckers, S.A. (1998): 'Use of approximate entropy measurements to classify ventricular tachycardia and fibrillation', *J Electrocardiol.*, **31**(supple.), pp. 101–105.
46. Pincus, S.M., Cummins, T.R., and Haddad, G.G. (1993): 'Heart rate control in normal and aborted-SIDS infants', *Am J Physiol.*, **264**, pp. R638–R646.
47. Fleisher, L.A., Pincus, S.M., and Rosenbaum, S.H. (1993): 'Approximate entropy of heartrate as a correlate of postoperative ventricular dysfunction', *Anesthesiology*, **78**, pp. 683–692.
48. Godin, P.J., Fleisher, L.A., Eidsath, A., Vandivier, R.W., Preas, H.L., Banks, S.M., Buchman, T.G., and Suffredini, A.F. (1996): 'Experimental human endotoxemia increases cardiac regularity: results from a prospective, randomized, crossover trial', *Crit Care Med.*, **24**, pp. 1117–1124.
49. Peng, C.K., Havlin, S., Hausdorf, J.M., Mietus, J.E., Stanley, H.E., and Goldberger, A.L., (1996): 'Fractal mechanisms and heart rate dynamics', *Journal on Electrocardiology*, **28**(suppl), pp. 59–64.

50. Iyengar, N., Peng, C.K., Raymond, M., Goldberger, A.L., and Lipsitz, L.A. (1996): 'Age-related alterations in the fractal scaling of cardiac interbeat interval dynamics', *American Journal of Physiology*, **271**(4), pp. 1078–1084.
51. Echeverria, J.C., Woolfson, M.S., Crowe, J.A., Hayes-Gill, B.R., Croaker, G.D., and Vyas, H. (2003): 'Interpretation of heart rate variability via detrended fluctuation analysis and alphabeta filter', *Chaos*, **13**(2), pp. 467–75.
52. Theiler, J., Eubank, S., Longtin, A., Galdrikian, B., and Farmer, J.D. (1992): 'Testing for nonlinearity in time series: the method of surrogate data', *Physica D*, **58**, pp. 77–94.

17

Soft Tissue Biomechanics of the Left Ventricular Myocardium

E.Y.K. Ng, Dhanjoo N. Ghista, Reginald C. Jegathese and Jian-Jun Shu

Large ventricular pumping and perfusion are interrelated. This chapter presents a perspective of biomechanical modeling of fluid perfusion through Left Ventricular (LV) myocardium, based on its fundamental constitutive relations.

17.1 Introduction

Myocardial perfusion is the flow or forced passage of blood through the coronary arteries to the heart myocardium. If cardiac pumping efficiency (contractility) increases, then cardiac output increases, and also myocardial perfusion increases. Herein, we are presenting the biomechanics of the myocardium and of its blood perfusion. In so doing, we have attempted to provide a quantitative insight into the perfusion process, and of how the various parameters of the process influence it, by means of fundamental processes and equations of mass, momentum and energy balances.

There is a need for estimating how myocardial perfusion is influenced by LV wall stress, because it has a direct bearing on LV pumping efficiency. Myocardium is the functional (parenchymal) tissue that endows the heart with its ability to pump blood. Coronary circulation is the supply to the myocardium with its own network of vessels, the left and right coronary arteries, which originate at the base of the aorta and branch out to encircle the myocardium.

In this chapter the focus is on the application of porous medium theory to myocardial perfusion. The flow of fluid through a porous medium depends upon its permeability (and porosity), which in the case of the myocardium also depends on the myocardial stress, due to intra-ventricular pressure. Since these physical factors are strongly altered in disease states, models describing the heart as a stressed porous material may shed some light on the etiology of example heart failure. In such a model one has to analyze the combined continuous interaction of important parameters that relate to myocardial perfusion as compliance, contractility, resistance-to-filling, and resistance-to-flow.

Below, we first will review briefly the literature on soft tissue mechanics and physiological aspects of myocardial perfusion. Then we will present a model of myocardial perfusion based on flow through stressed porous media, and use it to analyze myocardial perfusion.

17.2 Previous Research on Soft Tissue Biomechanics

It is a major challenge to include variations in LV myocardial contractility, resistance and compliance simultaneously in one single model. Such a model is needed since time varying elastance has a major effect on coronary perfusion [1–3]. Many studies can be found focusing on certain specialized aspects, as indicated in Fig. 17.1.

Figure 17.2 describes how the LV myocardial perfusion is interdependent on those LV myocardial perfusion [4–6], calculation of pressure and flow distributions in the myocardium.

17.2.1 Model Developments for Analyzing Myocardial Perfusion

The major research approaches to myocardial perfusion can be summarized in Table 17.1. The theory of flow through stressed porous medium was presented by Darcy [7], Biot [8], Fung [9, 10] and Hunter [11]. Mow *et al* [12] have analyzed the soft tissue [13] behavior in an articular cartilage, cornea, intervertebral disc tissues, based on a continuum (biphasic) mixture theory, for stress and strain, flow and pressure inside soft tissues. Humphrey [14]

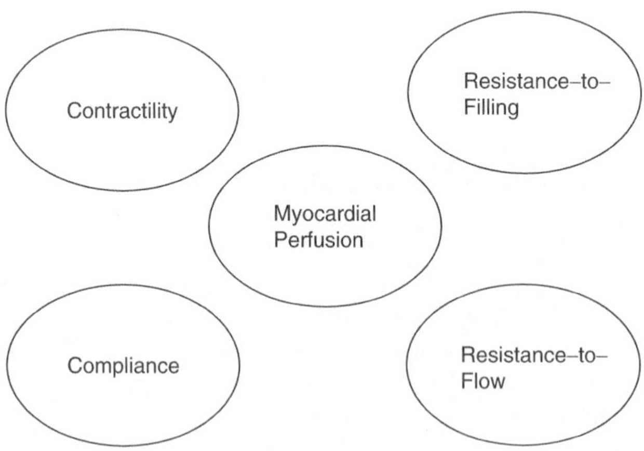

Fig. 17.1. Specialized areas for analyzing myocardial (soft tissue) properties

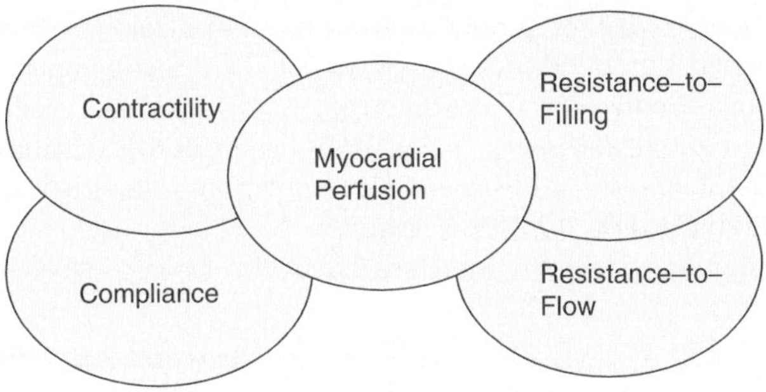

Fig. 17.2. Proposed integrated model for analyzing myocardial (soft tissue) properties

Table 17.1. Problems related to Myocardial Perfusion

Problems addressed	References
Continuum Mechanics	Fung [10], Hunter [11], Humphrey [39]
Constitutive Law	Criscione [24], Bischoff [16], Huyghe [15]
Soft Tissue mechanics	Meroi [26], Almeida [40], Ehlers [41], Gu [13]
Elastic properties	Oosterhout [17], Simon [42], Carcione [43]
Porous Media	Parker [29], Naili [44]
Permeability	Vaughn [14]
Fiber property	Pao [34], Nielsen [35], Stevens [45], Prinzen [36]
Poroelasticity	Coussy [33], Boer [30], Barry [31], Laible [46]
Myocardial Function	McCulloch [47], Winslow [48]
Myocardium Equivalent	Lipscomb [49]
Perfusion Mechanics	Vankan [27], Schonbein [37]
Perfusion Validation	Mazhari [50]
Myocardial Dysfunction	Lemmon [38]
Intramyocardial pump model	Bruinsma [20], Spaan [21]

has analyzed stresses and strains in soft biological tissues based on continuum mechanics [15], taking into account the nonlinear elasticity of microstructural tissue [16,17]. The work by Kajiya *et al* presents the function and the microcirculation of the coronary arteries [18,19]. Bruinsma *et al* [20] have provided a model wherein intramural vascular compliance and microvascular resistance are coupled to intramural pressure, in order to explain the perfusion differences to the heart on the basis of the intramyocardial pump model of Spaan *et al* [21]. Smith *et al* [22] have developed a model of the coronary circulation based on realistic anatomical data [23]. Therein, coronary blood flow and pressure distribution where calculated using Navier-Stokes equations to solve the mass and momentum conservation within this vascular structure. Criscione *et al* [24] developed a constitutive framework for the myocardium as a high-strain, parallel fibrous laminar structure, and analyzed the stresses and strains in it, in terms of the partial derivative function of the strain energy. FEA of poroelasticity equations (based on Biot's theory) has been carried out by Ferronato *et al* [25], for analyzing low speed flow through porous media [26]. Vanakn *et al* [27, 28] also analyzed blood perfusion through biological tissues, using FEA, their consideration of Reynolds number, Poisseuille-flow and extended Darcy's law has allowed for regional fluid flow analysis. For fluid flow through poroelastic deformable medium, theoretical and experimental analyses have also been performed by Parker *et al* [29].

A more recent survey carried out by de Boer [30] has emphasized the fundamentals of the porous media theory, where as Barry *et al* [31, 32] have analyzed flow-induced poroelastic deformation, using exact analytical models, their assumptions include homogeneous, isotropic and constant permeability, with stress being linearly dependent on the strain. The deformation of the porous tissue material is governed by elasticity theory, wherein the internal body force is due to the pressure gradient of the fluid. Naturally, poroelsticity of soft tissues [33] and fiber property [34–36] are important factors in influencing perfusion. In summary, the analysis of perfusion mechanics [37] provides an aid to the myocardial function and dysfunction [38].

17.2.2 Myocardial Material Properties and Their Influence on Perfusion

Figures 17.3 and 17.4 show the myocardium in a cross-section of the heart, with structural details [51]. The coronary artery blood supply is the entry condition for the myocardium whereas the venous return is the exit condition. The physiological parameters such as stroke volume, cardiac output, end diastolic volume, cardiac index, ejection fraction, LV wall stress and work together influence the LV pumping efficiency and myocardial perfusion [52].

The Young's modulus for biological tissue varies with the stress acting on it, in contrast to engineering materials wherein it remains constant.

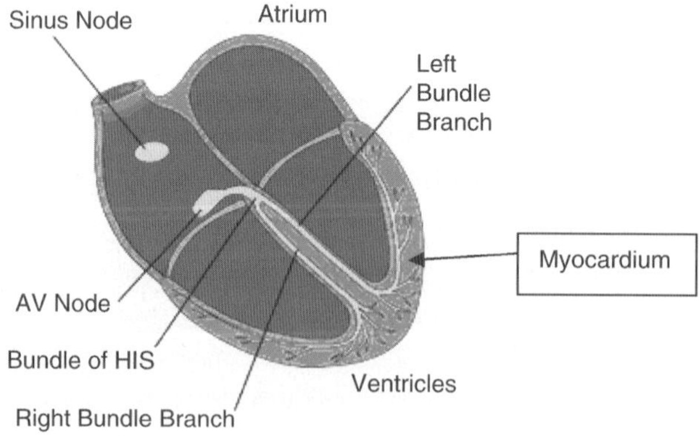

Fig. 17.3. Myocardium in the cross-section of the heart [44]

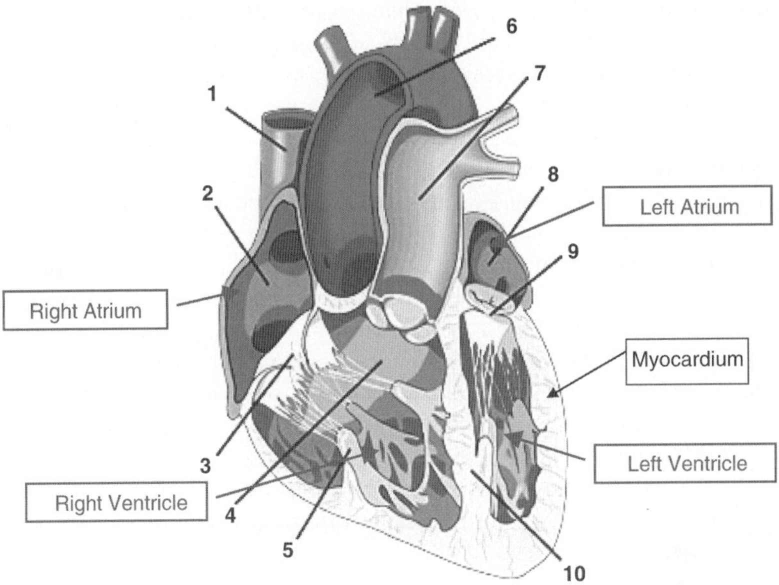

Fig. 17.4. Myocardium and detailed view in a cross-section of the heart [51]

By studying the morphological, structural and mechanical properties of soft tissues, the mechanical stresses in the myocardium can be analyzed. The stiffness (stress vs. strain) property of the myocardium varies throughout the cardiac cycle, but in a pressure load dependent manner. The challenge is to formulate the effective moduli of porous media based on strain energy considerations. In this regard the reader can refer to the classical introduction by Fung [9] in the 1980's on the material properties of living biological tissue.

In a recent study, Kakavas and Anifantis [53], indicated some of the critical parameters influencing the material properties of porous media, including micromechanical morphology, matrix material behavior, and the applied load range. The different material properties vary from linear elastic to incompressible hyper elastic. In another such study by Naili *et al* [44, 54, 55], consideration was given to the incompressible solid and fluid phases for analyzing the poroelastic behavior and micro structural parameters of deformable poroelastic medium, with specific applications for soft biological tissues like myocardium. Also, more extensive analysis on poroelastic behavior have been carried out by DiSilvestro *et al* [55], based on theory of mixtures applied to multiphase continuum. The use of poroelastic FEM [46] has enabled estimation of soft tissue material properties as a function of displacement or strain.

Accurate material property simulation [56] and identification [57] can enable perfusion analyses on poroelastic tissues [58] during active and passive states [59–61]. Bischoff *et al* [16] developed a constitutive model to characterize the material properties exhibiting hyper elastic orthotropic mechanical behavior, and compared the theoretical results with experimental values for uniaxial and biaxial test conditions. In summary, the material properties of the myocardium and its dynamic stiffness properties are factors that influence perfusion.

McCulloch *et al* [47] have analyzed the transmural stress and strain distributions and its effect on myocardial blood flow, ischemia and hypertrophy; therein, the need for three-dimensional strain distributions to completely describe the regional mechanics of myocardium has been emphasized. The importance of fibrous-sheet microstructure of myocardium has also been presented by Stevens *et al* [45] based on anatomically accurate FEMs; their focus was on fiber and sheet orientation of the myocardium. The optimized fiber orientations in the free wall, half-way between the apex and base, vary from approximately -60 degrees to 90 degrees, with the mid-wall fibers close to circumferential. The sheet orientations varied from -45 degrees at the subendocardium to -135 degrees at the subepicardium relative to a radically oriented vector. The transmural stress distributions were found to be within physiological range, when fiber and sheet orientation was optimized to minimize the stress.

Ehlers *et al* [41] have concentrated on the relation between finite dynamic equations based on Biot's theory and the concept of volume fractions. The variation in perfusion in these models occurs due to the effects of resistive forces, pressure gradient [62] and stresses affecting contractility [63], extensional and torsional strains [64]. In their work, Reza *et al* [63] modeled the stress tensor as the sum of active and passive stresses, based on the fiber coordinates. The active force generation was based on steady-state length-dependent tension developed in the fiber. The passive stresses were defined using an exponential strain-energy function. Figure 17.5 presents the distribution of stress in a thick-walled cylindrical model of the LV. This figure provides evidence as to why myocardial permeability is minimal in the endocardium and myocardial infarcts occur on the inner wall.

A significant contribution by Almeida *et al* [40, 65] has been the implementation and testing of a biphasic soft tissues with a transversely isotropic hyper-elastic solid phase, in their work, to solve the nonlinear biphasic governing equations, including the effect of strain-dependent permeability and a hyper-elastic transversely isotropic solid phase under finite deformation, the mixed-penalty and velocity-pressure FEM formulations have been used. Lipscomb *et al* [49] have done interesting research by developing a myocardial tissue equivalent, by combining quantitative analysis of mechanical and biological properties, the feasibility of their methodology has been demonstrated from the results of the contraction experiments. In order to understand the structure-function relationship of the complex behavior of soft biological tissues, a visual exploration tool has been simulated by Winslow *et al* [48].

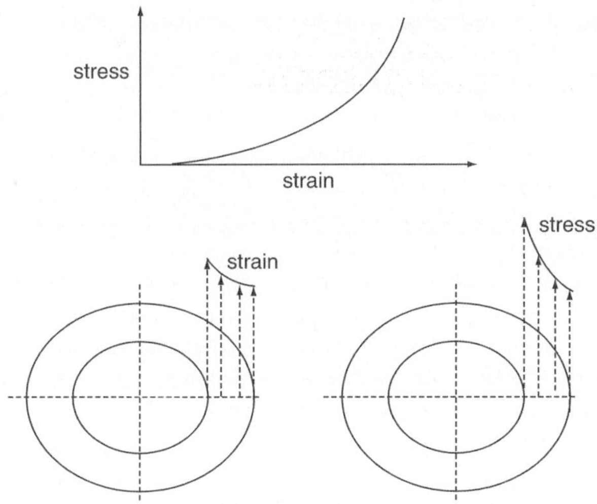

Fig. 17.5. Non-uniform strain and stress distribution in a thick-walled blood vessel. Stress and strain are higher in the inner wall than outer wall [9]

17.2.3 Myocardial Continuum Mechanics Approach for Analyzing Perfusion

Earlier works on continuum mechanics, initiated by Humphrey [39], focused on the behavior of soft tissue based on theories of elasticity, porous media and porous mixtures. The basic conservation laws by Darcy [7] and Biot [8] have defined the theory of deformation for a porous viscoelastic anisotropic solid. In earlier works of Seagrave [88], Fung [9] and Mow [12], some progress was made but in a very problem specific way. Humphrey [89] has attempted to provide some in-depth analysis to address a more general approach.

The theoretical importance of active contractile forces, during the growing stage of cardiovascular system, has been indicted by Taber [90, 91] and is characterized by growth, volumetric changes , remodeling, and morphogenesis. The theory of porous media was further extended by Huyghe *et al* [15, 92] to analyze the fluid and structure interaction between tissue stress and blood perfusion in the cardiac muscle, for which they applied a FE model.

The relation between ventricular wall stress distribution, and coronary blood flow [93] in cardiac hypertrophy has been analyzed by Costa *et al* [94]. Their analysis focused on large elastic deformations of ventricular myocardium, using three-dimensional FEM based studies on cylindrical and spherical coordinates. In other work by Costa *et al* [95] prolate spherical coordinates were used to analyze a fully converged non-symmetric three-dimensional higher-order FE model of the passive LV. Their main emphasis was how to use high-order cubic Hemite interpolation functions to improve computational efficiency and accuracy of numerical stress and strain solutions.

Table 17.2. Numerical methods for the analysis of myocardium

Main Objective	Problems addressed	References
Mathematical Model	Flow	Amini [96], Perktold [97]
Numerical Method	Poroelasticity	Mercer [98], Barry [31,32], Rappitsch [99], Ng [5,6]
Finite Element Analysis	Finite Strain	Costa [94, 95], Levenston [106], Ng [100]
		Gallagher [101], Vankan [28], Kwan [102]
		Laible [46], Simon [104]
Error Analysis	Soft tissue mechanics	Waldman [107], Afif [108], Donzelli [109]

Table 17.2 presents the works on mathematical models [96, 97], numerical methods [98, 99], finite element analysis [91, 101, 102] and error analysis. The study based on porohyperelastic constitutive theory, using Eulerian and Lagrangian forms of description, has been carried out by Simon *et al* [42,103–105] using Finite Element Method (FEM). An augmented Lagrangian formulation to enforce the saturation and incompressibility constraint-related to soft biological tissue has been used by Levenston *et al* [106]. Also pointed out are errors during measurement of strain distribution affect stresses and myocardial properties [107–109].

We also need to mention the numerical method proposed by Hon *et al* [110], using a simple meshless collocation algorithm to approximate the solution of the governing system of continuity, momentum and constitutive equations for the triphasic model. The macroscopic characteristics of the heart structure has been analyzed by Masood *et al* [111] to study LV contractility and function, with transmural gradient of perfusion observed across the heart wall.

Humphrey's work [39] on theoretical framework and characteristics of soft tissues, based on continuum biomechanics (finite elasticity, membrane theory, viscoelasticity, mixture theory, growth and remodeling and thermomechanics). Specifically, the paper has dealt with solid-fluid coupling, to study the elasto-dynamics of an intracranial secular aneurism, distended by a pulsatile blood pressure while surrounded by cerebrospinal fluid.

17.2.4 LV Diagnostics

When we discuss the LV, the question of LV diagnostic methods naturally arises. The importance of non-invasive diagnostic methods has been analyzed extensively in the books developed by Ghista *et al* [66, 67]. These works deal with (i) cardiac assessment technology, involving echocardiography, dyysler echocardiography, apexcardiography, and phonocardiography, and (ii) diagnostic processes, involving mathematical modeling of ECG-based diagnosis, compartmental analysis of the circulatory system by radio cardiogram, information calculus in non-dimensional parametric characterization of pulse-wave propagation, and indices for characterizing LV pumping efficiency.

In the work by Senda [71], a non-invasive method of measuring LV contractility in terms of *max (dp/dt)* of the LV by Doppler echocardiography has been used. A new index of cardiac contractility proposed is given by, max(dp/dt) = ρc max(du/dt), where ρ is the blood density, c the pulse wave velocity and u the flow velocity in the aorta. This method relies on imaging technologies and provides a reasonable approximation, of data obtained from in-situ experiments. Xiao *et al* [72] developed a computational method providing the relationship between arterial pressure and flow velocity. When applied to the aorta, it can indirectly provide a measure of LV contractility. The key parameters in their method were characteristic length, wall stiffness, heart rate, peak LV elasticity, end diastolic volume, and systemic vascular resistance.

The movement of soft tissue and fluid occurs simultaneously in the LV, and that complicated the resulting cardiac fluid dynamics, as indicated as a major problem by McQueen *et al* [73]. The unsteady behavior of biological fluid, [74–77] in blood vessels [78, 79] occurs during stages such as LV ejection [80, 81]. Simulation models [82, 83] and computational models [84] allow one to analyze the fluid dynamics [85] and fluid-structure interaction [86] behavior. Many studies have been carried out for the analysis of fluid inside LV chamber, but very few on the perfusion phenomena based on porous material theory.

The responses of the tissue forces and the physiological conditions during LV diastolic function have been analyzed by Lemmon *et al* [87]. They have solved the fluid mass and momentum conservation equations, using semi-implicit method for pressure equations. They have developed a computational model for the analysis of the blood-tissue interaction, and applied it to the analysis of thin-walled models of the heart. They have adopted an implicit moving boundary method for fluid-structure simulation of the interaction of the blood and cardiac structures in a beating heart.

17.2.5 Muscle Perfusion Modeling

Regional perfusion [50] varies over the different regions and with diastolic time fraction, and is further altered by disease processes such as myocardial infarction and ischemia [112–115]. For load acting on a tissue, an earlier study by Donkelaar [116] has shown the relation between perfusion and tissue pressure during contraction on skeletal muscle. A similar behavior in myocardial tissue has been analyzed and observed with the adoption of vascular waterfall theory [117], demonstrating rise in local venous resistance when intramuscular pressure exceeds the local intravascular pressure. This theory was somewhat challenged by the intramyocardial pump model, which explained that the intramural vascular volume never reached a steady state within a heart beat [21].

In order to quantify and visualize [118–120] myocardial regions at risk, Halman *et al* [120] presented a model relating myocardial perfusion to function; this involves combination of cardiac data obtained by tomographic and angiographic methods. The use of computer graphics in model-based and image-based methods may be very helpful in these types of studies, especially when it comes to patient specific modeling [122–125, 128].

The ventricular wall stress distribution, which is a principal factor governing myocardial properties [93] and coronary blood flow (and cardiac hypertrophy) has been analyzed by Costa *et al* [94]. Their analysis has involved large elastic deformations of resting and active ventricular myocardium, using three-dimensional FEM based on cylindrical and spherical coordinates. The analysis has been carried out for incompressible, nonlinear elastic, fibrous anisotropic materials, wherein the nonlinear mechanical properties of myocardium were described by a hyperelastic strain energy function. The governing equations were derived using general curvilinear coordinates, capable of handling cylindrical and spherical polar coordinates.

Another work by Costa *et al* [95] has used prolate spherical coordinates to analyze a fully converged non-symmetric 3D high-order FE model of the passive LV. Their main emphasis was to use high-order cubic Hermite interpolation functions, to improve computational efficiency and accuracy of numerical stress and strain solutions. Then, there is the work of Donkelaar *et al* [116] who computed spatial fluid flow within a blood compartment driven by a spatial gradient in fluid pressure. They found that during contraction the organization of vessels in the skeletal muscle was important for capillary perfusion.

Figure 17.6a shows the finite element mesh of the medial gastrocnemius muscle. Therein, thin lines indicate contractile elements, thick lines represent aponeurosis elements and dotted lines indicate the muscle fiber direction. The arrows indicate the locations where the arterial (P_a = 15 kPa) and venous (P_v = 0.6 KPa) blood pressures are described. Figure 17.6b shows the capillary flow through the muscle at rest (A, B) and during contraction (C, D) in simulation SIM_1 (A, C) and SIM_2 (B, D). At rest perfusion is identical (A, B) in capillary flow. During contraction, flow decreases in the distal muscle in simulation SIM_1 (C) but not in SIM_2 (D). Figure 17.6c shows the pressure distribution in skeletal muscle.

In Hron's approach to the muscle tissue consists of muscle fibers and capillary blood vessels aligned in one direction [126,127]. The extra-cellular matrix and interstitial fluid are limited in their movements. The model was tested by experiments of a muscle specimen with biaxial stretching, whereby perfusion occurs simultaneously through the capillaries. Figure 17.7a shows the schematic view of this biaxial stretching experiment Fig. 17.7b presents the undeformed and deformed configuration in two dimensions. Figure 17.7c and Fig. 17.7d show the finite element grid of the solid and its fluid velocity field.

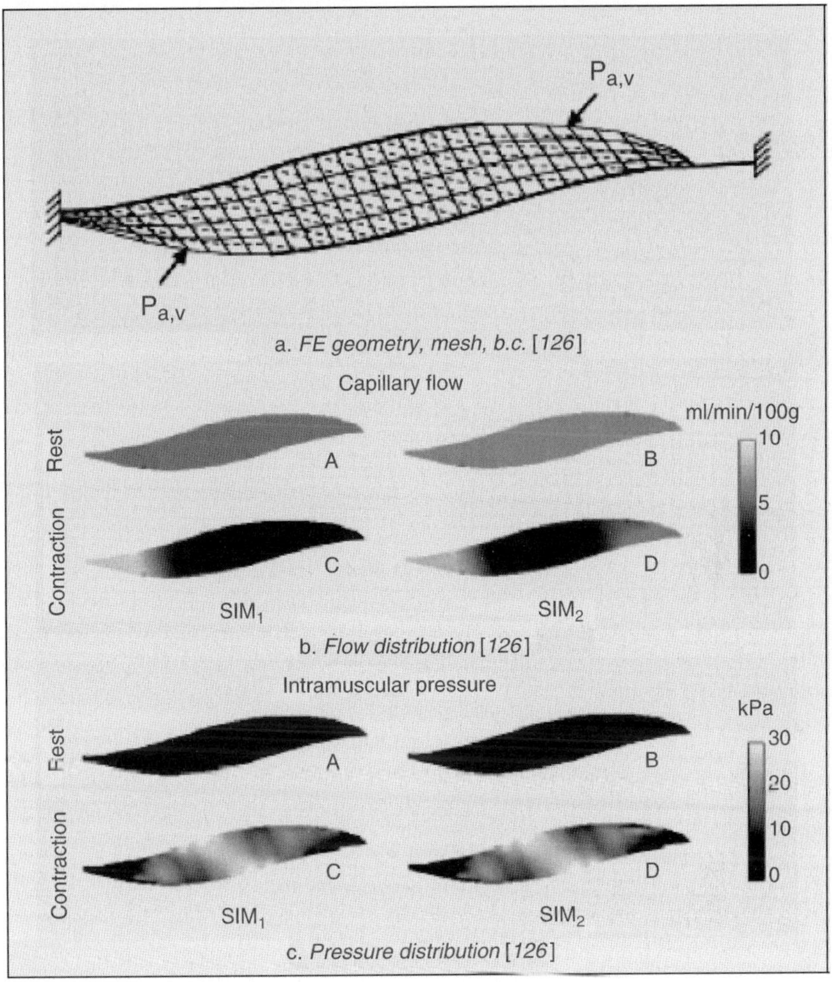

Fig. 17.6. Perfusion analysis on skeletal muscle during contraction [116]

17.3 Integrative Model

We have developed an analytical model of the myocardium to analyze LV perfusion mechanisms and distribution, and this model includes compliance, contractility, resistance-to-filling, and resistance-to-flow. The model is based on flow through a stressed porous medium, based on Darcy's law for flow through porous tissue.

To simulate a tissue in active and passive state, one needs to simulate for both stressed condition (as a function of pressure force at either end) as well as stress free condition due to the resistance to flow. The governing (Navier-Stokes) equations, adapted for the analyses, are well established and being

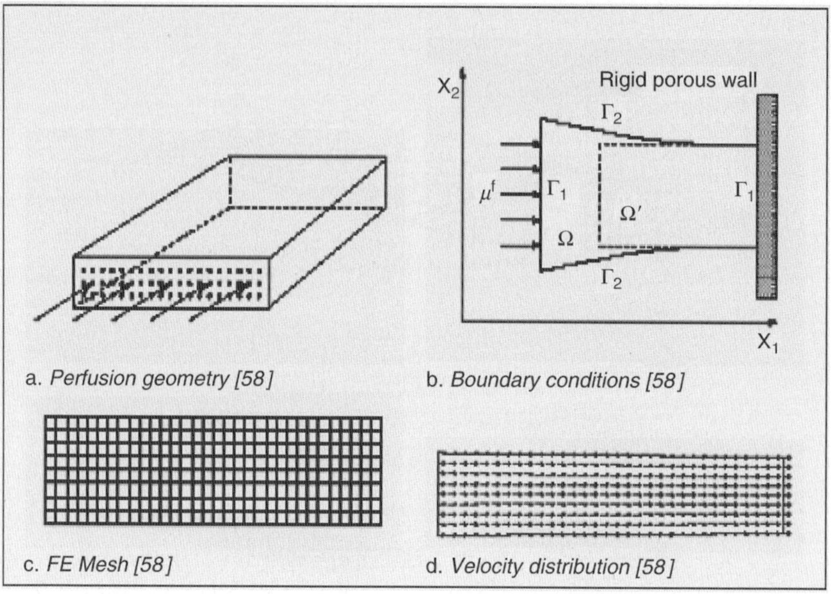

Fig. 17.7. Perfusion through soft tissue capillaries consists of muscle fibers and capillary blood vessels aligned in one direction [126]

widely used in soft tissue mechanics. Equations (17.1 to 17.8) consider the phenomena of flow through a porous medium as well as its elastic properties. Table 17.3 summarizes the steps in analyzing the poroelastic solid-flow interaction.

Continuity equation:

$$\nabla \bullet (\phi_s \boldsymbol{u}_s + \phi_f \boldsymbol{u}_f) = 0 \tag{17.1}$$

$$\phi_s + \phi_f = 1 \tag{17.2}$$

Momentum equations:

$$\nabla \bullet \boldsymbol{\sigma}_s + \boldsymbol{b}_s = 0 \tag{17.3}$$

$$\nabla \bullet \boldsymbol{\sigma}_f + \boldsymbol{b}_f = 0 \tag{17.4}$$

Constitutive equations:

$$\boldsymbol{\sigma}_s = -\phi_s p\mathbf{I} + \tilde{\boldsymbol{\sigma}}_s \tag{17.5}$$

$$\boldsymbol{\sigma}_f = -\phi_f p\mathbf{I} \tag{17.6}$$

Effective solid stress:

$$\tilde{\boldsymbol{\sigma}}_s = B_s \mathrm{tr}\,(\mathbf{T}_s)\,\mathbf{I} + 2\mu_s \mathbf{e}_s \tag{17.7}$$

Table 17.3. A Typical Flow Chart of Poroelastic Fluid-Structure Analysis

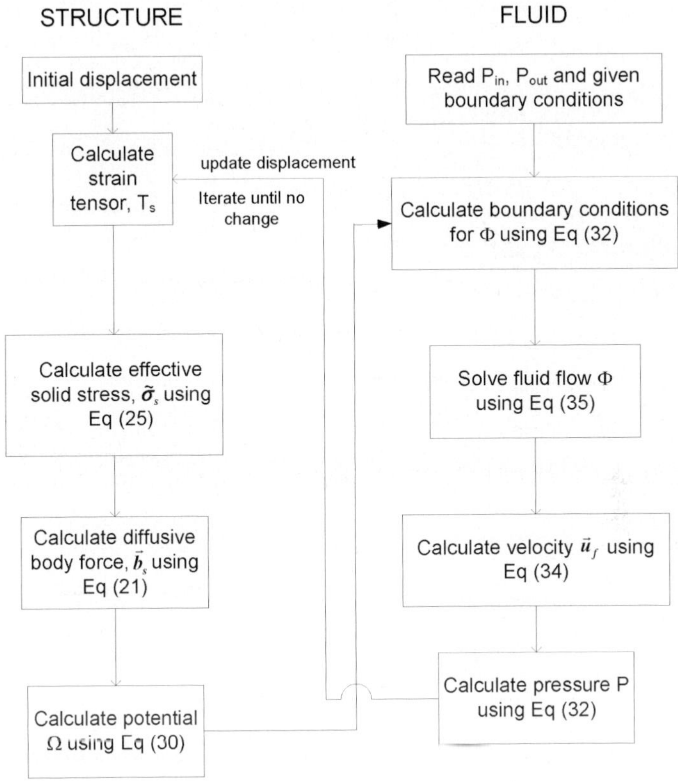

Darcy's law:

$$\vec{\boldsymbol{b}}_s - \vec{\boldsymbol{b}}_f = \frac{\mu_f}{\kappa}\left(\phi_f \vec{u}_f - \phi_s \vec{u}_s\right) \tag{17.8}$$

where the subscripts s and f refer to the incompressible, elastic solid phase and incompressible, viscous fluid (blood) phases respectively; ∇: gradient operator; ϕ: volumetric concentration; $\vec{u}$: velocity vector of solid; $\boldsymbol{\sigma}$: partial stress; $\vec{b}$: diffusive body force vector; p: hydrostatic pressure; $\mathbf{I}$: identity tensor; $\tilde{\boldsymbol{\sigma}}_s$: effective solid stress tensor; B_s: elastic bulk modulus; $\mathrm{tr}\,(\bullet)$: trace operator that yields first invariant of its tensorial argument; $\mathbf{T}_s$: solid strain tensor; μ_s: elastic shear modulus; $\mathbf{e}_s = \mathbf{T}_s - \frac{1}{3}\mathrm{tr}\,(\mathbf{T}_s)\,\mathbf{I}$: deviatoric component of solid strain tensor; μ_f: fluid dynamic viscosity; κ: permeability.
In our situation, we assume that $\phi_f >> \phi_s$ (*i.e.* $\phi_f \to 1, \phi_s \to 0$). The Eqs. (17.1) to (17.8) may be rewritten as:

Continuity equation:

$$\nabla \bullet \vec{u}_f = 0 \tag{17.9}$$

Momentum equations:

$$\nabla \bullet \boldsymbol{\sigma}_s + \vec{b}_s = 0 \tag{17.10}$$
$$\nabla \bullet \boldsymbol{\sigma}_f + \vec{b}_f = 0 \tag{17.11}$$

Constitutive equations:

$$\boldsymbol{\sigma}_s = \tilde{\boldsymbol{\sigma}}_s \tag{17.12}$$
$$\boldsymbol{\sigma}_f = -p\mathbf{I} \tag{17.13}$$

Effective solid stress:

$$\tilde{\boldsymbol{\sigma}}_s = B_s \mathrm{tr}\left(\mathbf{T}_s\right)\mathbf{I} + 2\mu_s \mathbf{e}_s \tag{17.14}$$

Darcy's law:

$$\vec{u}_f = \frac{\kappa}{\mu_f}\left(\vec{b}_s - \vec{b}_f\right) \tag{17.15}$$

Inserting Eqs. (17.10) to (17.13) into Eq. (17.15), the Darcy's law becomes:

$$\vec{u}_f = -\frac{\kappa}{\mu_f}\left(\nabla p - \vec{b}_s\right) \tag{17.16}$$

or

$$\vec{u}_f = -\frac{\kappa}{\mu_f}\left(\nabla p + \nabla \bullet \tilde{\boldsymbol{\sigma}}_s\right) \tag{17.17}$$

In Eq. (17.16), since the body force $\vec{b}_s$ is a field force ($\nabla \times \vec{b}_s = 0$ in the Eq. (17.21)), $\vec{b}_s$ can be rewritten as $\vec{b}_s = \rho_f \nabla\Omega$, where ρ_f is the fluid density and Ω is the potential.

The above equations enable us to analyze the behavior of (i) incompressible, elastic solid phase and (ii) incompressible, viscous fluid phase. The equations [127] can handle 2D and 3D geometric shapes as presented in Fig. 17.8. The loadings and the boundary conditions have to be appropriately taken into account. For analysis, the bulk blood velocity is given by Darcy's law, under the assumption that the pores are small and interconnected.

$$V = -\frac{\kappa}{\mu}\left(\nabla p - \rho\nabla\Omega\right) \tag{17.18}$$

where Ω – potential for body forces b_i.

$$b_i = \partial\Omega \Big/ \partial x_i \tag{17.19}$$

Rewriting V gives,

$$V = -\frac{\kappa}{\mu}\rho g\left(\nabla\left(\frac{p}{\rho g}\right) - \frac{1}{g}\nabla\Omega\right) = -\frac{\kappa\rho g}{\mu}\nabla\phi \tag{17.20}$$

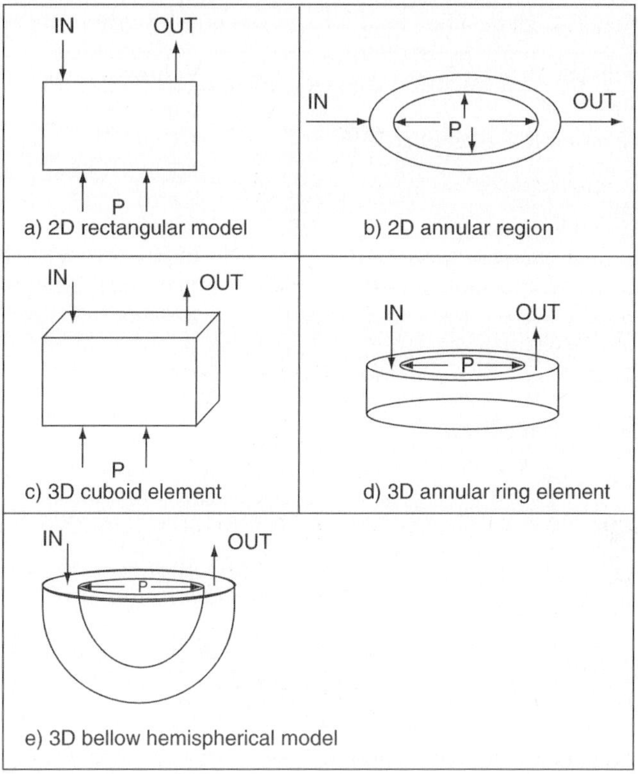

[IN: Fluid Inlet; OUT: Fluid Outlet; P: LV chamber pressure]

Fig. 17.8. Various LV myocardium perfusion models approximations with load conditions (IN: blood inlet; OUT: blood outlet; P: LV chamber pressure)

The pressure-head is given as,

$$\phi = \frac{p}{\rho g} - \frac{\Omega}{g} \tag{17.21}$$

Darcy's law is generally expressed in terms of hydraulic conductivity,

$$C = \frac{-\kappa \rho g}{\mu} \tag{17.22}$$

which gives (from Eqs. 17.20 to 17.22):

$$V_i = -C\frac{\partial \phi}{\partial x_i} \tag{17.23}$$

For an incompressible blood, $\nabla \bullet \mathbf{V} = 0$, and the pressure-head ϕ satisfies

$$\nabla \bullet (C\nabla\phi) = 0 \tag{17.24}$$

Table 17.4. Parameters used for analysis of myocardium tissue [18]

Parameter description	Value
Density	$1000\,\mathrm{kg/m^3}$
Acceleration due to gravity (g)	$9.81\,\mathrm{m/s^2}$
Dynamic Viscosity (μ)	$0.00015\,\mathrm{N.s.m^{-2}}$
Permeability (k)	$1 \times 10^{-15}\,\mathrm{m^2}$
LV Wall Thickness (t)	10 mm
Mean Inlet Pressure (p_{in})	100.0 mmHg (13.3 kPa)
Mean Outlet Pressure (p_{out})	30.0 mmHg (3.9 kPa)
Mean LV Chamber Pressure (p_1)	10.0 mmHg (1.3 kPa)

For an anisotropic medium,

$$V_i = -C_{ij}\frac{\partial \phi}{\partial x_j} \tag{17.25}$$

For this analysis, the last estimated value for V_i gives the velocity. Table 17.4 summarizes the parameters used for the analysis of myocardial tissue [18].

17.3.1 Analysis

The flow analysis of annular shaped LV model has been carried out by considering the inlet and outlet of blood flow at distant locations as well as at adjacent locations. Figure 17.9a presents a LV wall segment cross-sectional slice, with inlet and outlet of blood flow indicated. In a normal human heart, there exists three arterial supplies (inlet) and three venous returns (outlet). This can be modeled using three sets of inlet and outlet to match the physical model. For our simulation purpose, we need to analyze only one sector of the model, since the geometry exhibits symmetry. Two different sets of simulations are conducted. In the first set of simulation, we perform both stress-free and stressed boundary conditions with inlet and outlet as shown in Fig. 17.9b. For the second set of simulation, we define a sector with input adjacent to the output as shown in Figs. 17.9c and 17.9d.

Figure 17.10a shows the annular model and the numbering of boundary conditions (BCs) with inlet and outlet at a distance. Figure 17.10b shows the computed pressure distribution. Figures 17.11a and 17.11b depict the velocity distribution and streamline pattern, for the model in Fig. 17.9b. In general, the axial flow is dominant in the periphery of myocardial wall segment, but the normal flow is significant in the locations near the inlet and outlet due to the pressure variation effect at entry/exit. The pressure is uniform in the middle region, but non-uniform near the inlet and outlet, which is consistent with the velocity vector plots.

The geometry and numbering of BCs for the case with inlet adjacent to outlet are included in Fig. 17.12a. Figure 17.12b shows the pressure

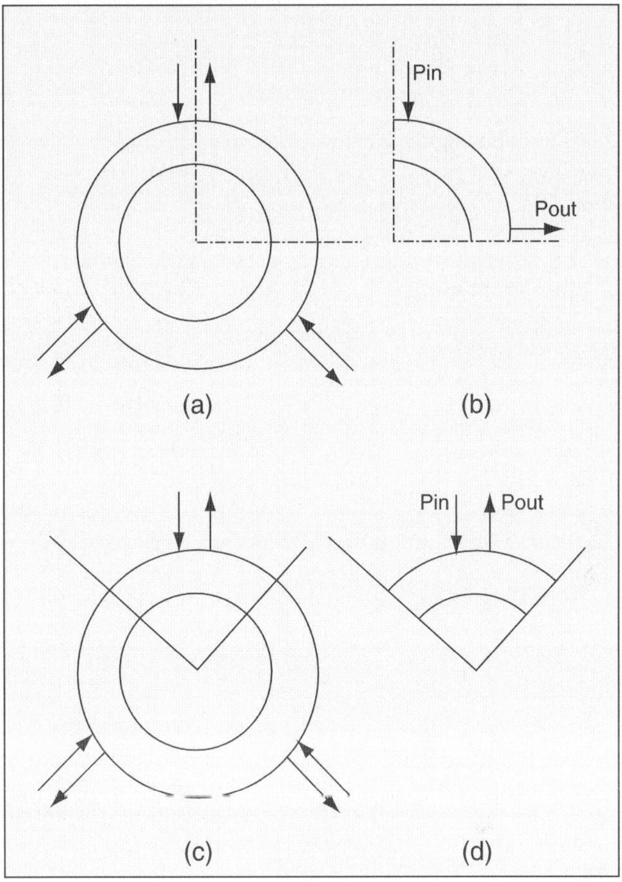

Fig. 17.9. LV sectional myocardial perfusion models, with different locations of inlet and outlet

distribution. Figure 17.13a shows the associated velocity distribution, and Fig. 17.13b gives the velocity streamline plot. The pressure is uniform far away from the entry/exit region, but non-uniform near inlet and outlet, which is consistent with the velocity vector plots. The poroelastic behavior as described above has been obtained using Darcy's law for soft tissues.

17.3.2 Applications

A failing heart assisted by an artificial vasculature allows it to eject at normal volumes, augmentates coronary perfusion, reduces wall tension, and improves cardiac mechanical and metabolic parameters as compared to conventional ventricular assist support. The porous models developed may be useful in the future to predict these improvements. Visualization of 3-D perfusion in a

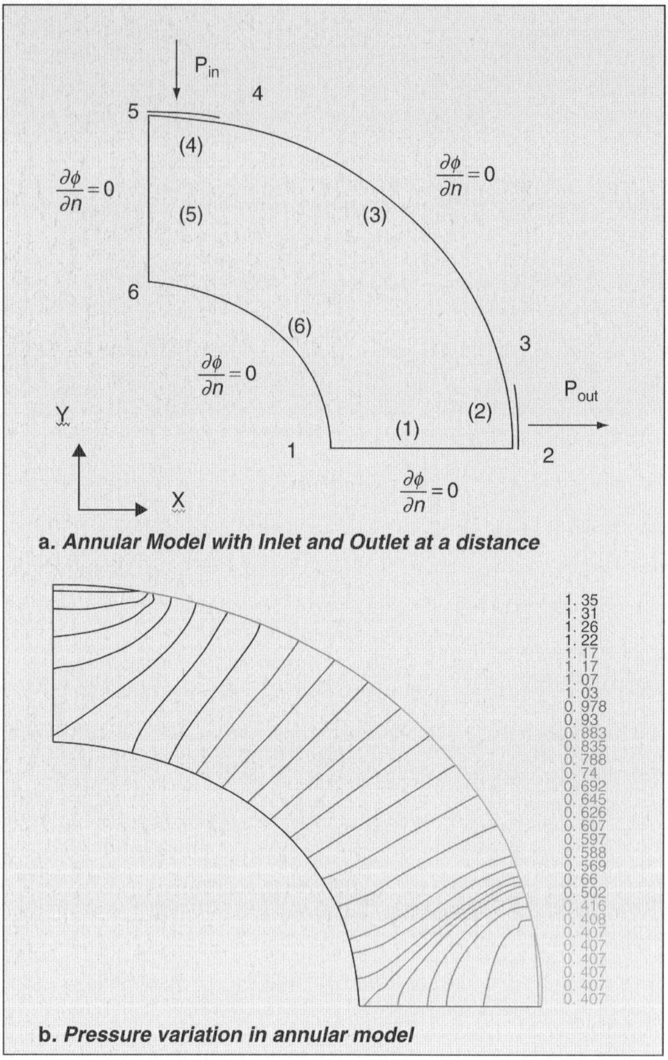

Fig. 17.10. LV annular myocardial wall model (with inlet, $p_{in} = 100\,\text{mmHg}$, and outlet, $p_{out} = 30\,\text{mmHg}$, at a distance), depicting pressure variation (kPa) inside the segment

patient's heart may be very helpful in clinical decision making. The construction of the LV as a mathematical FEM model, close to the shape of the actual LV may be suitable for simulating the deformation of the heart during the cardiac cycle. The results can be used to visualize the stress and strain across

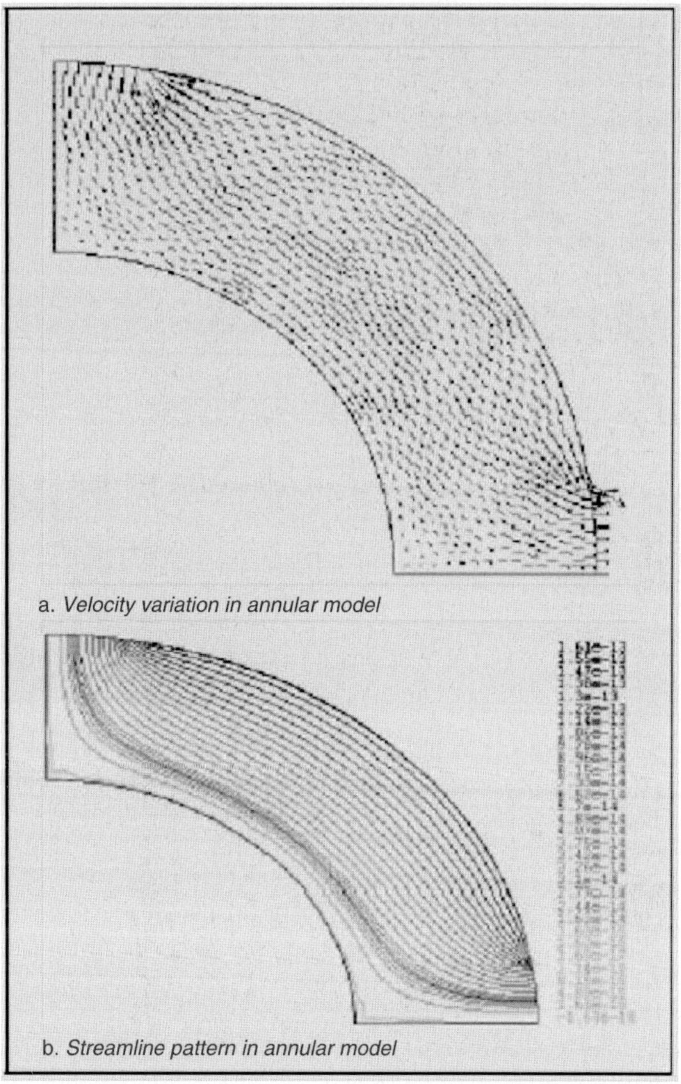

Fig. 17.11. LV annular myocardial wall segment model having flow configuration & conditions as in Fig. 17.10, a) velocity (max. vector of 0.9 m/s), b) streamline

the LV myocardium and simulate the diseased condition of the heart. This interactive virtual model is useful for wide range of simulation studies that can help in fields such as clinical treatment planning, manufacture of artificial hearts and to study heart diseases.

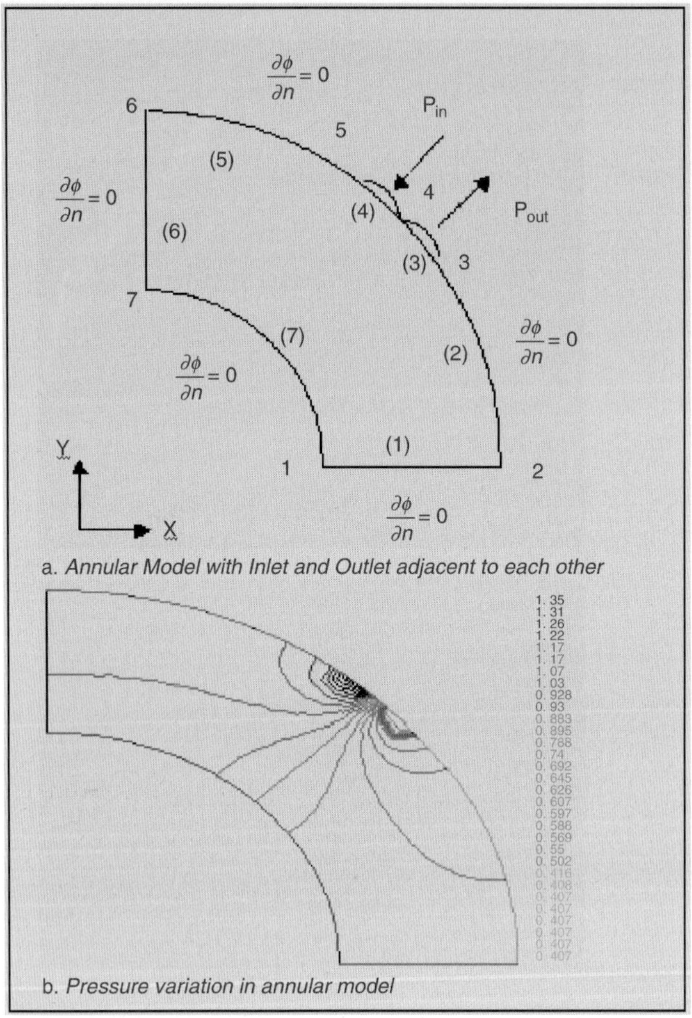

a. *Annular Model with Inlet and Outlet adjacent to each other*

b. *Pressure variation in annular model*

Fig. 17.12. LV annular myocardial model (with inlet, $p_{in} = 100\,\text{mmHg}$, and outlet, $p_{out} = 30\,\text{mmHg}$, adjacent to each other), depicting pressure variation (kPa) inside the segment

17.4 Conclusion

In this chapter, a review has been provided of literature and methods directed to understanding myocardial perfusion in porous materials. A specific biomechanical model has been derived to show a fluid perfusion through LV myocardium, based on its fundamental constitutive relations. Some results of the porous model have been presented to demonstrate the importance of location

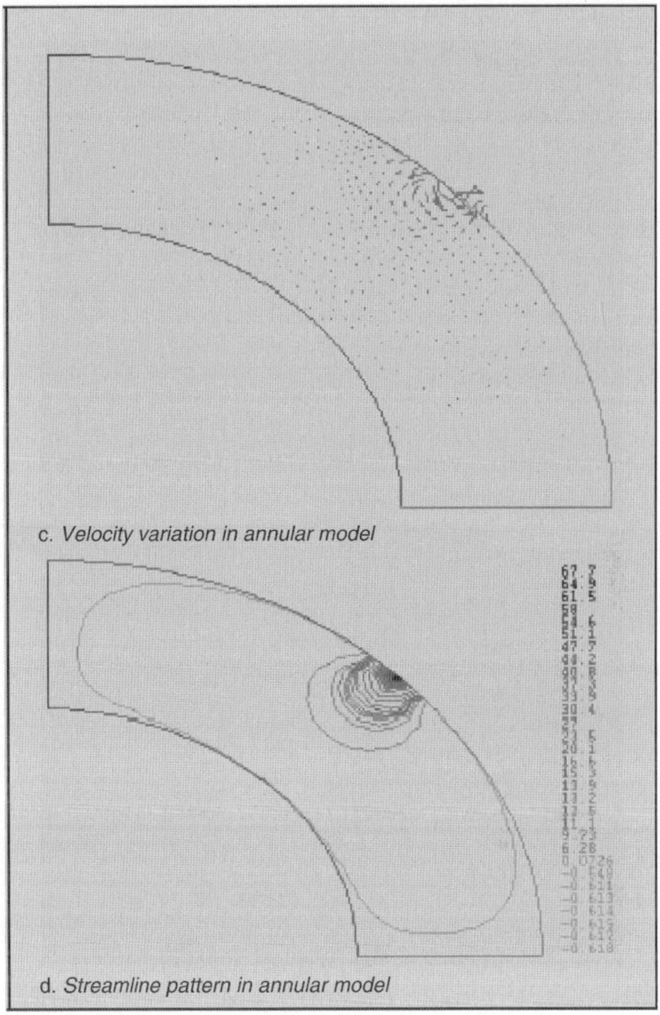

Fig. 17.13. LV annular myocardial model having flow configuration & conditions as in Fig. 17.12, a) velocity (max. vector of 0.88 m/s), b) streamline

of inlet and outlet points to its vascular bed. This porous model has been provided to analyze the effects of local stress and strain (Eqs. 17.7 & 17.14) on perfusion of the capillary bed.

In order to improve on the benchmark of numerical methods one should also include in the model analytical approaches to model elements as much as possible. The challenge will be to account appropriately for true tissue properties. We foresee that in the end such a model will be very useful not only in understanding the biological heart but also engineered heart muscle.

References

1. Vis, M.A., Bovendeerd, P.H., Sipkema, P. and Westerhof, N. (1997): 'Effect of ventricular contraction, pressure, and wall stretch on vessels at different locations in the wall', *Am. J. Physiol.*, **272**, pp. H2963–75.
2. Spaan, J.A. (1995): 'Mechanical determinants of myocardial perfusion', *Basic Res. Cardiol.*, **90**(2), pp. 89–102.
3. Smith, N.P., Pullan, A.J. and Hunter, P.J. (2000): 'Generation of an anatomically based geometric coronary model', Ann. Biomed. Eng., **28**(1), pp. 14–25.
4. Jegathese, C.R., Ng, E.Y.K. and Dhanjoo, N.G. (2004): 'Analysis of left ventricular myocardial properties', *Journal of Mechanics in Medicine and Biology*, **4**(2), pp. 173–185.
5. Ng, E.Y.K., Dhanjoo, N.G. and Cyril, R.J. (2005): 'Numerical approach to fluid-structure analysis of soft biological tissue', *Journal of Mechanics in Medicine and Biology*, **5**(1), pp. 11–28.
6. Ng, E.Y.K., Jegathese, C.R. and Dhanjoo, N.G. (2005): 'Perfusion studies of steady flow in poroelastic myocardium tissue', *International Journal of Computer Methods in Biomechanics and Biomedical Engineering*, **8**(6), pp. 349–358.
7. Darcy, H. (1856): 'Les Fontaines Publiques de la Ville de Dijon', Victor Delmont, Paris.
8. Biot, M.A. (1956): 'Theory of deformation of a porous viscoelastic anisotropic solid', *J. of Applied Physics*, **33**, pp. 1482–1498.
9. Fung, Y.C. (1981): 'Biomechanics: Mechanical properties of living tissues', *Springer Verlag*, New York, 1981.
10. Fung, Y.C. (1994): 'First course in continuum mechanics', *Prentice Hall*, 1994.
11. Hunter, P.J. and Smaill, B.H. (1988): 'The analysis of cardiac function: A continuum approach', *Proc. Biophys. Molec. Biology*, **52**, pp. 101–164.
12. Mow, V.C. and Guo, X.E. (2002): 'Mechano-electrochemical properties of articular cartilage: their inhomogeneties and anisotropies', *Annu. Rev. Biomed. Eng.*, **4**, pp. 175–209.
13. Gu, W.Y., Lai, W.M. and Mow, V.C. (1998): 'Charged-hydrated soft tissues containing multi-electrolytes: Passive transport and swelling behaviors', *Journal of Biomechanical Engineering*, **120**, pp. 169–180.
14. Vaughn, J., Czipura, A. and Humphrey, J.D. (2002): 'Measurements of finite-strain dependent permeability of biomembranes', *J. of Biomech Engg.*, **35**, pp. 287–291.
15. Huyghe, J.M. and Campen, D.H.V. (1995): 'Finite deformation theory of hierarchically arranged porous solids-II. Constitutive behavior', *Int. J. Engg. Sci.*, **33**(13), pp. 1873–1886.
16. Bischoff, J.E., Arruda, E.A. and Grosh, K. (2002): 'A microstructurally based orthotropic hyperelastic constitutive law', *ASME J. of Applied Mechanics*, **69**, pp. 570–579.
17. Oosterhout, M.F.M., Willigers, H.M.M., Reneman, R.S. and Prinzen, F.W. (1995): 'Fluorescent microspheres to measure organ perfusion: validation of simplified sample processing technique', *Am. J. Physiology*, **269**, pp. H725–H733.
18. Kajiya, F. and Goto, M. (1999): 'Integrative physiology of coronary microcirculation', *Japanese Journal of Physiology*, **49**, pp. 229–241.

19. Kajiya, F., Hiramatsu, O., Goto, M. and Ogasawara, Y. (2001): 'Mechanical characteristics of coronary circulation', *Journal of Mechanics in Medicine and Biology*, **1**(2), pp. 67–77.
20. Bruinsma, P., Arts, T., Dankelman, J. and Spaan, J.A. (1988): 'Model of the coronary circulation based on pressure dependence of coronary resistance and compliance', *Basic Res. Cardiol.* pp. 510–524.
21. Spaan, J.A., Breuls, N.P. and Laird, J.D. (1981): 'Diastolic-systolic coronary flow differences are caused by intramyocardial pump action in the anesthetized dog', *Circ. Res*. 1981, **49**(3), pp. 584–93.
22. Smith, N.P., Pullan, A.J. and Hunter, P.J. (2002): 'An anatomically based model on transient coronary blood flow in the heart', *SIAM J Appl. Math.*, **62**(3), pp. 990–1018.
23. Kassab, G.S., Rider, C.A., Tang, N.J. and Fung, Y.C. (1993): 'Morphometry of pig coronary arterial trees', *Am J Physiol.*, **265**(1 Pt 2), pp. H350–365.
24. Criscione, J.C., Mc Culloch, A.D. and Hunter, W.C. (2002): 'Constitutive framework optimized for myocardium and other high-strain, laminar materials with one fiber family', *Journal of Mechanics and Physics of Solids*, **50**(8), pp. 1681–1702.
25. Ferronato, M., Gambolati, G. and Teatini, P. (2001): 'Ill-conditioning of finite element poroelasticity equations', *Int. J. Solids Structures*, **38**, pp. 5995–6014.
26. Meroi, E.A., Natali, A.N. and Schrefler, B.A. (1999): 'A porous media approach to finite deformation behavior in soft tissues', *Computer Methods in Biomechanics and Biomechanical Engineering,* **2**, pp. 157–170.
27. Vankan, W.J., Huyghe, J.M., Janssen, J.D. and Huson, A. (1996): 'Poroelasticity of saturated solids with an application to blood perfusion', *Int. J. Engg. Sci.*, **34**(9), pp. 1019–1031.
28. Vankan, W.J., Huyghe, J.M., Janssen, J.D. and Huson, A. (1997): 'Finite element analysis of blood flow through biological tissue', *Int. J. Engg. Sci.*, **35**(4), pp. 375–385.
29. Parker, K.H., Mehta, R.V. and Caro, C.G. (1987): 'Steady flow in porous, elastically deformable materials', *J. of Applied Mechanics*, **54**, pp. 794–800.
30. De Boer, R. (2000): 'Contemporary progress in porous media theory', *Appl. Mech. Rev.*, **53**(12), pp. 323–370.
31. Barry, S.I. and Mercer, G.N. (1998): 'Exact solution for flow and deformation within finite poroelastic media', *Poromechanics*, pp. 345–350.
32. Barry, S.I., Mercer, G.N. and Zoppou, C. (1997): 'Approximate and analytical solutions for deformation of finite porous filters', *ASME Journal of Applied Mechanics*, **64**, pp. 929–934.
33. Coussy, O. (1995): 'Mechanics of porous continua', *Wiley, New York.*
34. Pao, Y.C. and Erik, R.L. (1994): 'Quantification of instantaneous extension and contraction of in vivo myocardial fiber', *Advances in Bioengineering*, **BED-28**, pp. 7–8.
35. Nielsen, P.M.F., Legrice, I.J., Smaill, B.H. and Hunter, P.J. (1991): 'Mathematical model of geometry and fibrous structure of the heart', *Am. J. Physiology,* **260**, pp. H1365–H1378.
36. Prinzen, F.W., Arts, T., Hokes, A.P.G. and Reneman, R.S.(1989): 'Discrepancies between myocardial blood flow and fiber shortening in the ischemic border zone as assessed with video imaging of epicardial deformation', *Pflugers Arch.*, **415**, pp. 220–229.

37. Scdmid-Schonbein, G.W. (1999): 'Biomechanics of microcirculatory blood perfusion', *Annu. Rev. Biomed.* Eng., **1**, pp. 73–102.
38. Lemmon, J.D. and Yoganathan, A. (2000): 'Computational modeling of left heart diastolic function: Examination of ventricular dysfunction', *Journal of Biomechanical Engineering*, **122**, pp. 297–303.
39. Humphrey, J.D. (2003): 'Continuum biomechanics of soft biological tissues', *Proc. R. Soc. Lond. A.*, **459**, pp. 2029.
40. Almeida, E.S. and Spilker, R.L. (1997): 'Mixed and penalty finite element models for nonlinear behavior of biphasic soft tissues in finite deformation: Part I–alternate or simulations', *Comput. Methods in Biomechanics and Biomechanical Engg.*, **1**, pp. 25–46.
41. Ehlers, W. and Markert, B. (2001): 'A linear viscoelastic biphasic model for soft tissues based on the theory of porous media', *Journal of Biomechanical Engineering*, **123**, pp. 418–424.
42. Simon, B.R., Kaufman, M.V., McAfee, M.A. and Baldwin, A.L. (1993): 'Determination of material properties for soft tissues using a porohyperelastic constitutive law', *Advances in Bioengineering*, **BED-26**, pp. 7–10.
43. Carcione, J.M.(2001): 'Energy balance and fundamental relations in dynamic anisotropic poro-viscoelasticity', *Proc. R. Soc. Lond. A*, **457**, pp. 331–348.
44. Naili, S., Oddou, C. and Geiger, D. (1998): 'Deformable porous medium: Application to soft biological tissues', *Poromechanics*, pp. 445–450.
45. Stevens, C. and Hunter, P.J. (2001): 'Investigation of feedback mechanisms for optimal fibrous-sheet structure with a computational model of the heart', *Bioengineering Conference ASME*, **50**, pp. 801–802.
46. Laible, J.P., Pflaster, D., Simon, B.R., Krag, M.H., Pope, M. and Haugh, L.D. (1994): 'A dynamic material parameter estimation for soft tissue using a poroelastic finite element model', *J. Biomed. Eng.*, **116**, pp. 19–29.
47. McCulloch, A.D. and Omens, J.H. (1991): 'Non-homogeneous analysis of three-dimensional transmural finite deformation in canine ventricular myocardium', *J. Biomech.*, **24**, pp. 539–548.
48. Winslow, R.L., Baumgartner, W., Scollan, D.F., Bhanot, G., Greenstein, J.L., Gresh, D.L., Yung, C.K. and Rogowitz, B.E. (2001): 'Mapping, modeling and visual exploration of structure-function relationships in the heart', *IBM Systems Journal*, **40**(2), pp. 342–359.
49. Lipscomb, M.L., Barocas, V.H., Leinwand, L.A. and Haubold, K.W. (2000): 'Development of a three dimensional in-vitro myocardial equivalent', *Southern Bioengineering Conference, Blacksburg.*
50. Mazhari, Omens J.H., Waldman L.K. and McCulloch A.D (1998): 'Regional myocardial perfusion and mechanics: A model based method of analysis', *Annals of Biomedical Engg.*, **26**, pp. 743–755.
51. Williams (Eds.), P.L. (1995): 'Gray's anatomy', *Churchill Livingstone.*
52. Katz, A.M. (1992): 'Physiology of the heart', *Raven Press Ltd.*, New York.
53. Kakavas, P.A. and Anifantis, N.K. (2003): 'Effective module of hyperelastic porous media at large deformation', *Acta Mechanica*, **160**, pp. 127–147.
54. Oddou, C., Naili, S., Burck, F.D. and Djerad, S.E. (1993): 'Unsteady rheology and poroelasticity of myocardial tissue', **BED-42**, *Bioengineering Conference ASME*, pp. 52–55.
55. Disilvestro, M.R., Zhu, Q. and Suh, J.K.F. (2001): 'Biphasic poroviscoelastic simulation of the unconfined compression of the articular cartilage: II-Effect of variable strain rates', *Journal of Biomechanical Engineering*, **123**, pp. 52–55.

56. Ottensmeyer, M.P., Salisbury, B.U.E. and Kenneth, J. (2000): 'Input and output for surgical simulation: Devices to measure tissue properties in-vivo and a hapticinterface for laparoscopy simulators', *Proceedings of Medicine Meets Virtual Reality,* WESTWOOD, J.D., Helene M. Hoffman, Greg T. Mogel, Richard A. Robb and Don Stredney. (2000): (Eds.), Newport Beach, CA, IOS Press, pp. 236–242.
57. Vorp, D.A., Rajagopal, K.R., Smolinski, P.J. and Borovetz, H.S. (1995): 'Identification of elastic properties for orthotropic vascular segments in Distension', *Journal of Biomechanics*, **28**, pp. 501–512.
58. Cheng, A.H.D. and Detornay, E. (): 'On singular integral equations and fundamental solutions of poroelasticity', *Int. J. Solids Structures*, **35**(34–35), pp. 4521–4555.
59. Horowitz, A., Lanir, Y., Yin, F.C., Perl, M., Sheinman, I. and Strumpf, R.K. (1998): 'Structural three-dimensional constitutive law for the passive myocardium', *ASME J. Biomech. Eng.*, **19**, pp. 200–207.
60. Yin, F.C.P., Chanc, C. and Judd, R.M. (1996): 'Compressibility of perfused passive myocardium', *Am. J. Physiology*, **271**, pp. H1864–H1870.
61. Guccione, J.M., Mcculloch, A.D. and Waldman, L.K. (1991): 'Passive material properties of intact ventricular myocardium determined from a cylindrical model', *ASME J. Biomech. Eng.*, **113**, pp. 42–55.
62. Vierendeels, J.A., Riemslagh, K., Dick, E. and Verdonck, P.R. (2000): 'Computer simulation of intraventricular flow and pressure gradients during diastole', *Journal of Biomechanical Engineering*, **122**(6), pp. 667–674.
63. Mazhari, R. and Mc Culloch, A.D. (1999): 'Three dimensional mechanics of myocardial contraction: Mechanism of transverse systolic stress', *Proceedings of ASME* 4th *Summer Bioengineering Conference*, pp. 43–44.
64. Criscione, J.C., Lorenzen-Schmidt, I., Humphrey, J.D. and Hunter, W.C. (1999): 'Mechanical contribution of endocardium during finite extension and torsion experiments on papillary muscles', *Annals Biomed Engg.*, **27**, pp. 123–130.
65. Almeida, E.S. and Spilker, R.L. (1998): 'Finite element formulations for hyperelastic transversely isotropic biphasic tissues', *Comput. Methods Appl. Mech. Engg.*, **151**, pp. 513–538.
66. Ghista, D.N. and Mihoczy, L. (1989): 'Noninvasive cardiac assessment technology', *Advances in Cardiovascular Physics*, (D.N. Ghista, editor), **6**, Karger, Basel, New York.
67. Ghista, D.N., Van Vollenhoven, E., Yang, W.-J. and Revl, H. (1979): 'Foundations of noninvasive cardiovascular diagnostic processes', *Advanced in Cardiovascular Physics*, (D.N. Ghista, editor), **4**, Karger, Basel, New York.
68. Ray, G., Ghista, D.N. and Sandler, H. (1979): 'Left ventricular biomechanical analyses for the development of indices characterizing normal-diseased myocardium and left ventricular pumping efficiency', *Adv. Cardiovasc. Phys.*, Karger, Basel, **4**, pp. 161–178.
69. Mirsky, I., Ghista, N. and Sandler, H. (1974): 'Cardiac mechanics: Physiological, clinical, and mathematical considerations', *John Wiley and Sons.*
70. Osterhues, H.H., Hombach, V. and Moss, A.J. (2000): 'Advances in noninvasive electrocardiographic monitoring techniques', *International Society for Holter and Noninvasive Electrocardiography*, Kluwer and Academic.

71. Senda, S., Sugawara, M., Matsumoto, Y., Kan, T. and Matsuo, H. (1992): 'A non-invasive method of measuring Max (dp/dt) of the left ventricle by Doppler echocardiography', *J. of Biomechanical Engineering*, **114**, pp. 15–19.
72. Xiao, X., Kaazempur-Mofrad, M.R., Ozawa, E.T. and Kamm, R.D. (2001): 'Non-invasive assessment of cardiovascular health: a computational/ experimental study', **BED-50**, *Bioengineering Conference ASME*, pp. 67–68.
73. McQueen, D.M. and Peskin, C.S. (2000): 'A three-dimensional computer model of the human heart for studying cardiac fluid dynamics', *Computer Graphics*, pp. 55–60.
74. Nicoud, F. and Schonfeld, T. (2002): 'Integral boundary conditions for unsteady biomedical CFD applications', *Int. J. for Numerical Methods in Fluids*, **40**, pp. 457–465.
75. Power, H. (1995): 'Bio-fluid mechanics', Computational Mechanics Publications.
76. Majumdar, J. (1992): 'Bio-fluid mechanics', *World Scientific*, Singapore.
77. Nereem, R. (1992): 'Vascular fluid mechanics, arterial wall and atherosclerosis', *ASME J. Biomech. Eng.*, **114**, pp. 274–282.
78. Pedley, T.J. (1990): 'The fluid mechanics of large blood vessels', *Cambridge University Press*, U.K.
79. Wooten, D.M. and Ku, D.N. (1999): 'Fluid mechanics of vascular systems, diseases, and thrombosis', *Annu. Rev. Biomed. Eng.*, **1**, pp. 299–329.
80. Georgiadis, J. G., Wang, M. and Pasipoularides, A. (1992): 'Computational fluid dynamics of left ventricular ejection', *Annals of Biomedical Engineering*, **20**, pp. 81–97.
81. Taylor, T.W., Okino H. and Yamaguchi, T. (1994): 'Three-dimensional analysis of left ventricular ejection using computational fluid dynamics', *Journal of Biomechanical Engineering*, **116**, pp. 127–130.
82. Liepsch, D. (2002): 'An introduction to biofluid mechanics-basic models and applications', *J of Biomechanics*, **35**, pp. 415–435.
83. Penrose, J.M.T. and Staples, C.J. (2002): 'Implicit fluid-structure coupling for simulation of cardiovascular problems', *Int. J. for Numerical Methods in Fluids*, **40**, pp. 467–478.
84. Peskin, C.S. and McQueen, D. (1989): 'A three-dimensional computational method for blood flow in the heart-I Immersed elastic fibers in a viscous incompressible fluid', *J. of Comput. Phys.*, **81**(2), pp. 372–405.
85. Verdonck, P. (1998): 'Intra and extracorporeal cardiovascular fluid dynamics: Vols. I and II', *WIT Press*, U.K.
86. Bathe, M. and Kamm, R.D. (2001): 'Fluid-structure interaction analysis in biomechanics', *First MIT Conference on Computational Fluid and Solid Mechanics*, U.S.A.
87. Lemmon, J.D. and Yoganathan, A.P. (2000): 'Three-dimensional computational model of left heart diastolic function with fluid-structure interaction', *Journal of Biomechanical Engineering*, **122**, pp. 109–117.
88. Seagrave, R.C. (1971) 'Biomedical applications of heat, and mass transfer', *Iowa State Univ. Press*, AMES, 66.
89. Humphrey, J.D. (2001): 'Cardiovascular solid mechanics: Cells, tissues, and organs', *Springer, NY*.
90. Taber, L.A. (2001): 'Investigating a mechanical growth law for the heart', *ASME Bioengg. Conference*, **BED-50**.

91. Taber, L.A. (2001): 'Biomechanics of cardiovascular development', *Annu. Rev. Biomed. Eng.*, **3**, pp. 1–25.
92. Huyghe, J.M. and Campen, D.H.V. (1995): 'Finite deformation theory of hierarchically arranged porous solids-I. Balance of mass and momentum', *Int. J. Engg. Sci.*, **33** (13), pp. 1861–1871.
93. Costa, K.D., Holmes, J.W. and McCulloch, A.D. (2001): 'Modeling cardiac mechanical properties in three dimensions', *Phil. Trans. R. Soc. Lond. A*, **359**, pp. 1233–1250.
94. Costa, K.D., Hunter, P.J., Rogers, J.M., Guccione, J.M., Waldman, L.K. and McCulloch, A.D. (1996): 'A three dimensional finite element method for elastic deformation of ventricular myocardium: I-cylindrical and spheroidal polar coordinates', *Journal of Biomechanical Engineering*, **118** (4), pp. 452–463.
95. Costa, K.D., Hunter, P.J., Rogers, J.M., Guccione, J.M., Waldman, L.K. and McCulloch, A.D. (1996): 'A three dimensional finite element method for large elastic deformation of ventricular myocardium: II-Prolate spheroidal coordinates', *J. Biomech. Eng.*, **118** (4), pp. 464–472.
96. Amini, A.A. and Prince, J.L. (2001): 'Measurement of cardiac deformations from MRI: Physical and mathematical models', *Kluwer Academic Publishers.*
97. Perktold, K. and Rappitsch, G. (1994): 'Mathematical modeling of local arterial flow and vessel wall mechanics', In Crolet, J., Ohayon, R. (1994): 'Computational Methods for Fluid Structure Interaction', *Pitman Research Notes in Mathematics*, Longman, **306**, pp. 230–245.
98. Mercer, G.N. and Barry, S.I. (1999): 'Flow and deformation in poroelasticity: II Numerical Method', *Math. And Comput. Modeling*, **30**, pp. 31–38.
99. Rappitsch, G., Perktold, K. and Pernkopf, E. (1997): 'Numerical modeling of shear-dependence mass transfer in large arteries', *International Journal for Numerical Methods in Fluids*, **25**, pp. 847–857.
100. Ng, E.Y.K. and Chua, L.T. (2000): 'Mesh independent prediction of skin burns injury', *J. of Medical Engineering and Technology*, **24**(6), pp. 255–261.
101. Gallagher, R.H. (1982): 'Finite elements in biomechanics', *John Wiley.*
102. Kwan, Huyghe, J.M., Janssen, J.D. and Huson, A. (1990): 'Finite element analysis of blood flow through biological tissue', *Int. J. Engg. Sci.*, **35**(4), pp. 375–385.
103. Simon, B.R. (1992): 'Multiphase poroelastic finite element models for soft tissue structures', AMR, 45(6), pp. 191–218.
104. Simon, B.R., Kaufman, M.V., McAfee, M.A. and Baldwin, A.L. (1993) 'Finite element models for arterial wall mechanics', *Journal of Biomechanical Engineering*, **115**, pp. 489–496.
105. Simon, B.R., Kaufman, M.V., McAfee, M.A. and Baldwin, A.L. (1996): 'Porohyperelastic theory and finite element models for soft tissues with application to arterial mechanics', *Mechanics of Poroelastic Media*, Ed. SELVADURAI, A.P.S. (1996): *Kluwer Academic Publishers.*
106. Levenston, M.E., Frank, E.H. and Grodzinsky, A.J. (1998): 'Variationally derived 3-D field finite element formulation for quasi-static poroelastic analysis of hydrated biological tissues', *Comput. Methods. Appl. Mech. Engg.*, **156**, pp. 231–246.
107. Waldman, L.K. and McCulloch, A.D. (1993): 'Non-homogeneous ventricular wall strain: analysis of errors and accuracy', *J. Biomech. Engg.*, **115**, pp. 497–502.

108. Afif, M. and Amaziane, B. (2002): 'Convergence of finite volume schemes for a degenerate convection-diffusion equation arising in flow in porous media', *Comput. Methods Appl. Mech. Engg.*, **191**(46), pp. 5265–5286.
109. Donzelli, P.S., Spilker, R.L., Bachmann, P.L., Niu, Q. and Shephard, M.S. (1992): 'Automated adaptive analysis of the biphasic equations for soft tissue mechanics using a posteriori error indicator', *Int. J. for Numerical Methods in Engineering*, **34**, pp. 1015–1033.
110. Hon, Y.C., Lu, M.W., Xue, W.M. and Zhou, X. (2002): 'Numerical algorithm for triphasic model of charged and hydrated soft tissues', **19**, pp. 1–15.
111. Masood, S. and Yang, G.Z. (2001): 'Macroscopic structure and physiology of the normal and diseased heart', *Technical Report*, Department of Computing, Imperial College, London.
112. Tad, W.T. (1996): 'The effects of cardiac infarction on realistic three dimensional left ventricular blood ejection', *Journal of Biomechanical Engineering*, **118**, pp. 106–110.
113. Leuven, S.L.V., Waldman, L.K. and McCulloch, A.D. (1994): 'Gradients of epicardial strain across the perfusion boundary during acute myocardial ischemia', *Am. J. Physiology*, **267**, pp. H2348–H2362.
114. Hoffman, J.I. and Spaan, J.A. (1990): 'Pressure-flow relations in coronary circulation', *Physiol. Rev.*, **70**(2), pp. 331–390.
115. Fokkema, D.S., Vanteeffelen, J.W., Dekker, S., Vergroesen, I., Reitsma, J.B. and Spaan, J.A. (2005): 'Diastolic time fraction as a determinant of subendocardial perfusion', *Am. J. Physiol. Heart Circ. Physiol.*, **288**(5), pp. H2450–H2456.
116. Van Donkelaar, C.C., Huyghe, J.M., Vankan, W.J. and Drost, M.R. (2001): 'Spatial interaction between tissue pressure and skeletal muscle perfusion during contraction', *Journal of Biomechanics*, **34**, pp. 631–637.
117. Downey, J.M. and Kirk, E.S. (1975): 'Inhibition of coronary blood flow by a vascular waterfall mechanism', *Circ Res.*, **36**(6), pp. 753–760.
118. Robb, R.A. (1999): '3-D visualization in biomedical applications', *Annu. Rev. Biomed. Eng.*, **1**, pp. 377–399.
119. Park, J. and Park. S.I. (2000): 'Strain analysis and visualization: left ventricle of a heart', *Computers and Graphics*, **24**, pp. 701–714.
120. Johannes and Behr. (2004): 'Modeling, visualization, and interaction techniques for diagnosis and treatment planning in cardiology', *Computers and Graphics*, **24**, pp. 741–753.
121. Halman, M., Sideman, S., Lessick, J. and Beyar, R. (1994): 'Relating coronary perfusion to myocardial function using three-dimensional reconstruction of heart and coronary arteries', *Med. & Biol. Eng. & Comput.*, **32**, pp. S144–S150.
122. Ridgeway, J.P., Radjenovic, A., Higgins, D.M., Kassner, A. and Sivananthan, M. (2001): 'A fast multi-slice multiple-point method for determination of TI relaxation time for use in quantitative myocardial perfusion measurement', *Proc. Intl. Soc. Magn. Reson. Med.*, 1905.
123. Radjenovic, A. (2001): 'Global measurement of the myocardial perfusion index:application to the assessment of the incomplete Gd-DTPA clearance in rest/stress studies of cardiac perfusion', *Proc Intl. Soc. Magn. Reson. Med.*, **229**.

124. Song, S.M., Leahy, R.M.H., Brundage, D.P.B. and Napel, S. (1994): 'Determining cardiac velocity fields and intraventricular pressure distribution from a sequence of ultra fast CT cardiac images', *IEEE Transactions on Medical Imaging*, **13**(2), pp. 386–397.
125. O'Dell, W.G. and McCulloch, A.D. (2000): 'Imaging three-dimensional cardiac function', *Annu. Rev. Biomed. Eng.*, **2**, pp. 431–56.
126. Hron, J. (2000): 'Mathematical description of soft tissue mechanics', *Proceedings of Algoritmy*, pp. 37–46.
127. Hron, J., Malek, J. and Rajagopal, K.R. (2001): 'Simple flows of fluids with pressure dependent viscosities', *Proc R Soc Lond A*, **457**, pp. 1603–1622.
128. Spaan, J.A.E., Ter Wee, R., Van Teeffelen, J.W.G.E., Streekstra, G., Siebes, M., Kolyva, C., Vink, H., Fokkema, D.S. and Vanbavel, E. (2005): 'Visualisation of intramural coronary vasculature by an imaging cryomicrotome suggests compartmentalisation of myocardial perfusion areas', A Medical and Biological Engineering and Computing **43**(4), pp. 431–435.

18

Wavelets and its Application in Cardiology

Jayachandran E S, Paul Joseph K

Wavelet theory is associated with building a model for a signal or a system with a set of special signals called wavelets. Wavelets are "small waves" which are oscillatory with fast decaying to zero in nature. Fourier Transform is a kind of signal representation which makes use of infinitely supported sinusoidal signals where the time localization of the signal is not achieved. Short Time Fourier Transform (STFT) or a windowed Fourier Transform provides some degree of localization in time, but the resolution is fixed. But wavelet transform provides time and frequency localization at various levels. Here wavelets are used to approximate a signal. Each element in a wavelet set is constructed from a single function called the "mother wavelet". Each element is a scaled and translated version of the mother wavelet. Thus the given signal is broken down into scaled and translated forms of wavelet, the process being called wavelet decomposition or wavelet transformation [1–4]. The reconstruction of the signal from the Wavelet transform is the Inverse Wavelet Transform.

Such an analysis and representation is useful in the processing of seismic signals, image processing, signal compression, acoustics, control systems, bio-signal analysis etc. where time localization is essential. This chapter deals with a brief theory of the wavelet transforms and a few applications specific to the analysis of ECG signals.

18.1 The Fourier Transform

For a signal f(t), a square integrable function i.e. $f(t) \in L2(R)$, the Fourier Transform (FT) is defined by

$$F(\omega) = \int_{-\infty}^{\infty} f(t)\, e^{-j\omega t}\, dt \tag{18.1}$$

thus decomposing the signal into its frequency components. The inverse relationship is

$$f(t) = 1/2\pi \int_{-\infty}^{\infty} F(\omega)\, e^{j\omega t}\, d\omega \qquad (18.2)$$

Equation 18.2 can be interpreted as expressing f as a combination of harmonic waves, $e^{-j\omega t}$. Here we can achieve well localization in frequency. But we do not have any information on the time at which a particular frequency component appears. A windowed FT or STFT provides both time and frequency localization. For example, a large value of the transform near a time 't' and a frequency 'ω' implies that the signal 'f' contains a large component frequency 'ω' near that time 't'.

18.2 Short Time Fourier Transform

For obtaining the frequency contents of a signal at some desired location in time 'b', the given f(t) is windowed by multiplying with a suitable window function w(t) shifted to exist near 'b'. i.e. $f_w(t) = f(t)\, w(t-b)$. Then the FT is taken [Fig. 18.1].

$$\{STFT_w\ f\}(b, \zeta) = \int_{-\infty}^{\infty} f(t)\, w^*_{b,\zeta}(t) dt \qquad (18.3)$$

Where

$$w_{b,\zeta}(t) = w(t - b)\, e^{j\zeta t}$$

By changing 'b' we can slide the window to cover the entire signal. Thus the frequency content of the signal at different localization in time 'b' can be found. This is termed as 'Running Window Fourier transform'.

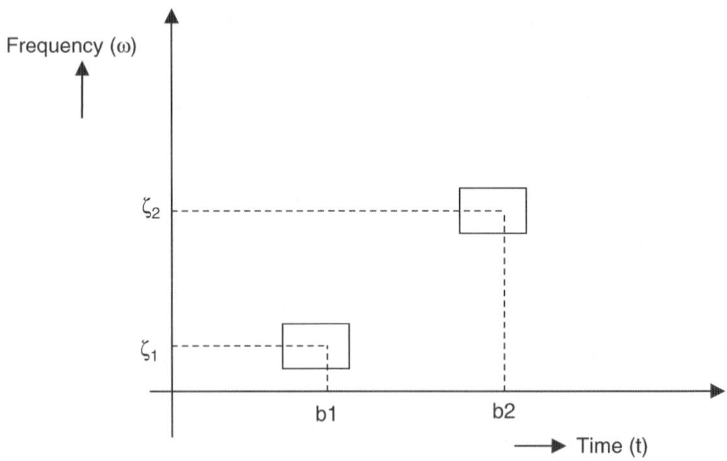

Fig. 18.1. Time – frequency localization

The Inverse STFT can be expressed as

$$f(t) = 1/2\pi \int_{-\infty}^{\infty} e^{j\zeta t}\, d\zeta \int_{-\infty}^{\infty} \{STFT\}\, w^*(t - b)\, db \qquad (18.4)$$

If we impose the time-frequency localization in a time-frequency plane, we get the plot as shown in Fig. 18.2.
See that the time window width is fixed and hence the resolution[1].

18.3 Continuous Wavelet Transform (CWT)

In wavelet theory, the scaling and translation operators act simultaneously on the mother wavelet function. (Affine operation). [The name wavelet, meaning little wave, came from the study of short duration seismic acoustic wave packets.] Performing affine operation on the mother wavelet creates a set of scaled translated versions of original mother wavelet. This is called a wavelet set. The mother wavelet is called the kernel of the wavelet transform.
The Continuous Wavelet Transform (CWT) of a function f(t) with respect to a mother wavelet $\Psi(t)$ is given by

$$[W_\Psi f(t)](a, b) = \int_{\infty}^{\infty} f\,(t)(1/\sqrt{a})\Psi^*((t - b)/a)dt = W_\Psi f(a, b) \quad ; a \neq 0 \quad (18.5)$$

Here a function 'f' in 't' is transformed into another function in 'a' and 'b'. A wavelet coefficient $W_\Psi\, f(a, b)$ at a particular scale and translation represents how well the signal 'f' and the scaled and translated mother wavelet match; or the coefficient represents the "degree of correlation" between the functions at a particular scale and translation.
The following conditions are to be satisfied for the function $\Psi(t)$ to behave as a wavelet.

Average value $\int \Psi(t)\, dt = 0;$ C-1

$\int |\Psi\,(t)|^2\, dt - 0;$ energy signal; C-2

$$C_\Psi = \int_{-\infty}^{\infty} |\Psi(\omega)|^2/|\,\omega|\, d\omega < \infty \qquad \text{C-3}$$

$\neq 0$: admissibility condition for wavelet.

If we define $\Psi_{a,b}\,(t)$ as the translated and scaled version of $\Psi\,(t)$, then

$$\Psi_{a,b}\,(t) = (1/\sqrt{|a|})\ \Psi\ ((t - b)/a)$$

The multiplication with $1/\sqrt{|\,a\,|}$ is to satisfy the second condition C-2. (Normalizing — $\|\Psi\| = 1$)
So $W_\Psi\ f(a, b) = \langle f(t),\ \Psi_{a,b}(t)\rangle$ CWT as the inner product. The concept of CWT is shown in figure below (Fig. 18.2).

[1] **A function can not be limited in time and frequency simultaneously. The figure of merit of a time-frequency window is its time frequency width product which is bounded below by uncertainty principle,* $\Delta t \Delta\omega \geq 1/2$.

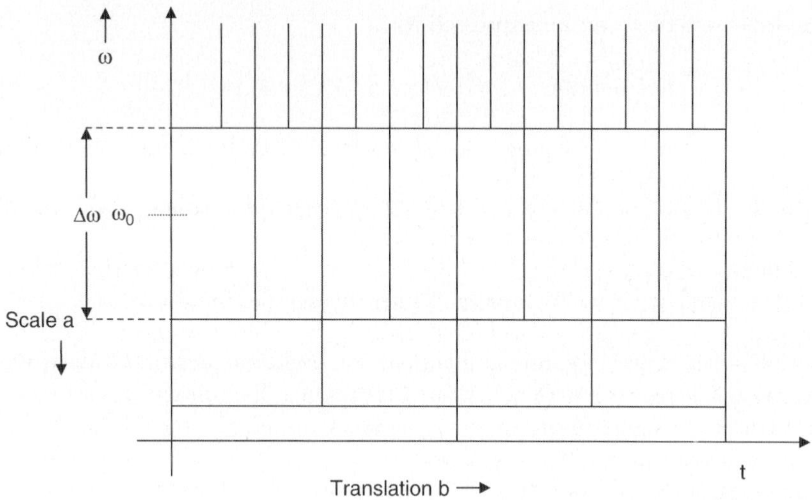

Fig. 18.2. Time frequency tiling

18.3.1 The Inverse Continuous Wavelet Transform (ICWT)

The expression for ICWT is obtained from the resolution of identity given by

$$\int_{-\infty}^{\infty} (1/a^2) \int_{-\infty}^{\infty} \langle f_1(t),\ \Psi_{a,b}(t)\rangle\langle\Psi_{a,b}(t),\ f_2(t)\rangle \text{ db da} = C_\Psi\langle\ f_1(t),\ f_2(t)\rangle \quad (18.6)$$

where $f_1(t), f_2(t),\ \Psi(t) \in\ L^2(R)$ and $\Psi(t)$ satisfies the admissibility condition. Letting $f_1(t) = f(t)$ and $f_2(t) = \delta(t\text{-}T)$ will yield the ICWT

$$f(t) = (C_\Psi)^{-1}\ \int a^{-2} \int\ W_\Psi f(t)(a, b)\Psi_{a,b}\ (t) \text{ db da} \quad (18.7)$$

Note that the CWT produces a 3 dimensional plot with a, b and $W_\Psi f$ on the respective axes. It is a surface with 'a' and 'b' as independent variables. The signal energy in the wavelet transform representation is given by

$$E(f) = (C_\Psi)^{-1} \int a^{-2} \int |\ W_\Psi\ f(t)(a, b)|^2\ \Psi_{a,b}\ (t) \text{ db da} \quad (18.8)$$

18.4 Discrete Wavelet Transform (DWT)

In practice CWT is less often employed. Most of the computer implementations utilize the Discrete Wavelet Transform or DWT. Here the independent variable 't' of the function is not discretized; rather the transform variables 'a' and 'b' are discretized. The transform then is often referred to as Continuous time wavelet series.

The scaling parameter 'a' is discretized as $a = p^j$, $j \in Z$. and $p \neq 0$.
If $p = 2$, it is called dyadic sampling.
Then the wavelet function will take the form

$$\Psi_{j,k} = 2^{-j/2}\ \Psi(2^{-j}\ t - k) \text{ where } k = 2^{-j}\ b \quad j, k \in Z \qquad (18.9)$$

Then the DWT will be a 2 dimensional sequence of numbers, W(j, k). The time–scale resolution is shown in Fig. 18.2.

The size of each rectangle in the Fig. 18.2 is decided by a particular scaled and translated wavelet. Each rectangle represents the simultaneous time–scale resolution corresponding to each scaled and translated wavelet. For a fixed scale, the resolution cell size is identical for all translations. For each block along the frequency axis, the $Q = \omega_0/\Delta\omega$ remains the same. Moving up in the scale results in higher time resolution and poor frequency resolution. Going down results in higher frequency resolution and poor time resolution. Also each resolution cell represents a discrete wavelet coefficient in the wavelet domain [1–4]. It is equivalent to decomposing the signal into different frequency bands at different time intervals, by passing the signal through a bank of constant Q filters.

18.5 Multi Resolution Analysis (MRA)

The Multi Resolution analysis (MRA) is the most important part of the theory of wavelets. This forms the building blocks for the construction of scaling functions and wavelet functions. Meyer and Mallat developed the idea of MRA. Here the function consisting of slow and rapid variations is decomposed, level by level, (scaling) into an approximation part and a detailed part. The approximation part is taken care off by the scaling function and the detailed part is taken care off by the wavelet function. For achieving the approximation, which is equivalent to a low pass action, we should have the scaling function which should satisfy the requirement

$$\int \Phi(t)\ dt = c, \text{ a constant or } c = 1; \Phi(t) \in\ L^2(\mathbf{R})$$

The approximation operation done by $\Phi(t)$ at various scales generates a nested sequence of function spaces.

$$\ldots\ldots V_3\ \mathbf{C}\ V_2\ \mathbf{C}\ V_1\ \mathbf{C}\ V_0\ \mathbf{C}\ V_{-1}\ldots\ldots\ldots\ldots\ldots\ldots\ L^2(\mathbf{R})$$

Where V_j: vector space of functions on $L^2(R)$.
Let $\Phi(t)$ and its integer translates $\Phi(t - k)$, $k \in Z$ form a set of basis vectors to span the linear space V_0. Since $\Phi(t)$ has the low pass characteristics, $\Phi(2t)$ and its integer translates $\Phi(2t - k)$, $k \in Z$ form a set of basis vectors to span the space V_{-1}. Every vector in V_0 now can be represented as a linear combination of $\Phi(2t)$ and its translates.

$$\Phi(t) = \sum_{k} c_k \, \Phi(2t - k) \tag{18.10}$$

This is a 2 scale difference equation and is termed as a '*refinement equation*' for the scaling function. Note that the function $\Phi(t)$ is represented as the combination of its own scaled translates $\Phi(2t - k)$. The vectors in V_{-1} which are orthogonal to the vectors in the approximation space V_0 will lie in the detail space W_0. Obviously the vectors in W_0 are generated by wavelet function $\Psi(t)$ acting on V_{-1}. Again since W_0 **C** V_{-1}, the wavelet function $\Psi(t)$ can be represented as a combination of the basis vectors of V_{-1}.

$$\Psi(t) = \sum_{k} d_k \Phi(2t - k) \tag{18.11}$$

For an orthogonal wavelet decomposition,

$$V_{-1} = V_0 + W_0 \text{ (direct sum)}$$

at the first decomposition level.

$$V_0 = V_1 + W_1 \text{ (direct sum)}$$

Generally

$$V_{j-1} = V_j + W_j \quad \text{(direct sum)}$$

and

$$V_j \perp W_j,$$

where V_j and W_j are the approximation and detail spaces at the jth decomposition level.
If our signal space under consideration is V_0, then after j levels of decomposition,

$$V_0 = V_j + W_j + W_{j-1} + W_{j-2} \ldots\ldots.. + W_1 \quad \text{(direct sum)}$$

This is usually pictorially represented as given below (Fig. 18.3).

Thus for an orthogonal wavelet system the following equations hold good.

$$\langle \Phi(t), \Phi(t-k) \rangle = \delta_{0,k} \text{ ie} \int \Phi(t)\Phi(t-k)\, dt = \delta_{0,k} = 1, k = 00, k \neq 0 \tag{18.12}$$

$$\langle \Phi(t), \Psi(t-k) \rangle = 0 \tag{18.13}$$

$$\langle \Psi(t), \Psi(t-k) \rangle = \delta_{0,k} \tag{18.14}$$

Now consider equations 10 and 11. Since the decomposition to V_0 from V_{-1} is a low pass operation, the coefficients c_k of the 2 scale difference equation (10) should form a low pass filter. Similarly since the wavelet function represents the detailed space, the coefficients d_k of the 2 scale difference equation (11) should form a high pass filter.

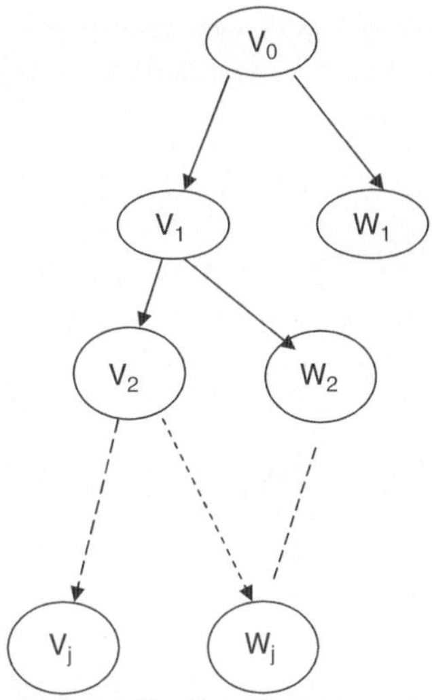

Fig. 18.3. Vector space decomposition of signal in space V_0 with wavelet function up to j^{th} level. $V_1, V_2, \ldots.. V_j$: approximation spaces; $W_1, W_2, \ldots W_j$: detail spaces

From the orthogonality relations it can be shown that

$$|H(\omega)|^2 + |H(\omega+\pi)|^2 = 1 \tag{18.15}$$
$$|G(\omega)|^2 + |G(\omega+\pi)|^2 = 1 \tag{18.16}$$

where

$$H(\omega) = 1/2 \sum_k c_k \, e^{-j\omega k} \quad \text{DTFT of sequence}\{c_k\}$$

$$G(\omega) = 1/2 \sum_k d_k \, e^{-j\omega k} \quad \text{DTFT of sequence } \{d_k\}$$

These equations are the *"power complementarity relations"* in filter bank theory [2].
The approximation ability of the wavelet system is still decided by the regularity of the wavelet and scaling functions. In other words it depends on the number of vanishing moments of the wavelet function.

$$\int t^k \Psi(t)\, dt = 0; \text{ k vanishing moments.} \tag{18.17}$$

All these conditions and equations will form the basic requirement for the construction of scaling and wavelet functions and the computation of the wavelet transforms. Mallat algorithm provides one method for the computation of the wavelet transform.

The above mentioned method is for constructing orthogonal wavelet system. There are semi orthogonal and bi-orthogonal systems which are very much useful in specific applications.

18.6 Some Applications

ECG analysis usually involves the detection of QRS complex and determining beat to beat time and frequency parameters. For clinical purposes it is also usually required to detect T wave as in the case of QT interval analysis and ST segment analysis. Detecting and separating these waves can be a difficult task, especially in the case where the bandwidths of the QRS waves change from beat to beat or overlap with the bandwidth of the T, or even P waves. Furthermore the morphologies of the waves making up the ECG may differ greatly from normal ECGs. Additionally noise may degrade the signal to such a degree that it is difficult to distinguish the ECG. Furthermore, baseline wander may make the detection of T waves and their offsets more difficult as the spectrum of the baseline may overlap with that of the T waves. Most frequency components of the baseline wander are usually below 0.5 Hz although these may extend to even higher frequencies. The American Heart Association (AHA) recommends that if a high-pass filter were used to remove the baseline wander, the cut-off frequency of such filter should not exceed 0.05 Hz [15]. A filtering stage prior to the detection of the T wave and its offset does not guarantee avoiding the spectrum of the baseline interfering with the spectrum of the T waves, and in some cases also with that of broad P waves. Various methods are being used for the analysis of ECG, *wavelet* being one among them. Since Wavelet transform can give good time and frequency information simultaneously, it provides a better tool for ECG analysis overcoming the limitations of frequency analysis by filtering. More over the ability of the wavelet to correlate or approximate with the ECG signal makes it extensively useful in the morphological analysis of ECG. A properly selected wavelet can produce wavelet transform coefficients at different scales corresponding to various frequency components of ECG.

The popular way of frequency analysis of the ECG signal has been the Fourier transform usually performed by the fast Fourier transform (FFT) algorithm. Since the wavelet transform (WT) uses wavelet functions that have time-widths adapted to each frequency as window, it is believed to be better than FFT for analyzing the time-frequency characteristics limited to QRS waves [5–7]. Takeshi *et al* have developed a method based on the wavelet transform for analyzing the frequency power spectrum during the QRS interval of patients with abnormalities such as the myocardial infarction (MI), the bundle branch block or the paced beat IVCA: intraventricular conduction abnormalities with or without MI [8].

Vincet *et al* have compared the respective yields of Fourier and wavelet transforms in analyzing heart rate variability during dynamic changes in autonomous nervous system balance induced by atropine and propranolol [9]. Fourier and wavelet transforms were applied to sequences of heart rate intervals in six subjects receiving increasing doses of atropine and propranolol. At the lowest doses of atropine administered, heart rate variability increased, followed by a progressive decrease with higher doses. With the first dose of propranolol, there was a significant increase in heart rate variability, which progressively disappeared after the last dose. Wavelet transform gave significantly better quantitative analysis of heart rate variability than Fourier transform during autonomous nervous system adaptations induced by both agents and provided novel temporally localized information.

Nicholas *et al* have used wavelet packet analysis in denoising the digital recording of the heart sounds in the detection of the heart murmurs [10]. Cavitation is known to cause blood element damage and may introduce gaseous emboli into the cerebral circulation, increasing the patient's risk of stroke. Discovering methods to reduce the intensity of cavitation induced by mechanical heart valves (MHVs) has long been an area of interest. Wavelet analysis approach for analyzing MHV cavitation is studied [11].

Reduction in overall heart rate variability (HRV) associated with aging is determined by a decreased amplitude of heart rate oscillations at all frequency levels, including high frequency (HF) oscillations attributed to respiratory sinus arrhythmia, low frequency (LF) oscillations attributed to Meyer waves and very low frequency (VLF) oscillations of an uncertain origin, presumably linked among others to thermoregulation. Wavelet transform have been used to find the age-related changes of heart rate variability within independent frequency components [12].

Baseline wander elimination is considered as a classical problem. Behzad *et al*, have presented a wavelet based search algorithm using the energy of the signal in different scales to isolate baseline wander from ECG signal [13]. Myocardial infarction (MI) is known to elicit activation of the autonomic nervous system. Reperfusion, induced by thrombolysis, is thus expected to bring about a shift in the balance between the sympathetic and vagal systems, according to the infarct location. In this study, time-dependent spectral analysis of HRV using the wavelet transform was used for explaining the patterns of cardiac rate control during reperfusion [14]. Some applications of wavelet for ECG signal analysis is discussed in the following section very briefly.

18.6.1 Decomposition of ECG signal

Wavelet Transform (WT) is made use of to detect the QRS complexes and their onsets as well as the T waves and their offsets by decomposing the ECG into different frequency bands controlled by a scaling parameter and looking for the presence of the QRS complexes/T waves appearing at certain frequencies. The method used to implement the wavelet transform makes use of low pass and high pass filter banks. (The mother wavelets used in ECG analysis should tend to resemble the ECG's Morphology). One method is to use a quadratic spline mother wavelet. This mother wavelet is then used to create a set of wavelets which are constructed by scaling and translating the mother wavelet. The wavelet transform used can be considered to be an analysis filter as it breaks a signal down into various components, also termed detailed signals, by low-pass filtering the signal at four different frequency bands before high-pass filtering the signal to obtain the detailed signals. These detailed signals can be analyzed or operated on, instead of the original signal. The wavelet transform generates a series of maximum-minimum pairs associated with the PQRST waves that are distinct and different from artifacts. The maximum-minimum pairs vary for each ECG wave and also for each scale. Thresholds are needed to determine whether a maximum-minimum pair corresponds to a QRS complex or a T wave. Such detailed signals are represented by the transformed ECG signal at scales 2 to 4. Scales 2 and 3 were found to characterize the ECG signals well enough for the detection of the QRS complexes and their onsets whereas scales 2 and 4 were used for the detection of the T wave and its offset [15].

The ECG signal is discretised. The wavelet transform is computed by Mallat algorithm. The discrete time wavelet transform maps discrete finite energy sequences to a 2-D discrete grid of coefficients. A specific structure can be implemented, where time dilation is accomplished by simply dropping every other sample of the signal under analysis (down sampling) for every stage in the wavelet decomposition. The signal is convolved with the impulse response of the high-pass filter. The scale parameter controls the rate of decay of the mother wavelet, i.e. controls the size of the "analysis window" dictated by the mother wavelet thus specifying the required shape in the ECG at specified time points. The following figure (Fig. 18.4) represents the algorithm for wavelet decomposition.
A quadratic spline function is used by C Gamo *et al* [15] as the wavelet for the analysis. Its frequency characteristics and the frequency response of the corresponding filters are given below.

$$\Psi(\omega) = i\,\omega \frac{\sin^4(\omega/4)}{(\omega/4)^4}$$

$$H(\omega) = e^{j\omega/2} \cos^2(\omega/2)$$

$$G(\omega) = 4\, e^{j\omega/2} \sin(\omega/2)$$

With these equations the wavelet transform is computed for various scales using Mallat Algorithm. The onset and offsets of the characteristic wave patterns were identified by the modulus maxima and minima of WT at corresponding points. This method detects Q, R, S and T waves. The time localization property of the transform allows taking measurements on the ST, QT intervals thus making analysis easy and correct.

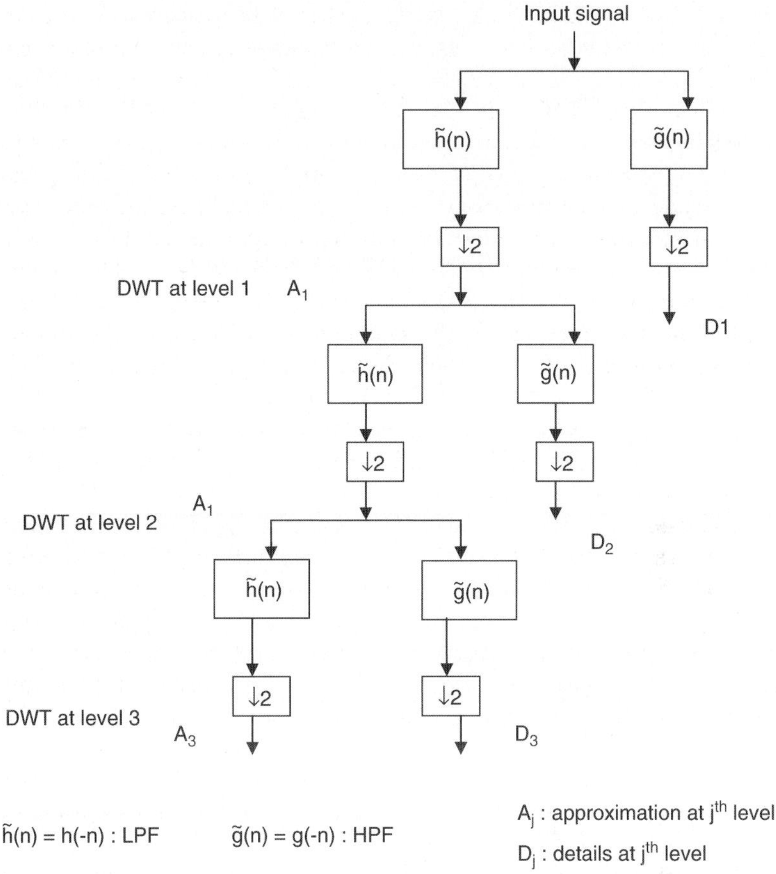

Fig. 18.4. Illustration of Mallat Algorithm

18.6.2 Detection of Myocardial Ischemia

Myocardial Ischemia is reflected in the ECG as the T wave amplitude changes ST deviation and even alterations in the terminal portions of QRS complex (Fig. 18.5). The most important ECG change associated with ischemia is the ST segment elevation or depression with depression being most common. Also this can be along with T wave amplitude changes and even T wave inversion. A long term monitoring of ECG data (usually 24hrs.) is required for the analysis of the ECG signal where automatic detection is proved useful. The wavelet transform method of analysis can be used to find the various characteristic points on ECG and enables to take measurements thus making the analysis easy and consistent.

Ranjith *et al* [16] used a quadratic spline wavelet which resembles an ECG to find the wavelet transform coefficients at different levels. From these coefficients ECG characteristic points are found out. The characteristic

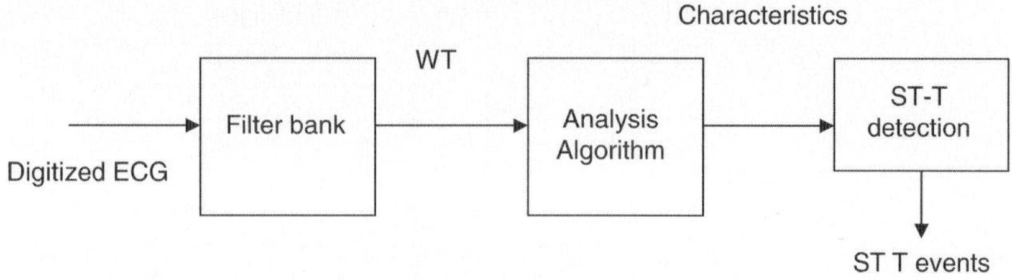

Fig. 18.5. Block schematic of the process

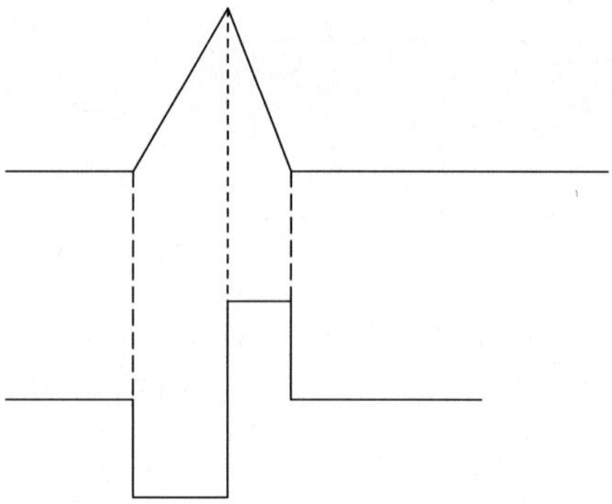

Fig. 18.6. Maxima and minima of WT

point–wavelet transform relationship can simply be explained by a small example. For the wave in the Fig. 18.6, WT value at a scale 2^1 is plotted.

The wave's rising edge corresponds to a negative minima and the trailing edge corresponds to a positive maxima. The moduli of these maxima and minima corresponding to the same edge are named as modulus maxima line.

For detection of R peak, the modulus maxima pair is located for the lowest scale (maximum resolution), by fixing a threshold for detection. The maxima minima pairs for other scales are located within the neighbourhood of these maxima minima pairs. If the amplitudes of these are consistent compared to that for the lower scale or it is increasing then the corresponding modulus maxima minima pair is treated as one that corresponds to a true R peak. From the modulus maxima pair of the R wave the beginning and end of first modulus maxima before and after the modulus maximum pair are detected within a time window. These correspond to QRS onset and offset points.

For T and P waves, since they have low frequency, WT at a coarser scale 2^4 is used to locate. T wave is detected in a window after the R wave and P wave in a window before the R wave.

The T wave creates a pair of modulus maxima with a different sign on WT at scale 2^4 within a time window after the detected R peak. The onset, peak and offset of P wave can also be detected in a window before the R peak.

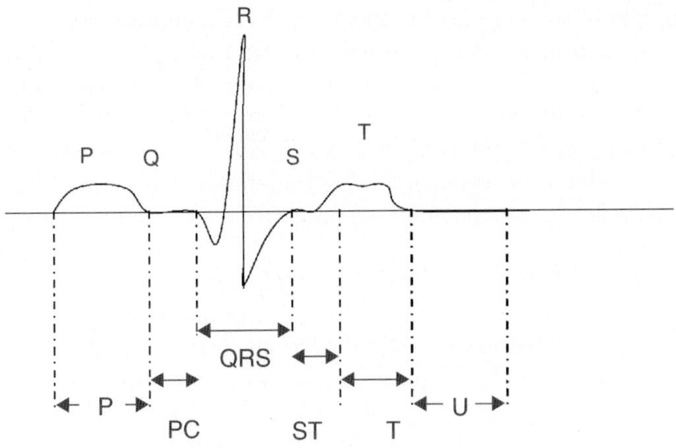

Fig. 18.7. ECG characteristic points

For detecting T wave, a reference level is created by drawing a line segment between 2 or more onsets of P waves (or offsets) on ECG characteristic points (Fig. 18.7). The deviation of the ST segment from this reference line is calculated. Relative amplitude of T wave from this reference line can also be found out. With this information we can check for the occurrence of an ischemic episode. The two performance indices for ST-T change detection algorithm are i] ST sensitivity (ST_{se}), an estimate of the likelihood of detecting an ischemic episode and ii] ST positive predictivity (ST + P), likelihood that the detection is a true ischemia event. With these performance measures, Ranjith et al. could conclude that the WT method was superior over the conventional Fourier Transform methods in the event detection. But the method suffered from the larger number of computations required compared to the FT or ANN methods.

18.6.3 De-noising ECG

ECG signals extracted from the body will be contaminated with different types of noise whose presence will prevent a true diagnosis. ECG signals can be contaminated from motion artifacts, 50 Hz signal interference from power lines and random white noise signals. The purpose of de-noising is to reduce the noise level in the ECG signal and to prevent signal distortion. The ability of WT to analyse a signal with time and frequency resolutions makes it an efficient tool for noise removal. Agante *et al* suggested [17] a method for the removal of the random noise and 50 Hz noise using wavelet transforms. After the selection of a suitable wavelet the noise removal process involves a soft thresholding operation on the coefficients of the ECG wavelet decomposition.

The wavelet decomposition is done by Mallat algorithm using a bank of filters as explained in the previous sections. The soft thresholding method is usually applied to the detail components of the decomposition tree (Fig. 18.4). The detail level bearing the noise will vary according to the noise type and sampling frequency. For white noise, most part of the noise coefficients are in the D_1 level while the power line signals lie in the D_3 level usually for a 500 Hz sampling rate. If the coefficient value is less than a threshold level it is considered to be a noise coefficient and therefore set to zero. After applying the soft thresholding operation the signal is reconstructed by climbing up the decomposition tree.

White noise typically has a normal distribution with zero mean and variance σ^2. A threshold value for this noise is given by [17] $t_n = \sigma\sqrt{[2\log(n)]}$ where σ is the standard deviation and n is the signal strength. The efficiency of the process is dependant on the signal pattern studied and the type of wavelet used. Agante *et al* [17] have found that the bi-orthogonal wavelet gave a better performance in denoising QR and QRS morphologies.

18.6.4 Classification of Arrhythmias

The loss of the natural rhythm of heart is called arrhythmia. There are more than 60 arrhythmias being identified and used for the diagnosis. Hence an automatic detection and classification prove useful. The feature extraction properties of the wavelets can be utilized along with an ANN classifier to identify some arrhythmias (Fig. 18.8).
Krishna *et al* [18] proposed a method for this and is given in the block diagram below.

Wavelet coefficients were found out by computing the wavelet transform for different scales. Most of the energy of the ECG signal lies between 0.5 Hz and 40 Hz. The wavelet coefficients at various resolution levels corresponding to this frequency range is selected for classification. Krishna *et al* [18] have selected 23 DWT coefficients which were termed as feature vector.
The segment of ECG signal 100 ms before and 150 ms after R wave is considered for analysis. The feature vector is computed and given to the NN for classification. A feed forward NN with 2 hidden layers is selected which used sigmoid activation functions and trained with back propagation algorithm.

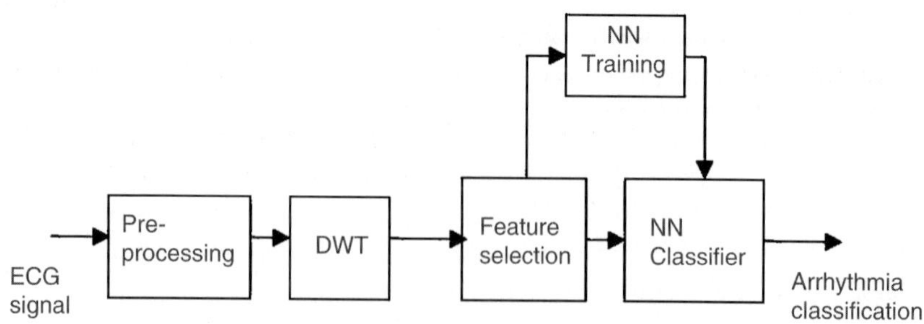

Fig. 18.8. ECG arrhythmia classifier

It is then tested to check the classification efficiency and the method is able to identify 12 arrythmias with about 95% accuracy.

18.7 Conclusion

We have discussed few applications of wavelet transform in the analysis of ECG morphology. The applications of wavelets in EMG and EEG analysis along with ECG are emerging fast and diverse. A lot of research is going on in the ECG and EEG data compression with wavelets. Continuous monitoring of bio-signals requires huge memory spaces in the computers associated with data storage. Wavelets are used for compressing the data by virtue of its multi resolution analytical capability. Wavelets have got many applications in medical image analysis and compression. Since even a minute variations in ECG can be indicative of the overall health of a person, the analysis of ECG never ends and wavelets can be an efficient tool for more exact analysis, thus helping doctors in diagnosis and treatment.

References

1. Young, R.K. (1993): 'Wavelet, Theory and Applications', *Kluwer academic Publishers.*
2. Vetterli, M., Kovasevic, J. (1995): 'Wavelets and subband Coding', *Pentice Hall.*
3. Goswami, Chan,(1999): 'Fundamentals of Wavelets-Theory Algorithms and applications' *Willey-interscience publications.*
4. Cohen, A., Kovacevic, J. (1996): 'Wavelet, A mathematical background', Invited paper, *Proceedings of IEEE*, **84**(4), pp. 514–522.
5. Morlet, D., Peyrin, F., Desseigne, P., and Rubel, P. (1993): 'Wavelet analysis of high-resolution signal-averaged ECGs in postinfarction patients', *J Electrocardiol.*, **26**, pp. 311–320.
6. Zenda, N., Tsutsumi, T., Sato, M., Takeyama, Y., Harumi, K., and Wei, D. (2000): 'Computer simulation of notches on initial part of QRS complex in patients with anterior myocardial infarction', In 2000 *Electrocardiology*, Ambroggi, L.D. (2001): 'Casa Editrice Scientifica Internazionale', Roma, pp. 117–120.
7. Shimojima, H., Tsutsumi, T., Yanagisawa, F., Komukai, M., Zenda, N., Higashi, Y., Takeyama, Y., Okamoto, Y. (2003): 'Application of wavelet transform for analysis of QRS complex in intraventricular conduction abnormalities', *Inter J of Bioelectromag.*, **5**(1), pp. 279–281.
8. Takeshi, T., Daisuke, W., Hisa, S., Yukei, H., and Yoichi, T. (2004): 'Analyzing Time-Frequency Power Spectrum Limited in QRS Complex Based on the Wavelet Transform', *International Journal of Bioelectromagnetism,* **6**(1), pp. 1–5.
9. Vincent, P., Jean, M.G., Serge, M., Anestis, A., Thierry, B., Frédéric, R., Frédéric, C., Luc, Q., Jean, R.L., and Jean, C.B. (1999): 'Wavelet transform to quantify heart rate variability and to assess its instantaneous changes', *J Appl Physiol.*, **86**, pp. 1081–1091.

10. Andrisevic, N., Ejaz, K., Rios-Gutierrez, F., and Alba-Flores, R. (2005): 'Detection of Heart Murmurs Using Wavelet Analysis and Artificial Neural Networks', *Journal of Biomechanical Engineering*, **127**(6), 899–904.
11. Herbertson, L.H., Reddy, V., Manning, K.B., and Welz, J.P. (2006): 'Wavelet Transforms in the Analysis of Mechanical Heart Valve Cavitation', *Journal of Biomechanical Engineering*, **128**(2), pp. 217–222.
12. Vigo, D.E., Guinjoan, S.M., Scaramal, M., Siri, L.N., and Cardinali, D.P. (2005): 'Wavelet transform shows age-related changes of heart rate variability within independent frequency components', *Auton Neurosci.*, **123**(1–2), pp. 94–100.
13. Mozaffary, B., and Tinati, M.A. (2004): 'ECG Baseline Wander Elimination using Wavelet Packets", *Transactions On Engineering, Computing And Technology*, pp. 22–24.
14. Toledo, E., Gurevitz, O., Hod, H., Eldar,M., and Akselrod, S. (2003): 'Wavelet analysis of instantaneous heart rate: a study of autonomic control during thrombolysis', *Am J Physiol Regul Integr Comp Physiol.*, **284**, pp. R1079–R1091.
15. Gamo, C., Gaydecki, P., Zaidi, A., and Fitzpatrick, A. (2000): 'An implementation of wavelet transform for ecg analysis', *IEE International Conference on 'Advances in Medical Signal and Information Processing*, **476**, pp. 32–40.
16. Ranjith, P., Baby, P.C., and Joseph, P. (2003): 'ECG Analysis using Wavelet Transforms-Application to Myocardial Ischemia' , *ITBM-RBM*, **24**(1), pp. 44–47.
17. Agante, P.M., and Marques, J.P. (1999): 'ECG noise filtering using wavelets with soft thresholding methods', *IEEE Computers in Cardiology*, **26**, pp. 535–538.
18. Prasad, K.G., Sahambi, J.S. (2003): 'Classification of ECG Arrhythmia using MRA and Neural Networks', *TENCON 2003*, Conference on Convergent Technologies for Asia Pacific Region, **1**, pp. 227–231.

19

1/f Fluctuation of Heart Rate in Postoperative and Brain-Dead Patients

Nakajima K, Tamura T, Sasaki K, Maekawa T

The power spectra of heart rate fluctuations in normal subjects show that they are inversely proportional to the frequency [1] and are thus called 1/f fluctuations. Such 1/f fluctuations are a well known phenomenon that occurs when an electric current passes through a vacuum tube and a semiconductor. In biology, 1/f fluctuations have been observed in cellular membranes [2], brain waves [3], and during concentration modulation of action potentials [4]. The 1/f fluctuations of heart rate have been used as a novel index of autonomic function and it provides a qualitative estimate of the state of the cardiovascular control system.

Short-term analyses of power spectra such as the 256 beats of RR intervals or during a 1 hour period have shown that low-frequency heart rate variability (i.e., less than 0.1 Hz) is under both sympathetic and parasympathetic nervous control, whereas higher frequency variability is under parasympathetic control [5–9]. Recently, analyses of heart rate variability from 24-hour electrocardiogram recordings are reported as development of digital holter electrocardiograph [10–12].

Nonrandom heart rate variability has also been studied using long-term monitoring. Such variability can be analyzed in two ways, by evaluating the signal in the time or in the frequency domain. In the time domain, the heart rate in normal subjects shows a sinusoidal rhythm called a circadian rhythm, which has a period of 20 h to 28 h. However, spectra are more frequently calculated in the frequency domain.

A few studies have analyzed the rhythms and fluctuations in postoperative patients [13, 14]. In this study, the heart rates of patients who had been admitted to an intensive care unit (ICU) were studied, and the variability in both the time domain and frequency domain were analyzed and evaluated to differentiate between survivors and no survivors, including brain-dead patients [15].

19.1 Data Acquistion

The 17 postoperative subjects chosen for this study included 2 brain-dead patients who had required intensive care. The data analyzed are summarized in Table 19.1. Of the patients, 9 were returned to the surgical wards, on average 10.8 ± 4.6 (mean $\pm$ SD) days postoperatively, and 2 of the patients were treated by hypothermia. Eight patients, 2 of whom had been brain dead, died in the ICU at 20.8 ± 18.2 postoperative days. One patient with brain death was hospitalized for more than 60 days. No patients required artificial ventilation. For comparison with the parameters of the patients, the physical parameters of a normal subject were also investigated. Informed consent was obtained from the patients or their relatives and from the normal control subject. The control subject was paid to participate in the study.

In the ICU, the electrocardiographic (ECG) results, blood pressures, and body temperatures of the patients were recorded with a monitor (BMS-8500, Nihon Koden, Tokyo, Japan). The heart rate was obtained from the R-R interval of the ECG and calculated automatically by a monitor based on R-wave triggering. The data were transferred to a workstation (CWS-8100, Nihon Koden) and stored on a hard disk every minute, along with the instantaneous heart rate value at each 1-minute interval. Data that spanned periods of 6 to 12 days were selected and transferred to another workstation and analyzed at 1-minute intervals. The data were collected from the second postoperative day until one day before leaving the ICU, except for one brain-dead patient, who was in the ICU for more than 60 days.

Acquisition of Control Data

The normal subject was a healthy student. He was admitted to the ICU and stayed there for three consecutive days. During the day, he stayed in bed, watched television, played video games, and engaged in conversation with visitors and staff. His heart rate was measured by an electrocardiograph system (BMS-8500, Nihon Koden).

Table 19.1. Patient Characteristics [15]

Subject by treatment outcome	Number of patients	Age, mean $\pm$ SD (range in years)	ICU stay, mean $\pm$ SD (range in days)
All surviving surgical patients	9	56.9 ± 17.3 (20 to 74)	10.8 ± 4.6 (3 to 17)
Surgery plus hypothermia treatment	2	53, 65	15, 17
Nonsurviving ICU patients	8	54.5 ± 14.5 (28 to 69)	20.8 ± 18.2 (10 to 60)
Nonsurviving brain-dead ICU patients	2	28, 32	60, 6

19.2 Data Analysis

The main objective of this study was to determine whether the circadian rhythm and the power spectrum of the heart rate can be used to distinguish survivors from non-survivors. Cosine fitting was carried out to demonstrate circadian cycles. Our hypothesis was that the frequency spectrum of the heart rate in normal subjects and survivors would be the inverse of the power law spectrum, that is, a 1/f-like spectrum, and that the non-survivors would show a white-noise-like spectrum before death.

Data were analyzed as follows. The data lost during calibration and disconnection of the sensors for various treatments were edited, and linear interpolation was attempted. To minimize the effect of transient artifacts, the moving average of a triangular weighting function was used to smooth the data. A triangular weighting function W was used to produce the following equation. When the nth number of data was calculated, data from the previous and proceeding 4-minute periods were summed with the weighted value

$$W(n) = \frac{1}{25} \sum_{i=-4}^{4} (5 - |i|)X(n+i) \tag{19.1}$$

where n is the nth number of data. The i represents previous and proceeding data of n with a 1-minute interval. The time course of the heart rate during a 24 h period was displayed on a video terminal and printed out by the recorder.

In this study, interpretation of missing data was made over time intervals of 5 to 30 minutes, with previous data used for interpolation. The interpolation of data was different between patients since the terminal patient was relatively stable and few data needed to be interpolated, while postoperative patients for the first few days in the ICU often underwent examinations and treatments so that interpolation of data was often required 3 to 5 times a day.

Cosine fitting of the heart rate time course was carried out for the normal subject and one of the brain-dead patients using the following equation:

$$Y = M + A\cos\left(2\pi\frac{n}{1440} - \phi\right) \tag{19.2}$$

where Y is the estimated value at data number n that corresponds to monitored time. M is the MESOR, which is the midline estimation static of rhythm, defined as the average value of the rhythmic function; A is the amplitude, defined as one-half the peak-to-trough variation; $2\pi(n/1440)$ is the angular frequency, which is $2\pi/\tau$ where τ is the fitted period. In our case, the number 1440 represents the total data for 24 h with a 1-minute sampling interval. ϕ is the acrophase (the approximated crest time) as a lag from a defined reference time point (midnight of the first day of measurement). The MESOR, amplitude, and acrophase were determined using the least squares method. The simplex method was used to optimize the cosine function [16]. The estimates of amplitude and acraphase follow a bivariate normal distribution characterized by an F statistic, and thus the 95% confidence limits for amplitude and acrophase were obtained. The cosine fitting is applied to time series of data anticipated to the period of 24 h in a normal subject. For the brain-dead patient, the cosine curves differing in period length by 10 minutes were fitted by the method of least squares to determine the most prominent rhythm within the period domain from 12 h to 48 h. Fitted cosine curves that had the greatest coefficient of determination were tested by analysis of variance with a significance level under 0.05($p < 0.05$). When the analysis of variance showed that the fitting cosine curve was statistically significant, we selected the values of MESOR, amplitude, and acrophase for the brain-dead patient.

Power spectral density of heart rate fluctuation was obtained as follows.

(1) Autocorrelation function was obtained from 1440 minutes law heart rate data. It produced 720 points data.
(2) 1440 points data were obtained using the symmetry property of the autocorrelation function.
(3) Fast Fourier transform of 1024 points was achieved for the 1440 points data. First 1024 points were used from 1440 points, which means the time of the first data was 0:00 o'clock.

The linear regression line between the log scale of the power spectral densities and that of the frequencies was calculated, and the fluctuations related to the frequency were evaluated from the slope of the linear regression line. The power spectra were evaluated for periods of 1 and 6 to 12 days, depending on the length of stay of a patient in the ICU.

19.3 Results

First, the results for the normal subject are presented, followed by those for the postoperative patients who recovered or died in the ICU. The latter include the results obtained from the two brain-dead patients.

Normal Subject

Figure 19.1 shows the heart rate time course and power spectrum on the second and the third days of the study. The heart rate was higher during the day than at night. The power spectra show the 1/f relationship. The time domain data of the two successive days were fitted by a 24 h cosine curve, which is a typical rhythm for a normal subject. Table 19.2 shows the results of the cosinor analysis of the heart rate data. The daily cycle peak shown in Fig. 19.2 that corresponded to 1.157×10^{-5} Hz, that is 24 h, was insignificant, but the 1/f relationship is clearly depicted.

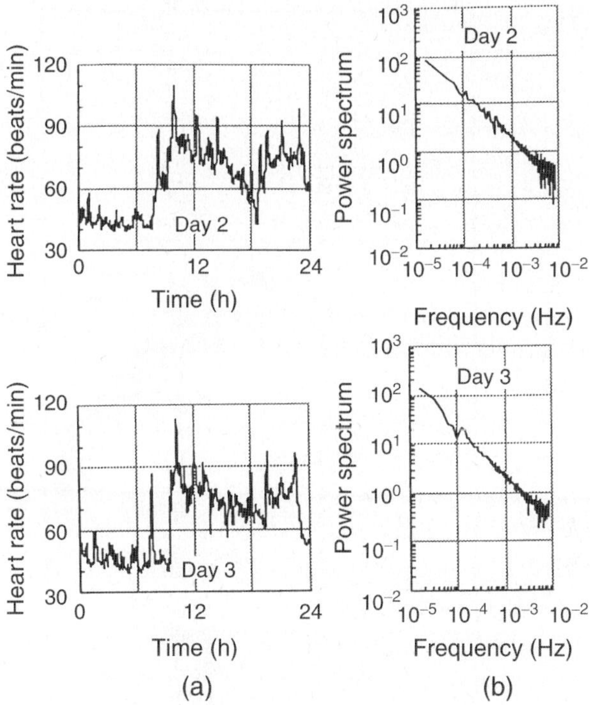

Fig. 19.1. Heart rate time courses (**a**) and power spectra (**b**) of a normal subject on the second and the third days in the intensive care unit

Table 19.2. Circadian Rhythmic Aspects of Heart Rate in Normal and Brain-Dead Patients [15]

	MESOR (mean ± SE)[a]	Amplitude(A)[a]	Acrophase(ϕ)[a]
Normal subject	63.8 ± 0.6	16.0 ± 0.7 (15.2, 16.7)	$-238 \pm 4°(-234, -242)$
Brain-dead subjects	102 ± 0.8	5.3 ± 0.2 (5.1, 5.5)	$-172 \pm 5°(-167, -177)$

Units of MESOR and amplitude are in beats per minute; acrophase is given in degrees with 15degrees = 1 h clock time as a negative value, a delay from local midnight, as a reference. Numbers in parentheses are the 95% confidence limits for amplitude and acrophase as derived by cosinor analysis using $\tau = 24$ h as the fitted approximation.
[a]95% confidence limits.

Patients

Survivors

Figure 19.3a shows the heart rate time courses from the second to the sixth postoperative days; the latter was 1 day before being retuned to the surgical

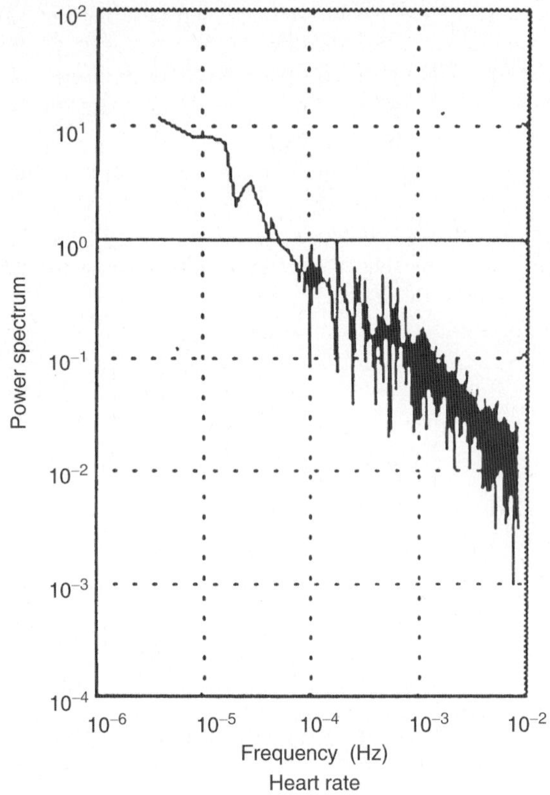

Fig. 19.2. Heart rate power spectrum, calculated for two consecutive days (48 h), of the normal subject [15]

ward for a postoperative patient who recovered. The heart rates including high frequency fluctuation were unstable from the second to the fifth days, but stabilized after the fifth day. The heart rate variation was approximately 25 beats/min. The power spectra (Fig. 19.3b) show the 1/f relationships with frequencies lower than 10^{-3} Hz from the second to the sixth days. However, from the second to the fourth days, the frequency components higher than 10^{-3} Hz showed broadband spectra, which corresponded to short-term variations in the heart rate. The total power spectrum for 6 consecutive days was calculated. The heart rate showed 1/f relationships with frequencies lower than 10^{-3} Hz, and a clear circadian peak (1.157×10^{-5} Hz) was observed.

Nonsurvivor

Figure 19.4 shows the results for a patient who died in the ICU on the 21st postoperative day. Figure 19.4a shows the time courses of heart rate from the second to fourth days and from the 18th to the 20th days (the latter was 1 day before the patient died). The heart rates were unstable on both of these days. The corresponding power spectra are shown in Fig. 19.4b. While the 1/f relationship was discernible, on the 20th day the frequency was consistently higher than 10^{-4} Hz, such that the frequency fluctuations were considered to be random. The total power spectrum for 6 days was also calculated and showed a 1/f relationship, with several spikes of approximately 10^{-5} Hz. A large peak that corresponded to the daily cycle was also observed.

The 1-day time courses and power spectra of two brain-dead patients are shown in the remaining figures. Figure 19.5 shows the heart rate time courses from the 45th to the 47th and the 57th to the 59th days for a patient who was pregnant and was treated to maintain intrauterine stability for growth of the fetus. The data shown were obtained after the fetus had been delivered by cesarean section. Her heart rate appeared to have a circadian rhythm. She also had relatively high heart rates due to the continuous administration of a catecholarnine. The cosine-fitted curve for the 57th and the 58th days is

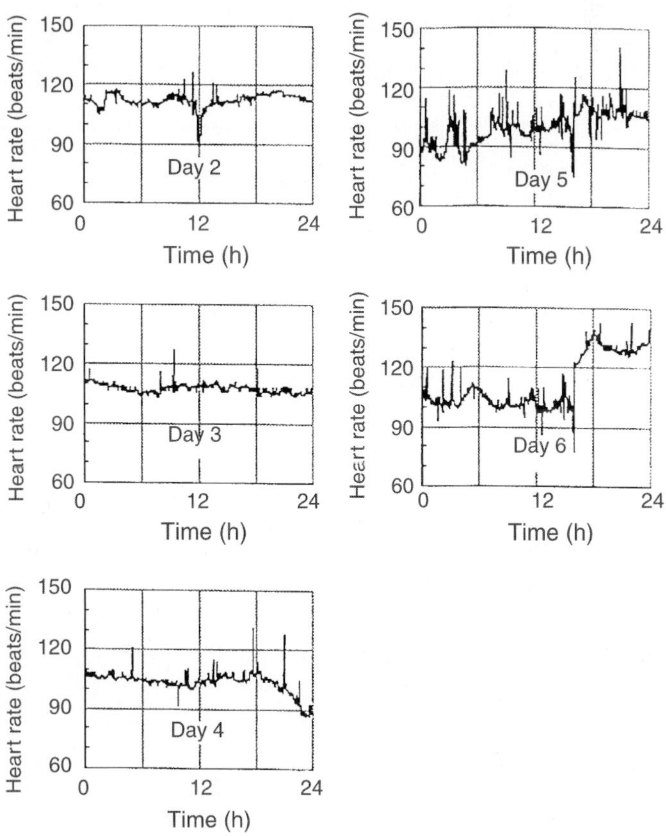

Fig. 19.3a. Heart rate time courses of a patient surviving on the 2nd and 6th postoperative days in the intensive care unit

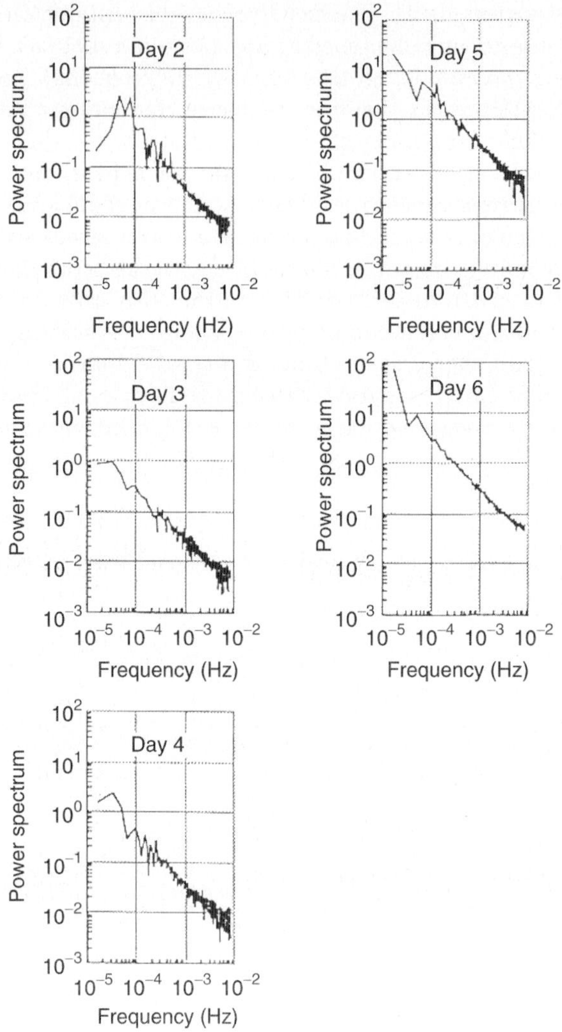

Fig. 19.3b. Power spectra of a patient surviving on the 2nd and 6th postoperative days in the intensive care unit

shown in Fig. 19.6, and the variable parameters are presented in Table 19.2. A significant ($p < 0.05$) circadian rhythm was validated by the fit of a 24 h cosine curve. In other words, the most prominent period of the rhythm was 24 h in this brain-dead patient. The comparison of circadian characteristics of heart rate indicates a statistically significant difference between the normal subject and brain-dead patient ($p < 0.001$). Compared with the normal subject the amplitude (A) was very small, with total heart rate variation being around 10–11 beats/min. The difference in circadian amplitude and acrophase between the two groups was statistically significant ($p < 0.001$). The corresponding power spectra are presented in Fig. 19.5b and show a 1/f relationship. The power density and variation of each frequency were small compared with those of the normal subject. In the total power spectra over a 12-day period, the 1/f relationship was observed in a frequency band of 10^{-5} to 10^{-2} Hz (Fig. 19.7). Figure 19.6 shows the Cosine 24h fitted heart rate curve for the brain-dead patient described in the Fig. 19.5.

The heart rate time course data for the other brain-dead patient are shown in Fig. 19.8a. The judgment of brain death was made on the fourth day, and the patient died on the seventh day. The heart rate on the fourth day was unstable due to artifacts and various treatments. The corresponding power spectra are shown in Fig. 19.8b. The power spectrum on the fourth day was relatively broad in the frequency range higher than 2×10^{-3} Hz. Despite the

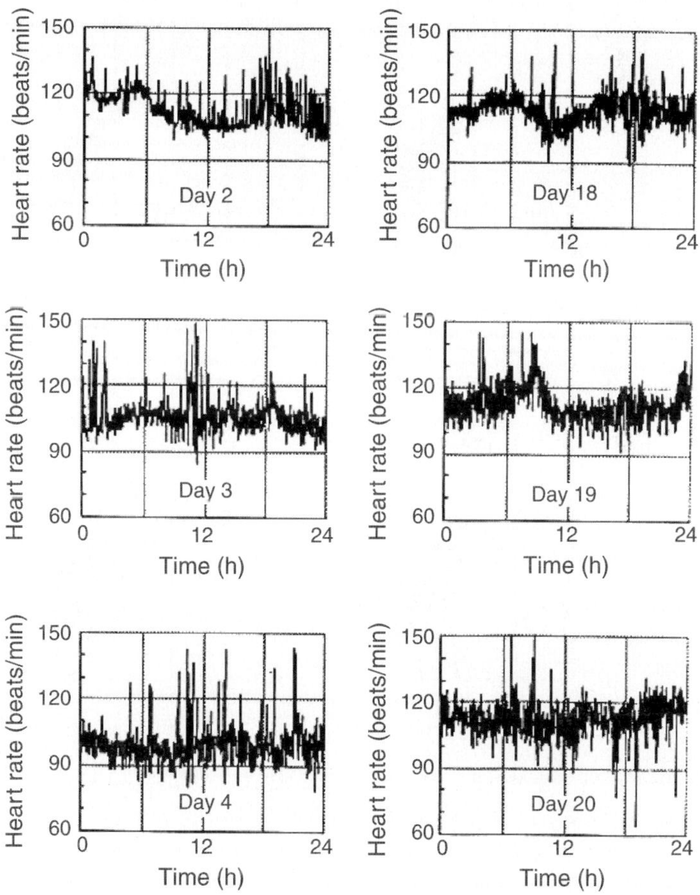

Fig. 19.4a. Heart rate time courses of a deceased (day 21) patient from the 2nd to 4th and from the 18th to 20th postoperative days in the intensive care unit

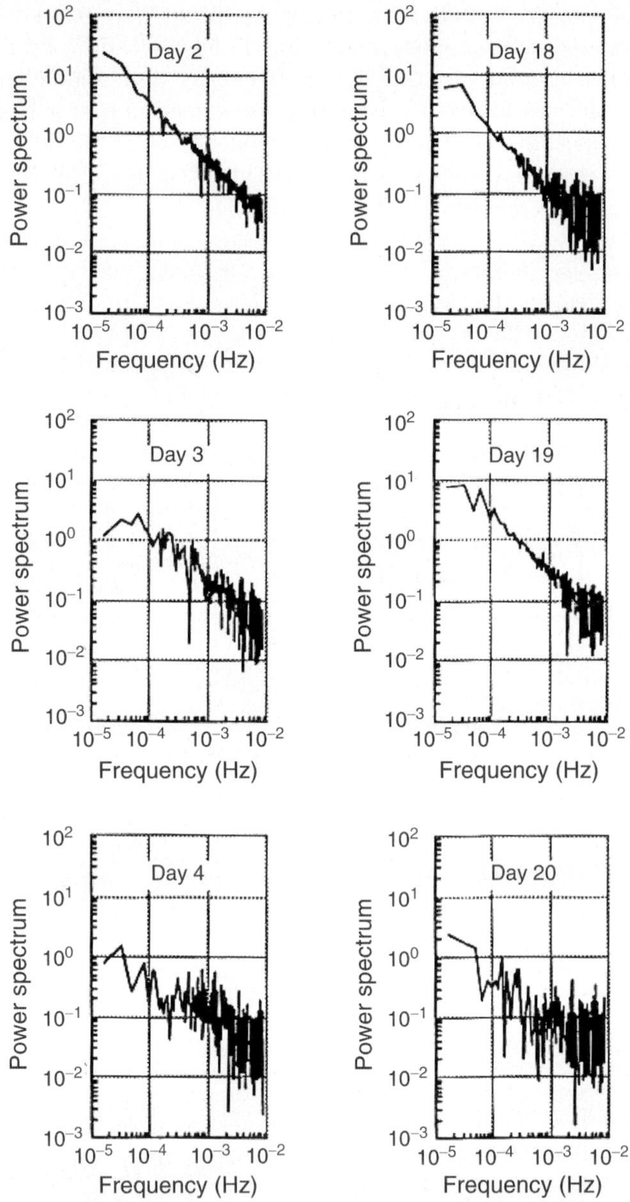

Fig. 19.4b. Power spectra of a deceased (day 21) patient from the 2nd to 4th and from the 18th to 20th postoperative days in the intensive care unit

sympathetic and parasympathetic nerves having been severed, the 1/f relationship persisted. In the total power spectrum over 6 days, the 1/f relationship and a peak that corresponded to a daily cycle were observed (Fig. 19.9). The frequency fluctuations of this patient were larger than those in the other brain-dead patient. Table 19.3 presents the mean and the dispersion parameters of the 1/f slope for all patients, including the two brain-dead patients. The 1/f slope values varied between 0.83 and 1.10.

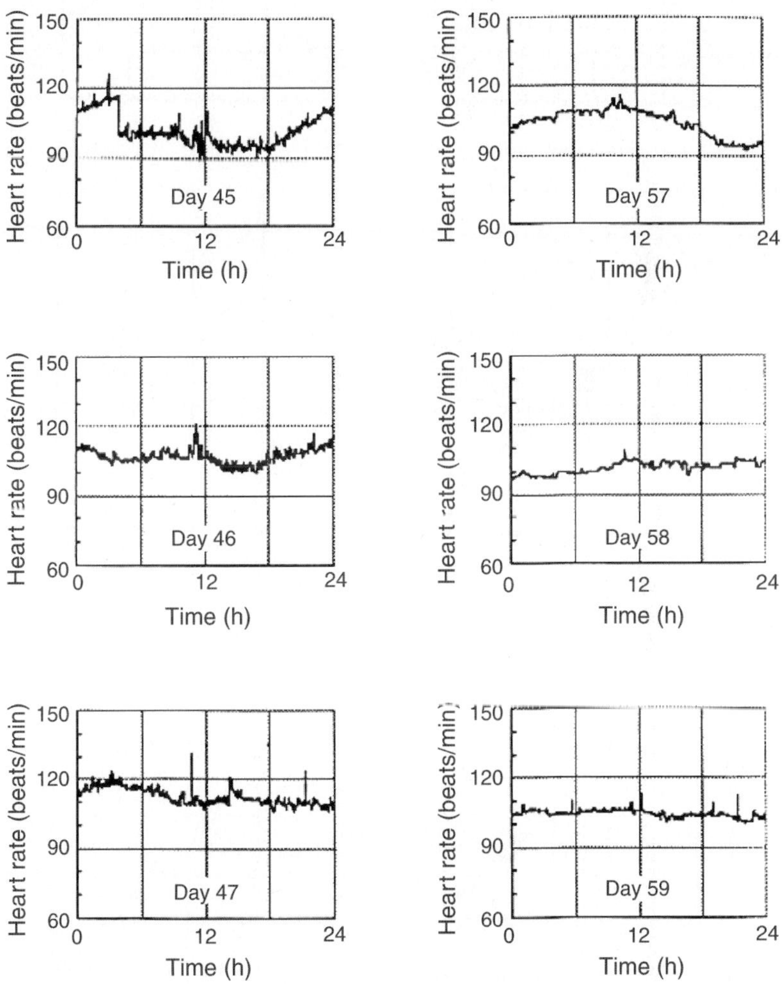

Fig. 19.5a. Heart rate time courses of a brain-dead patient from the 45th to 47th day and from the 57th to 59th day in the intensive care unit [15]

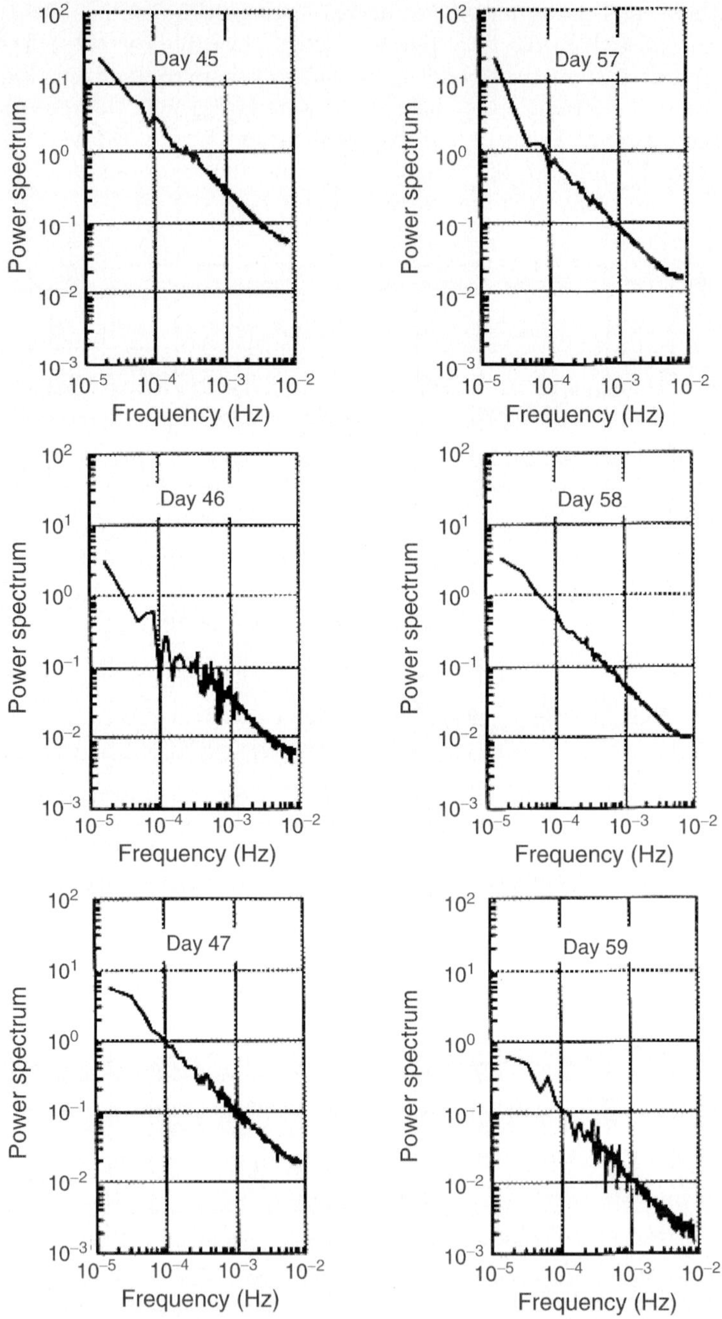

Fig. 19.5b. Power spectra of a brain-dead patient from the 45th to 47th day and from the 57th to 59th day in the intensive care unit [15]

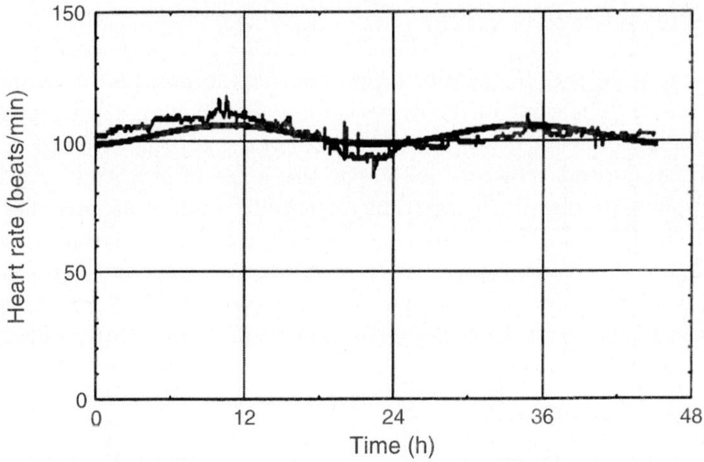

Fig. 19.6. Cosine 24 h fitted heart rate curve for the brain-dead patient described in the Fig. ?? legend [15]

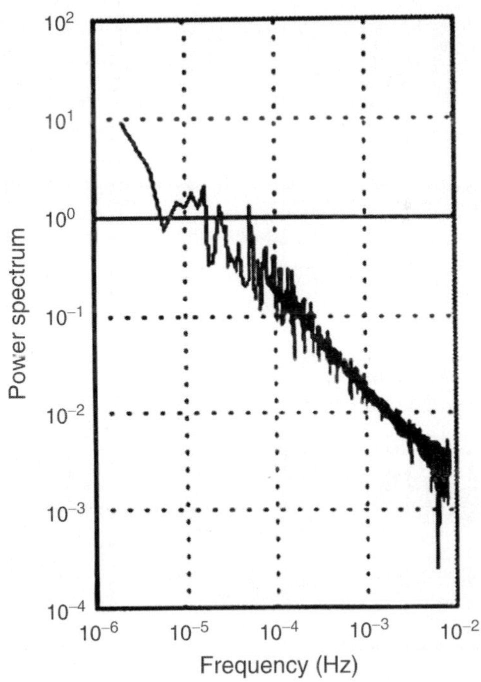

Fig. 19.7. Power spectrum of a consecutive 12 days after the cesarean section

19.4 Discussion

Long-term heart rate fluctuations were observed in postoperative and brain-dead patients. The slope of the regression line between log-frequency and log-power demonstrated a 1/f relationship in all of the subjects studied, including the two brain-dead patients. Although this type of study usually requires that patients be classified according to specific disease groups, we focused only on survivors versus nonsurvivors, without consideration of disease. We concluded that surviving patients showed essentially a 1/f relationship, like normal subjects in previous studies of ours and others [1, 17, 18]. In contrast, nonsurvivors showed a white-noise-like power spectrum shortly before death.

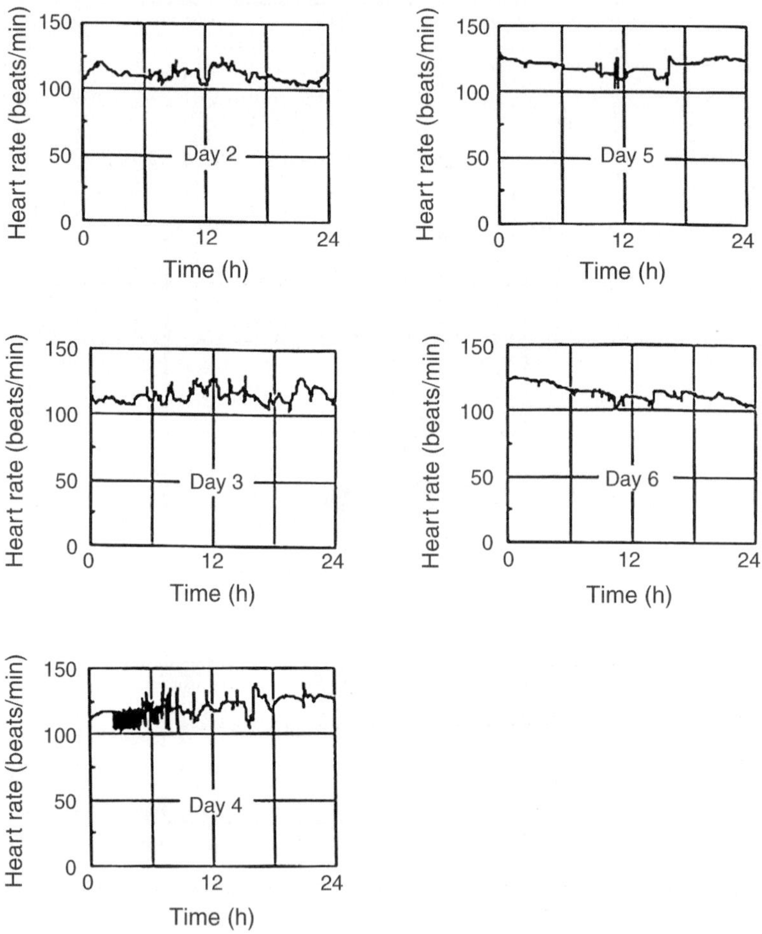

Fig. 19.8a. Heart rate time courses of a brain-dead patient from the 2nd to 6th days in the intensive care unit

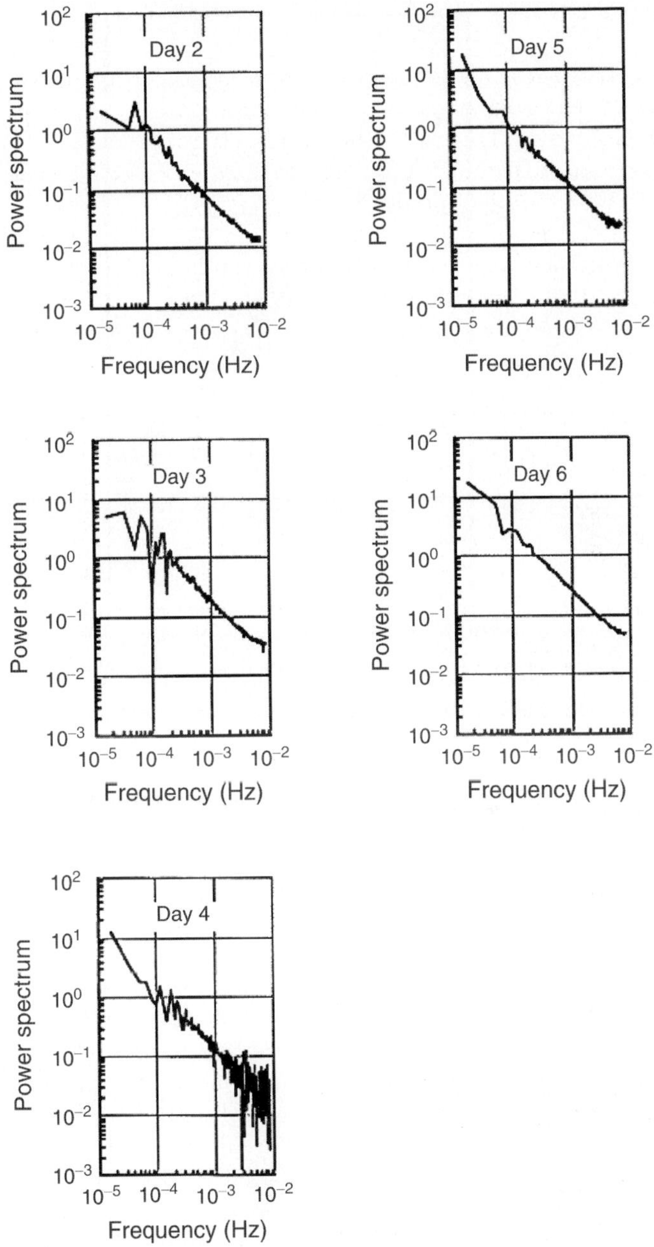

Fig. 19.8b. Power spectra of a brain-dead patient from the 2nd to the 6th days in the intensive care unit

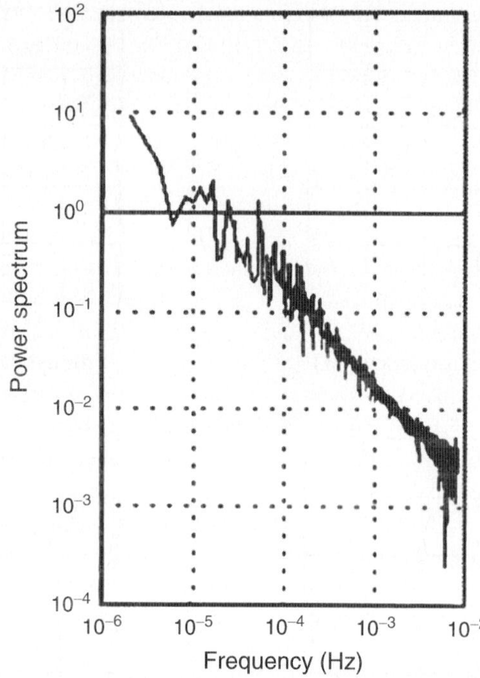

Fig. 19.9. Power spectrum of a consecutive six days

Table 19.3. 1/f Slope of ICU Patients [15]

Result	Number of subjects	1/f Slope	
		Mean ± SD	Range
Recovery	9	0.99 ± 0.07	0.83–1.10
Death	8	1.02 ± 0.03	0.98–1.10

There is no significant difference between two results of patient by t-test ($p > 0.1$).

In our study, data were obtained only at 1-minute intervals. Therefore, the spectral peak corresponding to the heart beat and respiratory rate could not be clearly visualized in the frequency domain. Further study is required using beat-by-beat heart rate recordings to accomplish this.

The heart rate fluctuations that occurred in our postoperative patients showed a 1/f relationship at frequencies lower than 10^{-3} Hz. No remarkable differences were observed between the fluctuation patterns at a frequency range lower than 10^{-3} Hz in the patients who returned to the ward and those who were brain dead. Differences were seen in the amplitude of the time domain and power density of the frequency domain between the normal subject and the brain-dead patients. The heart rate variation in the normal subject was approximately 32 beats/min between the day and night, whereas in the brain-dead patients, it was around 10–11 beats/min. Several studies have shown less-distinct circadian variation as an index of mortality postinfarction [7, 19]. Our results are consistent with their findings.

Random spectra in high-frequency regions constituted a critical clinical sign. The slope of frequencies higher than 10^{-3} Hz differed according to the patient's condition. The frequency distribution in postoperative patients who recovered showed a 1/f slope, whereas that of those who died in the ICU formed broadband spectra, especially shortly before death. The broadband spectrum on the fourth day, as shown in Fig. 19.7, produced results that somewhat resembled those presented in Fig. 19.3. These results indicate that the broadband spectrum is indicative of impending death.

In the stable condition, even in a brain-dead patient, the frequency domain showed a 1/f slope. However, when the rhythm suddenly became uncontrollable, the frequency distribution showed broadband, white-noise-like signals. We carried out long-term analyses and only recorded heart rates every minute, and it was therefore not possible to determine whether short-term heart rate variations had occurred. The frequencies of 10^{-2} and 10^{-3} Hz corresponded to cycles of 1.6 and 16.6 minutes, respectively. Changes in patient condition were evident with rhythms lasting less than 20 minutes. Therefore, both long- and short-term time domain analyses, corresponding to ultra- and very-low-frequency power spectra, are important for fully ascertaining the condition of a patient.

Myers *et al* [20] compared the ECG recordings of patients and normal subjects. They concluded that the power spectra in the range of 0.15 to 0.50 Hz differed significantly between patients and normal subjects. Bigger *et al* [7] studied the frequency heart rate variability and found the strongest predictors of mortality were alterations in the ultra-law-frequency and very-low-frequency power domain (<0.04 Hz). They concluded that rhythms of autonomic or greater period, rather than shorter ones indicative of sympathetic and parasympathetic influence on the heart after myocardial infarction, were associated with mortality. Although low- and high-frequency power spectra could not be calculated with our system, the present results indicate the power spectra in the frequency range of ultra-low and very-law-frequency power show different patterns in surviving and deceased patients.

In brain-dead patients, whose heart beats were controlled by intrinsic cardiac automaticity without cerebral nervous system control, clear 1/f relationships were observed. In brain-dead subjects in general, brain stem and spinal cord reactions had disappeared. When circulation is maintained artificially with a respirator and by medication, local spinal cord and sympathetic nerve reactions persist, and the heart rate rhythm is maintained. Conci *et al* [21] reported heart rate variability before and after brain-dead patients. Although they evaluated short-term, 512 seconds data, heart rate fluctuations of before and after brain-dead, their data supported our results of the 1/f relationship in heart rate obtained in brain-death patients [15].

Although further investigations are required before definite conclusions can be drawn, 1/f fluctuation appears to be a universal phenomenon that is independent of nervous activity control. The analysis of long-term heart rate variability provides a more complete understanding of the cardiac regulatory mechanisms in postoperative patients.

Circadian rhythms documented by the fit of a cosine function, and thus the power spectrum of cosine function, showed a 1/f relationship. However, we obtained a 1/f (ranged from 0.83 to 1.10 in Table 19.3) relationship. The reasons why power spectral density showed 1/f, not $1/f^2$, may be the result of a smoothing effect due to physical treatments and infusion of nutrients and secondary to the interpolation of missing data.

References

1. Kobayashi M, Musha T, "1/f fluctuation of heart beat period", IEEE Trans Biomed Eng., BME-29, 1982, 456–57.
2. Verveen AA, Derksen HE, "Fluctuation phenomena in nerve membrane", Proc IEEE 56, 1968, 906–16.
3. Yamamoto M, Nakahama H, Shima K, Kodama T, Mushiake H, "Markov-dependency and spectral analysis on spike counts in mesencephalic reticular neurons during sleep and attentive states", Brain Res., 366, 1986, 279–289.
4. Musha T, Kosugi Y, Matsumoto G, Suzuki M, "Modulation of the time relation of action potential impulses propagating along an axon", IEEE Trans Biomed Eng., BME-28, 1981, 616–623.
5. Akselrod S, Gordon D, Ubel FA, Shannon DC, Barger AC, Cohen RJ, "Power spectrum analysis of heart rate fluctuation: a quantitative probe of beat-to-beat cardio vascular control", Science, 213, 1981, 220–222.
6. Pomeranz BR, Macaulay JB, Caudill MA, Kutz I, Adam D, Gordon D, Kilborn KM, Barger AC, Shannon DC, Cohen RJ, Benson H, "Assessment of autonomic function in humans by heart rate spectral analysis", Am J Physiol., 248, (Heart Circ Physiol 17), 1985, H151–H153.
7. Bigger JT. Fleiss JL, Steinman RC, Rolnitzky LM, Kleiger RE, Rottman JN, "Frequency domain measures of heart period variability and mortality after myocardial infarction", Circulation, 85, 1992, 164–171.
8. Stein PK, Rich MW, Rottman JN, Kleiger RE, "Stability of index heart rate variability in patients with congestive heart failure", Am Heart J, 129, 1995, 975–981.
9. Task Force of the European Society of Cardiology and the North American Society of Pacing and Electrophysiology, "Heart rate variability standards of measurement, physiological interpretation, and clinical use", Circulation, 93, 1996, 1043–1065.
10. Hoffmann J, Grimm W, Menz V, Maisch B, "Cardiac autonomic tone and its relation to nonsustained ventricular tachyarrhythmias in idiopathic dilated cardiomyopathy", Clin Cardiol., 23, 2000, 103–108.
11. Karcz M, Chojnowska L, Zareba W, Ruzyllo W, "Prognostic significance of heart rate variability in dilated cardiomyopathy", Int J Cardiol., 87, 2003, 75–81.
12. Alter P, Grimm W, Vollrath A, Czerny F, Maisch B, "Heart rate variability in patients with cardiac hypertrophy–relation to left ventricular mass and etiology", Am Heart J., 151, 2006, 829–836.

13. Narumi J, Suma K, Kaneka H, Takeuchi Y, Inoue K, Shiroma K, Koyama Y, "Postoperative deep body temperature rhythm", Jpn J Surg., 17(3), 1987, 209–212.
14. Dworschak M, Lorenzl N, Rapp HJ, "Diminished circadian rhythm of heart rate variability after aortic surgery appears to be independent of myocardial ischemia", Jpn Heart J., 44, 2003, 645–654.
15. Tamura T, Maekawa T, Nakajima K, Sadamitsu D, Tateishi A, "Long-term heart rate fluctuations in postoperative and brain-dead patients", Chronobiol Int., 15(6), 1998, 633–646.
16. Nelder JA, Mead R, "A simple method for function minimization", Computer J., 7, 1965, 308–313.
17. Goldberger AL, West BJ, "Application of nonlinear dynamics to clinical cardiology", Ann New York Acad Sci., 504, 1987, 195–213.
18. Aoyagi N, Ohashi K, Tomono S, Yamamoto Y, "Temporal contribution of body movement to very long-term heart rate variability in humans", Am J Physiol Heart Circ Physiol., 278, 2000, H1035–H1041.
19. Kleiger RE, "Heart rate variability and mortality and sudden death post infarction", J Cardiol Electrophysiol., 6, 1995, 365–367.
20. Myers GA, Martin GJ, Magid NM, Barnett PS, Schaad JW, Weiss JS, Lesch M, Singer DH, "Power spectral analysis of heart rate variability in sudden cardiac death: comparison to other methods", IEEE Trans Biomed Eng., BME-33, 1986, 1149–1156.
21. Conci F, Di Rienzo M, Castiglioni P, "Blood pressure and heart rate variability and baroreflex sensitivity before and after brain death", J Neurol Neurosurg Psychiatry, 71(5), 2001, 621–631.

20

Stress During Speech Therapy

Toshiyo Tamura, Ayako Maeda, Masaki Sekine, Yuji Higashi and Toshiro Fujimoto

20.1 Speech Therapy

Speech therapists are specialists in the diagnosis and treatment of a variety of speech, voice, and language disorders. They treat people who are unable to make speech sounds or cannot make them clearly, such as those who exhibit a stutter, fluency and rhythm problems, inappropriate pitch, a harsh voice, and speech quality problems. The most widespread and obvious speech disorder is stuttering, which is frequently caused by anxiety. The speech therapist establishes a program of speech exercises to reduce the disability, and if necessary, enlists the aid of a psychologist or psychiatrist. Other speech disorders may result from hearing loss, stroke, cerebral palsy, mental disability, or brain injury. Speech therapists keep careful records of the evaluation and progress of patients, often developing and implementing individualized treatment programs based on input from physicians, psychiatric social workers, and psychologists. As speech disorders are usually related to neurological, psychological, and physical conditions, speech therapists can work as a member of a team that might include other healthcare specialists, such as neurologists and psychiatrists.

An important part of a speech therapist's work is to counsel and support individuals and families with regard to speech disorders and ways of coping with the associated stress. Therapists also work with families to establish treatment techniques for use at home and to assist in modifying behavior that impedes communication. Although a speech therapist's job is not physically demanding, it requires patience and compassion, as progress may be slow and halting. Speech therapy is a painstaking process that can be as rewarding as it is frustrating. Tremendous attention to detail and sharp focus are necessary in evaluating a patient's progress. Overall, speech therapists must be able to understand and empathize with the emotional strains and stresses attributable to speech problems, from both the patient's and family members' perspectives. Therapists must monitor patient progress, eliminate ineffective programs, and introduce new, more effective programs. The most stressful task for patients is the speech therapy itself, and the therapist must understand the stress related to the program [1,2].

In the present study, we investigated the stress experienced by patients during a therapy program.

20.2 Heart Rate Variability and Stress

This study examined the heart rate variability (HRV) as a marker of both dynamic and cumulative stress load, although the meaning of HRV has not been completely defined. As a marker of *dynamic* load, HRV appears to be sensitive and responsive to acute stress. Under laboratory conditions, mental load, such as that produced by making complex decisions or receiving rehabilitation therapy, has been shown to lower HRV. As a marker of *cumulative* wear and tear, HRV has been shown to decline with age, although the resting heart rate does not change significantly with advancing age. The decline in HRV has been attributed to a decrease in efferent vagal tone and reduced beta-adrenergic responsiveness. By contrast, regular physical activity, which slows the aging process, has been shown to raise HRV, presumably by increasing vagal tone. The HRV has been assessed by manually calculating the mean R-R interval and its standard deviation as measured on short-term (*e.g.*, 5 min) electrocardiograms; smaller standard deviations in the R-R interval indicated lower HRV values.

Measurements assessed using frequency domain parameters are represented as milliseconds squared (ms^2). Low frequencies (LF) of 0.04–0.15 Hz reflect mixed sympathetic and parasympathetic activity, the latter being prevalent during slow breathing. The normalized low frequency (LF norm) is determined by applying a mathematical ratio to LF; LF norm reflects changes in sympathetic activity (the flight-fright response). Mental stress increases LF activity. High frequencies (HF) of 0.15–0.4 Hz reflect parasympathetic activity and correspond to the time between two heartbeats (N-N variation) caused by respiration, *i.e.*, respiratory sinus arrhythmia. Deep, even breathing activates the parasympathetic system and increases the amplitude of HF. Mental stress decreases HF activity. The normalized high frequency (HF norm) is determined by applying a mathematical ratio to HF and reflects the change in parasympathetic regulation. A high LF/HF ratio (normalized units) indicates dominant sympathetic activity, whereas a low ratio indicates dominant parasympathetic activity. After metal stress, a decrease in the ratio reflects a change in parasympathetic regulation.

20.3 Calculating the RR Interval

The heart rate variability signal can be derived by monitoring the time interval between successive R-R events. In this way, it differs fundamentally from a signal derived by sampling a continuous process and requires special consideration before analysis. Although the HRV signal is a sequence of R-waves, one strategy is to assume that there is an underlying continuous process behind the events. With this approach, the R-wave events are considered "snapshots" of the continuous process, and reconstruction techniques are used for estimations. The advantage of this method is that the continuous process, once derived, may be sampled equidistantly in time and analyzed using standard techniques. The drawback is the risk of adding spurious information to the HRV measurements via the reconstruction algorithm. It is also unclear whether there is a sound physiological basis for a heart rate signal defined for all points in time.

In this study, we adopted an approach in which the HRV signal was defined in terms of R-R intervals rather than sampled time. Thus, the HRV signal was sampled equidistantly in terms of R-R intervals, not time, and could be processed using methods that require equidistant samples. It is important to note that the spectral content of such a signal refers to the "interval" power spectra and not the conventional power spectra of a time-sampled process.

To apply an interval power spectral representation of the HRV, the data must be sampled evenly at a known frequency. Many reports using spectral analysis have failed to address this topic, and there is no universal method for solving this problem. In the present study, we adopted a linear interpolation technique to resample the HRV signal in preparation for spectral analysis.

20.4 Experimental Methods

20.4.1 Subjects

Seven subjects (age, 65 ± 11 years old; three males and four females) participated in the therapy. The ethics committee of Fujimoto Hayasuzu Hospital approved this experiment, and written informed consent was obtained from each subject. The Table 20.1 summarizes the age, gender, symptoms, and number of months of therapy for each subject. The disability level of each subject

Table 20.1. Details of type and level of disorder, age, gender and duration of therapy

Subject #	Disorder	Level	Age	Gender	Therapy period (months)
1	Motor	Mild	70	Male	24
2	Sensory	Intermediate	71	Male	118
3	Motor	Intermediate	55	Female	23
4	Sensory	Intermediate	57	Male	5
5	Sensory	Intermediate	64	Female	16
6	Sensory	Severe	84	Female	11
7	Motor	Severe	52	Female	22

was judged against the impact that the communication difficulty had on the person's ability to communicate, in both understanding and expression, as judged by the family/caregiver as well as by the subject. The level of disability was severe in two subjects, intermediate in four, and mild in one.

20.4.2 Protocol

The subject sat resting for 3 min, and individual therapy lasted for 20–30 min depending on the subject. Then the subject sat for 3 min after the therapy. ECG monitoring began during the resting condition. The therapy followed the general protocol used at our hospital and consisted of exercises involving naming, repetition, comprehension, and sentence completion; following instructions; and conversations on topics of the subject's choice. Communication methods were also included, but this was not the major focus of treatment. A session began with simple tasks, progressed to more difficult tasks, and ended with simple tasks.

20.5 Results

All seven subjects exhibited high LF/HF values during difficult tasks. They sighed and made facial expressions that fit the LF/HF values. Figure 20.1 shows three typical LF/HF values. Six out of seven LF/HF values were increased significantly. During the sessions, six of eight subjects had high LF/HF values in the first and second halves of the session. One subject had a higher LF/HF in the second half, and one showed relatively little change throughout the session. Difficult exercises, including verbal fluency, naming, and explanation, resulted in relatively high LF/HF values. The tendency was unique for each patient.

20.6 Program with Relaxation

We assumed that intermittent relaxation would lead to a stress-free condition. For comparison, two subjects, one with mild aphasia (Broca's aphasia) and one with severe aphasia (Wernicke's aphasia), performed the exercise with relaxation.

Broca's aphasia patients is characterized by non-fluent speech, few words, short sentences, and many pauses. Words are produced with great effort and often sound distorted; prosody is flat and monopitched. This makes the speech sound telegraphic in nature because of the deletion of factor words and disturbances in word order. Moreover, the repetition of words and phrases is impaired. However, auditory comprehension of conversational speech is relatively intact. There is often an accompanying right hemiparesis involving the face, arm, and leg.

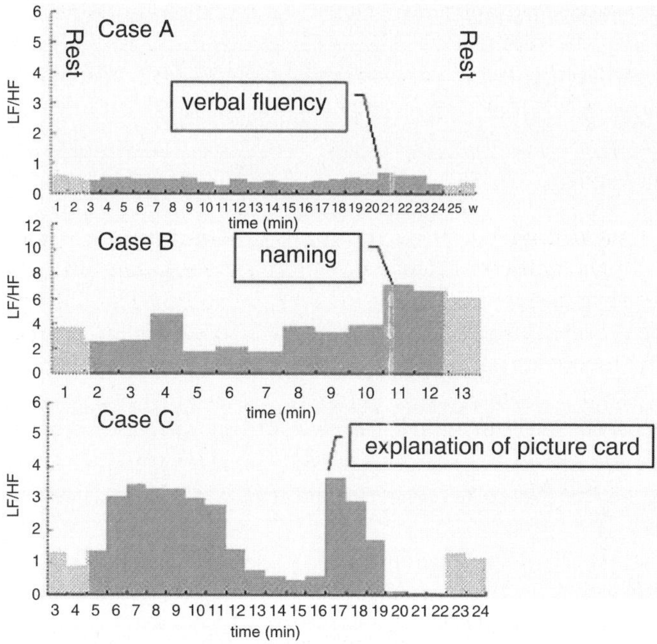

Fig. 20.1. Three typical examples of LF/HF values in subjects during speech therapy

Wernicke's aphasia syndrome involves the loss of the abilities to comprehend spoken language, to read silently, and to write, as well as the distortion of articulate speech. Hearing is intact. Speech may be fluent with a natural language rhythm, but the result has neither understandable meaning nor syntax. Despite the loss of comprehension, word memory is preserved, and words are often chosen correctly. Alexia, agraphia, acalculia, and paraphasia are frequently associated. The disorder is attributable to cortical lesions in the posterior portion of the left first temporal convolution.

These two patients underwent therapy with and without relaxation. The experiments were performed on 4 consecutive days, alternating daily between programs without and with relaxation. The relaxation included deep breathing and stretching of the trunk and neck.

Figures 20.2 and 20.3 show the relaxation effect in the subjects with mild and severe aphasia, respectively, during the therapy program. In the subject with mild aphasia, the LF/HF values changed daily and decreased after relaxation, as shown in the red circle. By contrast, the subject with severe aphasia showed no significant change with relaxation; the stress level in the severe aphasia patient may have been too great to allow relief.

The relationship between different aphasia and tendency of LF/HF is not clear. We need further experiments to conduct the results.

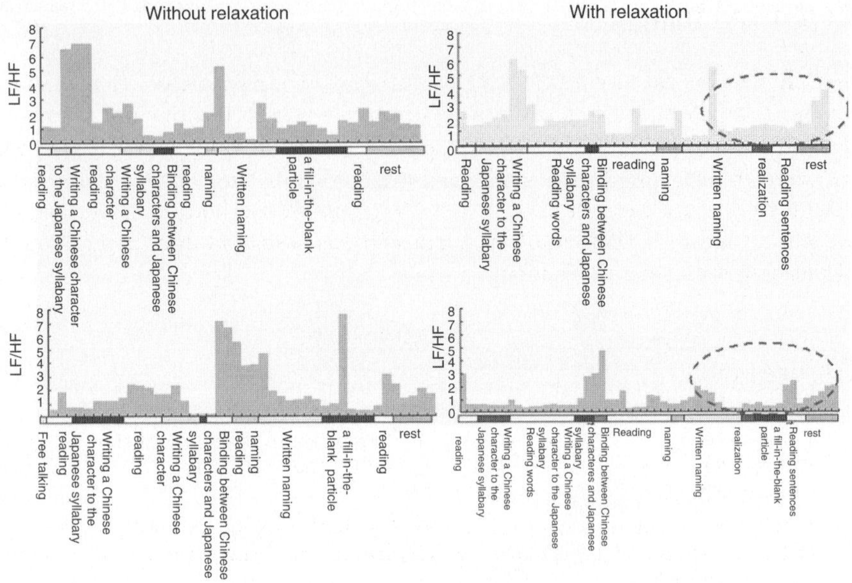

Fig. 20.2. The effect of relaxation during speech therapy in a subject with mild aphasia

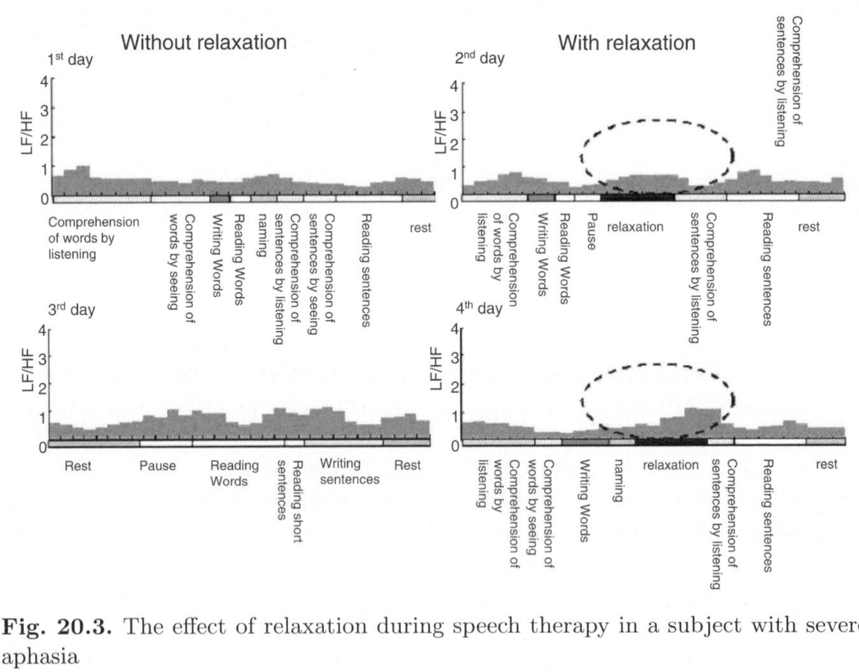

Fig. 20.3. The effect of relaxation during speech therapy in a subject with severe aphasia

20.7 Conclusion

Speech therapy is used in aphasia patients after stroke. After their sudden illness, the patients lose the ability to produce words and obviously suffer a decline in the quality of life. The stress associated with speech therapy was deceased in the mild aphasia subject by introducing intermittent relaxation. Monitoring heart rate variability may be a useful method for monitoring a patient's condition. This must be verified by assessments in a larger number of patients with severe aphasia.

References

1. Von Bonin, D., Fruhwirth, M., Heuser, P., Moser, M. (2001): 'Effects of speech therapy with poetry on heart rate variability and well-being'. *Forsch Komplementarmed Klass Naturheilkd*, **8**(3):144–160.
2. Laganaro, M., Vacheresse, F., Frauenfelder, U.H. (2002): 'Selective impairment of lexical stress assignment in an Italian-speaking aphasic patient'. *Brain Lang.*, **81**(1–3):601–609.

Color Plates

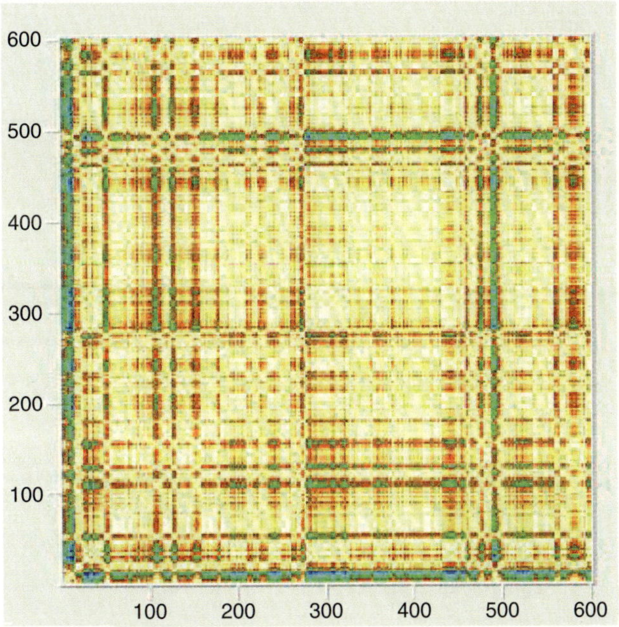

Fig. 5.6. Recurrence plot of normal heart rate

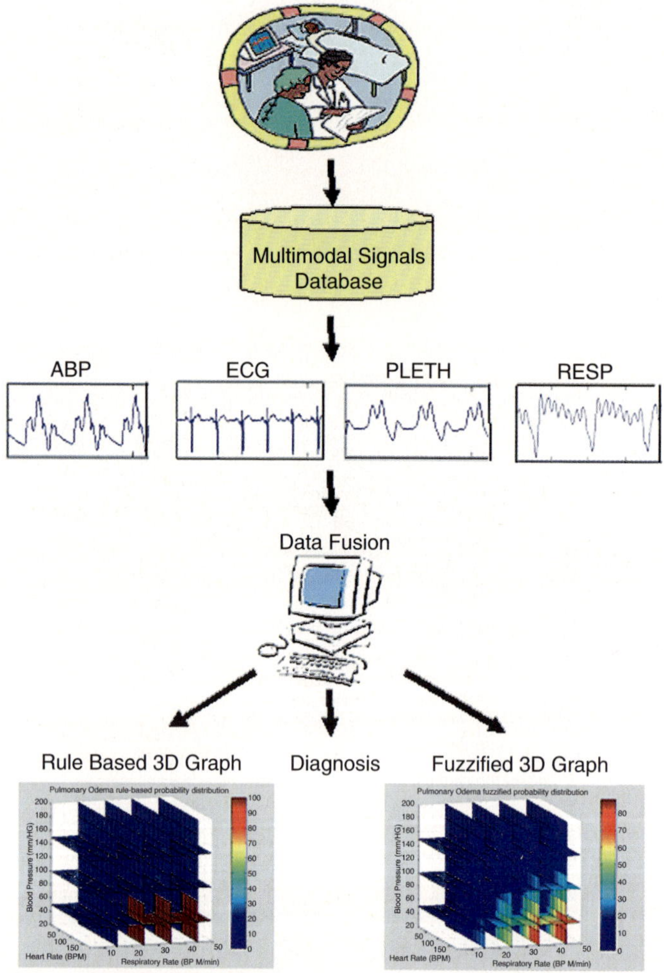

Fig. 6.6. Overview of the Patient state diagnosis system using data fusion

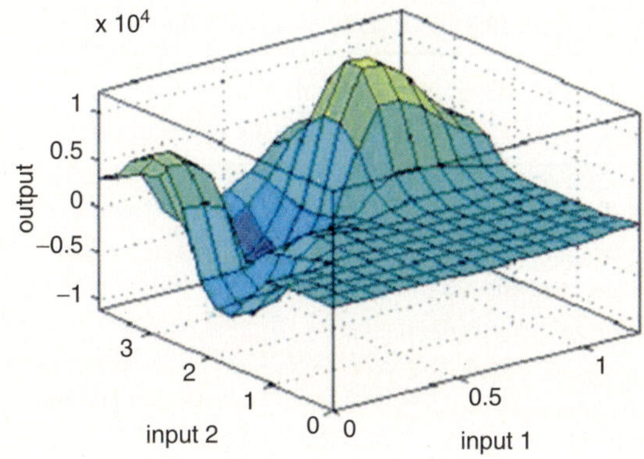

Fig. 9.6(a). Final decision surface for input 1(SD1/SD2) and input 2 (spectral entropy)

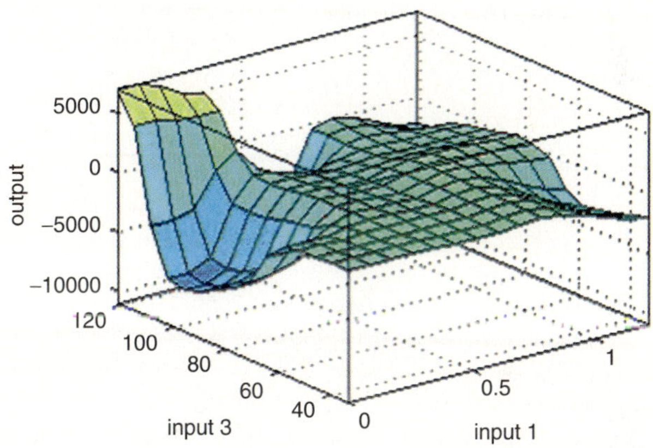

Fig. 9.6(b). Final decision surface for input 1(SD1/SD2) and input 3 (-slope)

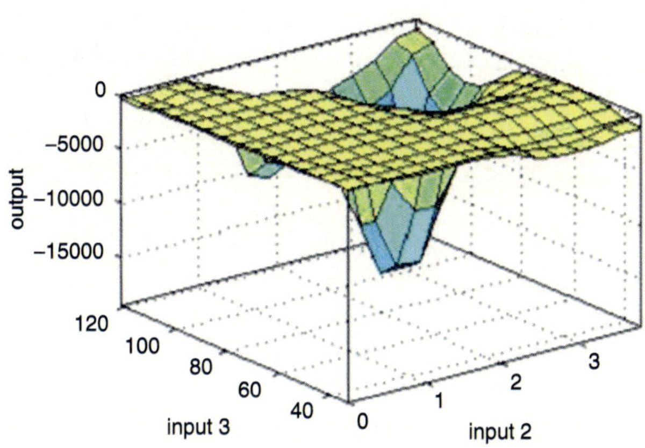

Fig. 9.6(c). Final decision surface for input 3 (-slope) and input 2 (spectral entropy)

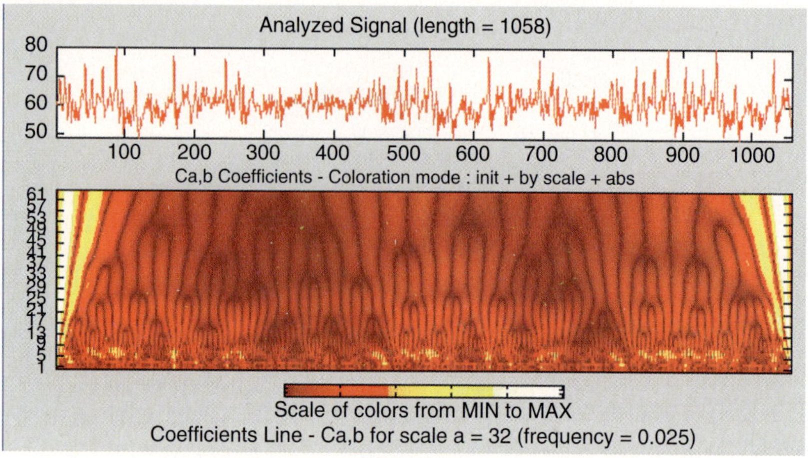

Fig. 16.1. CWT plot of normal heart rate

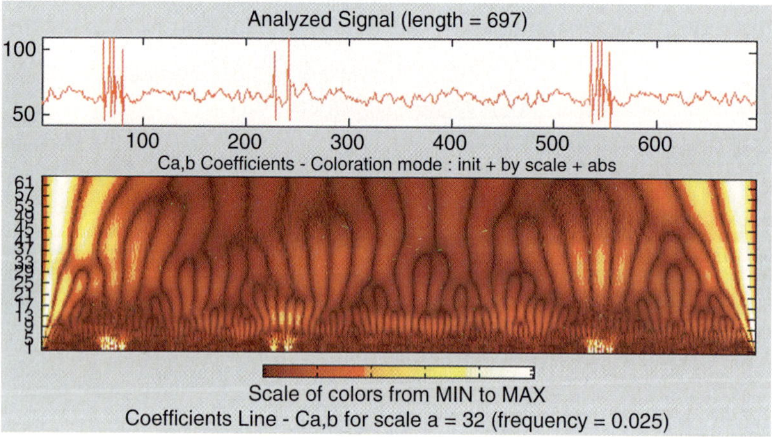

Fig. 16.3. CWT plot of ectopic beat

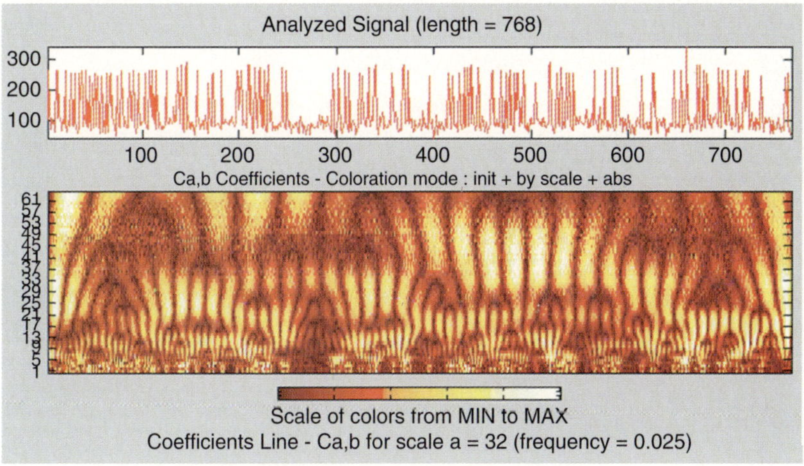

Fig. 16.5. CWT plot of AF heart rate

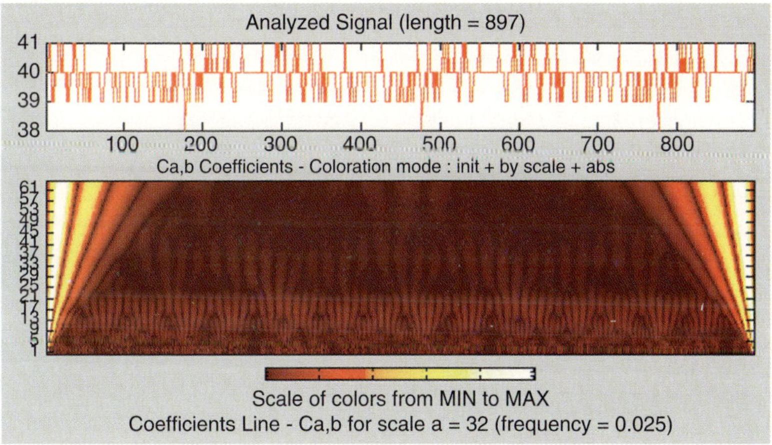

Fig. 16.7. CWT plot of CHB heart rate

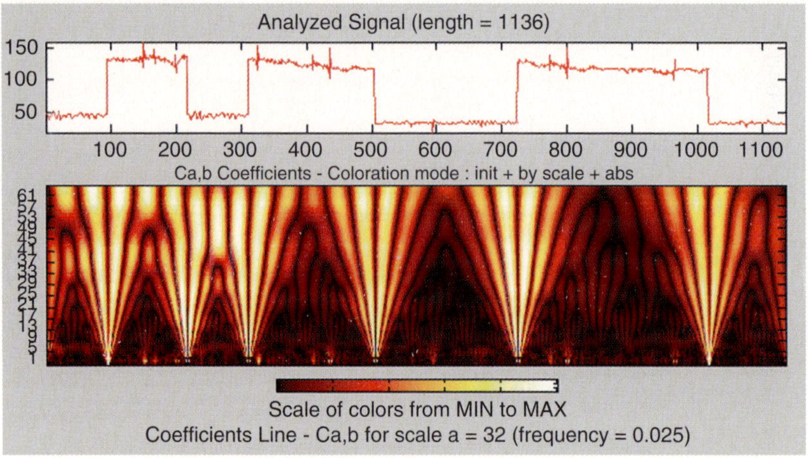

Fig. 16.9. CWT plot of SSS III heart rate

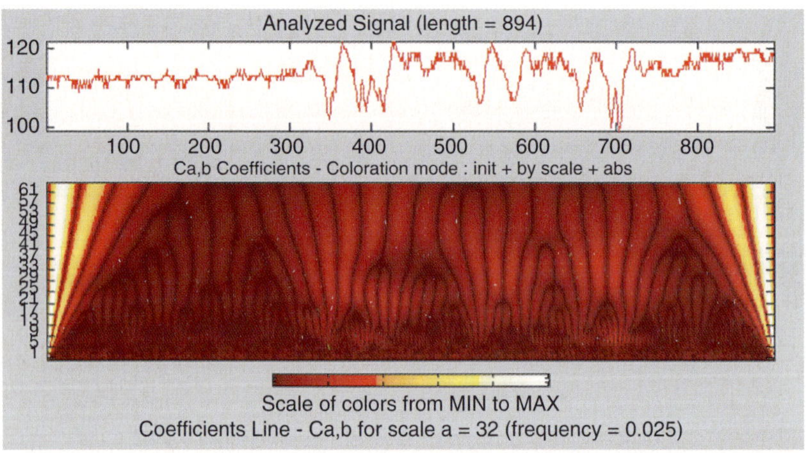

Fig. 16.11. CWT Poincare plot of Ischemic/Dilated Cardiomyopathy heart rate

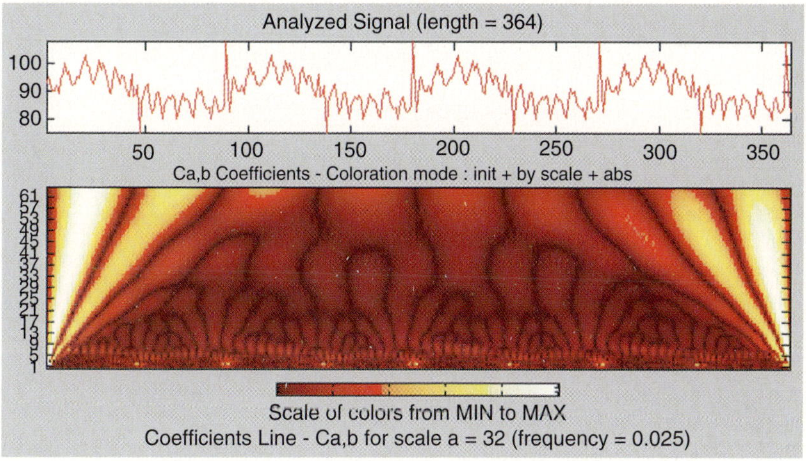

Fig. 16.13. CWT plot of LBBB heart rate

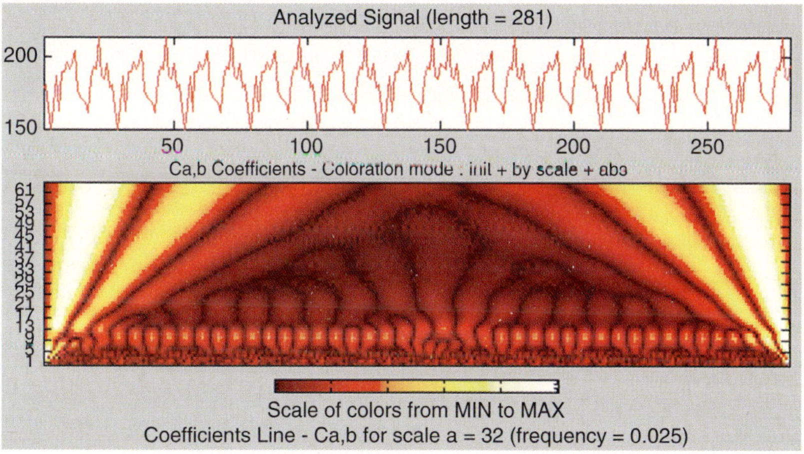

Fig. 16.15. CWT plot of VF heart rate

Index

Printing: Krips bv, Meppel
Binding: Stürtz, Würzburg